ELECTRICAL
SAFETY
HANDBOOK

ELECTRICAL SAFETY HANDBOOK

John Cadick, P.E.

Cadick Corporation, Garland, Texas

Mary Capelli-Schellpfeffer, M.D., M.P.A.

CapSchell, Inc., Chicago, Illinois

Dennis K. Neitzel, C.P.E.

AVO Training Institute, Inc., Dallas, Texas

Third Edition

McGRAW-HILL

New York   Chicago   San Francisco   Lisbon   London   Madrid
Mexico City   Milan   New Delhi   San Juan   Seoul
Singapore   Sydney   Toronto

The **McGraw·Hill** Companies

Library of Congress Cataloging-in-Publication Data

Cadick, John.
   Electrical safety handbook / John Cadick, Mary Capelli-Schellpfeffer, Dennis Neitzel.
   3rd ed.
       p.   cm.
   Includes index.
   ISBN 0-07-145772-0
   1. Electric engineering—Safety measures. 2. Electricity—Safety measures. 3. Industrial
safety. I. Capelli-Schellpfeffer, Mary. II. Neitzel, Dennis. III. Title.

TK152.C22   2005
621.319′028′9—dc22   2005051099

2 3 4 5 6 7 8 9 0 DOC/DOC 0 1 0 9 8 7 6

ISBN 0-07-145772-0

*The sponsoring editor for this book was Steve Chapman and the production
supervisor was Pamela A. Pelton. It was set in Times Roman by International
Typesetting and Composition. The art director for the cover was Anthony Landi.*

Printed and bound by RR Donnelley.

McGraw-Hill books are available at special quantity discounts to use as premiums and
sales promotions, or for use in corporate training programs. For more information,
please write to the Director of Special Sales, McGraw-Hill Professional, Two Penn
Plaza, New York, NY 10121-2298. Or contact your local bookstore.

To my wife, Sheryl, for her love and support.

John Cadick

To my wife, Brenda Neitzel, who always believed in me and who encouraged me to continue my education and strive to be the best that I could be; to the U.S. Air Force for giving me my start in an electrical career, to all of my employers, who gave me countless opportunities to learn and progress; and to John Cadick, who believed in me enough to ask me to contribute to this book.

Dennis Neitzel

In dedication especially to Michael Allen, the father of Sarah, Benjamin, Amelia, and Natalie, with my hope that each will continue to learn the most they can from books.

Mary Capelli-Schellpfeffer

CONTENTS

Foreword xvii

Preface xix

Acknowledgments xxi

Chapter 1. Hazards of Electricity   1.1

Introduction / 1.1
Glossary / 1.1
Hazard Analysis / 1.2
Shock / 1.2
    Description / 1.2
    Influencing Factors / 1.3
Arc / 1.7
    Definition and Description / 1.8
    Arc Energy Release / 1.9
    Arc Energy / 1.12
    Arc Energy Input / 1.12
    Arcing Voltage / 1.13
    Arc Surface Area / 1.13
    Incident Energy / 1.14
    Arc Burns / 1.16
Blast / 1.16
Affected Body Parts / 1.17
    General / 1.17
    Skin / 1.19
    The Nervous System / 1.20
    Muscular System / 1.21
    The Heart / 1.22
    The Pulmonary System / 1.23
Summary of Causes—Injury and Death / 1.23
    Shock Effect / 1.23
    Arc Flash Effect / 1.24
    Causes of Injury / 1.24
    Causes of Death / 1.25
Protective Strategies / 1.25
References / 1.26

Chapter 2. Electrical Safety Equipment   2.1

Introduction / 2.1
Glossary / 2.1
General Inspection and Testing Requirements for Electrical
  Safety Equipment / 2.2

Flash and Thermal Protection / *2.2*
   A Note on When to Use Thermal Protective Clothing / *2.2*
   Thermal Performance Evaluation / *2.3*
   Clothing Materials / *2.4*
   Non-Flame-Resistant Materials / *2.5*
   Flame-Resistant Materials / *2.6*
   Work Clothing / *2.9*
   Flash Suits / *2.11*
   Head, Eye, and Hand Protection / *2.13*
Head and Eye Protection / *2.14*
   Hard Hats / *2.14*
   Safety Glasses, Goggles, and Face Shields / *2.15*
Rubber-Insulating Equipment / *2.15*
   Rubber Gloves / *2.16*
   Rubber Mats / *2.20*
   Rubber Blankets / *2.21*
   Rubber Covers / *2.22*
   Line Hose / *2.25*
   Rubber Sleeves / *2.26*
   In-Service Inspection and Periodic Testing of Rubber Goods / *2.29*
Hot Sticks / *2.44*
   Description and Application / *2.44*
   When to Use / *2.47*
   How to Use / *2.47*
   Testing Requirements / *2.47*
Insulated Tools / *2.48*
   Description and Application / *2.48*
   When to Use / *2.49*
   How to Use / *2.49*
Barriers and Signs / *2.49*
   Barrier Tape / *2.49*
   Signs / *2.49*
   When and How to Use / *2.50*
Safety Tags, Locks, and Locking Devices / *2.51*
   Safety Tags / *2.51*
   Locks and Multiple-Lock Devices / *2.52*
   Locking Devices / *2.53*
   When/Where to Use Lockout Tagout / *2.53*
Voltage-Measuring Instruments / *2.54*
   Safety Voltage Measurement / *2.54*
   Proximity Testers / *2.55*
   Contact Testers / *2.56*
   Selecting Voltage-Measuring Instruments / *2.57*
   Instrument Condition / *2.59*
   Low Voltage Voltmeter Safety Standards / *2.60*
   Three-Step Voltage Measurement Process / *2.60*
   General Considerations for Low-Voltage Measuring Instruments / *2.62*
Safety Grounding Equipment / *2.63*
   The Need for Safety Grounding / *2.63*
   Safety Grounding Switches / *2.64*
   Safety Grounding Jumpers / *2.65*
   Selecting Safety Grounding Jumpers / *2.70*
   Installation and Location / *2.74*
Ground Fault Circuit Interrupters / *2.74*
   Operating Principles / *2.74*
   Applications / *2.75*
Safety Electrical One-Line Diagram / *2.78*
The Electrician's Safety Kit / *2.78*
References / *2.79*

Chapter 3. Safety Procedures and Methods   3.1

Introduction / 3.1
The Six-Step Safety Method / 3.1
   Think—Be Aware / 3.2
   Understand Your Procedures / 3.2
   Follow Your Procedures / 3.2
   Use Appropriate Safety Equipment / 3.2
   Ask If You Are Unsure, and Do Not Assume / 3.2
   Do Not Answer If You Do Not Know / 3.3
Pre-Job Briefings / 3.3
   Definition / 3.3
   What Should Be Included? / 3.3
   When Should Pre-Job Briefings Be Held? / 3.3
Energized or De-Energized? / 3.3
   The Fundamental Rules / 3.3
   A Hot-Work Decision Tree / 3.5
   After the Decision Is Made / 3.6
Safe Switching of Power Systems / 3.6
   Introduction / 3.6
   Remote Operation / 3.7
   Operating Medium-Voltage Switchgear / 3.7
   Operating Low-Voltage Switchgear / 3.11
   Operating Molded-Case Breakers and Panelboards / 3.15
   Operating Enclosed Switches and Disconnects / 3.17
   Operating Open-Air Disconnects / 3.18
   Operating Motor Starters / 3.20
Energy Control Programs / 3.23
   General Energy Control Programs / 3.23
   Specific Energy Control Programs / 3.24
   Basic Energy Control Rules / 3.24
Lockout-Tagout / 3.26
   Definition and Description / 3.26
   When to Use Locks and Tags / 3.26
   Locks without Tags or Tags without Locks / 3.26
   Rules for Using Locks and Tags / 3.27
   Responsibilities of Employees / 3.27
   Sequence / 3.28
   Lock and Tag Application / 3.28
   Isolation Verification / 3.28
   Removal of Locks and Tags / 3.28
   Safety Ground Application / 3.29
   Control Transfer / 3.31
   Nonemployees and Contractors / 3.31
   Lockout-Tagout Training / 3.31
   Procedural Reviews / 3.32
Voltage-Measurement Techniques / 3.32
   Purpose / 3.32
   Instrument Selection / 3.32
   Instrument Condition / 3.33
   Three-Step Measurement Process / 3.34
   What to Measure / 3.34
   How to Measure / 3.36
Placement of Safety Grounds / 3.37
   Safety Grounding Principles / 3.37
   Safety Grounding Location / 3.38
   Application of Safety Grounds / 3.38
   The Equipotential Zone / 3.43

Removal of Safety Grounds / *3.44*
Control of Safety Grounds / *3.44*
Flash Hazard Calculations and Approach Distances / *3.46*
Introduction / *3.46*
Approach Distance Definitions / *3.46*
Determining Shock Hazard Approach Distances / *3.46*
Calculating the Flash Hazard Minimum Approach Distance
(Flash Protection Boundary) / *3.49*
Calculating the Required Level of Arc Protection (Flash Hazard Calculations) / *3.51*
Introduction / *3.51*
The Lee Method / *3.52*
Methods Outlined in NFPA 70E / *3.52*
IEEE Standard Std 1584-2002 / *3.53*
Software Solutions / *3.55*
Required PPE for Crossing the Flash Hazard Boundary / *3.55*
A Simplified Approach to the Selection of Protective Clothing / *3.56*
Barriers and Warning Signs / *3.56*
Illumination / *3.61*
Conductive Clothing and Materials / *3.61*
Confined Work Spaces / *3.62*
Tools and Test Equipment / *3.62*
General / *3.62*
Authorized Users / *3.62*
Visual Inspections / *3.63*
Electrical Tests / *3.63*
Wet and Hazardous Environments / *3.63*
Field Marking of Potential Hazards / *3.65*
The One-Minute Safety Audit / *3.65*
References / *3.66*

Chapter 4. Grounding of Electrical Systems and Equipment   **4.1**

Introduction / *4.1*
Electric Shock Hazard / *4.1*
General Requirements for Grounding and Bonding / *4.2*
Definitions / *4.2*
Grounding of Electrical Systems / *4.3*
Grounding of Electrical Equipment / *4.6*
Bonding of Electrically Conductive Materials and Other Equipment / *4.6*
Performance of Fault Path / *4.8*
Arrangement to Prevent Objectionable Current / *4.8*
Alterations to Stop Objectionable Current / *4.8*
Temporary Currents Not Classified as Objectionable Current / *4.8*
Connection of Grounding and Bonding Equipment / *4.8*
Protection of Ground Clamps and Fittings / *4.9*
Clean Surfaces / *4.9*
System Grounding / *4.9*
Purposes of System Grounding / *4.9*
Grounding Service-Supplied Alternating-Current Systems / *4.9*
Conductors to Be Grounded—Alternating-Current Systems / *4.11*
Main Bonding Jumper / *4.11*
Grounding Electrode System / *4.12*
Grounding Electrode System Resistance / *4.14*
Grounding Electrode Conductor / *4.14*
Grounding Conductor Connection to Electrodes / *4.16*
Bonding / *4.18*

Equipment Grounding / *4.19*
    Equipment to Be Grounded / *4.19*
    Grounding Cord- and Plug-Connected Equipment / *4.19*
    Equipment Grounding Conductors / *4.21*
    Sizing Equipment Grounding Conductors / *4.22*
    Use of Grounded Circuit Conductor for Grounding Equipment / *4.22*
    Ferroresonance / *4.25*
Summary / *4.27*

Chapter 5. Electrical Maintenance and Its Relationship to Safety   5.1

Introduction / *5.1*
The Safety-Related Case for Electrical Maintenance / *5.1*
    Overview / *5.1*
    Regulatory / *5.2*
    Relationship of Improperly Maintained Electrical Equipment to the
      Hazards of Electricity / *5.2*
    Hazards Associated with Electrical Maintenance / *5.3*
The Economic Case for Electrical Maintenance / *5.3*
Reliability Centered Maintenance (RCM) / *5.4*
    What is Reliability Centered Maintenance? / *5.5*
    A Brief History of RCM / *5.5*
    RCM in the Industrial and Utility Arena / *5.5*
    The Primary RCM Principles / *5.6*
    Failure / *5.8*
    Maintenance Actions in an RCM Program / *5.8*
    Impact of RCM on a Facilities Life Cycle / *5.9*
    Conclusion / *5.10*
The Eight Step Maintenance Program / *5.10*
    Introduction / *5.11*
    Step 1—Plan / *5.11*
    Step 2—Inspect / *5.11*
    Step 3—Clean / *5.12*
    Step 4—Tighten / *5.12*
    Step 5—Lubricate / *5.12*
    Step 6—Test / *5.13*
    Step 7—Record / *5.13*
    Step 8—Evaluate / *5.13*
Frequency of Maintenance / *5.14*
    Determining Testing Intervals / *5.14*
    Summary / *5.14*
Maintenance Requirements for Specific Equipment and Locations / *5.14*
    General Maintenance Requirements / *5.14*
    Substations, Switchgear, Panel Boards, Motor Control Centers, and
      Disconnect Switches / *5.15*
    Fuse Maintenance Requirements / *5.16*
    Molded-Case Circuit Breakers / *5.16*
    Low-Voltage Power Circuit Breakers / *5.18*
    Medium Voltage Circuit Breakers / *5.20*
    Protective Relays / *5.21*
    Rotating Equipment / *5.23*
    Portable Electric Tools and Equipment / *5.23*
    Personal Safety and Protective Equipment / *5.24*
Conclusion / *5.24*
References / *5.24*

Chapter 6. Regulatory and Legal Safety Requirements and Standards   6.1

Introduction / 6.1
The Regulatory Bodies / 6.1
    The American National Standards Institute (ANSI) / 6.1
    The Institute of Electrical and Electronic Engineers (IEEE) / 6.3
    National Fire Protection Association (NFPA) / 6.3
    American Society for Testing and Materials (ASTM) / 6.4
    American Society of Safety Engineers (ASSE) / 6.5
    The Occupational Safety and Health Administration (OSHA) / 6.6
    Other Electrical Safety Organizations / 6.12
The National Electrical Safety Code (NESC)—ANSI C-2 / 6.12
    General Description / 6.12
    Industries and Facilities Covered / 6.13
    Technical/Safety Items Covered / 6.13
The National Electrical Code (NEC)—ANSI/NFPA 70 / 6.14
    General Description / 6.14
    Industries and Facilities Covered / 6.15
    Technical and Safety Items Included / 6.15
Electrical Equipment Maintenance—ANSI/NFPA 70B / 6.15
    General Description / 6.15
    Industries and Facilities Covered / 6.16
    Technical and Safety Items Covered / 6.16
Standard for Electrical Safety in the Workplace—ANSI/NFPA 70E / 6.16
    General Description / 6.16
    Industries and Facilities Covered / 6.17
    Technical Safety Items Covered / 6.18
The American Society for Testing and Materials (ASTM) Standards / 6.19
Occupational Safety and Health Administration (OSHA) Standards / 6.19
    Overview / 6.19
    General Industry / 6.19
    Construction Industry / 6.22

Chapter 7. Accident Prevention, Accident Investigation, Rescue, and First Aid   7.1

Accident Prevention / 7.1
    Individual Responsibility / 7.1
    Installation Safety / 7.1
    Power System Studies / 7.3
First Aid / 7.7
    General First Aid / 7.8
    Resuscitation (Artificial Respiration) / 7.12
    Heart-Lung Resuscitation / 7.12
Rescue Techniques / 7.14
    General Rescue Procedures / 7.14
    Elevated Rescue / 7.18
    Confined-Space Rescue / 7.30
    Ground-Level Rescue / 7.31
Accident Investigation / 7.37
    Purpose / 7.37
    General Rules / 7.38
    Data Gathering / 7.38
    Accident Analysis / 7.40

Chapter 8. Medical Aspects of Electrical Trauma 8.1

Introduction / 8.1
Statistical Survey / 8.1
  Non-Occupational Electrical Trauma / 8.4
  Fatality and Injury Related Costs / 8.4
Electrical Events / 8.6
  Electrocution and Electrical Fatalities / 8.7
Medical Aspects / 8.8
  Non-Electrical Effects in Electrical Events / 8.10
  Stabilization and Initial Evaluation / 8.13
  Medical and Surgical Intervention / 8.14
Rehabilitation Focus and Return to Work Planning / 8.16
  Reentry to Employment Settings / 8.16
  Plateau in Recovery / 8.16
References / 8.18

Chapter 9. Low-Voltage Safety Synopsis 9.1

Introduction / 9.1
Low-Voltage Equipment / 9.1
  Extension Cords / 9.2
  Plug- and Cord-Connected Equipment / 9.3
  Current Transformers / 9.5
Grounding Low-Voltage Systems / 9.5
  What Is a Ground? / 9.5
  Bonding Versus Grounding / 9.6
  Voltage Hazards / 9.6
  System Grounds / 9.8
  Equipment Grounds / 9.11
  Ground Fault Circuit Interrupters / 9.14
Safety Equipment / 9.14
  Overview / 9.14
  Hard Hats / 9.14
  Eye Protection / 9.15
  Arc Protection / 9.18
  Rubber Insulating Equipment / 9.18
  Voltage-Testing Devices / 9.18
Safety Procedures / 9.20
  General / 9.20
  Approach Distances / 9.20
  Voltage Measurement / 9.20
  Locking and Tagging / 9.21
  Closing Protective Devices After Operation / 9.21
Electrical Safety Around Electronic Circuits / 9.21
  The Nature of the Hazard / 9.21
  Special Safety Precautions / 9.22
Stationary Battery Safety / 9.23
  Introduction / 9.23
  Basic Battery Construction / 9.24
  Safety Hazards of Stationary Batteries / 9.25
  Battery Safety Procedures / 9.25

Chapter 10. Medium- and High-Voltage Safety Synopsis 10.1

Introduction / 10.1
High-Voltage Equipment / 10.1
  Current Transformers / 10.1

Grounding Systems of Over 1000 V / 10.3
    What Is a Ground? / 10.3
    Bonding Versus Grounding / 10.4
    Voltage Hazards / 10.5
    System Grounds / 10.5
    Equipment Grounds / 10.7
Safety Equipment / 10.7
    Overview / 10.7
    Hard Hats / 10.8
    Eye Protection / 10.8
    Arc Protection / 10.9
    Rubber Insulating Equipment / 10.9
    Voltage-Testing Devices / 10.9
Safety Procedures / 10.12
    General / 10.12
    Approach Distances / 10.12
    Voltage Measurement / 10.13
    Locking and Tagging / 10.13
    Closing Protective Devices After Operation / 10.13

Chapter 11. Human Factors in Electrical Safety   11.1

Introduction / 11.1
    Background / 11.1
Power Systems and Human Factors / 11.3
    Visualization / 11.4
    Cognitive Ergonomics / 11.4
Summary / 11.8
References / 11.8

Chapter 12. Safety Management and Organizational Structure   12.1

Introduction / 12.1
Electrical Safety Program Structure / 12.1
Electrical Safety Program Development / 12.2
    The Company Electrical Safety Team / 12.2
    Company Safety Policy / 12.4
    Assessing the Need / 12.4
    Problems and Solutions / 12.4
    Program Implementation / 12.5
    Examples / 12.5
    Company Safety Procedures / 12.7
    Results Assessment / 12.7
Employee Electrical Safety Teams / 12.8
    Reason / 12.8
    Method / 12.8
Safety Meetings / 12.9
    Who Attends / 12.9
    What Material Should Be Covered / 12.9
    When Meetings Should Be Held / 12.10
    Where Meetings Should Be Held / 12.10
    How Long Meetings Should Be / 12.10
    Evaluation of Safety Meetings / 12.10
Outage Reports / 12.11
Safety Audits / 12.11

Description / *12.11*
Purposes / *12.12*
Procedure / *12.12*
The Audit Team / *12.14*
Audit Tools / *12.14*
Follow-Up / *12.14*
Internal versus External Audits / *12.14*

Chapter 13. Safety Training Methods and Systems   13.1

Introduction / *13.1*
   Safety Training Definitions / *13.1*
   Training Myths / *13.2*
   Conclusion / *13.3*
Elements of a Good Training Program / *13.3*
   Classroom Training / *13.3*
   On-the-Job Training (OJT) / *13.5*
   Self-training / *13.6*
   Conclusion / *13.6*
On-the-Job Training / *13.6*
   Setup / *13.6*
   Implementation / *13.8*
   Evaluation / *13.8*
   Conclusion / *13.9*
Training Consultants and Vendors / *13.9*
   Canned Programs and Materials / *13.9*
   Tailored Programs / *13.10*
   Training Analysis / *13.10*
   Evaluating Training Vendors and Consultants / *13.11*
   Conclusion / *13.11*
Training Program Setup—A Step-By-Step Method / *13.12*
   Introduction / *13.12*
   Background / *13.12*
   A Plan / *13.14*
   Analyze / *13.14*
   Design / *13.16*
   Develop / *13.17*
   Implementation / *13.18*
   Evaluation / *13.18*
   Modification / *13.19*

Index   **I.1**

FOREWORD

Electricity has been recognized as dangerous since it began to be used for street lighting in large cities in the northeastern United States in the late 1800s. People received severe shocks and were frequently electrocuted as they made contact with energized electrical equipment. Inadequate electrical installation often resulted in fires, and standard installation methods did not exist.

At that time, even though fires and electrocutions were occurring, the use of electricity quickly expanded to other parts of the United States. Electrical designs and installations varied widely from one facility to another. Injury data and economic losses illustrated that both fire and electrocution were hazards, and insurance companies recognized the importance of standardization. If an installation standard could be developed, both electrocution and fire could be reduced.

The system of voluntary electrical standards that currently exists was developed after fire, and electrocution became recognized electrical hazards. The system of codes and standards that guides installations consists of documents generated by standards-developing organizations, third-party inspection of electrical equipment, and enforcement by inspecting organizations. The system does a good job of keeping the public safe from both electrocution and fires, provided the electrical equipment doors are latched, adequately maintained, and the equipment is operating normally.

In recent years, the community has begun to recognize that in addition to fire and electrocution, arc-flash and arc-blast hazards also result in injury. The knowledge base about these hazards is expanding but is not yet complete. The community knows that as the distance between a worker and an electrical hazard decreases, the degree of exposure increases. Workers must understand that they are exposed to these hazards until an electrically safe work condition has been established, as explained in this handbook.

Before an electrically safe work condition exists, workers are exposed to *all* hazards associated with electrical energy, in many different ways:

- Electrical equipment, devices, and components have a life expectancy, and control devices sometimes malfunction. When a failure occurs, a worker is expected to identify the problem, repair the problem, and restore the equipment to normal service.

- To extend the life of electrical equipment, the equipment must be maintained. Although the electrical energy sometimes is removed before a worker begins a maintenance task, those tasks often are executed while the source of electricity is energized.

- Equipment and circuits sometimes are modified to add new devices or circuits. Short-term employees may be expected to work in an environment that includes exposure to energized electrical circuits and components. Consultant and service employees are frequently exposed to energized electrical equipment and circuits.

- When a problem exists that causes equipment to operate other than normally, a worker might open a door or remove a cover and expose an energized electrical conductor or component. In many cases, the worker might troubleshoot while the circuit is energized. Components and conductors might be added within a piece of equipment while the equipment or parts of the equipment remain energized.

- After correcting a problem, workers sometimes create further hazardous conditions: leave an equipment door ajar, with latches open; replace covers with a minimum number of screws; and remove devices, leaving penetrations through a door.

In recent years, the electrical community has begun to understand that workers are exposed to many different electrical hazards. Workers certainly should not be unnecessarily exposed to hazards. Workers should understand how and when they could be exposed to a hazard and how to assess the hazard and risk of injury. They must also understand how to select and use work practices that minimize or eliminate risk of injury and how to select and wear protective equipment that will minimize or eliminate that risk.

Employees must be trained to understand the concepts discussed in each chapter. The authors of this handbook are warriors in the fight for electrical safe workplaces. This new edition provides critical, up-to-date information for employers about how to avoid injury. All of the information is here in the handbook, and work practices, procedures, and the overall electrical safety program should embody the ideas discussed herein. I highly recommend this valuable resource to employers as they join the growing effort to ensure that their workers remain safe. Providing a safe work place is not only an economical asset, it is the right thing to do.

Ray A. Jones, P.E.
NFPA 70E Technical Committee Chair

PREFACE

It seems like only a few days since the second edition of this handbook was completed; however, in the five years that have passed, amazing things have happened in the world of safety. First, the 2000 edition of NFPA 70E has become the 2004 edition with a multitude of changes. Second, the electrical world seems to have become much more aware of the hazards of electricity and has started to embrace many, if not all, of the more modern safety requirements, such as flash-hazard evaluations. Finally, companies throughout the world are starting to gear up to provide enhanced safety programs for their personnel and work in teams to achieve that goal.

The *Electrical Safety Handbook* has continued to receive remarkably broad acceptance in the electrical safety world, perhaps because it is the only independent reference source for all of the various aspects of electrical safety. With this in mind, we have expanded virtually all of the previous chapters in this edition as well as added a brand new chapter covering the safety aspects of electrical maintenance. We truly hope that the increased detail in the old chapters and the addition of the new chapter will be met with the same enthusiasm as the previous edition.

Chapters 1, 2, and 3 continue to serve as the central core of the book, by presenting the case for electrical safety (Chapter 1), a broad coverage of electrical safety equipment (Chapter 2), and a detailed coverage of electrical safety procedures (Chapter 3). The changes in Chapter 3 should be of special interest to the reader since we have updated the arc energy calculations to be consistent with new industry innovations as introduced in IEEE Standard Std 1584 and NFPA 70E.

Chapter 4 has been revised to include references to the 2005 edition of the National Electrical Code (NEC). Additional information has also been added that includes ferroresonance. Chapter 4 continues to provide a detailed overview of the general requirements for grounding and bonding electrical systems and equipment. This chapter also provides some needed explanations, illustrations, and calculations necessary for applying the requirements of NEC Article 250 as well as OSHA 29 CFR 1910.304(f). It should be emphasized, however that this chapter is not intended to replace or be a substitute for the requirements of the current NEC or OSHA regulations. Always utilize the most current standards and regulations when designing, installing, and maintaining the grounding systems within a facility.

Chapter 5, new to this edition, introduces a broad coverage of safety-related maintenance concepts. The chapter is not intended to be a maintenance reference, rather it introduces the economic and safety-related reasons for performing maintenance on an electrical power system. We also introduce the current philosophies on good maintenance including discussions of topics such as reliability-centered, predictive, preventive, and condition-based maintenance. Finally, the chapter covers the eight broad steps of a good maintenance program and introduces a cross section of maintenance and testing procedures for a variety of electrical equipment.

Chapter 6 (Chapter 5 in the second edition), updates the previous coverage on the consensus and mandatory standards and regulations in the workplace. Also included are explanations of several of the general industry OSHA regulations, which are quoted directly from the Federal Registers preamble as well as the Directorate of Compliance documents.

The specific information reprinted from OSHA has been updated to the most recent versions as of the date of this publication. As before, the reader should always refer to the OSHA publications (available at www.osha.gov) for the most recent information.

Chapter 7 (Chapter 6 in the second edition) has been expanded and rewritten in several key ways. New information, addressing the relationship of proper engineering and safety, has been added showing the reader how they can engineer safety hazards out of existence. Also, the section on first aid and CPR has been brought more up-to-date.

In Chapter 8, a statistical survey of electrical injuries and fatalities is added. Details about the possible medical scenarios following exposure to electrical hazards are included to assist the reader in appreciating how survivors of a multivictim electrical incident may share an event, but not necessarily the same effects to their bodies. Graphs and tables, new to the handbook's third edition, give quick summaries of enhanced information. Figures are presented to explain the basic principle that injury results with the transformation and transfer of energy.

Chapters 9 and 10 (Chapters 8 and 9 in the second edition) have been updated with new information taken from Chapters 2 and 3. As before, these two chapters serve as a quick reference to the reader for low-, medium-, and high-voltage safety.

Added to the topic of human factors now found in Chapter 11 is a historical perspective. Insights regarding human factors engineering developed within the civilian nuclear power industry are also included.

Chapter 12 (Chapter 11 in the second edition) has been updated to include information garnered from the collective experience of the authors since the second edition was published. Of special importance is the material covering the interrelationships among management, labor, and legal counsel.

Chapter 13 (Chapter 12 in the second edition) has a significant new section covering a method to develop a training that will provide a more consistent and compliant safety program. By using tried and tested educational methods, the new sections show the reader in a step-by-step manner how to create their training program.

John Cadick, P.E.

Mary Capelli-Schellpfeffer, M.D., M.P.A.

Dennis K. Neitzel, C.P.E.

ACKNOWLEDGMENTS

The authors gratefully acknowledge contributions, permissions, and assistance from the following individuals and organizations:

Kathy Hooper, ASTM; Robert L. Meltzer, ASTM; Laura Hannan, TEGAM Incorporated; Raymond L. Erickson, Eaton Corporation, Cutler-Hammer Products; H.G. Brosz, Brosz and Associates Archives; P. Eng, Brosz and Associates Archives; Ellis Lyons, W.H. Salisbury and Company; Tom Gerard, TIF Instruments, Inc., Miami, Florida; LaNelle Morris, Direct Safety Company, Phoenix, Arizona; Tony Quick, Santronics Inc.; Ed DaCruz, Siebe North, Inc.; Benjamin L. Bird, CIP Insulated Products; E.T. Thonson, AVO Biddle Instruments, Blue Bell, Pennsylvania; Terry Duchaine, Ideal Industries, Inc., Sycamore, Illinois; Mary Kay S. Kopf, DuPont Fibers; Stephen Gillette, Electrical Apparatus Division of Siemens Energy and Automation, Inc.; Mary Beth Stahl, Mine Safety Appliances Company; Debbie Prikryl, Encon Safety Products; Craig H. Seligman, NOMEX® III Work Clothing, Workrite Uniform Company, Oxnard, California; Kareem M. Irfan, Square D Company, A.B. Chance Co., AVO Multi-Amp Institute; Dr. Brian Stevens, Ph.D., Phelps Dodge Mining Company; Jason Saunders, Millennium Inorganic Chemicals; Cindy Chatman, Bulwark; Alan Mark Franks; Sandy Young; Bruce McClung; Dr. Raphael C. Lee, M.D., Sc.D., and Zoe G. Foundotos.

John Cadick

AVO Training Institute, Inc. (a Subsidiary of Megger); Erico, Inc. (Cadweld); Ronald P. O'Riley, *Electrical Grounding* (Delmar Publishers); National Fire Protection Association, *2005 National Electrical Code Handbook* (NFPA).

Dennis Neitzel

Dr. Capelli-Schellpfeffer's research was supported in part by grant R01 OH04136-02 from the U.S. Center for Disease Control and Prevention (CDC) and the National Institute of Occupational Safety and Health (NIOSH). Her comments do not represent official agency views.

Mary Capelli-Schellpfeffer

ELECTRICAL
SAFETY
HANDBOOK

CHAPTER 1
HAZARDS OF ELECTRICITY

INTRODUCTION

Modern society has produced several generations who have grown accustomed to electricity. This acclimatization has been made easier by the fact that electricity is silent, invisible, odorless, and has an "automatic" aspect to it. In the late 1800s, hotels had to place signs assuring their guests that electricity is harmless. By the late 1900s, signs had to be hung to remind us that electricity is a hazard. In fact, the transition of electricity from a silent coworker to a deadly hazard is a change that many cannot understand until it happens to them. Because of these facts, the total acceptance of an electrical safety procedure is a requirement for the health and welfare of workers.

Understanding the steps and procedures employed in a good electrical safety program requires an understanding of the nature of electrical hazards. Although they may have trouble writing a concise definition, most people are familiar with electric shock. This often painful experience leaves its memory indelibly etched on the human mind. However, shock is only one of the electrical hazards. There are two others—arc and blast. This chapter describes each of the three hazards and explains how each affects the human body.

Understanding the nature of the hazards is useless unless protective strategies are developed to protect the worker. This chapter also includes a synopsis of the types of protective strategies that should be used to protect the worker.

GLOSSARY

Arc (electric)	The heat and light energy release that is caused by the electrical breakdown of and subsequent electrical discharge through an electrical insulator, such as air.
Arc energy input	The total amount of energy delivered by the power system to the arc. This energy will be manifested in many forms including light, heat, and mechanical (pressure) energy.
Arc incident energy	The amount of energy delivered by an electric arc to the clothing or body of a worker. This amount of energy will be somewhat less than the arc energy based on factors in the workplace.
Arc-flash	See Arc
Arc-resistant switchgear	Metal-clad switchgear which features strengthened mechanical construction as well as pressure relief systems. Arc-resistant switchgear is designed to minimize the probability of an

	arc-flash as well as contain the energy in the event that one occurs.
Blast (electric)	The explosive effect caused by the rapid expansion of air and other vaporized materials that are a superheated by the sudden presence of an electric arc.
Contractor muscle	A muscle whose contraction bends or closes a joint. The bicep is a flexor muscle.
Electrocution	Death caused by the passage of electricity through the body. Death caused by electric shock.
Extensor muscle	A muscle whose contraction extends or stretches a body part. The tricep is an extensor muscle.
Fibrillation	Rapid and inefficient contraction of muscle fibers of the heart caused by disruption of nerve impulses.
Flash	See Arc
Horny layer	The commonly applied name for the stratum corneum layer of the epidermis. The stratum corneum is called the horny layer because its cells are toughened like an animal's horn.
Plasma	A high-temperature, electrically ionized gas. Because of the high temperatures and electrical characteristics of a plasma, it is usually identified as a fourth state of matter. The others including solid, liquid, and gas.
Shock circuit	The path that electric current takes through the body. If the shock circuit includes critical organs, severe trauma is more likely than if it does not.
Shock (electric)	The physical stimulation or trauma that occurs as a result of electric current passing through the body.

HAZARD ANALYSIS

The division of the electrical power hazard into three components is a classic approach used to simplify the selection of protective strategies. The worker should always be aware that electricity is the single root cause of all of the injuries described in this and subsequent chapters. That is, the worker should treat electricity as the hazard and select protection accordingly.

SHOCK

Description

Electric shock is the physical stimulation that occurs when electric current flows through the human body. The distribution of current flow through the body is a function of the resistance of the various paths through which the current flows. The final trauma associated with the electric shock is usually determined by the most critical path called the *shock circuit*. The symptoms may include a mild tingling sensation, violent muscle contractions, heart arrhythmia, or tissue damage. Detailed descriptions of electric current trauma are included in Chap. 8. For the purposes of this chapter, tissue damage may be attributed to at least two major causes.

Burning. Burns caused by electric current are almost always third-degree because the burning occurs from the inside of the body. This means that the growth centers are destroyed. Electric-current burns can be especially severe when they involve vital internal organs.

Cell Wall Damage. Research funded by the Electric Power Research Institute (EPRI) has shown that cell death can result from the enlargement of cellular pores due to high-intensity electric fields.[1] This research has been performed primarily by Dr. Raphael C. Lee and his colleagues at the University of Chicago. This trauma called electroporation allows ions to flow freely through the cell membranes, causing cell death.

Influencing Factors

Several factors influence the severity of electrical shock. These factors include the physical condition and responses of the victim, the path of the current flow, the duration of the current flow, the magnitude of the current, the frequency of the current, and the voltage magnitude causing the shock.

Physical Condition and Physical Response. The physical condition of the individual greatly influences the effects of current flow. A given amount of current flow will usually cause less trauma to a person in good physical condition. Moreover, if the victim of the shock has any specific medical problems such as heart or lung ailments, these parts of the body will be severely affected by relatively low currents. A diseased heart, for example, is more likely to suffer ventricular fibrillation than a healthy heart.

Current Duration. The amount of energy delivered to the body is directly proportional to the length of time that the current flows; consequently, the degree of trauma is also directly proportional to the duration of the current. Three examples illustrate this concept:

1. Current flow through body tissues delivers energy in the form of heat. The magnitude of energy may be approximated by

$$J = I^2 Rt \qquad (1.1)$$

   where J = energy, joules
   I = current, amperes
   R = resistance of the current path through the body, ohms
   t = time of current flow, seconds

   If sufficient heat is delivered, tissue burning and/or organ shutdown can occur. Note that the amount of heat that is delivered is directly proportional to the duration of the current (t).

2. Some portion of the externally caused current flow will tend to follow the current paths used by the body's central nervous system. Since the external current is much larger than the normal current flow, damage can occur to the nervous system. Note that nervous system damage can be fatal even with relatively short durations of current; however, increased duration heightens the chance that damage will occur.

3. Generally, a longer duration of current through the heart is more likely to cause ventricular fibrillation. Fibrillation seems to occur when the externally applied electric field overlaps with the body's cardiac cycle. The likelihood of this event increases with time.

Frequency. Table 1.1 lists the broad relationships between frequency and the harmful effects of current flow through the body. Note that at higher frequencies, the effects of

TABLE 1.1 Important Frequency Ranges of Electrical Injury

Frequency	Regimen	Applications	Harmful effects
DC–10 kHz	Low frequency	Commercial electrical power, soft tissue healing; transcutaneous electrical stimulation	Joule heating; destructive cell membrane potentials
100 kHz–100 MHz	Radio frequency	Diathermy; electrocautery	Joule heating; dielectric heating of proteins
100 MHz–100 GHz	Microwave	Microwave ovens	Dielectric heating of water
10^{13}–10^{14} Hz	Infrared	Heating; CO_2 lasers	Dielectric heating of water
10^{14}–10^{15} Hz	Visible light	Optical lasers	Retinal injury; photochemical reactions
10^{15} Hz and higher	Ionizing radiation	Radiotherapy; x-ray imaging; UV therapy	Generation of free radicals

Joule (I^2t) heating become less significant. This decrease is related to the increased capacitive current flow at higher frequencies.

It should be noted that some differences are apparent even between DC (zero Hz) and standard power line frequencies (50 to 60 Hz). When equal current magnitudes are compared (DC to AC rms), DC seems to exhibit two significant behavioral differences:

1. Victims of DC shock have indicated that they feel greater heating from DC than from AC. The reason for this phenomenon is not totally understood; however, it has been reported on many occasions.
2. The DC current "let-go" threshold seems to be higher than the AC "let-go" threshold.

In spite of the slight differences, personnel should work on or near DC power supplies with the same level of respect that they use when working on or near AC power supplies. This includes the use of appropriate protective equipment and procedures.

Note: Unless otherwise specifically noted, the equipment and procedures suggested in this handbook should be used for all power frequencies up to and including 400 Hz.

Voltage Magnitude.   Historically, little attention was paid to the effect that voltage magnitude has on an electrical trauma. It was assumed that a 200-V source would create the same amount of physical trauma that a 2000-V source would—assuming that the current magnitude is the same. In fact, higher voltages can be more lethal for at least three reasons:

1. At voltages above 400 V the electrical pressure may be sufficient to puncture the epidermis. Since the epidermis provides the only significant resistance to current flow, the current magnitude can increase dramatically.
2. The degree of electroporation is higher for greater cellular voltage gradients. That is, the higher voltages cause more intense fields, which in turn increase the severity of the electroporation.
3. Higher voltages are more likely to create electrical arcing. While this is not a shock trauma per se, it is related to the shock hazard since arcing may occur at the point of contact with the electrical conductor.

Current Magnitude. The magnitude of the current that flows through the body obeys Ohm's law, that is,

$$I = \frac{E}{R} \tag{1.2}$$

where I = current magnitude, amperes (A)
E = applied voltage, volts (V)
R = resistance of path through which current flows, ohms (Ω)

In Fig. 1.1 the worker has contacted a 120-V circuit when an electric drill short-circuits internally. The internal short circuit impresses 120 V across the body of the worker from the hand to the feet. This creates a current flow through the worker to the ground and back to the source. The total current flow in this case is given by the formula

$$I = \frac{E}{R_1 + R_2} \tag{1.3}$$

Variable R_2 is the resistance of the earth and for the purposes of this analysis may be ignored. Variable R_1 is the resistance of the worker's body and includes the skin

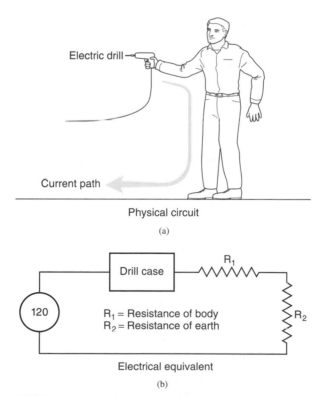

Physical circuit

(a)

Electrical equivalent

(b)

FIGURE 1.1 Electric shock current path.

TABLE 1.2   Nominal Resistance Values for Various Parts of the Human Body

Condition (area to suit)	Resistance Dry	Wet
Finger touch	40 kΩ–1 MΩ	4–15 kΩ
Hand holding wire	10–50 kΩ	3–6 kΩ
Finger-thumb grasp*	10–30 kΩ	2–5 kΩ
Hand holding pliers	5–10 kΩ	1–3 kΩ
Palm touch	3–8 kΩ	1–2 kΩ
Hand around 1½-inch (in) pipe (or drill handle)	1–3 kΩ	0.5–1.5 kΩ
Two hands around 1½-in pipe	0.5–1.5 kΩ	250–750 Ω
Hand immersed	—	200–500 Ω
Foot immersed	—	100–300 Ω
Human body, internal, excluding skin	—	200–1000 Ω

* Data interpolated.
Source: This table was compiled from Kouwenhoven and Milner. Permission obtained from estate of Ralph Lee.

resistance, the internal body resistance, and the resistance of the shoes where they contact the earth.

Typical values for the various components can be found in Tables 1.2 and 1.3. Assume, for example, that a worker shown in Fig. 1.1 is wearing leather shoes and is standing in wet soil. This person is perspiring heavily and has an internal resistance of 200 Ω. From Tables 1.2 and 1.3 the total resistance can be calculated as

500 Ω (drill handle) + 200 Ω (internal) + 5000 Ω (wet shoes) = 5700 Ω

From this information the total current flow through the body for a 120-V circuit is calculated as

$$I = \frac{120}{5700} = 21.1 \text{ milliamperes(mA)} \tag{1.4}$$

Table 1.4 lists the approximate effects that various currents will have on a 68-kilogram (kg) human being. The current flow of 21.1 mA is sufficient to cause the worker to go into

TABLE 1.3   Nominal Resistance Values for Various Materials

Material	Resistance*
Rubber gloves or soles	>20 MΩ
Dry concrete above grade	1–5 MΩ
Dry concrete on grade	0.2–1 MΩ
Leather sole, dry, including foot	0.1–0.5 MΩ
Leather sole, damp, including foot	5–20 kΩ
Wet concrete on grade	1–5 kΩ

* Resistances shown are for 130-cm² areas.
Source: Courtesy of Ralph Lee.

TABLE 1.4 Nominal Human Response to Current Magnitudes

Current (60 Hz)	Physiological phenomena	Feeling or lethal incidence
<1 mA	None	Imperceptible
1 mA	Perception threshold	Mild sensation
1–3 mA		Painful sensation
3–10 mA		
10 mA	Paralysis threshold of arms	Cannot release hand grip; if no grip, victim may be thrown clear (may progress to higher current and be fatal)
30 mA	Respiratory paralysis	Stoppage of breathing (frequently fatal)
75 mA	Fibrillation threshold 0.5%	Heart action discoordinated (probably fatal)
250 mA	Fibrillation threshold 99.5% (≥5-s exposure)	
4 A	Heart paralysis threshold (no fibrillation)	Heart stops for duration of current passage. For short shocks, may restart on interruption of current (usually not fatal from heart dysfunction)
≥5 A	Tissue burning	Not fatal unless vital organs are burned

Notes: (1) This data is approximate and based on a 68-kg (150-lb) person. (2) Information for higher current levels is obtained from data derived from accident victims. (3) Responses are nominal and will vary widely by individual.
Source: Courtesy of Ralph Lee.

an "electrical hold." This is a condition wherein the muscles are contracted and held by the passage of the electric current—the worker cannot let go. Under these circumstances, the electric shock would continue until the current was interrupted or until someone intervened and freed the worker from the contact. Unless the worker is freed quickly, tissue and material heating will cause the resistances to drop, resulting in an increase in the current. Such cases are frequently fatal.

The reader should note that the values given in this example are for illustration only. Much lower values can and do occur, and many workers have been electrocuted in exactly this same scenario.

Parts of the Body. Current flow affects the various bodily organs in different manners. For example, the heart can be caused to fibrillate with as little as 75 mA. The diaphragm and the breathing system can be paralyzed, which possibly may be fatal without outside intervention, with less than 30 mA of current flow. The specific responses of the various body parts to current flow are covered in later sections.

ARC

Caution: The calculation and formulas in this section are shown to illustrate the basic concepts involved in the calculation of arc parameters including current, voltage, and energy. The calculation of actual values for specific field conditions is a complex, safety-related procedure, and should be done only under the direction of experienced engineers.

Definition and Description

ANSI/IEEE Std 100-1988 defines arc as: "A discharge of electricity through a gas, normally characterized by a voltage drop in the immediate vicinity of the cathode approximately equal to the ionization potential of the gas."[2]

A similar definition, perhaps more useful in the discussion of electrical safety is given in the glossary of this handbook as: "The heat and light energy release that is caused by the electrical breakdown of and subsequent electrical discharge through an electrical insulator such as air."

Electric arcing occurs when a substantial amount of electric current flows through what previously had been air. Since air is a poor conductor, most of the current flow is actually occurring through the vapor of the arc terminal material and the ionized particles of air. This mixture of super-heated, ionized materials, through which the arc current flows, is called a *plasma*.

Arcs can be initiated in several ways:

• When the voltage between two points exceeds the dielectric strength of the air. This can happen when overvoltages due to lightning strikes or switching surges occur.

• When the air becomes superheated with the passage of current through some conductor. For example, if a very fine wire is subjected to excessive current, the wire will melt, superheating the air and causing an arc to start.

• When two contacts part while carrying a very high current. In this case, the last point of contact is superheated and an arc is created because of the inductive flywheel effect.

Electric arcs are extremely hot. Temperatures at the terminal points of the arcs can reach as high as 50,000 kelvin (K). Temperatures away from the terminal points are somewhat cooler but can still reach 20,000 K. Although the specific results of such temperatures will vary depending on things such as distance from the arc, ambient environmental conditions, and arc energy; anecdotal evidence supported by experimental results developed by the Institute of Electrical and Electronics Engineers (IEEE) clearly shows the following:

• The heat energy of an electrical arc can kill and injure personnel at surprisingly large distances. For example, second-degree burns have been caused on exposed skin at distances of up to 12 feet (ft) or (3.6 meters [m]) and more.

• Virtually all types of clothing fibers can be ignited by the temperatures of electrical arcs. Clothing made of non-flame resistant fibers will continue to burn after the arc source has been removed and will continue to cause serious physical trauma. Table 1.5 shows the ignition temperature of various fabrics and identifies those that will support combustion after the arc energy is gone.

TABLE 1.5   Ignition Temperatures and Characteristics of Clothing Fibers

Fiber	Melt temperature	Decomposition temperature	Ignition temperature	Burning temperatures
Cotton/Rayon*	N/A	554/581	752/788	1562
Polyester	482	734	1040	1337
Wool	N/A	446	1112	Unknown
Nylon 6,6	490	653	990	1607
PBI	N/A	860	N/A	N/A
NOMEX	N/A	900	N/A	N/A

* FR treatment of cotton of rayon does not affect ignition temperatures.

All temperatures are expressed in °F. Please note that polyester ignites at a higher temperature and burns at a lower temperature than cotton. This shows the fallacy of using untreated cotton as an FR garment.

Source: Courtesy Bulwark Protective Clothing.

The amount of energy, and therefore heat, in an arc is proportional to the maximum available short circuit volt-amperes in the system at the point of the arc. Calculations by Ralph Lee indicate that maximum arc energy is equal to one-half the available fault volt-amperes at any given point.[3] Later research by Neal, Bingham, and Doughty show that while the maximum may be 50 percent, the actual value will usually be somewhat different depending on the degree of distortion of the waveform, the available system voltage, and the actual arc power factor.[4] The same research also shows that enclosing the arc to create a so-called "arc in the box" focuses the incident arc energy and increases its effect by as much as threefold.[4,5]

The arc energy determines the amount of radiated energy and, therefore, the degree of injury. The arc energy will be determined by the arc voltage drop and the arcing current. After the arc is established, the arc voltage tends to be a function of arc length; consequently, the arc energy is less dependent on the system voltage and more dependent on the magnitude of the fault current. This means that *even low voltage systems have significant arc hazard* and appropriate precautions must be taken. Figures 1.2 and 1.3 show the results of two experiments that were conducted with manikins exposed to electric arcs. As can be seen, both high and low voltages can create significant burns.

Arc Energy Release

Arc energy is released in at least three forms—light, heat, and mechanical. Table 1.6 describes the nature of these energy releases and the injuries that they cause. Note that light and heat tend to cause similar injuries and will, therefore, be treated as one injury source in later calculations. Also note that mechanical injuries are usually categorized as blast injuries, even though the ultimate cause is the electric arc.

To be conservative in arc energy release calculations, two assumptions must be made:

1. All arc energy is released in the form of heat measured in cal/cm^2 or J/cm^2. *The reader should remember that this assumption is made solely for the purpose of analyzing electric arc thermal injury. Other hazards such as shock and blast are considered separately.*
2. Every arc is fed by a sinusoidal source, thereby creating the maximum amount of energy release.

TABLE 1.6 Factors That Affect the Amount of Trauma Caused by an Electric Arc

Distance	The amount of damage done to the recipient diminishes by approximately square of the distance from the arc. Twice as far means one-fourth the damage. (Empirical evidence suggests that the actual value may be somewhat different because of the focusing effect of the surroundings.)
Temperature	The amount of energy received is proportional to the difference between the fourth power of the arc temperature and the body temperature ($T_a^4 - T_b^4$).
Absorption coefficient	The ratio of energy received to the energy absorbed by the body.
Time	Energy received is proportional to the amount of time that the arc is present.
Arc length	The amount of energy transmitted is a function of the arc length. For example, a zero length arc will transmit zero energy. Note that for any given system, there will be an optimum arc length for energy transfer.
Cross-sectional area of body exposed to the arc	The greater the area exposed, the greater the amount of energy received.
Angle of incidence of the arc energy	Energy is proportional to the sine of the angle of incidence. Thus, energy impinging at 90° is maximum.

(a)

(b)

(c)

FIGURE 1.2   Electric arc damage caused by a medium voltage arc. (*Courtesy Brosz and Associates.*)

(a)

(b)

(c)

FIGURE 1.3 Electric arc damage caused by 240 volt arc. (*Courtesy Brosz and Associates.*)

1.11

TABLE 1.7   Electrical Arc Injury, Energy Sources

Energy	Nature of injuries
Light	Principally eye injuries, although severe burns can also be caused if the ultra-violet component is strong enough and lasts long enough.
Heat	Severe burns caused by radiation and/or impact of hot objects such as molten metal.
Mechanical	Flying objects as well as concussion pressures.

Arc Energy

Several major factors determine the amount of energy created and/or delivered by an electric arc. Table 1.7 lists the major factors and their qualitative effect. The quantitative effects of electric arc are the subject of many on-going studies.

An individual's exposure to arc energy is a function of the total arc energy, the distance of the subject from the arc, and the cross-sectional area of the individual exposed to the arc.

Arc Energy Input

The energy supplied to an electric arc by the electrical system, called the arc input energys may be calculated using the formula

$$J_{arc} = \int_{o}^{t} V_{arc} \times I_{arc} \times \cos(\theta) \times dt \tag{1.5}$$

where J_{arc} = arc energy, joules
$\quad\quad V_{arc}$ = arc voltage, volts
$\quad\quad I_{arc}$ = arc current, amperes
$\quad\quad t$ = time, seconds

Research has shown that electric arcs are rarely perfect sinusoids; however, the perfect sinusoid creates the greatest arc power. Therefore, Eq. 1.5 can be solved as

$$J_{arc} = V_{arc} \times I_{arc} \times t \times \cos(\theta) \tag{1.6}$$

where θ = the angle between current and voltage

Arcing Current.   The actual arcing current varies as a function of several variables, and has been calculated or estimated in different ways. IEEE Standard Std 1584-2002, for example, gives two equations that may be used to calculate the arcing current.[7] Equation 1.7 is the formula used for electrical arcs in systems with voltages less than 1000 V and Eq. 1.8 is used for systems with voltages equal to or greater than 1000 V.

$$\log_{10}(I_a) = K + 0.662 \log_{10}(I_{bf}) + 0.0966V + 0.000526G$$
$$+ 0.5588V \log_{10}(I_{bf}) - 0.00304G \log_{10}(I_{bf}) \tag{1.7}$$

$$\log_{10}(I_a) = 0.00402 + 0.9831 \log_{10}(I_{bf}) \tag{1.8}$$

where I_a = arcing current (kA)
    K = a constant: (−0.153 for open configurations or −0.097 for box configurations)
    I_{bf} = the bolted, RMS symmetrical, three-phase fault current (kA)
    V = system phase-to-phase voltage (kV)
    G = the gap between the arcing conductors (mm)

Note that these equations are based on a specific model as developed for Std 1584-2002. The model includes the following:

- Three-phase voltages in the range of 208 to 15,000 V phase-to-phase
- Frequencies of 50 or 60 Hz
- Bolted fault current in the range of 700 to 106,000 A
- Grounding of all types and ungrounded
- Equipment enclosures of commonly available sizes
- Gaps between conductors of 13 to 152 mm
- Faults involving all three phases

Arcing Voltage

Arc voltage is somewhat more difficult to determine. Values used in power system protection calculations vary from highs of 700 V/ft (214.4 V/m) to as low as 300 V/ft (91.4 V/m). Two things are well understood:

1. Arc voltages start low and tend to rise. Periodically, the arc voltage will drop if the arc lasts long enough.[6]
2. Arc voltage is proportional to arc length. Therefore, from Eq. 1.6, arc power and energy are proportional to arc length.

Modern software programs used to calculate incident arc energy take different approaches to determine arc voltage. It should be noted that, at best, arc voltage calculation is only approximate for any given scenario.

Arc Surface Area

While the actual shape of an electrical arc may vary, all classic, realistic solutions start by assuming that an arc causes an approximately cylindrical plasma cloud with length L and radius r. This cylindrical structure will have a lateral surface area equal to $2\pi rL$. The area of the ends of the cylinder are ignored in this calculation since they are so small relative to the side of the arc. To simplify the calculation of energy density, the arc is assumed to form a sphere with a surface area equal to the cylinder, Fig. 1.4. Thus, the arc sphere will have a radius of

$$r_s = \tfrac{1}{2}\sqrt{2rL} \tag{1.9}$$

where r_s = radius of equivalent sphere
    r = radius of arc cylinder
    L = length of arc

FIGURE 1.4   Arc cylinder and equivalent arc sphere.

Incident Energy

The single most important of all arc energy calculations is the one that determines the energy transfer from the arc to the nearby body. This is called the incident energy. This information can be used to determine the necessary level of protective clothing required, and can also be used in the performance of a risk analysis.

The ultimate measure of tissue injury from electrical arc is the temperature to which the tissue rises during exposure. However, calculation of these temperatures starts with the calculation of the amount of energy, or heat flux (measured in calories per square centimeter), delivered to the skin. Many methods have been developed to calculate the incident energy—some more conservative than others. The following sections describe some of the methods that have been determined either empirically or theoretically. The reader should be aware that research into these areas is continuing at a frantic pace. Always refer to the most recent industry literature for the most up-to-date information.

The Lee Method.   Ralph Lee has predicted that the heat energy received by an object (or worker) can be calculated using Eq. 1.10.

$$Q_o = \frac{Q_s \times A_s}{4\pi \times r^2} \times t \qquad (1.10)$$

where Q_o = heat flux received by the object (cal/cm^2)
$\quad Q_s$ = heat flux generated by source (cal/s/cm^2)
$\quad A_s$ = surface area of arc sphere
$\quad r$ = distance from center of source to object (cm)
$\quad t$ = length of arc exposure

Using Eq. 1.8 as a starting point, Lee determined that the energy received by the worker is calculated using Eq. 1.11

$$E = 2.142 \times 10^6 \, VI_{bf} \left(\frac{t}{D^2} \right) \qquad (1.11)$$

where E = incident energy in J/cm^2
$\quad V$ = system voltage (phase-to-phase)

t = arcing time (seconds)
D = distance from arc point to person or object (mm)
I_{bf} = bolted fault current (kA)

Other Early Research. Research by Bingham and others[4,5,7] has yielded a slightly different result based primarily on empirical results. Using an experimental setup,[7] the researchers measured energy received from an electric arc at various distances. The arc was created using a 600-V source, and different configurations were used to simulate a completely open-air arc versus the so-called "arc-in-a-box." Using these experiments, they developed two equations to model the amount of energy received.

$$E_{MA} = 5271 D_A^{-1.9593} t_A (0.0016F^2 - 0.0076F + 0.8938) \qquad (1.12)$$

$$E_{MB} = 1038.7 D_B^{-1.4738} t_B (0.0093F^2 - 0.3453F + 5.975) \qquad (1.13)$$

where E_{MA} = maximum open-arc incident energy (cal/cm^2)
E_{MB} = maximum arc-in-a-box incident energy (cal/cm^2)
D = distance from the arc electrodes in inches ($D \geq 18$ in)
F = bolted fault current available in kA (16 to 50 kA range)
t_A = duration of the open-air arc (seconds)
t_B = duration of the arc-in-a-boc (seconds)

Note that these equations were developed with three constraints:

1. System voltage = 600 V
2. System available fault current $F - 16,000$ A $< F < 50,000$ A
3. Electrode distance $D \geq 18$ in (45.72 cm)

Using the model previously described (see "Arcing Current"), the IEEE Std 1584-2002 calculates incident energy by first calculating the normalized incident energy. The normalized energy is calculated for an arc time of 0.2 seconds and a distance of 610 mm. The empirically developed formula for this calculation is shown in Eq. 1.14.

$$\log_{10}(E_n) = K_1 + K_2 + 1.081[\log_{10}(I_a)] + 0.0011G \qquad (1.14)$$

where I_a = arc current calculated from Eq. 1.7
E_n = incident energy (J/cm^2) normalized for time and distance
K_1 = constant and is equal to -0.792 for open configurations and -0.555 for enclosed configurations
K_2 = constant and is equal to 0 for ungrounded and high resistance systems and -0.113 for grounded systems
G = gap between conductors (mm)

After the $\log_{10} E_n$ is calculated from Eq. 1.14, Eq. 1.15 is used to calculate E_n, and Eq. 1.16 is used to calculate the actual incident energy.

$$E_n = 10^{\log_{10} E_n} \qquad (1.15)$$

$$E = 4.184 C_f E_n \left(\frac{t}{0.2} \right) \left(\frac{610^x}{D^x} \right) \qquad (1.16)$$

where E = incident energy (J/cm^2)
C_f = calculation factor equal to 1.0 for voltages above 1 kV and 1.5 for voltage equal to or below 1 kV.

E_n = normalized incident energy calculated from Eqs. 1.14 and 1.15
t = arcing time in seconds
D = distance from the arc point to the exposed worker (mm)
x = distance exponent whose value is dependent on voltage. (See IEEE Std 1584-2002, Table 4 for values of x.)

Extensive research continues to be performed on the subject of incident arc energy. Practical applications used in the selection of protective equipment are covered in detail in Chap. 2.

Arc Burns

Arc burns are thermal in nature and, therefore, fall into one of the three classical categories:

- *First-degree burns.* First-degree burning causes painful trauma to the outer layers of the skin. Little permanent damage results from a first-degree burn because all the growth areas survive. Healing is usually prompt and leaves no scarring.

- *Second-degree burns.* Second-degree burns result in relatively severe tissue damage and blistering. If the burn is to the skin, the entire outer layer will be destroyed. Healing occurs from the sweat glands and/or hair follicles.

- *Third-degree burns.* Third-degree burns to the skin result in complete destruction of the growth centers. If the burn is small, healing may occur from the edges of the damaged area; however, extensive third-degree burns require skin grafting.

Refer to Chap. 8 for more detailed coverage of electrical arc trauma.

BLAST

When an electric arc occurs, it superheats the air instantaneously. This causes a rapid expansion of the air with a wavefront that can reach pressures of 100 to 200 lb per square foot (lb/ft^2)(4.79 to 9.58 kPa). Such pressure is sufficient to explode switchgear, turn sheet metal into shrapnel, turn hardware into bullets, push over concrete walls, and propel molten metal at extremely high velocities.

Blasts do not always occur. Sometimes an arc is not accompanied by a blast, but when it is, it can be lethal. Figure 1.5a to c shows physical evidence of the pressure exerted by an electric blast.

In Fig. 1.5a the interior of a medium-voltage cubicle can be seen. The severe scorching on the right-hand side of the interior and exterior of the cubicle is clear evidence of a significant arc-flash event. By looking closely at the cable terminations evidence of the contact points for the electric arc as it occurred can be seen.

Figure 1.5b shows the aisle between the switchgear where the arc flash occurred (right side) and the adjacent gear across the aisle. The metal panel in the aisle is the one that covered the cubicle where the arc flash event occurred. Note that the fully secured panel was blown completely off the faulted switchgear across the aisle and smashed into the adjacent switchgear.

Figure 1.5c is a close-up view of the adjacent switchgear showing where the metal panel, seen in the lower right side of the photo, is dented and crumpled in the cubicle cover.

(a)

FIGURE 1.5 (a) Interior of a medium-voltage cubical showing the results of an electrical arc and accompanying electrical blast; (b) external view of an aisle and adjucent switchgear for arc flash event shown in Fig. 1.5a; (c) closeup view of adjucent switchgear showing metal covers damaged by impact of panel blown across aisle by arc flash event.

Taken together, these three photos clearly illustrate the following two key points:

- Workers may not assume that they are safe from electrical arc flash events even though the access doors and panels are fully secured.
- Unless it is specially designed arc-resistant switchgear, metal-clad equipment might not withstand the explosive force of an electrical blast.

AFFECTED BODY PARTS

General

Detailed information on the medical aspects of electrical trauma is provided in Chap. 8. The following sections are for overview only.

(b)

(c)

FIGURE 1.5   *(Continued)*.

Skin

Definition and Description.   Skin is the outer layer that completely encloses and envelopes the body. Each person's skin weighs about 4 lb, protects against bacterial invasion and physical injury of underlying cells, and prevents water loss. It also provides the body with sensation, heat regulation, excretion (sweat), and absorbs a few substances. There are about 20 million bacteria per square inch on the skin's surface as well as a forest of hairs, 50 sweat glands, 20 blood vessels, and more than 1000 nerve endings. Figure 1.6 is a cross section of the upper layers of skin tissue.

The main regions of importance for electrical purposes are the epidermis, the sweat glands, and the blood vessels.

The epidermis, the top-most layer of skin, is 0.1 to 1.5 millimeters thick; however it is made up of five layers including the basal cell layer, the squamous cell layer, the stratum granulosum, the stratum lucidum, and the outermost layer called the *stratum corneum* or "horny layer."

The stratum corneum comprises 10 to 30 thin layers of dead cells which have been "pushed" up from the lower layers in the process of the normal growth process. It is called the horny layer because its cells are toughened like an animal's horn.

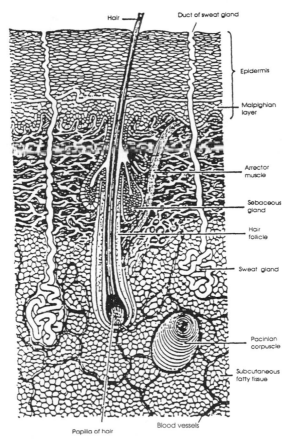

FIGURE 1.6   Typical skin cross section.

The stratum corneum is composed primarily of a protein material called *keratin*. Of all the skin layers, keratin exhibits the highest resistance to the passage of electricity. When areas of the epidermis such as the hands or feet are subjected to friction, the horny layer becomes thickened and toughened. Areas that are toughened in this manner are called corns or callouses.

The sweat glands and the blood vessels have relatively low resistance to the passage of electricity and provide a means of electrical access to the wet and fatty inner tissues. Most of the electrical resistance exhibited by the human body (See Table 1.2) comes from the stratum corneum. Internal resistance is typically in the area of 200 Ω.

Effects on Current Flow.   Since the body is a conductor of electricity, Ohm's law applies as it does to any other physical substance. The thicker the horny layer, the greater the skin's electrical resistance. Workers who have developed a thick stratum corneum have a much higher resistance to electricity than a child with an extremely thin layer. However, as Table 1.2 shows, even high skin resistance is not sufficient to protect workers from electric shock.

Skin resistance is also a function of how much skin area is in the circuit. Therefore, grasping a tool with the entire hand gives a much lower resistance than touching the tool with a finger. Also, any cut or abrasion penetrates the horny layer and significantly reduces the total resistance of the shock circuit. Moisture, especially sweat, greatly reduces the skin's resistance.

A remarkable thing occurs to the skin insulation when voltages above 400 V are applied. At these voltages the epidermis is punctured like any film insulation, and only the low-resistance inner layers are left. This is a major consideration for the many 480 V distribution systems commonly used today. Note that the epidermis may not puncture, but if it does, the current flow increases and shock injury is worse.

Burns.   Electrically caused burns can come from at least four different sources:

- Physical contact with conductors, tools, or other equipment that have been heated by the passage of electrical current flow. These types of burns are no different than burns received from any hot object.

- Current flow through the skin can cause burns from the I^2R energy. Since such burns occur from the inside-out, they are frequently third degree.

- Thermal or radiation burns are caused by the radiant energy of the electrical arc.

- Contact with superheated plasma caused by the vaporization of solid materials in the vicinity of an electrical arc.

Figure 1.7 shows a victim of burns caused by both the radiant arc energy and contact with the superheated plasma.

The Nervous System

Definition and Description.   The nervous system comprises the electrical pathways that are used to communicate information from one part of the human body to another. To communicate, electric impulses are passed from one nerve to another. For example, the heart beats when an electric impulse is applied to the muscles that control it. If some other electric impulse is applied, the nervous system can become confused. If the current is high enough, the damage can be permanent.

Shock.   As far as the nervous system is concerned, at least three major effects can occur when current flows through the body:

1. *Pain.*   Pain is the nervous system's method of signaling injury. When current flows through the nerves, the familiar painful, tingling sensation can result.

FIGURE 1.7 Thermal burns caused by high-voltage electric arc.

2. *Loss of control.* An externally applied current can literally "swamp" the normal nervous system electric impulses. This condition is similar to electrical noise covering an information signal in a telemetering or other communications system. When this happens, the brain loses its ability to control the various parts of the body. This condition is most obvious during the electrical paralysis, or electrical hold, that is described later in this chapter.

3. *Permanent damage.* If allowed to persist, electric current can damage the nervous system permanently. This damage takes the form of destroyed neurons and/or synapses. Since the nervous system is the communications pathway used to control the muscles, such damage can result in loss of sensation and/or function depending on the type of injury.

Muscular System

Definition and Description. The muscular system provides motor action for the human body. When the nervous system stimulates the muscles with electric impulses, the muscles contract to move the body and perform physical activity. Skeletal muscles are either flexor or extensor muscles. When a flexor muscle, such as the bicep, contracts, it bends or closes the joint. When an extensor muscle, such as the tricep, contracts, it unbends or opens the joint.

The heart and pulmonary system are also muscle related, and they will be covered in a later section.

Shock. Electrical shock can affect muscles in at least three significant ways:

1. *Reflex action.* Muscular contractions are caused by electric impulses. Normally these impulses come from the nervous system. When an externally induced current flows

through a muscle, it can cause the muscle to contract, perhaps violently. This contraction can cause workers to fall off ladders or smash into steel doors or other structures.

2. *Electrical paralysis.* Current magnitudes in excess of 10 mA may be sufficient to block the nervous system signals to the muscular system. Thus, when such an external current is flowing through the body, the victim may be unable to control his or her muscles. This means that the victim cannot let go—he or she is caught in an electrical hold. As the current continues, the heating and burning action can lower the path resistance and cause an increase in the current. If the current is not cut off or if the victim is not freed from the circuit, death will occur.

3. *Permanent damage.* If the current is high enough, the muscle tissue can be destroyed by burning. Currents of even less than 5 A will cause tissue destruction if they last long enough. Because such burning destroys the growth areas in tissue, the damage can be extremely slow to heal. Physical therapy and other extraordinary methods may be required to restore muscular function.

The Heart

Definition and Description. The heart is a fist-sized pump that beats more the 2.5 billion times in a 75 year lifetime. Figure 1.8 shows the structural layout of the heart. The atria and ventricles work together, alternately contracting and relaxing to pump blood through your heart.

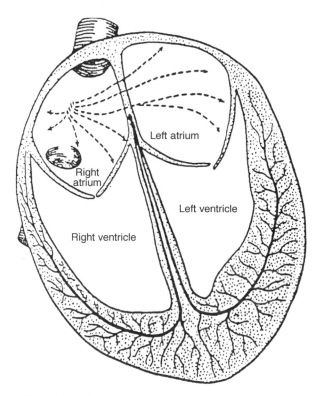

Left atrium

Right atrium

Left ventricle

Right ventricle

FIGURE 1.8 The heart.

The electrical system of your heart is the power source that makes this possible. Normally a heart beat starts in the sinus node, travels at approximately 7 ft/s (2.1 m/s) through the AV node, "HIS bundle," and right and left bundle branches. The resulting contraction sends blood flowing from the heart.

This sequence occurs with every beat (usually 60 to 100 times per minute). If the path is interrupted for any reason, even for a few minutes, changes in the heart rate and rhythm occur that can be fatal.

Shock.   When the heart's electrical system is disturbed for any reason, such as an outside current from an electric shock, changes in the heart's rate and rhythm occur. Such disruptions result in a large percentage of heart deaths.

The electric impulses in the heart must be coordinated to give a smooth, rhythmic beat. An outside current of as little as 60 to 75 mA can disturb the nerve impulses so that there is no longer a smooth, timed heartbeat. Instead the heart fibrillates—that is, it beats in a rapid, uncoordinated manner. When a heart is fibrillating, it flutters uselessly. Prolonged exposure to an outside current exceeding 75 mA is likely to result in death.

Like any muscle, the heart will become paralyzed if the current flowing through it is of sufficient magnitude. Oddly, paralysis of the heart is not often fatal if the current is removed quickly enough. In fact, such paralysis is used to an advantage in defibrillators. A defibrillator intentionally applies heart-paralyzing current. When the current is removed, the heart is in a relaxed state ready for the next signal. Frequently the heart restarts.

Burns.   Any internal organ, such as the heart, can be burned by current flows in excess of 5 A. Such burns are often fatal.

The Pulmonary System

Definition and Description.   With the exception of the heart, the pulmonary system is the most critical to human life. If breathing stops, which can happen with as little as 30 mA, all other functions cease shortly thereafter. When the lower diaphragm moves down, it creates a partial vacuum in the chest chamber. This in turn draws air into the sacs in the lungs. The oxygen is then passed to the bloodstream through the tiny capillaries. At the same time, carbon dioxide is returned to the air in the lungs. When the lower diaphragm moves up, the air is forced out of the lungs, thus completing the breathing cycle.

Shock.   Current flow through the midsection of the body can disrupt the nervous system impulses which regulate the breathing function. This disruption can take the form of irregular, sporadic breathing, or—if the current flow is sufficient—the pulmonary system may be paralyzed altogether. When such stoppage occurs, first aid is often required.

SUMMARY OF CAUSES—INJURY AND DEATH

Shock Effect

Table 1.4 summarizes the effects that electric shocks of varying amounts of current will have on a 68-kg (150-lb) person. Note that these effects are only approximate and vary between individuals.

Arc Flash Effect

Tables 1.6 and 1.8 summarize the effects that electric arc flash energy have on personnel. Note that these effects are still being studied and the full range of injury potential is not yet fully evaluated and documented.

Causes of Injury

Injury from electrical hazard can come from both direct and indirect sources:

- The reflex action caused by the passage of current flow can cause falls resulting in cuts, abrasions, or broken limbs.
- Nerve damage from shock or burns can cause loss of motor function, tingling, and/or paralysis.
- Burns, both thermal- and current-induced, can cause extremely long duration and intensely painful suffering. Third-degree burns may require skin grafting to heal.
- The light intensity, molten metal and/or burns to the eyes can cause blindness.
- The concussion of a blast can cause partial or complete loss of hearing.
- Current-induced burns to internal organs can cause organ dysfunction.
- The superheated plasma may be inhaled, causing severe internal burns.
- Metal vapors may be inhaled filling the lungs with toxic residues.

TABLE 1.8   Equipment and Procedural Strategies for Protection from the Three Electrical Hazards (See Chaps. 2 and 3 for detailed information.)

Hazard	Equipment strategy	Procedural strategy for all three hazards
Shock	• Rubber insulating equipment including gloves with leather protectors, sleeves, mats, blankets, line hose, and covers. • Insulated tools when working near energized conductors.	• De-energize all circuits and conductors in the immediate work area. • Develop and follow a lockout/tagout procedure. • Maintain a safe working distance from all energized equipment and conductors. • Use all specified safety equipment. • Follow all safety procedures and requirements. • Carefully inspect all equipment before placing it into service. This includes tools, test equipment, electrical distribution equipment, and safety equipment. • Make certain that all nonenergized equipment is properly grounded. This applies to both normal system grounding and temporary safety grounds. • Design or re-design systems to be intrinsically safe.
Arc	• Approved flash/flame-resistant work clothing. • Approved flash suits when performing work with a high risk of arcing. • Use hot sticks to keep as much distance as possible. • Wear eye protection. • Wear rubber gloves with leather protectors and/or other flashproof gloves.	
Blast	• Approved flash/flame-resistant work clothing. This will protect from splashed molten material. • Approved flash suits when performing work with a high risk of arcing. This will protect from splashed molten material. • Wear face shields.	

Causes of Death

If the electrical injury is severe enough, death can result.

- An electric shock–induced fall can cause fatal physical injuries.
- When the skin is severely burnt, large quantities of liquid are brought to the burnt areas to aid in the healing process. This creates a stress on the renal system and could result in kidney failure.
- Severe trauma from massive burns can cause a general systemic failure.
- Burnt internal organs can shut down—causing death. Thus, the more critical the organ that is burnt, the higher the possibility of death.
- The pressure front from the blast can cause severe injury to the lungs, called *blast-lung,* resulting in death.
- Heart failure can result from fibrillation and/or paralysis.

PROTECTIVE STRATEGIES

The types of strategies that may be employed to protect from each of the three electrical hazards are remarkably similar. Table 1.8 summarizes the types of protective strategies that may be used. Note that the information given in Table 1.8 is general. Specific equipment and procedures are covered in Chaps. 2 and 3.

Be aware that any given strategy may not be applicable in a given situation. For example,

- When troubleshooting equipment, de-energization may not be possible.
- De-energization may create an additional, unacceptable hazard. For example, if de-energization shuts down ventilation equipment in a hazardous area, workers may opt for working with energized equipment.
- Shutdown of an entire continuous process plan to work on or around one small auxiliary circuit may not be economically feasible.

If electrical equipment cannot be de-energized, a hierarchy of electrical safe work practices and procedures must be implemented as follows:

- Document the reason(s) for having to work on or near energized electrical conductors or circuit parts. Identify that de-energizing is infeasible.
- Obtain responsible manager's (employer's designated person) signature, verifying knowledge and recognizing the need to work on or near exposed energized electrical conductors or circuit parts.
- Perform a job briefing that includes a hazard-risk analysis.
- Prepare and follow a documented plan of work processes and tasks.
- If worker(s) must enter closer to the potential arc source(s) than the flash protection boundary, then ensure that the worker is a "qualified person" and is wearing appropriate personal protective clothing (PPC) and using other personal protective equipment (PPE).

REFERENCES

1. Raphael C. Lee, "Injury by Electrical Forces: Pathophysiology, Manifestations and Therapy," *Current Problems in Surgery,* vol. 34, no. 9, 1997.
2. Institute of Electrical and Electronics Engineers (IEEE), *Standard Dictionary of Electrical and Electronics Terms*, ANSI/IEEE Std 100-1988, p. 240.
3. Ralph H. Lee, "The Other Electrical Hazard: Electric Arc Blast Burns," *IEEE Trans. Industrial Applications,* vol. 1A-18, no. 3, 1982, p. 246.
4. Thomas E. Neal, Allen H. Bingham, and Richard I. Doughty, "Protective Clothing Guidelines for Electric Arc Exposure," IEEE Paper PCIC-96-34 presented at the Petroleum Chemical Industry Conference in Philadelphia, 1996.
5. Richard I. Doughty, Thomas E. Neal, Terrence A. Dear, and Allen H. Bingham, "Testing Update on Protective Clothing & Equipment for Electric Arc Exposure," IEEE Paper PCIC-97-35, 1997.
6. J. Lewis Blackburn, *Protective Relaying Principles and Applications,* 2nd ed., Marcel Dekker, New York, 1998.
7. Richard L. Doughty, Thomas E. Neal, and H. Landis Floyd, "Predicting Incident Energy to Better Manage the Electric Arc Hazard on 600-V Power Distribution Systems," IEEE Paper.
8. Institute of Electrical and Electronic Engineers (IEEE), *IEEE Guide for Performing Arc-Flash Calculations*, IEEE Std 1584–2002.

CHAPTER 2

ELECTRICAL SAFETY EQUIPMENT

INTRODUCTION

The safety aspects of any job or procedure are greatly enhanced by the use of proper tools meters, apparel, and other such equipment. This chapter outlines the construction and use of a variety of electrical safety equipment. Some of the equipment is used to actually perform work—items such as insulated tools or voltage-measuring devices fall into this category. Other safety products are used strictly to protect the worker, for example, flash suits and rubber goods.

Each specific piece of safety equipment is used to protect the worker from one or more of the three electrical safety hazards, and each piece of equipment should be employed when performing various types of jobs in the electric power system.

Always be certain that the clothing or apparel and the meters or tools you are using are designed and tested to match or exceed the incident energy level, and the voltage level for the application to which you will be exposed.

GLOSSARY

ATPV (Arc Thermal Performance Value)	". . .the incident energy on a fabric or material that results in sufficient heat transfer through the fabric or material to cause the onset of a second-degree burn based on the Stoll curve."[1]
E_{BT} (Energy Break-Through)	"the average of the five highest incident energy exposure values below the Stoll curve where the specimens do not exhibit breakopen."[1]
Flame Resistant	A fabric or material that burns very slowly and self-extinguishes upon the removal of an external flame.
Flame retardant	A chemical applied to or incorporated into a fabric at the time of production, which significantly reduces a fabric's flammability.
Flammable	Easily ignited and capable of burning rapidly.
Flash Boundary	The radial distance from an energized conductor at which the heat energy will equal 1.2 cal/cm^2 in the event that the worst case arc-flash event occurs.
Stoll Curve	A tabular or graphical relationship which quantitatively describes the relationship between heat energy and human skin.[2]

GENERAL INSPECTION AND TESTING REQUIREMENTS FOR ELECTRICAL SAFETY EQUIPMENT

Each of the types of electrical safety equipment described in this chapter has specific inspection and testing requirements. These requirements are identified in each of the individual sections. In addition to specific requirements, the following precaution should always be observed:

Always perform a detailed inspection of any piece of electrical safety equipment before it is used. Such an inspection should occur at a minimum immediately prior to the beginning of each work shift, and should be repeated any time the equipment has had a chance to be damaged.

Where possible, the guidelines for wearing and/or using the various types of safety equipment discussed in this chapter are based on existing industry standards. In any event, the guidelines used in this book should be considered *minimum*. Requirements for specific locations should be determined on a case-by-case basis using current industry standards.

FLASH AND THERMAL PROTECTION

The extremely high temperatures and heat content of an electric arc can cause extremely painful and/or lethal burns. Since an electric arc can occur at any time in electrical equipment that has not been placed in an electrically safe work condition, the worker must take precautions and wear protection when exposed to potential arc hazards. Note that these sections address equipment for electrical hazard. Fire protection equipment has slightly different requirements and is not covered.

Table 2.1 itemizes the type of equipment required to protect the worker from the thermal hazards of electric arc. The next sections describe the type of equipment used and will identify when and how to use that equipment.

A Note on When to Use Thermal Protective Clothing

The usage directions given in this chapter should be used as minimum guidelines only. Modern technology has enabled the calculation of actual incident arc energies. When these arc energies are compared to the Arc Thermal Performance Value (ATPV) discussed later in this chapter, the exact weight and type of thermal clothing can be determined.

TABLE 2.1 Equipment Used to Protect Workers from Arc Hazard

Area of body to be protected	Equipment used
Torso, arms, legs	Thermal work uniforms, flash suits
Eyes	Face shields, goggles, safety glasses
Head	Insulating hard hats, flash hoods
Hands	Rubber gloves with leather protectors, thermally resistant gloves

If workers are required to place any part of their body within the *flash boundary* distance of an energized electrical component, they must wear thermal protective clothing with an ATPV or E_{BT} equal to or greater than the amount of arc energy to which they might be exposed. Chapter 3 of this handbook provides specific methods for these calculations. The ATPV or E_{BT} for any given material is calculated using the procedures defined in ASTM Standard F 1959/F 1959M.[1]

Thermal Performance Evaluation

Flame Resistance (FR).   Virtually all clothing will ignite when exposed to a sufficient heat source. When the heat source is removed, normal clothing will continue to burn. Flame-resistant clothing may burn and char when it is exposed to a heat source, but it will not continue to burn after the heat source is removed.

The most common test for flame resistance is defined in Method 5903.1 of Federal Test Standard 191A (Flame Resistance of Cloth: Vertical). This test suspends a 12-inch-long specimen of fabric vertically in a holder. The fabric is enclosed and subjected to a controlled flame on the bottom edge of the fabric for 12 seconds. Table 2.2 lists the three sets of data that are recorded in this test.

Note that the results are gathered *after* the flame source has been removed. Note also that the *afterglow* is not included in most of the industry standards that reference this method.

Arc Thermal Performance Value (ATPV).   Research by Stoll & Chianta[2] developed a curve (the so-called Stoll curve) for human tolerance to heat. The curve is based on the minimum incident heat energy (in kJ/m^2 or cal/cm^2) that will cause a second-degree burn on human skin. Modern standards that define the level of thermal protection required are based on the Stoll curve. That is, clothing must be worn that will limit the degree of injury to a second-degree burn. This rating is called the *Arc Thermal Performance Value* (ATPV).

Energy Break-Through (E_{BT}).   Some types of flame resistant materials become brittle when exposed to very high temperatures. The combination of the brittleness and the concussion from the electrical blast can cause the material to fail and break open. ASTM has defined this value as the average of the five highest incident energy exposure values below the Stoll curve where the specimens do not exhibit breakopen.

Since the E_{BT} is defined for values below the Stoll curve, garments made from such materials can be safely used where exposures will be less than the energy break-through level.

TABLE 2.2   Test Data Drawn from Method 5903.1 of Federal Test Standard 191A

Test result measured	Description
Afterflame	The number of seconds (in tenths) during which there is a visible flame remaining on the fabric
Afterglow	The number of seconds (in tenths) during which there is a visible glow remaining on the fabric
Char length	The length of the fabric in tenths of an inch destroyed by the flame that will readily tear by application of a standard weight

ASTM and Other Standards. The American Society of Testing and Materials (ASTM) has three standards that apply to the thermal protective clothing to be worn by electrical workers. Table 2.3 identifies these standards.

ASTM Standard F 1506 specifies three requirements for workers' clothing in Part 6:

1. Thread, bindings, and closures used in garment construction shall not contribute to the severity of injuries to the wearer in the event of a momentary electric arc and related thermal exposure.
2. Afterflame is limited to 2.0 s or less and char length is limited to 6.0 in or less. Afterglow is mentioned but is not judged a serious hazard.
3. Garments must be labeled with the following information:
   a. Tracking identification code system
   b. Meets requirements of ASTM F 1506
   c. Manufacturer's name
   d. Size and other associated standard labeling
   e. Care instructions and fiber content

Standard F 1959 defines the technical specifications of measuring the Arc Thermal Performance Value. Note that garments only pass the 5903.1 method with 6.0 in or less char length and 2.0 s or less afterflame.

Standard F 2178 defines the technical specifications for measuring the arc rating for protective face shields used in arc protective clothing. The standard provides methods for arc thermal ratings only and specifically does not address molten materials, fragmentation materials, and other such flying debris.

Usage Standards. The principal standards for electrical worker thermal protection are OSHA 1910.269 and ANSI/NFPA 70E. Of these two, 70E is the most rigorous and provides the best level of protection, and it defines user thermal protection requirements on the basis of the ATPV. The usage requirements described in this handbook are based primarily on ANSI/NFPA 70E.

Clothing Materials

Materials used to make industrial clothing fall into two major categories, with several subcategories under each as follows:

1. Non-flame-resistant materials. When these materials are treated with a flame-retardant chemical, they become flame resistant.
   a. Natural fibers such as cotton and wool
   b. Synthetic fibers such as polyester, nylon, and rayon

TABLE 2.3 ASTM Standards Defining Electrical Worker Thermal Clothing

Standard number	Title
F 1506	Standard Performance Specification for Textile Materials for Wearing Apparel for Use by Electrical Workers Exposed to Momentary Electric Arc and Related Thermal Hazards
F 1959	Standard Test Method for Determining the Arc Thermal Performance Value of Materials for Clothing
F2178	Standard Test Method for Determining The Arc Rating of Face Protective Products

2. Flame-resistant materials
   a. Non-flame-resistant materials that have been chemically treated to be made flame-resistant
   b. Inherently flame-resistant materials such as PBI, Kermel, and Nomex

The following sections describe some of the more common fibers and identify their general capabilities with respect to thermal performance.

Non-Flame-Resistant Materials

Contrary to some misunderstandings, natural fibers such as cotton and wool *are not flame-resistant*. In fact, the only advantage that natural fibers exhibit over synthetics such as polyester is that they do not melt into the burn. Do not use natural fibers and expect to get the type of protection afforded by true flame-resistant materials.

Cotton. Cotton work clothing made of materials such as denim and flannel is a better choice than clothing made from synthetic materials. Cotton does not melt into the skin when heated; rather, it burns and disintegrates, falling away from the skin. Thick, heavy cotton material provides a minimal barrier from arc temperatures and ignites quickly. At best, cotton provides only minimal thermal protection.

Wool. Wool clothing has essentially the same thermal properties as cotton clothing. Note the following key points:

- Certain OSHA standards allow the use of natural cotton or wool clothing when a worker is exposed to electrical arcing hazards. The standards suggest weights of 11 oz/yd^2 or heavier.
- General industrial practice and industry standards have rejected the use of cotton and wool even though they are permitted. The reason for this is that these natural fibers are not flame-resistant. They will burn and, while they are not as bad as some synthetic fibers, they can aggravate the degree of injury.

Synthetic Materials. Untreated synthetic clothing materials such as polyester and nylon provide extremely poor thermal protection and should *never be used* when working in areas where an electric arc may occur. Some synthetic materials actually increase the danger of exposure to an electric arc. Synthetic materials have a tendency to melt into the skin when exposed to high temperatures. This melting causes three major difficulties.

1. The melted material forms a thermal seal which holds in heat and increases the severity of the burn.
2. Circulation is severely limited or cut off completely under the melted material. This slows healing and retards the flow of normal nutrients and infection-fighting white blood cells and antibodies.
3. The removal of the melted material is extremely painful and may increase the systemic trauma already experienced by the burn victim.

Synthetic-Cotton Blends. Synthetic-cotton blends such as polyester-cotton are used to make clothing that is easier to care for. Although slightly less vulnerable to melting than pure polyester, the blends are still extremely vulnerable to the heat of an electric arc and the subsequent plasma cloud. Such blends provide poor thermal protection and should not be used in areas where the hazard of electric arc exists.

Flame-Resistant Materials

Chemically Treated Materials.   Both natural and synthetic fibers can be chemically treated to render them flame resistant. Such materials are sometimes used in disposable, coverall-type clothing. While some chemical treatments (such as Borax and boric acid–salt combinations) may be temporary in nature, others are quite satisfactory and may last for the life of the garment.

Historically, chemically treated natural fibers did not exhibit as high an ATPV as synthetic materials when compared by weight. This difference has decreased over the last few years and some workers report that the natural fibers are more comfortable to wear in the climatic extremes of heat and cold.

Heavy weights of chemically treated natural fibers may provide superior protection against certain molten metals.

NOMEX* IIIA.   NOMEX is an aramid fiber made by the DuPont Company. It has a structure that thickens and carbonizes when exposed to heat. This unique characteristic allows NOMEX to provide excellent thermal protection.

NOMEX has been modified in the years since it was first introduced. NOMEX IIIA is made with an antistatic fiber and is, therefore, suitable for use in hazardous environments such as those with high concentrations of hydrocarbon gas.

Since the characteristics of NOMEX are inherent to the fiber, and not a chemical treatment, the thermal protection capabilities of NOMEX are not changed by repeated laundering.

Polybenzimidazole (PBI).†   PBI is a product of the Hoechst Celanese Corporation. It is similar to NOMEX in that it is a synthetic fiber made especially to resist high temperatures. PBI is non-flammable, chemically resistant, and heat stable. This heat stability makes it less prone to shrinking or embrittlement when exposed to flame or high temperatures.

PBI does not ignite, melt or drip in Federal Vertical; Flame Tests FSTM 5903 and FSTM 5905. PBI's characteristics are permanent for the life of the garment. Hoescht Celanese performed tests indicate that PBI has heat protection characteristics which are equal to or superior than other materials.

KERMEL.‡   KERMEL is a synthetic polyamide imide aramid fiber manufactured in France by Rhone-Poulenc. KERMEL fiber is only offered in fabrics blended with other fibers. KERMEL is blended with wool for dress uniforms, sweaters, and underwear, and with high-tenacity aramid for bunker gear and gloves. In the professional firefighter and work wear areas, KERMEL is offered in a 50/50 blend with FR viscose rayon.

Like other synthetic flame-retardant materials, KERMEL is flame-resistant and does not drip or melt when heated.

Material Comparisons.   Table 2.4 illustrates the relative properties of the various types of clothing materials. The information given in Table 2.4 is drawn from general industry experience and/or manufacturer experiments.

---

* NOMEX is a registered trademark of the E. I. Dupont de Nemours Company.
† PBI is a trademark of the Hoescht Celanese Corporation.
‡ KERMEL is a registered trademark of Rhone-Poulenc.

TABLE 2.4 Clothing Material Characteristics

Generic name	Fiber	Manufacturer	Moisture Regain*	Tenacity g/den[†]	Comments
Aramid (meta)	NOMEX®	DuPont	5.5	4.0–5.3	• Long chain synthetic polyamide fiber. · Excellent thermal stability. Will not melt and drip. · Excellent chemical and abrasion resistance. · Fair colorfastness to laundering and light exposure.
Aramid (para)	KEVLAR® TWARON® Technora	DuPont Akzo (Netherlands) Teijin (Japan)	4.3 4.0	21–27 22.6	• Blended with Nomex for fabric integrity in high temperature exposures. · Fair abrasion resistance. · Sensitive to chlorine bleach, light, and strong mineral acids.
Polyamide imide	KERMEL® (France)	Rhone-Poulenc	3.4	4.0–4.5	• Long chain synthetic polyamide fiber. · Excellent thermal stability. Will not melt and drip. · Excellent chemical and abrasion resistance. · Fair colorfastness to laundering and light exposure.
Basofil	BASOFIL®	BASF	5.0	2.0	• A melamine fiber formed when methylol compounds react to form a three dimensional structure of methylene ether and methylene bridges. · Resistant to many solvents and alkalis. Moderately resistant to acids. · Will not shrink, melt or drip when exposed to a flame.
Modacrylic	Protex	Kaneka (Japan)	2.5	1.7–2.6	• Long chain synthetic polymer fiber containing acrylonitrile units modified with flame retardants. · Excellent chemical resistance. · Fair abrasion resistance. · High thermal shrinkage.
FR Acrylic	Super Valzer	Mitsubishi (Japan)	2.5	1.7–2.6	• Long chain synthetic polymer fiber containing acrylonitrile units modified with flame retardants. · Excellent chemical resistance. · Fair abrasion resistance. · High thermal shrinkage.
PBI	PBI Gold	Celanese	15.0	2.8	• Polymer is a sulfonated poly (2.2-m-phenylene-5,5 bibenzim idazole). · Will not ignite, does not melt. · Excellent chemical resistance. · Dyeable in dark shades only.
Polyimide	P84*	Imitech (Austria)	3.0	4.3	• Long chain synthetic polyimide fiber. ◊ High thermal shrinkage. · Thermal properties inferior to Nomex.
FR Viscose FR Cotton	PFR Rayon FR Cotton	Lenzing (Austria) Natural Fiber	10.0 8.0	2.6–3.0 2.4–2.9	• Man-made cellulosic, properties similar to cotton. · Fiber contains flame retardants. · Flame retardant treated in fabric form. · Poor resistance to acids. · Fair abrasion resistance. · Relatively poor colorfastness to laundering and light exposure. · Wear properties similar to untreated cotton.
Vinal	VINEX® FR9B®	Westex	3.0	3.0	• Fabric blended of 85% Vinal/15% rayon. · Fiber composed of vinyl alcohol units with acetal crosslinks. · Sheds aluminum splash. · Very sensitive to shrinkage from wet and dry heat.
FR Polyester	TREVIRA® FR Polyester	Trevira	.4	4.5	• Polyester with proprietary organic phosphorus compound incorporated into the polymer chain. · Properties similar to regular polyester except as modified by flame retardants. · Melt point 9°C lower than regular polyester.
Polyamide	Nylon	DuPont Monsanto	6.0	6.0–8.0	• Long chain synthetic polyamide in which less than 85% of the aramide linkages are attached. · Blended with FR cotton to improve abrasion resistance. · Wear properties significantly better than untreated cotton.

* A measure of ability to absorb moisture. (Percent by weight of moisture gained from a bone dry state at 65% relative humidity.)
[†] A measure of strength and durability. (Tenacity is defined as force per unit linear density to break a known unit of fiber.)
Source: Courtesy Bulwark.

2.7

Thermal performance of NOMEX® III flame retardant cotton, flame retardant polyester/cotton blend, and untreated polyester/cotton blend.

FIGURE 2.1   Relative thermal performance of various materials. (*Courtesy E.I. Dupont de Nemours.*)

FIGURE 2.2   Thermal manikin used to generate the graph shown in Fig. 2.1. (*Courtesy E.I. DuPont de Nemours.*)

Figure 2.1 graphically illustrates the predicted, thermal performance of four of the commonly used materials. To generate these numbers, a specially instrumented manikin called Thermo-Man* is used (Fig. 2.2). The manikin is 6-ft, 1-in tall and has 122 heat sensors distributed over the entire body. The sensors measure the heat transmitted through the garments worn while a flash fire or electric arc is applied to the manikin. The heat data recorded by these sensors was used to generate Fig. 2.1.

More recent research and development such as by Doughty, Neal, Dear, Bingham, and Floyd[3,4] also use electrical arcs to develop the ATPV rating (see Fig. 2.3). In this system a controlled electrical arc is generated adjacent to a manikin with heat sensors installed. The arc is controlled by using the Faraday cage to minimize random horizontal movement.

Work Clothing

Construction.   Work clothing used for routine day-to-day electrical safety may be employed as secondary flash protection. Flame-retardant cotton, flame-retardant synthetic-cotton blend, NOMEX, PBI, or other flame-retardant materials are preferred. The clothing should meet the following minimum requirements:

1. Long sleeves to provide full arm protection
2. Sufficient weight for both thermal and mechanical protection

The suggested minimum is 4 ounces per square yard (oz/yd^2) if synthetics such as NOMEX or PBI are used and 7 to 8 oz/yd^2 if flame-retardant cloth or blends are used. Figure 2.4 shows several examples of flame resistant work clothing.

When to Use Thermally Protective Work Uniforms.   Thermally protective work uniforms should be required for all workers who are routinely exposed to the possibility of electric arc and/or flash. This applies especially to workers in the industries which have the added hazard of flash fire. At a minimum, all employees who are routinely exposed to

---

* Thermo-Man is a registered trademark of the E. I. DuPont de Nemours Company.

FIGURE 2.3   Electrical arc test setup used to determine Arc Thermal Performance Value. (*ATPV.*)

(a)   (b)

(c)   (d)

FIGURE 2.4   Thermal protective clothing. (*Courtesy Bulwark Protective Apparel.*)

480 V and higher should use the thermally protective materials. See Chap. 3 for more specific information.

Thermal protective clothing should be worn everytime the worker works inside the flash boundary. The clothing should be selected with an ATPV or E_{BT} sufficient for the maximum incident energy that may be created in the work area. See Chap. 3 for information on determining the flash boundary.

Care of Thermally Protective Work Uniforms. The following information is necessarily general in nature. Always refer to the manufacturer's care and laundering instructions for specific information. Work uniforms should be kept clean and free of contaminants. Contaminated work clothing can be extremely hazardous. Table 2.5 lists typical care and use precautions for thermal work clothing and flash suits.

Flash Suits

Construction. A flash suit is a thermal-protective garment made of a heavier-weight NOMEX, PBI, or other flame resistant material. This flash suit shown in Fig. 2.5 is made of 12 oz twill and 88 percent cotton or 12 percent nylonLined with 9 oz twill, 88 percent cotton or 12 percent nylon. The entire garment is flame resistant and has an ATPV rating of 40 cal/cm².

Flash suits are composed of a minimum of two parts—the face shield/hood (Fig. 2.6) and the jacket (Fig. 2.7). Flash suits designed for ATPV ratings of 40 cal/cm² and higher (as shown in Fig. 2.5) also include coveralls for lower body protection (Fig. 2.8). If FR pants are not used the jacket should be long enough to protect the legs.

Using Flash Suits. Flash suits should be worn anytime the worker works inside the flash boundary. The suit should be selected with an ATPV or E_{BT} sufficient for the maximum incident energy that may be created in the work area. See Chap. 3 for information on determining the flash boundary. The procedures listed in Table 2.6 are typical of those in which many companies require the use of flash suits; however, specific rules should be developed for each company. *Flash suits should always be used in conjunction with adequate head, eye, and hand protection.* Note that all workers in the vicinity of the arc potential should be wearing a flash suit.

TABLE 2.5 Care and Use Guidelines for Thermal Protective Clothing

---

- Clothing should not be allowed to become greasy and/or impregnated with flammable liquids.
- Launder according to manufacturer's instructions. Generally, home laundering in hot water with a heavy-duty detergent will be effective.
- Do not mix flame-resistant garments with items made of other materials in the same wash.
- Do not use bleaches or other treatments unless recommended by the manufacturer.
- Remember that laundering may degrade the chemical treatment on some flame-retardant materials. Observe manufacturer's recommendations as to how many washes constitute the life of the garment.
- Inspect work uniforms and flash suits before each use. If they are contaminated, greasy, worn, or damaged in any way, they should be cleaned or replaced as required.

FIGURE 2.6 Face shield for flash suit. (*Courtesy Bulwark Protective Apparel.*)

FIGURE 2.5   Flash suit. (*Courtesy Bulwark Protective Apparel.*)

Some facilities impose a current limitation on the rule as well. For example, one large petrochemical company requires flash suits for the procedures listed in Table 2.6, but only when the circuit breaker feeding the circuit has an ampacity of 100 A or greater. See Chap. 3 for more specific information, and methods that should be used to select flash suits.

FIGURE 2.7 Flash suit jacket. (*Courtesy Bulwark Protective Apparel.*)

FIGURE 2.8 Flash suit coveralls. (*Courtesy Bulwark Protective Apparel.*)

Head, Eye, and Hand Protection

When wearing flash suits, or whenever exposed to arc hazard, employees should wear full protection for the head, eyes, and hands. Head and eye protection will be provided if the employee is equipped with a flash suit. When not in a flash suit, however, employees should wear hard hats and eye shields or goggles. Hand protection should be provided by electrical insulating rubber gloves covered with leather protectors.

TABLE 2.6 Procedures which Require the Use of Flash Suits

- Operating open-air switches on circuits of 480 V and higher
- Open-door switching and racking of circuit breakers—480 V and higher
- Removing and installing motor starters in motor control centers—208 V and higher
- Applying safety grounds—480 V and higher
- Measuring voltage in any circuit which is uncertain or has exhibited problems—208 V and higher
- Working on or near any exposed, energized conductors—208 V or higher
- Anytime the incident energy may exceed a level that can be protected by day-to-day thermal-resistant work clothing.

HEAD AND EYE PROTECTION

Hard Hats

Construction and Standards.   In addition to wearing protection from falling objects and other blows, electrical workers should be equipped with and should wear hard hats that provide electrical insulating capabilities. Such hats should comply with the latest revision of the American National Standards Institute (ANSI) standard Z89.1 which classifies hard hats into three basic classes.

1. Class G* hard hats are intended to reduce the force of impact of falling objects and to reduce the danger of contact with exposed low-voltage conductors. They are prooftested by the manufacturer at 2200 V phase-to-ground.

2. Class E* hard hats are intended to reduce the force of impact of falling objects and to reduce the danger of contact with exposed high-voltage conductors. They are prooftested by the manufacturer at 20,000 V phase-to-ground.

3. Class C hard hats are intended to reduce the force of impact of falling objects. They offer no electrical protection.

Figure 2.9 shows two examples of class E (formerly class B) hard hats. Note that a hard hat must be a class G or class E hat to be used in areas where electrical shock may occur. The label for a class B (now class E) hard hat is shown in Fig. 2.10.

Use and Care.   Electrically insulating class G or E hard hats should be worn by workers any time there is a possibility they will be exposed to shock, arc, blast, mechanical blows, or injuries. Table 2.7 lists typical working conditions in which workers should be wearing such protection.

All components of the hard hat should be inspected daily, before each use. This inspection should include the shell, suspension, headband, sweatband, and any accessories. If dents, cracks, penetrations, or any other damage is observed, the hard hat should be removed from service. Class G and E hard hats may be cleaned with warm water and soap.

---

* Prior to 1998 the class G hat was labeled class A and class E hats were labeled class B.

Shockgard ®

insulating safety headgear
adjustable head sizes 6½ thru 7¾
This hat or cap complies with
ANSI Z89.1 1986 CLASS "B"
& E.E.I. AP-1. 1961 safety requirements for industrial head protection

MADE BY
MSA   MINE SAFETY APPLIANCES COMPANY
PITTSBURGH PENNSYLVANIA U S A 15230
AP783 Rev 8   993618

FIGURE 2.9   Hard hats suitable for use in electrical installations. (*Courtesy Mine Safety Appliances Company.*)

FIGURE 2.10   Hard hat label for a class E hard hat. (*Courtesy Mine Safety Appliances Company.*)

TABLE 2.7 Work Situations that Require Non-conductive Head Protection and Eye Protection

- Working close to exposed, overhead energized lines
- Working in switchgear, close to exposed energized conductors
- Any time that a flash suit is recommended (see Table 2.6)
- When any local rules or recognized standards require the use of nonconductive hard hats or eye protection
- Any time there is danger of head, eye, or face injury from electric shock, arc, or blast

Solvents and other harsh cleaners should be avoided. Always refer to the manufacturer's instructions for specific cleaning information.

Safety Glasses, Goggles, and Face Shields

The plasma cloud and molten metal created by an electric arc are projected at high velocity by the blast. If the plasma or molten metal enters the eyes, the extremely high temperature will cause injury and possibly permanent blindness. Electrical workers exposed to the possibility of electric arc and blast should be equipped with and should wear eye protection. Such protection should comply with the latest revision of ANSI standard Z87.1 and should be nonconductive when used for electric arc and blast protection.

Flash suit face shields (Figs. 2.5 and 2.6) will provide excellent face protection from molten metal and the plasma cloud. Goggles which reduce the ultraviolet light intensity (Fig. 2.11) are also recommended. Figure 2.12 is a photograph of a worker with an insulating hard hat and protective goggles.

FIGURE 2.11 Ultraviolet-resistant safety goggles. (*Courtesy Mine Safety Appliances Company.*)

Use and Care. Eye and face protection should be worn by workers any time they are exposed to the possibility of electric arc and blast. Table 2.7 lists typical situations where such protection might be required.

Face protection should be cleaned before each use. Soft, lint-free cloths and warm water will normally provide the necessary cleaning action; however, most manufacturers supply cleaning materials for their specific apparatus.

RUBBER-INSULATING EQUIPMENT

Rubber-insulating equipment includes rubber gloves, sleeves, line hose, blankets, covers, and mats. Employees should use such equipment when working in an area where the hazard of electric shock exists. This means anytime employees are working on or near an energized, exposed conductor, they should be using rubber-insulating equipment.

Rubber goods provide an insulating shield between the worker and the energized conductors. This insulation will save the workers' lives should they accidently contact

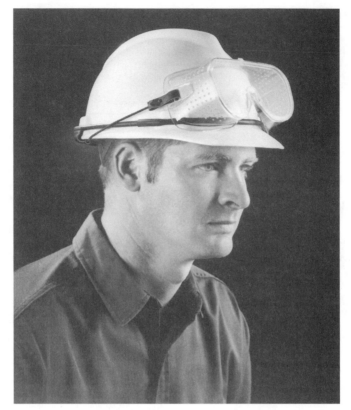

FIGURE 2.12 Hard hat and goggles. (*Courtesy Mine Safety Appliances Company.*)

the conductor. The American Society of Testing and Materials (ASTM) publishes recognized industry standards which cover rubber insulating goods.

Rubber Gloves

Description.   A complete rubber glove assembly is composed of a minimum of two parts—the rubber glove itself and a leather protective glove. In service, the leather protector fits over the outside of the rubber glove and protects it from physical damage and puncture. Sometimes the glove set will include a sheer, cotton insert that serves to absorb moisture and makes wearing the gloves more pleasant. Figure 2.13 shows a typical set of rubber gloves with leather protectors. *Caution:* Rubber gloves should never be used without their leather protectors except in certain specific situations as described later in this chapter.

Construction and Standards.   The ASTM publishes four standards which affect the construction and use of rubber gloves.

FIGURE 2.13 Typical rubber glove set. (*Courtesy W.H. Salisbury and Co.*)

1. Standard D 120 establishes manufacturing and technical requirements for the rubber glove.
2. Standard F 696 establishes manufacturing and technical requirements for the leather protectors.
3. Standard F 496 specifies in-service care requirements.
4. Standard F 1236 is a guide for the visual inspection of gloves, sleeves, and other such rubber insulating equipment.

Rubber gloves are available in six basic voltage classes from class 00 to class 4, and two different types: types I and II. Table 2.8 identifies each class, its maximum use voltage, and the root-mean-square (rms) and direct current (dc) voltages that are used to proof-test the gloves. Figure 2.14 shows the general design and dimensions of rubber gloves. Rubber gloves are available in the following lengths:

For class 00:	278 mm (11 in), 356 mm (14 in)
For class 0:	280 mm (11 in), 360 mm (14 in), 410 mm (16 in), and 460 mm (18 in)
For classes 1,2,3:	360 mm (14 in), 410 mm (16 in), 460 mm (18 in)
For class 4:	410 mm (16 in), 460 mm (18 in)

Note that all lengths have an allowable tolerance of ±13 mm (±½ in).

TABLE 2.8 Rubber Insulating Equipment Classifications, Use Voltages, and Test Voltages

Class of insulating blankets	Nominal maximum-use voltage* phase-phase, ac, rms, max	AC proof-test voltage, rms, V	DC proof-test voltage, avg, V
00	500	2,500	10,000
0	1,000	5,000	20,000
1	7,500	10,000	40,000
2	17,000	20,000	50,000
3	26,500	30,000	60,000
4	36,000	40,000	70,000

* *Note:* The ac voltage (rms) classification of the protective equipment designates the maximum nominal design voltage of the energized system that may be safely worked. The nominal design voltage is equal to (1) the phase-to-phase voltage on multiphase circuits or (2) the phase-to-ground voltage on single-phase grounded circuits. Except for Class 0 and 00 equipment, the maximum-use voltage is based on the following formula: Maximum-use voltage (maximum nominal design voltage) = 0.95 ac proof-test voltage − 2000.

(a) **Length Measurement on Standard Cuff Glove**

(b) **Length and Contour Measurements on Contour Cuff Gloves**

FIGURE 2.14 Dimension measurements for standard and contour cuff rubber gloves. (*Courtesy ASTM.*)

Table 2.9 lists minimum and maximum thicknesses for the six classes of gloves.

In addition to the voltage classes, rubber gloves are available in two different types: type I which is *not* ozone-resistant and type II which is ozone-resistant.

All rubber goods must have an attached, color-coded label subject to the minimum requirements specified in Table 2.10.

When to Use Rubber Gloves.   Rubber gloves and their leather protectors should be worn any time there is danger of injury due to contact between the hands and energized parts of the power system. Each of the work situations described in Table 2.7 should require the use of rubber gloves and their leather protectors.

Rubber gloves should also be worn anytime a worker's hands are closer to an energized conductor than the distance specified by the restricted approach boundary. See Chap. 3 for the definition and calculation methods for the restricted approach boundary.

TABLE 2.9 Rubber Glove Thickness Standard

| Class of glove | Minimum thickness | | | | Maximum thickness | |
| | In crotch | | Other than crotch | | | |
	mm	in	mm	in	mm	in
00	0.20	0.008	0.25	0.010	0.75	0.030
0	0.46	0.018	0.51	0.020	1.02	0.040
1	0.63	0.025	0.76	0.030	1.52	0.060
2	1.02	0.040	1.27	0.050	2.29	0.090
3	1.52	0.060	1.90	0.075	2.92	0.115
4	2.03	0.080	2.54	0.100	3.56	0.140

Source: Courtesy ASTM.

When to Use Leather Protectors. As stated earlier in this chapter, leather protectors should always be used over rubber gloves to provide mechanical protection for the insulating rubber. Furthermore, leather protectors should **never** be used for any purpose other than protecting rubber gloves.

Sometimes the need for additional dexterity may require that the leather protectors not be used. The various industry standards allow such an application in only three situations.

- Class 00   Up to and including 250 V, leather protectors may be omitted for Class 00 gloves. Such omission is only permitted under limited-use conditions when small-parts manipulation requires unusually good finger dexterity.
- Class 0   Leather protectors may be omitted under limited-use conditions when small-parts manipulation requires unusually good finger dexterity.
- Classes 1, 2, 3, 4   Under limited use conditions the leathers may be omitted. *However: When the leathers are omitted for these classes, the user must employ gloves rated at least one (1) voltage class higher than normal.* For example, if working in a 4160-Volt circuit without leather protectors, the worker must use Class 2 gloves.

Leather protectors should never be omitted if there is even a slight possibility of physical damage or puncture. Also, rubber gloves previously used without protectors shall not be used with protectors until given an inspection *and electrical retest.*

How to Use Rubber Gloves.   Rubber gloves should be thoroughly inspected and air-tested before each use. They may be lightly dusted inside with talcum powder or manufacturer-supplied powder. This dusting helps to absorb perspiration and eases putting them on and removing them. *Caution:* Do not use baby powder on rubber gloves. Some baby powder

TABLE 2.10   Labeling Requirements for Rubber Goods

- Color coding according to voltage class: class 00–beige. class 0–red, class 1–white, class 2–yellow, class 3–green, class 4–orange.
- Manufacturer's name
- Voltage class (00, 0, 1, 2, 3, 4)
- Type
- Size (gloves only)

products contain additives which can damage the glove and reduce its life and effectiveness. Cotton inserts are highly recommended for worker confort and convenience.

Rubber gloves should be applied before any activity that exposes the worker to the possibility of contact with an energized conductor. As previously stated, anytime the worker's hands must be closer to an energized conductor than the restricted approach boundary, rubber gloves must be worn. Be certain to wear the leather protector with the glove. Always check the last test date marked on the glove and do not use it if the last test was more than 6 months earlier than the present date.

Rubber Mats

Description. Rubber mats are used to cover and insulate floors for personnel protection. Rubber insulating mats should not be confused with the rubber matting used to help prevent slips and falls. This type of mat is sold by many commercial retail outlets and is not intended for electrical insulation purposes. Rubber insulating mats will be clearly marked and labeled as such.

Insulating rubber matting has a smooth, corrugated, or diamond design on one surface and may be backed with fabric. The back of the matting may be finished with cloth imprint or other slip-resistant material.

Construction and Standards. The ASTM standard D-178 specifies the design, construction, and testing requirements for rubber matting.

Rubber mats are available in five basic voltage classes, from class 0 to class 4, in two different types, and in three different subcategories. Table 2.8 identifies each class, its maximum use voltage, and ac rms and dc voltages that are used to test them. Table 2.11 identifies each of the types and special properties for insulating mats. Table 2.12 identifies the thickness requirements for insulating mats. Table 2.13 identifies the standard widths for insulating rubber matting.

Rubber mats must be clearly and permanently marked with the name of the manufacturer, type, and class. ASTM D-178 must also appear on the mat. This marking is to be placed a minimum of every 3 ft (1 m).

When to Use Rubber Mats. Employers should use rubber mats in areas where there is an ongoing possibility of electric shock. Because permanently installed rubber mats are subject to damage, contamination, and embedding of foreign materials, they should not be relied upon as the sole or primary source of electrical insulation.

TABLE 2.11 Types and Special Property Specifications of Insulating Rubber Matting

	Type I	Type II
Composition	Made of any elastomer or combination of elastomeric compounds, properly vulcanized	Made of any elastomer or combination of elastomeric compounds with one or more of the special properties listed by subcategory
Subcategories	None	A: Ozone resistant B: Flame resistant C: Oil resistant

TABLE 2.12 Thickness Specifications for
Insulating Rubber Mats

Class	Thickness		Tolerance	
	mm	in	mm	in
0	3.2	0.13	0.8	0.03
1	4.8	0.19	0.8	0.03
2	6.4	0.25	0.8	0.03
3	9.5	0.38	1.2	0.05
4	12.7	0.50	1.2	0.05

Source: Courtesy ASTM.

TABLE 2.13 Standard Widths for
Insulating Rubber Matting

24 ± 0.5 in	(610 ± 13 mm)
30 ± 0.5 in	(760 ± 13 mm)
36 ± 1.0 in	(914 ± 25 mm)
48 ± 1.0 in	(1220 ± 25 mm)

How to Use Rubber Mats.   Rubber mats are usually put in place on a permanent basis to
provide both electrical insulation and slip protection. Mats should be carefully inspected
before work is performed which may require their protection. Rubber mats should only be
used as a backup type of protection. Rubber blankets, gloves, sleeves, and other such personal apparel should always be employed when electrical contact is likely.

Rubber Blankets

Description.   Rubber blankets in Fig. 2.15 are rubber insulating devices that are used to cover
conductive surfaces, energized or otherwise. They come in a variety of sizes and are used anytime employees are working in areas where they may be exposed to energized conductors.

Construction and Standards.   The ASTM publishes three standards which affect the construction and use of rubber blankets.

1. Standard D 1048 specifies manufacturing and technical requirements for rubber blankets.
2. Standard F 479 specifies in-service care requirements.
3. Standard F 1236 is a guide for the visual inspection of blankets, gloves, sleeves, and
    other such rubber insulating equipment.

FIGURE 2.15 Typical insulating rubber blanket.

Rubber blankets are available in five basic voltage classes (0 to 4), two basic types (I and II), and two styles (A and B). Table 2.8 identifies each class, its maximum use voltage, and the ac rms and dc voltages that are used to proof-test the blankets. Table 2.14 lists standard blanket sizes, and Table 2.15 lists standard blanket thicknesses.

Type I blankets are made of an elastomer which is not ozone-resistant. Type II blankets are ozone-resistant. Both type I and type II blankets are further categorized into style A and style B. Style A is a nonreinforced construction, and style B has reinforcing members built in. The reinforcing members may not adversely affect the insulating capabilities of the blanket.

Blankets have a bead around the entire periphery. The bead cannot be less than 0.31 in (8 mm) wide nor less than 0.06 in (1.5 mm) high. Blankets may have eyelets to facilitate securing the blanket to equipment; however, the eyelets must not be metal.

Rubber blankets must be marked either by molding the information directly into the blanket or by means of an attached, color-coded label. The labeling is subject to the minimum requirements specified in Table 2.10. Figure 2.16 summarizes the voltage ratings for rubber goods and illustrates the labels which are applied by one manufacturer.

When to Use Rubber Blankets.   Rubber blankets should be used anytime there is danger of injury due to contact between any part of the body and energized parts of the power system. Rubber blankets may be used to cover switchgear, lines, buses, or concrete floors. They differ from mats because they are not permanently installed.

How to Use Rubber Blankets.   Rubber blankets should be thoroughly inspected before each use. They may then be draped over metal conductors or buses or hung to form insulating barriers.

Blankets should be applied before any activity which exposes the worker to the possibility of contact with an energized conductor. Always check the last test date marked on the blanket and do not use it if the last test was more than 1 year earlier than the present date.

Rubber Covers

Description.   Rubber covers are rubber insulating devices that are used to cover specific pieces of equipment to protect workers from accidental contact. They include several

TABLE 2.14   Standard Blanket Sizes—Length and Width

Without slot mm (in)	
457 by 910	(18 by 36)
560 by 560	(22 by 22)
690 by 910	(27 by 36)
910 by 910	(36 by 36)
910 by 2128	(36 by 84)
1160 by 1160	(45.5 by 45.5)
With slot mm (in)	
560 by 560	(22 by 22)
910 by 910	(36 by 36)
1160 by 1160	(45.5 by 45.5)

Source: Courtesy ASTM.

TABLE 2.15   Rubber Blanket Thickness Measurements

Class	Thickness	
	mm	in
0	1.6 to 2.2	0.06 to 0.09
1	2.6 to 3.6	0.10 to 0.14
2	2.8 to 3.8	0.11 to 0.15
3	3.0 to 4.0	0.12 to 0.16
4	3.2 to 4.3	0.13 to 0.17

Source: Courtesy ASTM.

ASTM Labeling Chart
for Salisbury Linemen's Rubber and SALCOR® Protective Equipment

Class Color	Proof Test Voltage AC / DC	Max. Use Voltage AC / DC *	Blanket, Line Hose, Sleeve, Covers Label	Glove Label	Conventional Work Position for Electrical Worker or Lineman	ASTM Specification Reference
00 Beige	2500 / 10,000	500 / 750 *	CLASS 00 TYPE I	W.H. SALISBURY & CO. ASTM D120 TYPE I CLASS 00 10	Ground, Structure or Basket	D120 Rubber Insulating Gloves
0 Red	5,000 / 20,000	1,000 / 1,500 *	CLASS 0 TYPE I	W.H. SALISBURY & CO. ASTM D120 TYPE I CLASS 0 10	Ground, Structure or Basket	D178 Rubber Insulating Matting
1 White	10,000 / 40,000	7,500 / 11,250 *	CLASS 1 TYPE I	W.H. SALISBURY & CO. ASTM D120 TYPE I CLASS 1 10	Structure or Basket	D1048 Rubber Insulating Blankets
2 Yellow	20,000 / 50,000	17,000 / 25,500	CLASS 2 TYPE I	W.H. SALISBURY & CO. ASTM D120 TYPE I CLASS 2 10	Electrically Isolated Basket or Platform	D1049 Rubber Insulating Covers
3 Green	30,000 / 60,000	26,500 / 39,750 *	CLASS 3 TYPE I	W.H. SALISBURY & CO. ASTM D120 TYPE I CLASS 3 10	Electrically Isolated Basket or Platform	D1050 Rubber Insulating Line Hose
4 Orange	40,300 / 70,000	36,000 / 54,000 *	CLASS 4 TYPE I	W.H. SALISBURY & CO. ASTM D120 TYPE I CLASS 4 10	Electrically Isolated Basket or Platform	D1051 Rubber Insulating Sleeves

ASTM Specification Reference (continued):

- D120 Rubber Insulating Gloves
- D178 Rubber Insulating Matting
- D1048 Rubber Insulating Blankets
- D1049 Rubber Insulating Covers
- D1050 Rubber Insulating Line Hose
- D1051 Rubber Insulating Sleeves
- F478 In-Service Care of Line Hose & Covers
- F479 In-Service Care of Insulating Blankets
- F496 In-Service Care of Gloves & Sleeves
- F696 Leather Protectors for Insulating Gloves
- F1236 Inspection Guide for Rubber Products

Type I on all labels designates natural rubber.
Type II designates SALCOR UV and ozone resistant equipment.

SALISBURY
847.679.6700

10409

* Max. use DC voltage is not part of any ASTM specification. Max. use DC voltages are valid in reference to IEC 903 only

FIGURE 2.16 ASTM labeling chart. (*Courtesy W.H Sallisbury and Co.*)

classes of equipment such as insulator hoods, dead-end protectors, line hose connectors, cable end covers, and miscellaneous covers. Rubber covers are molded and shaped to fit the equipment for which they are intended.

Construction and Standards.   The ASTM publishes three standards which affect the construction and use of rubber covers.

1. Standard D 1049 specifies manufacturing and technical requirements for rubber covers.

2. Standard F 478 specifies in-service care requirements.

3. Standard F 1236 is a guide for the visual inspection of blankets, gloves, sleeves, and other such rubber insulating equipment.

Rubber covers are available in five basic voltage classes (0 to 4), two basic types (I and II), and five styles (A, B, C, D, and E). Table 2.8 identifies each class, its maximum use voltage, and the ac rms and dc voltages that are used to proof-test the covers. Many varieties of rubber covers are available (Fig. 2.17). Their size and shape are determined by the equipment that they are designed to cover. Refer to ASTM standard D 1049 for complete listings of the various standard covers.

Type I covers are made of a properly vulcanized, cis-1,4-polyisoprene rubber compound which is not ozone-resistant. Type II covers are made of ozone-resistant elastomers. Both type I and II covers are further categorized into styles A, B, C, D, and E. Table 2.16 describes each of the five styles.

FIGURE 2.17   Typical insulating rubber covers. (*Courtesy W.H. Salisbury and Co.*)

TABLE 2.16 Styles of Rubber Covers

Style	Description
A	Insulator hoods
B	Dead end protectors
C	Line hose connectors
D	Cable end covers
E	Miscellaneous covers

Rubber covers must be marked either by molding the information directly into the cover or by means of an attached, color-coded label. The labeling is subject to the minimum requirements specified in Table 2.10.

When to Use Rubber Covers. Rubber covers should be used anytime there is danger of an injury due to contact between any part of the body and energized parts of the power system.

How to Use Rubber Covers. Rubber covers should be thoroughly inspected before each use. They may then be applied to the equipment which they are designed to cover. Any covers that appear to be defective or damaged should be taken out of service until they can be tested.

Covers should be applied before any activity which exposes the worker to the possibility of contact with an energized conductor. Covers which are used to connect line hose sections should always be used when multiple line hose sections are employed.

Line Hose

Description. Rubber insulating line hoses are portable devices used to cover exposed power lines and protect workers from accidental contact. Line hose segments are molded and shaped to completely cover the line to which they are affixed.

Construction and Standards. The ASTM publishes three standards which affect the construction and use of rubber line hose.

1. Standard D 1050 specifies manufacturing and technical requirements for rubber line hose.

2. Standard F 478 specifies in-service care requirements.

3. Standard F 1236 is a guide for the visual inspection of blankets, gloves, sleeves, and other such rubber insulating equipment.

Rubber line hose is available in five basic voltage classes (0 to 4), three basic types (I, II, and III), and four styles (A, B, C, and D). Table 2.8 identifies each class, its maximum-use voltage, and the ac rms and dc voltages that are used to proof-test them.

Type I line hose is made of a properly vulcanized, cis-1,4-polyisoprene rubber compound which is not ozone-resistant. Type II line hose is made of ozone-resistant elastomers. Type III line hose is made of an ozone-resistant combination of elastomer and thermoplastic polymers. Type III line hose is elastic. All three types are further categorized into styles A, B, C, and D. Table 2.17 lists the characteristics of each of the four styles.

Rubber line hose must be marked either by molding the information directly into the hose or by means of an attached, color-coded label. The labeling is subject to the minimum requirements specified in Table 2.10.

When to Use Rubber Line Hose. Rubber line hose should be used any time personnel are working on or close to energized lines or lines that could be energized.

TABLE 2.17   Characteristics of the Four Styles of Line Hose

Style	Description
A	Straight style, constant cross section
B	Connect end style. Similar to straight style with connection at one end
C	Extended lip style with major outward extending lips
D	Same as style C with a molded connector at one end

How to Use Rubber Line Hose.   Line hoses should be thoroughly inspected before each use. They may then be applied to the lines which they are designed to cover. Any line hose that appears to be defective or damaged should be taken out of service until it can be tested.

Line hose should be applied before any activity which exposes the worker to the possibility of contact with an energized conductor. When more than one section of line hose is used, connecting line covers should be employed. The line hose should completely cover the line.

Rubber Sleeves

Description.   Rubber sleeves (Fig. 2.18) are worn by workers to protect their arms and shoulders from contact with exposed energized conductors. They fit over the arms and complement the rubber gloves to provide complete protection for the arms and hands. They are especially useful when work must be performed in a cramped environment.

FIGURE 2.18   Insulating rubber sleeves. (*Courtesy W.H. Salisbury and Co.*)

Construction and Standards. The ASTM publishes three standards which affect the construction and use of rubber sleeves.

1. Standard D 1051 specifies manufacturing and technical requirements for rubber sleeves.
2. Standard F 496 specifies in-service care requirements.
3. Standard F 1236 is a guide for the visual inspection of blankets, gloves, sleeves, and other such rubber insulating equipment.

Insulating sleeves are available in five basic voltage classes (0 to 4), two basic types (I and II), and two styles (A and B). Table 2.8 identifies each class, its maximum use voltage, and the ac rms and dc voltages that are used to proof-test them.

Type I sleeves are made of properly vulcanized, cis-1,4-polyisoprene rubber compound which is not ozone-resistant. Type II sleeves are made of ozone-resistant elastomers.

Style A sleeves are made in a straight, tapered fashion (Fig. 2.19). Type B sleeves are of a curved elbow construction (Fig. 2.18).

Rubber sleeves are manufactured with no seams. They have a smooth finish and self-reinforced edges. Sleeves are manufactured with holes used to strap or harness them onto the worker. The holes are nominally $\frac{5}{16}$ in (8 mm) in diameter and have nonmetallic, reinforced edges.

Rubber sleeves must be marked clearly and permanently with the name of the manufacturer or supplier, ASTM D 1051, type, class, size, and which arm they are to be used on (right or left). Such marking shall be confined to the shoulder flap area and shall be nonconducting and applied in such a manner as to not impair the required properties of the sleeve.

FIGURE 2.19 Style A, straight taper rubber insulating sleeves. (*Courtesy ASTM.*)

TABLE 2.18   Standard Thickness for Rubber Insulating Sleeves

Class of sleeve	Minimum		Maximum	
	mm	in	mm	in
0	0.51	0.020	1.02	0.040
1	0.76	0.030	1.52	0.060
2	1.27	0.050	2.54	0.100
3	1.90	0.075	2.92	0.115
4	2.54	0.100	3.56	0.140

Source: Courtesy ASTM.

A sleeve shall have a color-coded label attached which identifies the voltage class. The labeling is subject to the minimum requirements specified in Table 2.10. Table 2.18 shows standard thicknesses for rubber sleeves, and Table 2.19 lists standard dimensions and tolerances.

When to Use Rubber Sleeves.   Rubber sleeves should be used any time personnel are working on or close to energized lines or lines that could be energized. They should be considered any time rubber gloves are being worn and should be required for anyone working around or reaching through energized conductors.

How to Use Rubber Sleeves.   Rubber sleeves should be inspected before each use. They may be worn to protect the worker from accidental contact with energized conductors. Be certain to check the last test date marked on the sleeve. If the date is more than 12 months earlier than the present date, the sleeve should not be used until it has been retested.

TABLE 2.19   Standard Dimensions and Tolerances for Rubber Insulating Sleeves

Style	Size	Dimensions*							
		A		B		C		D	
		mm	in	mm	in	mm	in	mm	in
Straight taper	regular	667	$26\frac{1}{4}$	394	$15\frac{1}{2}$	286	$11\frac{1}{4}$	140	$5\frac{1}{2}$
	large	724	$28\frac{1}{2}$	432	17	327	$12\frac{7}{8}$	175	$6\frac{7}{8}$
	extra large	762	30	483	19	337	$13\frac{1}{4}$	175	$6\frac{7}{8}$
Curved elbow	regular	673	$26\frac{1}{2}$	394	$15\frac{1}{2}$	311	$12\frac{1}{4}$	146	$5\frac{1}{4}$
	large	705	$27\frac{1}{2}$	406	16	327	$12\frac{7}{8}$	175	$6\frac{7}{8}$
	extra large	749	$29\frac{1}{2}$	445	$17\frac{1}{2}$	327	$12\frac{7}{8}$	178	7

* Tolerances shall be as follows:
    A—±13 mm ($\pm\frac{1}{2}$ in)
    B—Minimum allowable length
    C—±13 mm ($\pm\frac{1}{2}$ in)
    D—±6 mm ($\pm\frac{1}{4}$ in)
Source: Courtesy ASTM.

In-Service Inspection and Periodic Testing of Rubber Goods

Field Testing.   Rubber goods should be inspected before each use. This inspection should include a thorough visual examination and, for rubber gloves, an air test. Table 2.20 is a synopsis of the inspection procedures defined in ASTM standard F 1236. Rolling is a procedure in which the rubber material is gently rolled between the hands or fingers of the inspector. This procedure is performed on both the inside and outside of the material. Figures 2.20 and 2.21 illustrate two types of rolling techniques.

Rubber gloves should be air tested before each use. First, the glove should be inflated with air pressure and visually inspected, and then it should be held close to the face to feel for air leaks through pinholes. Rubber gloves can be inflated by twirling, by rolling, or by using a mechanical glove inflater. Note that "rolling" in this situation is not the same as the rolling discussed earlier in the description of rubber goods inspection.

To inflate a glove by twirling, grasp the side edges of the glove (Fig. 2.22a), gently stretch the glove until the end closes and seals (Fig. 2.22b), and then twirl the glove in a

TABLE 2.20   Inspection Techniques for Rubber Insulating Equipment

Gloves and sleeves	Inspect glove and sleeve surface areas by gently rolling their entire outside and inside surface areas between the hands. This technique requires gently squeezing together the inside surfaces of the glove or sleeve to bend the outside surface areas and create sufficient stress to inside surfaces of the glove or sleeve to highlight cracks, cuts, or other irregularities. When the entire outside surface area has been inspected in this manner, turn the glove or sleeve inside-out and repeat the inspection on the inside surface (now the outside). If necessary, a more careful inspection of suspicious areas can be achieved by gently pinching and rolling the rubber between the fingers. Never leave a glove or sleeve in an inside-out condition. Stretch the thumb and finger crotches by pulling apart adjacent thumb and fingers to look for irregularities in those areas.
Blankets and mats	Place rubber blankets on a clean, flat surface and roll up tightly starting at one corner and rolling toward the diagonally opposite corner. Inspect the entire surface for irregularities as it is rolled up. Unroll the blanket and roll it up again at right angles to the original direction of rolling. Repeat the rolling operations on the reverse side of the blanket.
Line hose	Examine the inside surfaces of the insulating line hose by holding the hose at the far end from the lock and placing both hands side-by-side palms down around the hose. With the slot at the top and the long free end of the hose on the left, slowly bend the two ends of the hose downward while forcing the slot open with the thumbs. The hose should be open at the bend, exposing the inside surface. Slide the left hand about a foot down the hose and then, with both hands firmly gripping the hose, simultaneously move the left hand up and the right hand down to pass this section over the crown of the bend for inspection. Slide the right hand up the hose to the left hand. Hold the hose firmly with the right hand while the left hand again slides another foot down the hose. Repeat the inspection and, in this way, the entire length of hose passes through the hands from one end to the other

Source: This material is reproduced verbatim from ASTM standard F-1236. It appears with the permission of the American Society of Testing and Materials.

(a)

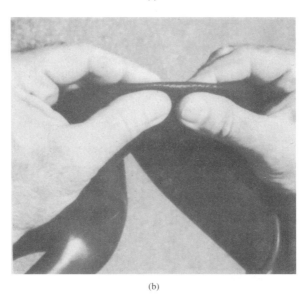

(b)

FIGURE 2.20   Inspection of gloves and blanket by rolling. (a) Hand
rolling; (b) pinch rolling. (*Courtesy ASTM.*)

rotating motion using the rolled edges of the glove opening as an axis (Fig. 2.22c). This will
trap air in the glove and cause it to inflate.

Gloves that are too heavy to inflate by twirling may be inflated by rolling. To do
this, lay the glove on a flat surface, palm up, and press the open end closed with the
fingers (Fig. 2.23a). Then while holding the end closed, tightly roll up about $1\frac{1}{2}$ in of the

FIGURE 2.21 Inspection of rubber blanket by rolling. (*Courtesy ASTM.*)

(a)

FIGURE 2.22 Inflating rubber gloves by twirling. (a) Grasping; (b) stretching; (c) twirling. (*Courtesy ASTM.*)

(b)

(c)

FIGURE 2.22 (*Continued*)

(a)

(b)

FIGURE 2.23 Inflating rubber gloves by rolling. (a) Pressing; (b) rolling. (*Courtesy ASTM.*)

gauntlet (Fig. 2.23*b*). This will trap air in the glove and cause it to inflate. Gloves may also be inflated by commercially available mechanical inflaters (Fig. 2.24). After the gloves are inflated, they may be visually inspected (Fig. 2.25) and then checked by holding them close to the face to listen and feel for air leaks through pinholes (Fig. 2.26).

Description of Irregularities in Rubber Goods. If damage or imperfections are discovered during inspections, the rubber goods may need to be removed from service. The following are typical of the types of problems that may be discovered.

1. *Abrasions and scratches* (Fig. 2.27). This is surface damage that normally occurs when the rubber material makes contact with an abrasive surface.
2. *Age cracks* (Fig. 2.28). Surface cracks that may look like crazing of glazed ceramics and may become progressively worse. Age cracks are due to slow oxidation caused by exposure to sunlight and ozone. It normally starts in areas of the rubber that are under stress.
3. *Chemical bloom* (Fig. 2.29). A white or yellowish discoloration on the surface. It is caused by migration of chemical additives to the surface.

FIGURE 2.24   Mechanical inflation. (*Courtesy ASTM.*)

FIGURE 2.25   Visually inspecting inflated rubber gloves. (*Courtesy ASTM.*)

FIGURE 2.26 Inspecting rubber gloves by listening and feeling for air leaks through pinholes. (*Courtesy ASTM.*)

FIGURE 2.27 Scratches. (*Courtesy ASTM.*)

FIGURE 2.28   Age cracks. (*Courtesy ASTM.*)

4. *Color splash* (Fig. 2.30).   This is a spot or blotch caused by a contrasting colored particle of unvulcanized rubber that became embedded into the finished product during the manufacturing process.
5. *Cuts* (Fig. 2.31).   These are smooth incisions in the surface of the rubber caused by contact with sharp-edged objects.

FIGURE 2.29   Chemical bloom. (*Courtesy ASTM.*)

FIGURE 2.30   Color splash. (*Courtesy ASTM.*)   **FIGURE 2.31**   Cuts. (*Courtesy ASTM.*)

6. *Depressions or indentations* (Fig. 2.32).   A shallow recess which exhibits a thinner rubber thickness at the bottom of the depression than in the surrounding areas.

7. *Detergent cracks.*   Cracks that appear on the inside surface of the glove or sleeve. The cracks form around a spot of detergent residue that was not removed during the cleaning and rinsing of the form prior to the dipping process.

8. *Embedded foreign matter* (Fig. 2.33).   This is a particle of nonrubber that has been embedded in the rubber during the manufacturing process. It normally shows up as a bump when the rubber is stretched.

9. *Form marks.*   This is a raised or indented section on the surface of the rubber. It is caused by an irregularity in the form that was used to mold the product.

10. *Hard spot* (Fig. 2.34).   A hardened spot caused by exposure to high heat or chemicals.

11. *Mold mark* (Fig. 2.35).   A raised or indented section caused by an irregularity in the mold.

12. *Scratches, Nicks, and Snags* (Figs. 2.27, 2.36, 2.37).   Angular tears, notches, or chip-like injuries in the surface of the rubber caused by sharp objects such as wire, pointed tools, staples, or other similar sharp-edged hazards.

FIGURE 2.32 Depression or indentations. (*Courtesy ASTM.*)   **FIGURE 2.33** Embedded foreign material. (*Courtesy ASTM.*)

(a)

(b)

FIGURE 2.34 Hard spot. (*Courtesy ASTM.*)

(c)

(d)

FIGURE 2.34   (*Continued*)

(a)   (b)

FIGURE 2.35 Mold mark. (*Courtesy ASTM.*)

13. *Ozone cracks* (Fig. 2.38).   A series of interlacing cracks that start at stress points and worsen as a result of ozone-induced oxidation.

14. *Parting line or flash line.*   A ridge of rubber left on finished products. They occur at mold joints during manufacturing.

15. *Pitting* (Fig. 2.39).   A pockmark in the rubber surface. It is often created by the rupturing of an air bubble close to the surface during manufacturing.

FIGURE 2.36 Nick. (*Courtesy ASTM.*)   **FIGURE 2.37** Snag. (*Courtesy ASTM.*)

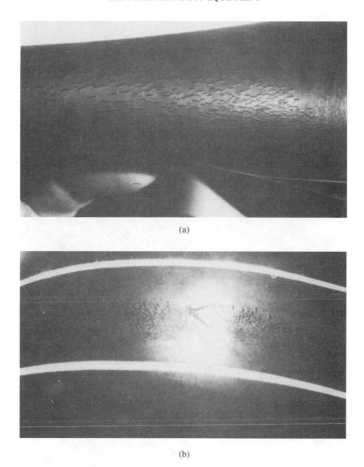

(a)

(b)

FIGURE 2.38   Ozone cracks. (*Courtesy ASTM.*)

FIGURE 2.39   Pitting. (*Courtesy ASTM.*)

FIGURE 2.40 Protuberance. (*Courtesy ASTM.*)

16. *Protuberance* (Fig. 2.40).   A bulge or swelling above the surface of the rubber.

17. *Puncture* (Fig. 2.41).   Penetration by a sharp object through the entire thickness of the product.

18. *Repair marks* (Fig. 2.42).   An area of the rubber with a different texture. Usually caused by repair of the mold or form.

19. *Runs.*   Raised flow marks which occur on rubber glove fingers during the dipping process.

20. *Skin breaks.*   Cavities in the surface of the rubber. They have filmy ragged edges and smooth interior surfaces. They are usually caused by embedded dirt specks during the manufacturing process.

FIGURE 2.41 Puncture. (*Courtesy ASTM.*)

FIGURE 2.42 Repair mark. (*Courtesy ASTM.*)   **FIGURE 2.43** Tear. (*Courtesy ASTM.*)

21. *Soft spots.*   Areas of the rubber which have been soft or tacky as a result of heat, oils, or chemical solvents.

22. *Tears* (Fig. 2.43).   A rip through the entire thickness of the rubber. Usually caused by forceful pulling at the edge.

Electrical Testing.   Rubber insulating equipment should be electrically tested periodically. Table 2.21 summarizes the requirements and/or recommendations for such testing. Electrical testing of rubber goods is a relatively specialized procedure and should be performed only by organizations with the necessary equipment and experience.

TABLE 2.21 In-service Electrical Tests for Rubber Insulating Goods

Product	Maximum test interval, months	ASTM standard	Notes
Gloves	6	F 496	Tested, unused gloves may be placed into service within 12 months of the previous tests without retesting.
Mats	—	D 178	There is no regulatory requirement for inservice electrical testing of mats. They are tested by the manufacturer when new.
Blankets	12	F 479	Tested, unused blankets may be placed into service within 12 months of the previous tests without being retested.
Covers	—	F 478	Covers should be retested when in-service inspections indicate a need.
Line hose	—	F 478	Line hose should be retested when in-service inspections indicate a need.
Sleeves	12	F 496	Tested, unused sleeves may be placed into service within 12 months of the previous tests without being retested.

For detailed testing information refer to the ASTM standards referenced in Table 2.21. Note that all rubber goods are tested at the same voltages dependent upon their class (00, 0, 1, 2, 3, or 4). See Table 2.8 for a listing of ASTM-required test voltages.

HOT STICKS

Description and Application

Hot sticks are poles made of an insulating material. They have tools and/or fittings on the ends which allow workers to manipulate energized conductors and equipment from a safe distance. Hot sticks vary in length depending on the voltage level of the energized equipment and the work to be performed. Modern hot sticks are made of fiberglass and/or epoxiglass. Older designs were made of wood which was treated and painted with chemical-, moisture-, and temperature-resistant materials. Figure 2.44 is an example of a simple hot stick fitted with a tool suitable for operation of open-air disconnect switches.

Hot sticks can be fitted with a variety of tools and instruments. The most common fitting is the NEMA standard design shown in Fig. 2.45 as the *standard universal* fitting. This fitting allows a variety of tools and equipment to be connected to the hot stick. Figure 2.46 shows a voltage tester attached to a hot stick using a standard fitting.

Figure 2.45 also shows other attachments and extensions that can be used to increase the usefulness of hot sticks. In addition to the equipment shown in Fig. 2.45, hot sticks can also be equipped with wrenches, sockets, screwdrivers, cutters, saws, and other such tools.

Hot sticks can also be purchased in telescoping models (Fig. 2.47) and so-called shotgun models (Fig. 2.48). The telescoping type of hot stick is composed of several hollow, tubular sections which nest inside of each other. The topmost section is first extended and locked in place by means of a spring-loaded button which snaps into a hole. The user of the hot stick extends as many of the sections as are required to accomplish the job at hand. The telescoping hot stick makes very long hot stick lengths available which then collapse to a small, easy-to-carry assembly.

The shotgun hot stick (Fig. 2.48) has a sliding lever mechanism that allows the user to open and close a clamping hook mechanism at the end. In this way the user can attach the stick to a disconnect ring and then close it. After the switch is operated, the shotgun

FIGURE 2.44   Typical hot stick. (*Courtesy AB Chance Corp.*)

FIGURE 2.45   Various fittings, couplings, and tools for hot sticks. (*Courtesy AB Chance Corp.*)

mechanism is operated to open the hook. The shotgun stick gets its name from its similarity to the pump-action shotgun.

Figure 2.49 shows a hot stick kit with several sections and various tools. This type of package provides a variety of configurations which will satisfy most of the day-to-day needs for the electrician and the overhead line worker. The kit includes the following components:

1. Six 4-ft sections of an Epoxiglas snap-together hot stick

2. Aluminum disconnect head for opening and closing switches and enclosed cutouts

3. Nonmetallic disconnect head for use in indoor substations where buswork and switches are in close proximity

4. Clamp stick head for use with 6-in-long eye-screw ground clamps. This is used to apply and remove safety grounds

FIGURE 2.46   Voltage tools attached to a hot stick. (*Courtesy W.H. Salisbury and Co.*)

FIGURE 2.47 Telescoping hot stick—shown collapsed. (*Courtesy AB Chance Corp.*)

5. Tree trimmer attachment with $1\frac{1}{2}$ ft of additional stick and pull rope used to close jaws of the trimmer
6. Pruning saw
7. Pistol-grip saw handle for use when tree limbs can be reached and insulation is not required
8. Heavy-duty vinyl-impregnated storage case

FIGURE 2.48 Shotgun-type hot stick. (*Courtesy AB Chance Corp.*)

Electricians involved primarily in indoor work might wish to substitute other tools for the tree trimming and pruning attachments.

FIGURE 2.49 Typical hot stick kit for electricians and line workers. (*Courtesy AB Chance Corp.*)

When to Use

Hot sticks should be used to insulate and isolate the electrician from the possibility of electric shock, arc, or blast. Table 2.22 identifies the types of procedures for which hot sticks are recommended.

How to Use

The specifics of hot stick use will depend upon the task being performed and the location in which the worker is positioned. As a general rule, if hot sticks are being used, the worker should also wear other protective clothing. At a minimum, rubber gloves and face shields should be employed. However, many recommend that flash suits should also be worn, especially when safety grounds are being applied.

Before each use the hot stick should be closely inspected for signs of physical damage which may affect its insulating ability. If the hot stick is cracked, split, or otherwise damaged, it should be taken out of service.

Testing Requirements

ASTM Standard F 711 requires that manufacturers test hot sticks to very stringent standards before they are sold. Additionally, OSHA standards require that hot sticks be inspected and/or tested periodically. The following should be the minimum:

TABLE 2.22 Typical Procedures Requiring Use of Hot Sticks

Medium voltage and higher
- Voltage measurement
- Any repairs or modifications to energized equipment

All voltages
- Operation of disconnects and cutouts
- Application of safety grounds

---

1. Hot sticks should be closely inspected for damage or defects
   a. Prior to each use
   b. At least every two years
2. If any damage or defects are noted the hot stick should be repaired or replaced
3. Hot sticks should be electrically tested according to ASTM Standard F 711
   a. Anytime an inspection reveals damage or a defect
   b. Every two years

INSULATED TOOLS

Description and Application

Insulated tools, such as those shown in Fig. 2.50, are standard hand tools with a complete covering of electrical insulation. Every part of the tool is fully insulated. Only the minimum amount of metallic work surface is left exposed. Such tools are used to prevent shock or arc in the event that the worker contacts the energized conductor.

ASTM Standard 1505 defines the requirements for the manufacture and testing of insulated hand tools. Such tools are to be used in circuits of 1000 V ac and 1500 V dc. Such tools are covered with two layers of material. The inner layer provides the

FIGURE 2.50   Insulated tool set. (*Courtesy CIP Insulated Products.*)

electrical insulation and the outer layer provides mechanical protection for the electrical insulation.

When to Use

Insulated tools should be used anytime work is being performed on or near exposed, energized conductors. They should be inspected before each use.

How to Use

Insulated tools are used in the same way that ordinary hand tools are used, and all the same precautions should be observed. Avoid using the tools in any application which may damage the insulation.

BARRIERS AND SIGNS

Whenever work is being performed which requires the temporary removal of normal protective barriers such as panels or doors, barriers and signs should be used to warn personnel of the hazard.

Barrier Tape

Barrier tape is a continuous length of abrasion-resistant plastic tape. It should be a minimum of 2 in wide and should be marked or colored to clearly indicate the nature of the hazard to which employees will be exposed if they cross the tape line. Figure 2.51 shows a type of barrier tape suitable for marking and barricading an area where an electrical hazard exists.

Such tape should be yellow or red to comply with Occupational Safety and Health Administration (OSHA) standards. Generally red tape, or red tape with a white stripe, is the

FIGURE 2.51   Barrier tape styles suitable for electrical hazards. (*Courtesy Direct Safety Supply Co.*)

preferred color for such applications. Whatever color is chosen should be a standard design that is used consistently for the same application.

Signs

Warning signs should be of standardized design and easily read. They should be placed in such a way to warn personnel of imminent hazard. Figure 2.52 shows a type of sign suitable for use as an electrical hazard warning.

When and How to Use

General Requirements.   Table 2.23 lists the general procedures for using signs, barriers, and attendants to warn personnel of electrical hazards. Signs should be placed so that they are easily seen and read from all avenues of access to the hazardous area.

Temporary Hazard Barricades.   Temporary hazard barricades should be constructed with a striped barrier tape using the following procedure:

1. The tape should be placed so that it completely encloses the hazardous area.
2. The tape should be clearly visible from all directions of approach.
3. The tape should be at a level such that it forms an effective barrier. Approximately 3 ft is suitable.

FIGURE 2.52   Typical electrical hazard sign. (*Courtesy Ideal Industries, Inc.*)

TABLE 2.23 Summary of the Use of Warning Methods

Type of warning	When to use
Signs	Signs should be placed to warn employees about electrical hazards which may harm them
Barricades	Barricades should be used to prevent and limit employee access to work areas where they may be exposed to electrical hazards. Such barriers should be made of nonconductive material such as plastic barrier tape
Attendants	If signs and/or barricades cannot provide a level of safety sufficient to protect employees, attendants shall be placed to guard hazardous areas. Attendants shall be familiar with the nature and extent of the hazard so as to adequately warn other employees

4. Allow an area of sufficient size to give adequate clearance between the hazard and any personnel outside the hazardous area.

5. If test equipment is being used on equipment inside the hazardous area, the tape should be arranged so that the equipment can be operated outside the area.

6. Do not use the same style and color of tape for any purpose other than marking temporary hazards.

7. Such a barricaded area should be considered to be the same as a metal enclosure; that is, *access is not possible.*

8. After the hazard has been eliminated, remove the tape.

SAFETY TAGS, LOCKS, AND LOCKING DEVICES

Safety tags, locks, and locking devices are used to secure and mark equipment that has been taken out of service. They are applied in such a way that the equipment cannot be reenergized without first removing the tags and/or locks. For more information about lockout and tagout procedures refer to Chap. 3.

Safety Tags

Safety tags are applied to equipment to indicate that the equipment is not available for service. They are tags constructed of a durable, environment-proof material. They should be of standardized construction and include a warning that says Do Not Start, Do Not Open, Do Not Close, Do Not Operate, or other such warning. The tag must also indicate who placed it on the equipment and the nature of the problem with the equipment. Figure 2.53 shows tags which are suitable for such an application.

Some manufacturers supply tags with individual employee photographs on them. This type of tag helps to further identify who placed the tag and to personalize the installation.

Tags are to be applied using strong, self-locking fasteners. Nylon cable wraps are suitable for such an application. The fastener must have a breaking strength of not less than 50 lb.

FIGURE 2.53 Typical tags suitable for tagout purposes. (*Courtesy Ideal Industries, Inc.*)

Locks and Multiple-Lock Devices

Locks are used to prevent operation of equipment that has been de-energized. They must be strong enough to withstand all but the most forceful attempts to remove them without the proper key. If a lock can be removed by any means other than a bolt cutter or the key that fits it, the lock should not be used.

Standard padlocks are normally applied for lockout purposes (Fig. 2.54). Each employee should have a set of padlocks which can be opened only by his or her key. A master key may be kept for emergency situations which require that the lock be opened by someone other than the one who placed it. See Chap. 3 for situations in which safety locks may be removed by someone other than the individual who placed it.

Departments may also have "group" padlocks which are placed by shift personnel and which are keyed to a departmental key. For example, the operations department may have shift operators who remove and restore equipment from service. Equipment may be removed from service during one shift and later returned to service during another shift. In such situations the group lock will be placed by one individual and removed by another. Thus, the group locks will have master keys so that any authorized shift operator can place or remove them.

FIGURE 2.54 Typical padlocks suitable for lockout purposes. (*Courtesy Ideal Industries, Inc.*)

Multiple-Lock Devices. Sometimes several workers will need to place a lock on one piece of equipment. This often happens when several crafts are working in the area secured by the lock. In these circumstances, a multiple-lock device is used (Fig. 2.55a). To lock out a piece of equipment, the multiple-lock device is first opened and applied as though it were a padlock. Then the padlock is inserted through one of the holes in the device. The padlock prevents

(a)   (b)

FIGURE 2.55   Multiple-lock devices. (*Courtesy Ideal Industries, Inc.*)

opening of the multiple-lock device which, in turn, prevents operation of the equipment. The devices shown in Fig. 2.55a can accommodate up to six locks; however, multiple-lock devices can be "daisy chained" to allow as many locks as required.

The multiple-lock devices in Fig. 2.55b combine the multiple-lock device with a safety tag. The information required on the tag is written on the tag portion of the device which is then applied to the equipment.

Locking Devices

Some equipment, such as wall switches and molded case circuit breakers, do not readily accommodate locks. In these instances, when lockout is required, a locking device must be used. Figure 2.56a shows locking devices which may be placed over the handle of a molded case circuit breaker and clamped in place. The lock is then installed through the hole left for that purpose. The breaker cannot be operated until the device is removed, and the locking device cannot be removed until the padlock is open.

Figure 2.56b is a similar device which mounts on a standard wall switch. The locking device is first attached to the wall switch with the switch faceplate mounting screws. The switch is moved to the OFF position and the hinged cover of the device is closed. A padlock is placed through the flange supplied for that purpose.

Devices like the wall switch lockout device can be left in place permanently for switches that are frequently locked out. Although the device shown in Fig. 2.56b can be used to lock the switch in the ON position, this practice is not recommended.

When/Where to Use Lockout Tagout

Equipment should be locked out and tagged out when it is being serviced or maintained and an unexpected start-up could injure personnel who are working in the area. Thus tags and locks should be placed anytime an employee is exposed to contact with parts of the equipment which have been de-energized.

The locks and tags must be applied to all sources of power for the affected equipment. They must be applied in such a way that the equipment cannot be reenergized without first

(a)

(b)

FIGURE 2.56   Locking devices. (*Courtesy Ideal Industries, Inc.*)

removing the locks and/or tags. See Chap. 3 for more complete information on lockout/tagout procedures. Figure 2.57 shows locks, tags, and multiple-lock devices being applied to electrical switching equipment.

VOLTAGE-MEASURING INSTRUMENTS

Safety Voltage Measurement

Safety voltage measurement actually involves measuring for zero voltage. That is, a safety measurement is made to verify that the system has been de-energized and that no voltage is present. (See Chap. 3 for more details.) Because of this, the instruments that are used for

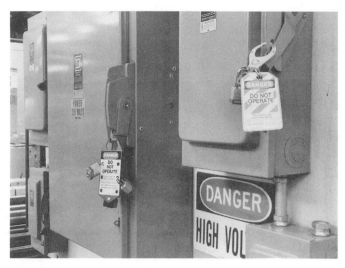

FIGURE 2.57 Typical application of locks, tags, and multiple-lock devices. (*Courtesy Ideal Industries, Inc.*)

safety voltage measurement need only be accurate enough to determine whether or not the nominal system voltage is present at the point of worker exposure.

The instruments discussed in the following sections are intended primarily for safety voltage measurements. Some of them are quite accurate. Contact the individual manufacturers for more specific information.

For various reasons, some authorities prefer a proximity tester over a contact tester, while others prefer a contact tester over a proximity tester. Whichever type is selected, it must be capable of indicating the presence of dangerous voltages and discriminating between nuisance static voltages and actual hazardous voltage source. See Chap. 3 for more information.

FIGURE 2.58 "AC Sensor" proximity voltage sensor for use on circuits up to 600 V alternating current. (*Courtesy Santronics, Inc., Sanford, NC.*)

Proximity Testers

Proximity testers do not require actual metal-to-metal contact to measure the voltage, or lack of voltage, in a given part of the system. They rely on the electrostatic field established by the electric potential to indicate the presence of voltage. Proximity testers will indicate voltage levels through insulation. They may not provide accurate results when cable is shielded.

Proximity testers are not accurate and do not indicate the actual level of the voltage that is present. Rather they indicate the presence of voltage by the illumination of a light and/or the sounding of a buzzer. Figures 2.58 through 2.60 are three different types of proximity detectors for various voltage levels.

FIGURE 2.59 "TIC" tracer proximity voltage sensor for use on circuits up to 35,000 V. (*Courtesy TIF Electronics.*)

FIGURE 2.60 Audiovisual proximity voltage tester for use in circuits from 240 V to 500 kV. (*Courtesy W.H. Salisbury and Co.*)

Figure 2.58 shows a simple neon light proximity tester. The end of the unit is plastic and sized to fit into a standard 120-V duplex receptacle. It requires two AAA flashlight cells to operate. When placed in proximity to an energized circuit, the red neon, located in the white plastic tip, glows.

Figure 2.59 shows a proximity tester that uses a combination of audio and visual indicators. When turned on, the unit emits a slow beeping sound which is synchronized to a small flashing light. If the unit is placed close to an energized circuit, the number of beeps per second increases. The light flashes increase in frequency along with the beeps. The higher the voltage and/or the closer the voltage source, the faster the beeps. When the voltage is very high, the beeping sound turns into a steady tone. This unit has a three-position switch. One position turns the unit off, and the two others provide low- and high-voltage operation, respectively. If high voltages are being measured, the unit should be attached to a hot stick. The manufacturer supplies a hot stick for that purpose.

Figure 2.60 shows a proximity tester that is similar in some ways to the one shown in Fig. 2.59. It has both a light and a tone; however, the light and tone are steady when a voltage is sensed above the unit's operational threshold. The unit in Fig. 2.60 also has a multiposition switch which has an OFF position, a 240-V test position, a battery test position, and multiple voltage level positions from 4200 V up to the unit's maximum. This unit is shown attached to a hot stick in Fig. 2.46. It should not be used to measure voltages without the use of a hot stick and/or appropriate rubber insulating equipment.

Contact Testers

Some personnel prefer the use of testers which make actual metal-to-metal contact with the circuit being energized. Such instruments are called contact testers. *Contact testers* may be simple indicators, but more often they are equipped with an analog or digital meter which indicates actual voltage level. Figures 2.61 through 2.64 illustrate

FIGURE 2.61 Safety voltage and continuity tester. (*Courtesy Ideal Industries, Inc.*)

various styles of contact testers that may be used for safety-related measurements.

Figure 2.61 shows one of the more popular models used for voltages up to 600 V alternating current or direct current. This unit is a solenoid type of instrument. That is, a spring-loaded solenoid plunger is connected to an indicator which aligns with a voltage scale. The distance that the plunger travels is proportional to the voltage level of the measured circuit. The voltage scale is read in volts. The instrument shown in Fig. 2.61 also indicates continuity and low voltage. It switches automatically between those functions. Note that this style of unit is one of the most popular styles in use for low-voltage measurements. Because of their solenoid mechanism, these styles are subject to small arcs when the contact is made with the measured circuit. This problem can be mitigated by using test leads equipped with fused resistors. Such resistors not only eliminate the arcing problem but also open in the event that an internal short circuit occurs in the meter.

Figure 2.62 shows a modern, digital readout safety voltmeter. This instrument is suitable for circuits up to 1000 V and is tested to 2300 V. It has an inherently high impedance; therefore, it is not prone to arcing when the leads make contact. The meter should only be used for voltage measurements. It has no continuity or ammeter scales.

Figure 2.63 shows a typical digital multimeter. These instruments are in common use by virtually all electricians and electrical and electronic technicians. Such instruments often have voltage ranges well above 1000 V; however, use of them in power circuits with voltages above 600 V is not recommended. Care should be taken when using this style of instrument in an electric power circuit.

The instrument shown in Fig. 2.64 is called a *phasing tester* because it is often used to phase two circuits—that is, to check that *A* phase in circuit 1 is the same as *A* phase in circuit 2 and so on. Such units can also be used for safety-related voltage measurements. The instrument is composed of two high-resistance elements in series with an analog instrument. The resistances are selected in such a way that the meter tracks accurately up to 16 kV. To extend the range of the instrument up to 48 or 80 kV, extension resistors can be added.

Selecting Voltage-Measuring Instruments

Voltage-measuring instruments must be selected based on a variety of criteria. The following sections describe each of the steps that should be used in the selection of voltage-measuring instruments.

(a)   (b)

FIGURE 2.62 Digital readout contact-type safety voltmeter. (*Courtesy Tegam, Inc.*)

Voltage Level. The instrument used must have a voltage capability at least equal to the voltage of the circuit to be measured. Make certain that the manufacturer certifies the instrument for use at that level.

Application Location. Some instruments are designed for use solely on overhead lines or solely in metal-clad switchgear. Make certain that the manufacturer certifies the instrument for the application in which it will be used.

Internal Short Circuit Protection. If the measuring instrument should fail internally, it must not cause a short circuit to appear at the measuring probes. Instruments with resistance leads and/or internal fuses should be employed.

Sensitivity Requirements. The instrument must be capable of reading the lowest voltage which can be present. This is from all sources such as backfeed as well as normal voltage supply.

FIGURE 2.63 Digital readout multimeter. (*Courtesy Fluke.*)

Circuit Loading. The instrument must be capable of measuring voltages that are inductively or capacitively coupled to the circuit. Therefore, it must have a high enough circuit impedance so that it does not load the circuit and reduce the system voltage to apparently safe levels.

Instrument Condition

Before performing voltage measurements, the measuring instrument must be carefully inspected to ensure that it is in good mechanical and electrical condition.

Case Physical Condition. The case and other mechanical assemblies of the instrument must be in good physical condition and not cracked, broken, or otherwise damaged. Any instrument with a broken case should be taken out of service and repaired or replaced.

Probe Exposure. Only the minimum amount of lead should be exposed on contact-type instruments. This minimizes the chance of accidently causing a short circuit when the lead contacts more than one conductor at a time. Low-voltage instruments such as those shown in Figs. 2.61 and 2.62 have spring-loaded plastic sleeves which cover the entire probe until pushed by the pressure during the measurement. These types of probes should always be used for measurement of voltages in power systems.

Lead Insulation Quality. The lead insulation should be closely inspected. If the insulation is frayed, scored, or otherwise damaged, the leads should be replaced before the instrument is used to measure voltage.

Fusing. If the instrument you are using is fused, the fuse should be checked to make certain that it is of the right size and capability. Instruments used to measure power system voltages should be equipped with high interrupting capacity fuses which will safely interrupt 200,000 A at rated voltage.

Operability. Make certain that the instrument is operable. If it uses batteries, they should be checked before the instrument is placed in service. Check the instrument on a known hot source before it is taken into the field. If it does not work, take it out of service until it can be repaired or replaced.

(a)

FIGURE 2.64 Phase tester extension handles. (*Courtesy AB Chance Corp.*)

(b)

FIGURE 2.64   (*Continued*)

Low Voltage Voltmeter Safety Standards

Background.   IEC 61010 is a standard that has been developed to define design and usage safety requirements for low voltage (less than 1000 V AC) meters. The standard applies primarily to contact type instruments and establishes requirements for meter categories based on the meter's ability to withstand voltage surges such as those experienced in a modern low voltage power system.

Categories.   Table 2.24 shows the four categories as defined in the standard. For industrial applications the minimum category used should be category III. If the work to be performed is in the incoming service, incoming substation, or other electrical systems close to the utility system, Category IV should be selected.
    It is also important to note that many meters are rated for 600 V while others are rated at 1000 V. The 1000 V rated meters are always the preferred choice from a safety standpoint. In many applications, a Category III meter rated at 1000 V is superior in surge withstand to a Category IV rated at 600 V.

Three-Step Voltage Measurement Process

Since a safety-related voltage measurement is normally made to make certain that the circuit is dead, the measuring instrument must be checked both before and after the actual

TABLE 2.24 Summary of IEC Meter Categories

Overvoltage category	Location of usage	Examples
Category I	Electronics and other types of cord connected equipment locations	• Any cord connected equipment where in the supply or the equipment itself has built-in transient suppression capabilities • In equipment that is inherently low energy
Category II	Single phase receptacle connected loads	• Appliances • Portable tools • Long branch circuits
Category III	Three-phase or single-phase distribution systems which are isolated from outdoor electric utility supplies by transformers, distance or other specific types of surge protection	• Indoor lighting circuits • Bus and feeders in industrial facilities • Motors, switchgear, and other such industrial equipment
Category IV	Facilities with direct paths to outdoor utility circuits and feeders	• Outdoor substation facilities • Electric meter equipment connected directly to the utility • Service entrance equipment • Overhead lines to isolated facilities

circuit is read. This ensures that a zero reading is, in fact, zero and not caused by a non-functional instrument. Note that this before-and-after check should be made in addition to the check that is given when the instrument is placed into service from the tool room or supply cabinet. This before-and-after measurement process is called the *three-step process.*

Step 1—Test the Instrument Before the Measurement. The measuring instrument should be applied to a source which is known to be hot. Ideally this source would be an actual power system circuit; however, this is not always possible, because a hot source is not always readily available—especially at medium voltages and higher.

Because of this problem manufacturers often have alternative means to check their instruments in the field. Some provide low-voltage positions on their instruments. Figure 2.65 shows the switch settings for the instrument of Fig. 2.60. This instrument may be verified in three different ways:

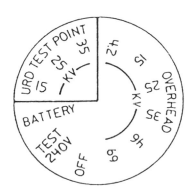

FIGURE 2.65 Selector switch positions for the meter shown in Fig. 2.60. (*Courtesy W.H. Salisbury and Co.*)

1. With the switch in the TEST-240v position, place the instrument head close to a live circuit in excess of 110 V.

2. With the switch in the TEST-240v position, rub the instrument head on cloth or clothing to obtain a static charge. The unit should indicate periodically.

3. Set the switch to the 35-KV overhead position and place the head close to a spark plug of a running engine.

After the instrument is verified, the BATTERY position can be used to verify the battery supply circuitry and the battery condition.

Figure 2.66(b) shows a test device and setup used to verify the type of instrument shown in Fig. 2.64. One lead from the test device (Fig. 2.66(a)) is plugged into a special jack mounted on the instrument. The other end is clipped to each of the measuring probes, one at a time. The reading on the instrument meter is used to determine the operability of the test instrument.

Of course, the best way to test an instrument is to check it on a known hot source using the scale position that will be used for the actual measurement.

Step 2—Measure the Circuit Being Verified.   The instrument should then be used to actually verify the presence or absence of voltage in the circuit. See Chap. 3 for a detailed description of how to make such measurements.

Step 3—Retest the Instrument.   Retest the instrument in the same way on the same hot source as was used in step 1. Table 2.25 summarizes the voltage-measuring steps discussed in this section. Because of the extreme importance of voltage measurement, this procedure is repeated in more detail in Chap. 3.

General Considerations for Low-Voltage Measuring Instruments

Most general-purpose multimeters are not designed for power system use. This is certainly true of many of the small units that are available from some major retail companies. Table 2.26 lists specific requirements for test instruments used in power system voltage measurement.

(a)   (b)

FIGURE 2.66   Field test unit for voltage testers similar to Fig. 2.64. (*Courtesy AB Chance Corp.*)

TABLE 2.25 Voltage Measuring Instrument Selection and Use*

Instrument selection	• Voltage level • Application location • Internal short circuit protection • Sensitivity requirements • Circuit loading
Instrument condition	• Case physical condition • Probe exposure • Lead insulation quality • Fusing • Operability
Three-step process	• Check the meter • Check the circuit • Check the meter

* See text for details.

SAFETY GROUNDING EQUIPMENT

The Need for Safety Grounding

Even circuits that have been properly locked and tagged can be accidently energized while personnel are working on or near exposed conductors. For example,

• If capacitors are not discharged and grounded, they could accidently be connected to the system.

• Voltages could be induced from adjacent circuits. Such voltages can be extremely high if the adjacent circuit experiences a short circuit.

• Switching errors could result in reenergizing of the circuit.

• An energized conductor can fall or be forced into the de-energized circuit, thereby energizing it and causing injury.

• Lightning strikes could induce extremely high voltages in the conductors.

TABLE 2.26 Checklist for Low-Voltage Measuring Equipment

Instrument should be certified for use in electric power systems
Leads should be of proper design
   Secured permanently to instrument
   Fully insulated probes with minimum metallic exposure
   Insulation rated for voltage of circuits to be measured
Internal fusing capable of interrupting fault current if short circuit occurs
High resistance measurement to minimize arcing at probe tips
Single-function (voltage readings) instrument preferred
   If a multimeter is used, the resistance and current measurement circuits must be adequately fused
   Auto-voltage detection is desirable
Instrument insulation levels should be rated for voltage of the circuits to be measured
Instrument should be the correct category for the application per IEC 61010.

Employees who are working on or near such exposed conductors could be severely injured from shock, arc, or blast which results when the conductors are accidently energized. Because of this, safety grounding equipment should be employed as one additional safety measure when employees must work near exposed conductors.

Whenever possible, safety grounds are applied to create a zone of equal potential around the employee. This means that the voltage is equal on all components within reach of the employee. Correct methods for applying safety grounds are discussed more fully in Chap. 3.

Safety Grounding Switches

Grounding switches are specially manufactured units designed to replace a circuit breaker in medium-voltage metal-clad switchgear. The switch rolls into the space normally occupied by the breaker, and the switch stabs connect to the bus and line connections which normally connect to the breaker stabs. Figures 2.67 through 2.69 show various views of a typical grounding switch.

Two different varieties of grounding switches are in common use. One type of switch has two sets of disconnect stabs: one set connects to the bus phase connections, and the other connects to the line connections. The other type of switch has only one set of stabs which can be moved to connect to either the bus or the line side of the switchgear connections.

For either variety of grounding switch, internal connections can be made to electrically ground the bus, the line, or both sides of the switch. Thus the switch grounds the circuit at the source.

Ground switches should be used very carefully and should not be relied on for 100 percent personal safety protection. Safety grounds (see next section) should be employed when secure personal protection is required.

FIGURE 2.67 Front view of a 15-kV grounding switch. (*Courtesy UNO-VEN Refinery, Lemont, IL.*)

FIGURE 2.68 Rear view of a 15-kV grounding switch. (*Courtesy UNO-VEN Refinery, Lemont, IL.*)

FIGURE 2.69 Closeup of jumper connections for 15-kV grounding switch. (*Courtesy UNO-VEN Refinery, Lemont, IL.*)

Safety Grounding Jumpers

General Description. Safety grounding jumpers (also called safety grounds) are lengths of insulated, highly conductive cable with suitable connectors at both ends. They are used to protect workers by short-circuiting and grounding de-energized conductors. Thus if a circuit is accidently energized, the safety grounds will short-circuit the current and protect the workers from injury. Safety grounds also drain static charges and prevent annoying or dangerous shocks. Figure 2.70 illustrates a typical set of safety grounds.

SAFETY GROUNDING JUMPERS
1. **Clamps: With Cable Strain Relief Sleeves**
2. **Ferrules: Per Cable size & Clamp mat'l.**
3. **Cable/s: Grounding: per system short circuit capacity.**
   Length minimum required to reach.

FIGURE 2.70 Typical safety ground jumpers with component parts marked. (*Courtesy W.H. Salisbury and Co.*)

Notice that the safety ground is composed of three major component parts: the clamp, the ferrule, and the cable. The entire assembly must be capable of withstanding the enormous thermal and magnetic forces that are applied when the circuit to which the jumper is applied becomes energized. The construction of safety grounds and their component parts are regulated by ASTM standard F 855.

Clamps.   Clamps serve to connect the grounding jumper to the cable or bus system that it is being used to ground. ASTM standard F 855 defines three types of clamps (I, II, and III). Each of the three types is available in two basic classes (A and B). Table 2.27 describes the basic types and classes of clamp.

Types I and II are used to connect the grounding jumper to the actual de-energized conductor. Type II has hot sticks which are permanently mounted to the clamp, and type I has only rings to which shotgun-type hot sticks can be connected. Type III clamps are used for connection to ground bus, tower steel, grounding clusters, and other such permanently grounded points. Figure 2.71a to c illustrates the three types of grounding clamp.

Each type of clamp can be purchased with either nonserrated (class A) or serrated (class B) jaws. Regardless of which type of jaw is purchased, the conductor surfaces should be thoroughly cleaned before application of the clamp. The class B jaws provide additional assurance that adequate electric contact will be made.

Clamps are available in seven ASTM grades depending on the short-circuit capability of the circuit to which they will be connected. Table 2.28 is taken from ASTM standard F 855 and identifies the seven grades and their associated electrical and mechanical characteristics.

Ferrules.   Ferrules are used to connect the jumper cable to the clamp. ASTM standard F 855 identifies six basic types of ferrule as described in Table 2.29. Figure 2.72a through d illustrates each of the six basic types.

In addition to types, ferrules are manufactured in seven grades depending upon the short circuit currents which they are designed to withstand. Table 2.30 shows the electrical and mechanical characteristics for the seven grades of ferrule. Table 2.31 shows the physical specifications for grounding cable ferrules. Of course, ferrules must be attached to the clamp with a mechanically and electrically secure type of termination. Table 2.32 lists the ASTM standard types of terminations between ferrules and grounding clamps.

Cables.   The cables used for grounding jumpers are specified in three basic types (I, II, and III) as illustrated in Table 2.33. Note that the insulation can be of any color; however, most manufacturers use yellow, black, or clear insulation. Clear insulation has the advan-

TABLE 2.27   Types and Classes of Grounding Jumper Clamps

Type I	Clamps for installation on de-energized conductors equipped with eyes for installation with removable hot sticks
Type II	Clamps for installation on de-energized conductors equipped with permanently mounted hot sticks
Type III	Clamps for installation on permanently grounded conductors of metal structures. These clamps are equipped with tee handles, eyes, or square or hexagon head screws, or both
Class A	Clamps equipped with smooth contact surfaces
Class B	Clamps equipped with serrated, cross-hatched, or other means used to abrade or bite through corrosion products on the conductor to which the jaw is clamped

Source: From ASTM Standard F 855.

(a)

(b)

(c)

FIGURE 2.71 ASTM type I, II, and III clamps. (a) Type I clamp; (b) Type II clamp; (c) Type III clamp. (*Courtesy AB Chance Corp.*)

tage of allowing easy visual inspection to detect signs of conductor fusing after severe short circuits. Clear insulation, however, is sometimes sensitive to ultraviolet degradation. Yellow insulation has the advantage of being easily visible and indicating a potentially hazardous condition. Black insulation, while acceptable, can be easily mistaken for a normal conductor. Clear or yellow insulation is the preferred choice for grounding jumper cable.

TABLE 2.28 Grounding Clamp Ratings

Grade	Grounding clamp torque strength, min				Short circuit properties*								Continuous current rating, A RMS, 60 Hz	Minimum cable size with ferrule installed equal or larger than
	Yield†		Ultimate		Withstand rating, symmetrical kA rms, 60 Hz			Ultimate rating/capacity,‡ symmetrical kA rms, 60 Hz						
	lbf·in	n·m	lbf·in	n·m	15 cycles (250 ms)	30 cycles (500 ms)	Copper cable size	6 cycles (100 ms)	15 cycles (250 ms)	30 cycles (500 ms)	60 cycles (1 s)	Maximum copper test cable size		
1	280	32	330	37	14.5	10	#2	30	19	13	9	2/0	200	#2
2	280	32	330	37	21	15	1/0	48	30	21	15	4/0	250	1/0
3	280	32	330	37	27	20	2/0	60	38	27	19	4/0	300	2/0
4	330	37	400	45	36	25	3/0	76	48	34	24	250 kcmil	350	3/0
5	330	37	400	45	43	30	4/0	96	60	43	30	250 kcmil	400	4/0
6	330	37	400	45	54	39	250 kcmil or 2 2/0	113	71	50	35	350 kcmil	450	250 kcmil or 2 2/0
7	330	37	400	45	74	54	350 kcmil or 2 4/0	159	100	71	50	550 kcmil 550 kcmil	550	350 kcmil or 2 4/0

Note: lbf·in = foot-pound-inch, n·m = newton-meter, ms = millisecond, kcmil = thousand circular mil.
* Withstand and ultimate short circuit properties are based on performance with surges not exceeding 20% asymmetry factor.
† Yield shall mean no permanent deformation such that the clamp cannot be reused throughout its entire range of application.
‡ Ultimate rating represents a symmetrical current which the clamp shall carry for the specified time.
Source: Courtesy of ASTM.

Typical
Type I
compression
ferrules

Typical
Type II
solder
ferrules

(a)

Typical
Type III
plain stud
shrouded
compression
ferrule for
grounding
cables

Type IV
threaded
stud
shrouded
compression
ferrule
for
grounding
cable

Type V
bolted
shrouded
compression
ferrule for
grounding
cable

(b)   (c)

FIGURE 2.72   ASTM types I, II, III, IV, V, and VI ferrules. (*Courtesy ASTM.*)

Type VI threaded
stud compression
ferrule

(d)

FIGURE 2.72 (*Continued*)

Grounding jumper cable sizes are determined by the amount of available short circuit current that is present in the part of the system to which they are being applied. Table 2.34 lists the maximum current ratings and the size of cable required by ASTM standards. Table 2.34 also identifies the current ratings for commercially manufactured, complete jumpers described in the next section.

Complete Jumpers. While some users make their own safety ground jumpers using the ASTM specifications and buying the individual components, most prefer to buy a completely preassembled jumper set as illustrated in Fig. 2.73. The jumper illustrated in Fig. 2.73 is designed for use on live-front pad-mount underground distribution (UD) switchgear and transformers. It is rated at 15,000 A for 30 cycles and 21,000

A for 15 cycles. It has a three-way terminal block assembly, four bronze body grounding clamps, and three 6-ft lengths of 2/0 clear jacketed copper cable.

Many types of jumpers are available from the manufacturer's stock, and many more types can be quickly assembled by the manufacturer for specific applications.

Selecting Safety Grounding Jumpers

The selection of grounding jumpers may be performed with a relatively simple four-step process. The following four sections are provided courtesy of W.H. Salisbury and Company.

Short Circuit Capacity. Determine the maximum short circuit capability at each typical grounding location. From this information the size of the cable can be determined from Table 2.34. (Table 2.34 is the same as Table 5 in ASTM standard F 855.) Note that in selecting the cable size, it is perfectly acceptable to use smaller sizes of cable such as 2/0 and then double the cables when required for higher short circuit duties.

TABLE 2.29 Types of Ferrules for Use on Safety Ground Jumpers

Type I	Compression ferrule. Cylindrical and made for installation on cable stranding by compression
Type II	Solder ferrule is tubular and made for installation on cable stranding with solder
Type III	Plain stud shrouded compression ferrule with a stepped bore that accepts the entire cable over the jacket
Type IV	Threaded stud shrouded compression ferrule with stepped bore that accepts entire cable over jacket and has male threads at forward end
Type V	Bolted shrouded compression ferrule with internal threads and a bolt at forward end
Type VI	Threaded stud compression ferrule with male threads at forward end

TABLE 2.30 Grounding Cable Ferrule Ratings

		Short circuit properties*—Symmetrical kA rms, 60 Hz						Continuous current rating, rms 60 Hz
		Withstand rating		Ultimate rating/capacity[†]				
Grade	Cable size	15 cycles (250 ms)	30 cycles (500 ms)	6 cycles (100 ms)	15 cycles (250 ms)	30 cycles (500 ms)	60 cycles (1 s)	
1	2	14.5	10	30	19	13	9	200
2	1/0	21	15	48	30	21	15	250
3	2/0	27	20	60	38	27	19	300
4	3/0	36	25	76	48	34	24	350
5	4/0	43	30	96	60	43	30	400
6	250 kcmil	54	39	113	71	50	35	450
7	350 kcmil	74	54	159	100	71	50	550

* Withstand and ultimate short circuit properties are based on performance with surges not exceeding 20% asymmetry factor.

[†] Ultimate rating represents a symmetrical current which the ferrule shall carry for the time specified.

Source: Courtesy ASTM.

TABLE 2.31 Grounding Cable Ferrule Physical Specifications

Type	Description	Shape	Minimum Specifications to be Supplied by Manufacturers in Addition to Cable Capacity and Material
I	Compression	O. D. after Compression	Installing Die Code O. D. after Installation
II	Solder	O. D.	O. D.
III	Plain Stud Shrouded Compression	O. D.	Installing Die Codes Stud O. D.
IV	Threaded Stud Shrouded Compression	Thread Size A 30°	Installing Die Codes Thread Size A
V	Bolted Shrouded Compression	O. D. 30°	Installing Die Codes Bolt O. D. or Thread Size
VI	Threaded Compression	Thread Size A 30°	Installing Die Codes Thread Size A

A Standard thread sizes are as follows:

½ in.–13 UNC ≃ M12 × 1.75, ⅝ in.–11 UNC ≃ M16 × 2.00, ¾ in.–10 UNC ≃ M20 × 2.50.

Note: Inspection or vent holes are optional for Types III, IV, V, and VI.

Source: Courtesy ASTM.

TABLE 2.32 Grounding Cable Ferrules and Compatible Grounding Clamp Terminations

CABLE FERRULES ASTM			MATCHING GROUNDING CLAMP CABLE TERMINATION ASTM		
ASTM Ferrule Type	Ferrule Description	Illustration of Ferrule Including Essential Interface Application Data Required	ASTM Termination Style	Termination Description	Essential Interface Application Data Required
I	Compression Ferrule	O. D. After Installation — Base Material	1.	Cable Retaining Eyebolt	
II	Solder Ferrule	O. D. — Base Material	2.	Eyebolt & Cable Support	Ferrule or Stud O. D. Accepted
			3	Plain Bore Clamp and Cable Support	
III	Plain Stud Shrouded Compression Ferrule	O. D. — Base Material	4.	Plain Bore Tubular With Screws and Cable Support	
IV	Threaded Stud Shrouded Compression Ferrule	Thread Size A — 30° Base Material	5. 6. 7.	Plain Bore Boss / Plain Bore Boss and Cable Support / Threaded Bore Boss	Tap Boss I. D. or Thread Size A and Material
V	Bolted Shrouded Compression Ferrule	O. D. or Thread Size — 30° Base Material	5. 6.	Plain Bore Boss / Plain Bore Boss and Cable Support	Tap Boss I. D. and Material
VI	Threaded Compression Ferrule	Thread Size A — 30° Base Material	6. 8. 9.	Plain Bore Boss and Cable Support / Threaded Bore Boss and Cable Support / Threaded Bore Clamp and Cable Support	Tap Boss Thread Size A and Material

A Standard thread sizes are as follows:
½ in.–13 UNC ≈ M12 × 1.75, ⅝ in.–11 UNC ≈ M16 × 2.00, ¾ in.–10 UNC ≈ M20 × 2.50.
Note: Inspection or vent holes are optional for Types III, IV, V, and VI.
Source: Courtesy ASTM.

Grounding Clamps. Grounding clamp types and styles must be determined for each conductor size and shape where grounding jumpers may be applied. If cable and/or bus is expected to be corroded and difficult to clean, serrated jaw clamps should be specified. Remember to specify clamps for both the conductor and the ground end of the grounding set.

TABLE 2.33 Cable Types for Use on Safety Ground Jumpers

Type I	Cables with stranded, soft drawn copper conductor with 665 or more strands of #30 American Wire Gauge (AWG) (0.0100 in) or #34 AWG (0.0063 in). Flexible elastomer jackets rated for installation and continuous service at temperatures from −40°F (−40°C) through +194°F (90°C)
Type II	Cables with stranded, soft drawn copper conductor with 133 wires or more for size #2 or 259 wires or more for size 1/0 and larger. Flexible elastomer jackets rated for installation and continuous service at temperatures from −13°F (−25°C) to +194°F (+90°C)
Type III	Cables with stranded, soft drawn copper conductor with 665 or more #30 AWG (0.0100 in). Flexible thermoplastic jackets rated for installation and continuous service at temperatures ranging from +14°F (−10°C) to +140°F (+60°C). *Caution:* The use of type III jacketed cables is restricted to open areas or spaces with adequate ventilation so that fumes which could be produced by overheating the jacket during a short circuit fault on the cable can be dispersed

TABLE 2.34 Grounding Jumper Ratings

| | Short circuit properties* | | | | | | |
| | Withstand-rating symmetrical kA rms, 60 Hz | | Ultimate capacity symmetrical kA rms, 60 Hz | | | | |
Grounding cable size (copper)	15 cycles (250 ms)	30 cycles (500 ms)	6 cycles (100 ms)	15 cycles (250 ms)	30 cycles (500 ms)	60 cycles (1 s)	Continuous current rating, rms, 60 Hz
#2	14.5	10	29	18	13	9	200
1/0	21	15	47	29	21	14	250
2/0	27	20	59	37	26	18	300
3/0	36	25	74	47	33	23	350
4/0	43	30	94	59	42	29	400
250 kcmil	54	39	111	70	49	35	450
350 kcmil	74	54	155	98	69	49	550

* Withstand and ultimate short circuit properties are based on performance with surges not exceeding 20% asymmetry factor.

Source: Courtesy ASTM.

Grounding Ferrules. Selection of the grounding ferrules is based on the cable size and material and the cable termination style of the grounding clamp(s). Aluminum or copper-plated ferrules should be used with an aluminum clamp and copper cable. Bronze clamps should be used only with copper ferrules and copper cables. Some manufacturers simplify the ferrule selection by employing only one ferrule and cable termination design.

Complete Jumper Specification. To simplify the selection of grounding sets, the user may first determine the short circuit capacity, conductor size and shape, and then refer the design of the jumper to the manufacturer. Most manufacturers have a variety of stock grounding jumpers and will design and build special systems on an as-needed basis.

FIGURE 2.73 Commercially manufactured safety ground jumper. (*Courtesy AB Chance Corp.*)

Installation and Location

The actual procedures and methods for the installation of safety grounds are discussed in detail in Chap. 3. In general the following rules apply:

1. Always make a voltage measurement to verify the system is dead before installing safety grounds.
2. Grounds should be installed in such a way that they equalize the voltage on all equipment within reach of the worker. This equipotential grounding method is illustrated in Chap. 3.
3. When possible, safety grounds should be applied on both sides of the work area. Generally, the more grounding points, the better.
4. Application of safety grounds is considered to be a procedure with more than a normal risk of electric arc. Always wear face shields, rubber gloves, and—whenever possible— flash suits.
5. Safety grounds should be applied on systems of any voltage level to add additional protection for personnel when the system is out of service and conductors are exposed. Systems of 480 V and higher should always be safety-grounded when locked and tagged. Lower-voltage systems may be safety-grounded on an as-needed basis.
6. Of course, if the nature of the work precludes the use of safety grounds, they should not be applied. For example, the measurement of insulation resistance cannot be done when safety grounds are installed.

GROUND FAULT CIRCUIT INTERRUPTERS

Operating Principles

Most 120-V circuits are fed from standard thermal-magnetic molded-case circuit breakers. The circuit is similar to that shown in Fig. 2.74. In this configuration, the only protection from overcurrent is the circuit breaker, which requires a minimum of 15 A to even begin operation. Clearly, if a human being contacts the 120-V circuit, the circuit breaker will not operate unless a minimum of 15 A flows. Even then, it will operate slowly. Since current levels as low as 10 to 30 mA can be fatal to human beings, such installations are clearly not effective for personnel protection.

FIGURE 2.74   Standard protection schemes for 120-V circuits.

FIGURE 2.75 Standard 120-V supply with ground fault circuit interrupter. (*Courtesy AB Chance Corp.*)

Figure 2.75 shows a system that has been in use since the late 1960s. This device, called a *ground fault circuit interrupter* (GFCI), has a current transformer which is applied to the hot and the neutral lead. The resulting output of the current transformer is proportional to the difference in the current between the two leads.

To understand its operation, consider what happens when a normal load is attached to the duplex outlet of Fig. 2.75. Under such circumstances, all the current that flows through the hot wire will go to the load and return to the source on the neutral wire. Since the currents on the hot wire and the neutral wire are equal in magnitude, the output from the GFCI current transformer will be zero and the GFCI will not operate.

Now consider what happens when a grounded person contacts the hot wire downstream of the GFCI. Under these circumstances current will flow from the hot wire through the person and will return on the ground wire. Because of this the currents on the hot wire and the neutral wire will not be equal. The current transformer will produce an output to the sensor which will, in turn, cause the GFCI breaker contacts to open.

This operation occurs instantaneously, that is, with no intentional time delay. The circuit is disconnected very quickly and the person is spared the long-duration shock. While a GFCI does not guarantee the complete safety of personnel, it is set sensitively enough to save the life of the person an overwhelming percentage of the time.

Ground fault circuit interrupters are set to trip when the current difference between the hot lead and the neutral lead differ by more than 5 ± 1 mA. They open typically in less than 25 milliseconds.

Applications

The National Electrical Code (NEC) specifies locations which require the use of GFCIs. These locations include bathrooms, swimming pool outlets, and temporary power supplies for construction work. Refer to the most recent edition of the NEC for a detailed list of these requirements.

At a minimum, GFCI receptacles should be used in portable power cords used for temporary construction power during electrical system and other types of industrial work. Figure 2.76 shows a typical GFCI suitable for such use.

Ground fault circuit interrupters should be tested before each use. In addition to being visually inspected to check for frayed insulation and a damaged case, the GFCI circuit

FIGURE 2.76   Portable ground fault circuit interrupter. (*Courtesy Direct Safety Supply Co.*)

should be tested for proper operation. This test can be easily performed using a simple test device like that shown in Fig. 2.77. This device plugs into the duplex receptacle. When the receptacle is energized, the user can observe the lights on the end which will display in a certain pattern if the wiring to the receptacle is correct. The user then presses the test button and an intentional 5-mA ground is placed on the hot wire. If the GFCI is functional, it will operate and open the circuit. The user can then reset the GFCI and put it into service.

FIGURE 2.77   Receptacle and GFCI tester. (*Courtesy Ideal Industries, Inc.*)

FIGURE 2.78 Typical safety electrical one-line diagram (SEOLD).

SAFETY ELECTRICAL ONE-LINE DIAGRAM

One-line diagrams are used in electric power systems for a variety of purposes including engineering, planning, short circuit analysis, and—most important—safety. The safety electrical one-line diagram (SEOLD) provides a roadmap for the electric power system. Figure 2.78 is an example of a typical SEOLD. Safety electrical one-line diagrams are used to ensure that switching operations are carried out in a safe, accurate, and efficient manner. *An accurate, up-to-date, legible SEOLD should be available for all parts of the electric power system.* Safety electrical one-line diagrams should be accurate, concise, and legible.

1. *Accuracy.* The SEOLD should be accurate and up to date. Regular revision reviews should be performed on the SEOLD, and any required changes or modifications should be implemented immediately. Personnel should be aware of the review process and should have easy access to suggesting changes on the SEOLD.

2. *Concise.* Because one-line diagrams are used for a variety of purposes, some facilities put a large amount of information on one diagram. Items such as cable lengths, sizes, and impedances; current transformer ratios; transformer impedances; protective relay logic circuits; and metering circuits are often found on one-line diagrams used for safety and operations. This practice is not acceptable. Table 2. 35 lists the only items that should be included on a one-line diagram. Note that the use of computer-aided drafting systems makes the development of job-specific one-line diagrams quite simple.

3. *Legibility.* All too often, SEOLDs are allowed to become illegible. They may be sun-bleached, or they may simply be the product of an original diagram that has been allowed to become too old. The SEOLD should be easy to read even in subdued lighting.

The SEOLD is the roadmap for the field electrician. Each electrical worker should have access to the SEOLD for day-to-day switching activities.

THE ELECTRICIAN'S SAFETY KIT

Each electrician should be supplied with a minimum list of electrical safety equipment. The list given in Table 2.36 should be considered the minimum. This represents the minimum requirement for an electrician's electrical safety. Other equipment should be added on an as-needed basis.

TABLE 2.35 Information Required to be Included on Safety Electrical One-Line Diagrams

Energy sources	All sources of energy in the power system should be included on the SEOLD. This includes the normal sources such as power company connections and generators, as well as possible sources of backfeed such as instrument transformers and foreign voltages brought in for remote-control purposes
Disconnect devices	The circuit breakers, switches, cutouts, fused disconnects, and any other devices which can be used to control the power system energy

Note: All energy sources and disconnect devices should be clearly marked with a unique identifying name and/or number. The device itself should be clearly marked with an identical name and/or number.

TABLE 2.36 Minimum Safety Equipment for Electricians

- Rubber insulating gloves with leather protectors and protective carrying bags. Glove voltage classes should be consistent with the voltages around which the electrician is expected to work. Recommend one set of class 0 or 00 gloves and one set suitable for the highest voltage which the electrician will encounter.
- Rubber insulating sleeves. Quantities and voltage classes consistent with the rubber gloves described above.
- Rubber insulating blankets. Quantities and voltage classes consistent with the systems around which the electrician will be working.
- Safety voltage tester. One each low-voltage plus one medium- or high-voltage unit.
- Insulated tools for below 1000 V.
- Padlocks, multiple-lock devices, and lockout tags.
- Hard hats, ANSI standard Z89.1 class B.
- Safety glasses with full side shields, ANSI standard Z87.1 glasses suitable for thermal protection.
- Warning signs—"Danger—High Voltage"—and others as required—red barrier tape with white stripe.
- Safety ground cables as required.
- Flame retardant work clothing—minimum 6 oz per yard.
- Flame retardant flash suit—10 oz per yard with full head protection and three-quarter length coat.
- Safety electrical one-line diagram.

Note: This listing is for electrical hazards only. Other protection such as ear protection, fire extinguishers, and flashlights should be supplied as required.

REFERENCES

1. American Society of Testing and Materials (ASTM), "Standard Test Method for Determining the Arc Thermal Performance Value of Materials for Clothing," ASTM Std F 1959/F 1959M R 99, Conshohocken, PA.

2. A.M. Stoll and M.A. Chianta, "Method and Rating system for Evaluation of Thermal Protection," *Aerospace Medicine,* **40:** 1232–1238, 1968.

3. Richard I. Doughty, Thomas E. Neal, Terrence A. Dear, and Allen H. Bingham, "Testing Update on Protective Clothing & Equipment for Electric Arc Exposure," IEEE Paper PCIC-97-35.

4. Richard L. Doughty, Thomas E. Neal, and H. Landis Floyd, "Predicting Incident Energy to Better Manage the Electric Arc Hazard on 600-V Power Distribution Systems," IEEE Paper.

CHAPTER 3
SAFETY PROCEDURES AND METHODS

INTRODUCTION

The way work is performed in or around an electrical power system is just as important as the safety equipment that is used. Proper voltage measurement can mean the difference between life and death. Standing in the right place during switching operations can mitigate the effects of an electric arc or blast and proper application of safety grounds can prevent an accidental reenergization from becoming a fatality.

This chapter summarizes many industry-accepted practices for working on or around energized electric power circuits. The methods and techniques covered in this section should be used as guidelines. Local work rules and regulatory standards always take precedence. Note that any and all safety procedures should be reviewed at least annually. System changes, employee reassignments, or accidents should be considered excellent reasons for modification of existing procedures or implementation of new procedures.

Safety is the one, truly personal concern in an electric power system. In the majority of electrical accidents, the injured victim was the so-called last link in the chain. The use of proper procedures and/or proper safety equipment could have prevented the accident. Equipment and procedures can be provided, but only the individual employee can make the decision to use them. Employees must be impressed with the knowledge that only they can make this final decision—a decision which may result in life or death.

Caution: This chapter presents several procedures that require technical calculations using a variety of mathematical and engineering applications. These calculations should be performed or evaluated only by qualified, experienced engineers or other such technical personnel with the requisite knowledge and skills.

THE SIX-STEP SAFETY METHOD

Table 3.1 lists six important steps to practicing safe behavior and will serve as the foundation for a personnel safety philosophy. If implemented, this method will greatly enhance the safe and efficient performance of electrical work.

Each individual is responsible for his or her own safety. The steps listed in Table 3.1 are all individual steps that can be taken by everyone who works on or around electric power circuits and conductors.

TABLE 3.1 Six-Step Safety Method

- Think—be aware
- Understand your procedures
- Follow your procedures
- Use appropriate safety equipment
- Ask if you are unsure, and do not assume
- Do not answer if you do not know

Think—Be Aware

Many accidents could have been prevented if the injured victim had concentrated on the safety aspects of the job. Thinking about personal or job-related problems while working on or near energized conductors is a one-way ticket to an accident. Always stay alert to the electrical hazards around the work area.

Understand Your Procedures

Every company has defined safety procedures which are to be followed. Each worker should be thoroughly familiar with all the safety procedures that affect his or her job. Knowledge of the required steps and the reasons for those steps can save a life. All employees should go through extensive safety training.

Follow Your Procedures

In the past, some facilities have allowed the violation of safety procedures in the name of production. Such actions have invariably proven to be costly in terms of human injury and/or death. Violation of safety procedures without good cause should be a discharge offense. What constitutes "good cause" must be decided on a local basis; however, excuses of lesser significance than immediate danger to life should not be acceptable.

Use Appropriate Safety Equipment

No matter how meticulous workers are, accidents do occasionally happen. Equipment failures, lightning strikes, switching surges, and other such events can cause shock, arc, or blast. Also, sometimes it becomes necessary for employees to work on or very close to energized conductors, which increases the chance of accidental contact.

Because of these reasons, appropriate safety equipment should be used any time workers are exposed to the possibility of one of the three electrical hazards. Remember that nothing is sadder than an accident report which explains that the dead or injured worker was not wearing safety equipment.

Ask If You Are Unsure, and Do Not Assume

Ignorance kills and injures many people each year. No one should ever get fired for asking a question—especially if it is a safety-related question. Anyone who is uncertain about a particular situation should be encouraged to ask questions which should then be answered by a qualified person immediately and to the fullest extent possible.

Do Not Answer If You Do Not Know

No one should answer a question if they are not certain of the answer. Self-proclaimed experts should keep their opinions to themselves.

PRE-JOB BRIEFINGS

Definition

A pre-job briefing (sometimes called a "tailgate meeting") is a meeting which informs all workers of the job requirements. In particular a pre-job briefing is used to alert workers to potential safety hazards. A pre-job briefing need not be a formal gathering; however, it is mandatory that all workers involved attend, and worker attendance should be documented.

What Should Be Included?

OSHA rules require that a pre-job briefing discuss, at a minimum, the following issues:

- **S**pecial precautions to be taken
- **H**azards associated with the job
- **E**nergy control procedures
- **P**rocedures and Policies
- **P**ersonal Protective Equipment

Note that the first letters of each of these bulleted items form the acronym **SHEPP**, which can be used to help remember the important issues that need to be discussed.

Pre-job briefings should be proactive meetings in which workers are informally quizzed to make certain that they fully understand the safety issues that they will face.

When Should Pre-Job Briefings Be Held?

- At the beginning of each shift
- At the beginning of any new job
- Any time that job conditions change
- When new personnel are introduced to an ongoing job

ENERGIZED OR DE-ENERGIZED?

The Fundamental Rules

All regulatory standards are quite clear in their requirements to de-energize a circuit before employees work on or near it. Stated simply:

All circuits and components to which employees may be exposed should be de-energized before work begins.

A few basic points will clarify this requirement:

- Production or loss of production is ***never*** an acceptable, sole reason to work on or near an energized circuit.
- Work that can be rescheduled to be done de-energized, should be rescheduled.
- De-energized troubleshooting is always preferred over energized troubleshooting.
- The qualified employee doing the work, must ***always*** make the final decision as to whether the circuit is to be de-energized. Such a decision must be free of any repercussions from supervision and management.

FIGURE 3.1   Hot-work flow chart.

TABLE 3.2 Examples of Additional Hazards

---

* Interruption of life-support systems
* Deactivation of emergency alarms
* Shutdown of ventilation to hazardous locations
* Removal of illumination from the work area

A Hot-Work Decision Tree

Figure 3.1 illustrates a method that may be used to determine the need to work on a circuit when it is energized. The numbers in the following explanation refer to the numbers assigned to each of the decision blocks shown in Fig. 3.1.

1. Work performed on or near circuits of less than 50 V to ground may *usually* be considered to be de-energized work. Note that if the circuit has high arcing capability, decision 1 should be answered as a "Yes."
2. If de-energizing simply changes the hazard from one type to another, or if it actually increases the degree of hazard. This decision should be answered "Yes." Table 3.2 lists the types of additional hazards that should be considered in answering this decision.
3. The need to keep production up is common to all industries—manufacturing, petrochemical, mining, steel, aluminum, and electrical power systems. However, many employers abuse the concept that "production must continue." The following points should clarify when production issues may be allowed to influence the decision to de-energize.
   a. Shutdown of a continuous process that will add extraordinary collateral costs may be a signal to work on the circuit energized. Table 3.3 lists examples of these types of collateral costs.
   b. Shutdown of a simple system which does not introduce the types of problems identified in Table 3.3 should always be undertaken rather than allowing energized work.
4. In some cases, the very nature of the work or the equipment requires that the circuit remain energized. Table 3.4 shows three of the most common examples of such work. Note, however that this work should still be de-energized if it is possible to do it that way. For example, troubleshooting a motor starter may be faster with the circuit energized; however, if it can be done de-energized it should be, even at the cost of a little more time.
5. If decisions 2 or 3 lead in the direction of energized work, the next decision should be rescheduling. If energized work can be done de-energized on a different shift or at a later time, it should be postponed. Many companies miss this elegantly simple alternative to exposing their personnel to hazardous electrical energy.
6. The final, and arguably the most important, decision of all is to determine whether the work can be done safely. If, in the opinion of the qualified personnel assessing the job, the work is simply too dangerous to do with the circuits energized, then it *must* be de-energized.

TABLE 3.3 Examples of Collateral Costs that May Justify Energized Work

---

* Excessive restart times in continuous process systems
* High product loss costs (in polyethylene process, for example, the product has to be physically dug out of process equipment after an unscheduled outage)

TABLE 3.4 Examples of Work that May Require Energization of Circuits

---

- Testing electrical circuits (to verify de-energization, for example)
- Troubleshooting complex controls
- Infrared scan

After the Decision Is Made

If the work must be done energized, all employees who work on or near energized conductors must be qualified to do the work, must use appropriate personal protective equipment, and must use appropriate safety-related work practices.

If the circuits are to be de-energized, the steps listed in Table 3.5 must be followed. Note that proper, safe procedures for each of the items in Table 3.5 are discussed later in this chapter.

SAFE SWITCHING OF POWER SYSTEMS

Introduction

The most basic safety procedure is to de-energize the parts of the system to which workers may be exposed. This procedure virtually eliminates the hazards of shock, arc, and blast. De-energizing, also called *clearing*, involves more than simply turning the switches off. To ensure maximum safety, de-energizing procedures that are precise for each situation should be written.

The following sections discuss the proper safety techniques for operation of various types of equipment and provide de-energizing and reenergizing procedures which may be used as the basis for the development of site-specific procedures. Please note that specific procedures may vary depending on the application and type of equipment. Refer to manufacturer's and/or local facility procedures for specific information. The methods given in these sections should be considered minimum requirements. These procedures assume that the device is being operated when one or both sides are energized.

Caution:

- Switching of electric power should only be carried out by qualified personnel who are familiar with the equipment and trained to recognize and avoid the safety hazards associated with that equipment.
- Non-load-interrupting devices—that is, devices which are not intended to interrupt any current—should never be used to interrupt current flow.

TABLE 3.5 Steps Required Before De-Energized Work May Commence

---

1. All energy control devices feeding the work area must be opened.
2. Locks and tags shall be placed on the energy control devices.
3. Voltage measurements shall be made at the point(s) of exposure to verify that the circuit is de-energized.
4. Safety grounds (if required) shall be placed to ensure the existence of an equipotential work zone.
5. The work area must be closely inspected by a qualified person to make certain that no energized parts remain. This critical step is often missed.

Remote Operation

The procedures on the following pages provide recognized safe work practices for manual operation of the various types of gear. Note that the *best* way to operate any electrical device is doing it remotely. If the equipment has supervisory or other type of remote control, it should always be operated from the remote position, with all personnel safely out of harm's way.

Operating Medium-Voltage Switchgear

General Description. Figures 3.2 through 3.6 illustrate various types of medium-voltage switchgear and the breakers used in them. In this style of gear, the circuit breaker rolls into the switchgear on wheels as shown in Fig. 3.3 or on a sliding type of racking mechanism as shown in Fig. 3.6. The opening and closing of the breaker is performed electrically using a front-panel-mounted, pistol-grip type of control. Turning the grip in one direction (usually counterclockwise) opens the breaker, and turning it in the other direction closes it.

The breaker connects to the bus and the line via a set of disconnects, visible at the top of Fig. 3.3 and on the right side of Fig. 3.6. When the breaker is open, it can be moved toward the front of the cubicle so it disconnects from the bus and line. This action is referred to as *racking* the breaker. The breaker may be completely removed from the switchgear, or it may be put into two or more auxiliary positions. Racking a circuit breaker is accomplished using removable cranks. These cranks may be of a rotary type or a lever bar that is used to "walk" the breaker from the cubicle.

Most switchgear have two auxiliary positions: the *test* position and the *disconnected* position. In the test position, the breaker is disconnected from the bus; however, its control power is still applied through a set of secondary disconnects. This allows technicians to operate the breaker for maintenance purposes. In the disconnected position, the breaker is completely disconnected; however, it is still in the switchgear.

FIGURE 3.2 Medium-voltage metal-clad switchgear. (*Courtesy Westinghouse Electric.*)

FIGURE 3.3 Circuit breaker used in switchgear shown in Fig. 3.2. (*Courtesy Westinghouse Electric.*)

FIGURE 3.4 Switchgear of Fig. 3.2 with door open showing circuit breaker cubicle. (*Courtesy Westinghouse Electric.*)

FIGURE 3.5 Metal-clad, medium-voltage switchgear for vacuum-type circuit breakers. (*Courtesy General Electric.*)

FIGURE 3.6 Medium-voltage, vacuum interrupter circuit breaker used in switchgear of Fig. 3.5. (*Courtesy General Electric.*)

TABLE 3.6 Recommended Minimum Safety Equipment for Operating Metal-Clad Switchgear

Closed door (arc-resistant gear only)
- Hard hat—ANSI type A, B, G, or E (as required by voltage level)
- Safety glasses with side shields
- Flame-resistant work clothing (select using arc-flash calculations)

Open door or non-arc resistant gear
- Hard hat—ANSI type A, B, E, or G
- Safety glasses with side shields
- Rubber gloves with leather protectors (class according to voltage level)
- Flame-resistant work clothing (select using arc-flash calculations)

Note: Closed door means that the front safety panels are closed and latched. If the front panels are not designed to contain blast and arc (such as with arc-resistant switchgear), then *open door* protective equipment should be worn.

For some types of switchgear, the front panel provides worker protection from shock, arc, and blast. This means that the switchgear is designed to contain arc and blast as long as the door is properly closed and latched. Such gear is called *arc-resistant* switch gear.

Closed-Door Operation.   Table 3.6 lists the recommended safety equipment to be used by operators when performing both closed-door and open-door switching on medium-voltage switchgear. Note that both the primary operator and backup operator should be wearing the recommended clothing (Fig. 3.7). The primary operator is the worker who

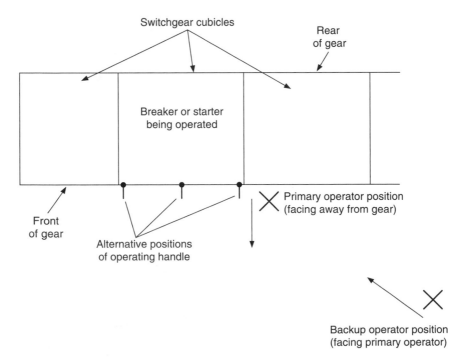

FIGURE 3.7   Proper position for operating electrical equipment (top view).

actually manipulates the handle which opens and/or closes the circuit breaker. The backup operator's responsibility is to back up the primary operator in the event there is a problem. The backup operator may be optional in some facilities.

To operate the switchgear with a closed door, the following steps apply:

1. The primary operator stands to the side of the cubicle containing the breaker to be operated. The side to which he or she stands should be determined by which side the operating handle is on. If the handle is in the middle, the operator should stand on the hinge or the handle side of the door depending on which side is stronger. (Refer to the manufacturer.)

2. The primary operator faces away from the gear. Some workers prefer to face the gear to ensure better control. This is okay but the preferred method is to face away.

3. The backup operator stands even farther from the cubicle, facing the primary operator.

4. The primary operator reaches across to the operating handle and turns it to open or close the breaker. Note that the primary operator continues to keep his or her face turned away from the gear. Some operators prefer to use a hot stick or a rope for this operation. This keeps the arms as far as possible from any hazard.

5. If the breaker can be racked with the door closed, and if the breaker is to be racked away, the primary operator inserts the racking handle. In this operation, the primary operator may have to face the breaker cubicle.

6. If lockout-tagout procedures are required, the primary operator places the necessary tags and/or locks.

Open-Door Operation. Refer to Table 3.6 for a minimum listing of safety equipment for this operation. If the door must be open for racking the breaker away from or onto the bus, the following steps should be observed:

1. The breaker is opened as described earlier under closed-door operation.

2. The primary operator opens the cubicle door and racks the breaker to the desired position.

3. If lockout-tagout procedures are required, the primary operator places the necessary tags and/or locks.

Operating Low-Voltage Switchgear

General Description. With some exceptions, the operation of low-voltage switchgear is very similar to the operation of medium-voltage gear. Figures 3.8 through 3.11 illustrate various types of low-voltage switchgear and the breakers used in them.

In this style of gear, the circuit breaker rolls into the switchgear on a sliding type of racking mechanism as in Fig. 3.11. The circuit breaker is tripped by the release of a powerful set of springs. Depending on the breaker the springs are released either manually and/or electrically using a front panel button and/or control switch. The springs may be charged manually and/or electrically, again depending on the breaker.

The breakers may be closed either with another set of springs or by means of a manual closing handle. The springs for spring-operated breakers may be charged either manually or electrically. For example, the breaker shown in Fig. 3.9c is a manually operated breaker. The large gray handle is cranked several times to charge the closing spring. When the spring is fully charged, the breaker may be closed by depressing a small, mechanical button on the face of the breaker. Tripping the breaker is accomplished by depressing another pushbutton. Other breakers have different means of opening and closing.

FIGURE 3.8 Low-voltage metal-clad switchgear. (*Courtesy General Electric.*)

Low-voltage breakers connect to the bus and the line via a set of disconnects, visible on the right side of Fig. 3.9*b*. When the breaker is open, it can be moved toward the front of the cubicle so that it disconnects from the bus and line. This action is referred to as *racking* the breaker. Racking the larger-size low-voltage breakers is accomplished using removable cranks. These cranks are usually of a rotary type. The breaker may be completely removed from the switchgear, or it may be put into two or more auxiliary positions. Smaller low-voltage breakers are racked by simply pulling or pushing them.

Like medium-voltage switchgear, most low-voltage breaker have two auxiliary positions (*test* and *disconnected*). In the test position, the breaker is disconnected from the bus; however, its control power and/or auxiliary switches are still applied through a set of secondary disconnects. This allows technicians to operate the breaker for maintenance purposes. In the disconnected position, the breaker is completely disconnected; however, it is still in the switchgear.

For some types of switchgear, the front panel provides worker protection from shock, arc, and blast. This means that the switchgear is designed to contain arc and blast as long as the door is properly closed and latched. Such gear is called *arc-resistant*.

Closed-Door Operation. The closed-door operation of low-voltage breakers is virtually identical to the closed-door operation of medium-voltage circuit breakers. Table 3.6 lists the recommended safety equipment to be used by operators when performing both closed-door

AKR-6D-75
(a)

AKR-5C-75
(b)

AKS-5A-50
(c)

FIGURE 3.9 Low-voltage circuit breaker used in the switchgear shown in Fig. 3.8. (*Courtesy General Electric.*)

FIGURE 3.10 Low-voltage metal-enclosed switchgear. (*Type R switchgear manufactured by Electrical Apparatus Division of Siemens Energy and Automation, Inc.*)

FIGURE 3.11   Low-voltage power circuit breaker racked completely away from the bus. (*Type RL circuit breaker manufactured by Electrical Apparatus Division of Siemens Energy and Automation, Inc.*)

and open-door switching on low-voltage switchgear. Note that both the primary operator and backup operator should be wearing the recommended clothing (Fig. 3.7).

The primary operator is the worker who actually manipulates the handle which opens and/or closes the circuit breaker. The backup operator's responsibility is to back up the primary operator in the event there is a problem. The backup operator may be optional in some facilities.

To operate the switchgear with a closed door, the following steps apply:

1. If the breaker requires manual spring charging, the primary operator may face the breaker to obtain the necessary leverage on the cranking handle.

2. After the springs are charged, the primary operator stands to the side of the cubicle containing the breaker to be operated. The side to which he or she stands should be determined by which side the operating handle is on. If the handle is in the middle, the operator should stand on the hinge or the handle side of the door depending on which side is stronger. (Refer to the manufacturer.)

3. The primary operator faces away from the gear. (preferred)

4. The backup operator stands even farther from the cubicle, facing the primary operator.

5. The primary operator reaches across to the operating buttons or handle and operates them to open or close the breaker. Note that the primary operator continues to keep his or her face turned away from the gear. Some operators prefer to use a hot stick or a rope for this operation. This keeps the arms as far as possible from any hazard.

6. If the breaker can be racked with the door closed, and if the breaker is to be racked in or out, the primary operator inserts the racking handle and turns it. Note that breakers which are racked manually cannot be racked with the door closed. In this operation the primary operator may have to face the breaker cubicle.

7. If lockout-tagout procedures are required, the primary operator places the necessary tags and/or locks.

Open-Door Operation. Refer to Table 3.6 for a minimum listing of safety equipment for this operation. If the door must be open for racking the breaker, the following steps should be observed:

1. The breaker is opened as described earlier under closed-door operation.

2. The primary operator opens the cubicle door and racks the breaker to the desired position.

3. If lockout-tagout procedures are required, the primary operator places the necessary tags and/or locks.

Operating Molded-Case Breakers and Panelboards

General Description. Molded-case circuit breakers are designed with a case that completely contains the arc and blast of the interrupted current, as shown in Fig. 3.12. Such breakers are permanently mounted in individual enclosures or panelboards (Fig. 3.13) along with many other such breakers.

Molded-case breakers have a three-position operating handle—open, closed, and tripped. When the operator opens the breaker, he or she does so by moving the operating handle to the open position. Likewise the close operation is accomplished by moving the handle to the close position.

FIGURE 3.12 Various molded-case circuit breakers. (*Courtesy Westinghouse Electric.*)

FIGURE 3.13 Panelboards equipped with molded-
case circuit breakers. (*Courtesy McGraw-Hill, Inc.*)

When the breaker trips via its internal automatic protective devices, the handle moves
to the tripped position. The trip position is normally an intermediate position between the
full-closed and full-open positions. After a trip operation, the breaker cannot be operated
until the handle is moved forcefully to the open position. This action resets the internal tripping mechanism and reengages the manual operating mechanism.

Operation.   *Caution:* Circuit breakers should not be used for the routine energizing and
de-energizing of circuits unless they are manufactured and marked for such purpose. They
may be used for occasional or unusual disconnect service.
    Table 3.7 lists the minimum recommended safety equipment to be worn when operat-
ing molded-case circuit breakers, and Fig. 3.14 illustrates the proper position for the operation. Notice that a backup operator is not required for this procedure; however, secondary
assistance is always a good practice. The procedure can be summarized as follows:

1. The operator stands to the side of the breaker and/or panel, facing the panel. The oper-
    ator may stand to either side, depending on the physical layout of the area.

2. The operator grasps the handle with the hand closest to the breaker.

3. The operator turns his or her head away from the breaker and then firmly moves the
    operating handle to the desired position.

4. If locks or tags are required, they are placed on the breaker using the types of equipment
    as described in Chap. 2.

TABLE 3.7 Recommended Minimum Safety Equipment for Operating Molded-
Case Circuit Breakers

- Hard hat—ANSI type A, B, E, or G
- Safety glasses with side shields
- Flame-resistant work clothing (select using arc-flash calculations)
- Hand protection—leather or flame-resistant gloves (do not need to be insulating)

Note: Operator turns his/her head when operating the circuit breaker.

FIGURE 3.14 Proper position for operating molded-case circuit breakers, enclosed switches, and motor controls (top view).

Operating Enclosed Switches and Disconnects

General Description. Figure 3.15 shows several basic types of enclosed switches. These devices are used to connect and/or disconnect circuits. Such devices may be load interrupting or non–load interrupting. If they are non–load interrupting, they must not be operated when current is flowing in the circuit. If you are uncertain as to whether the switches are load interrupting or not, look on the nameplate or check with the manufacturer. The presence of arc interrupters, such as those in Fig. 3.15*d*, is a good indication that the device is intended to interrupt load current.

These switches are operated by moving the handle. In some units, the handle is bolted or locked in place to prevent inadvertent operation. Enclosed switches have a mechanical interlock that prevents the case from being opened until the handle is in the open position. Qualified personnel may temporarily defeat the interlock if needed for maintenance or troubleshooting purposes. Such switches should not be operated with the door open when load current if flowing. Be very cautious about opening medium-voltage switches.

Operation. The basic operating procedure for such switches is similar to the procedure given for molded-case circuit breakers. *Caution:* Switches should not be used to interrupt load current unless they are intended for that purpose. Refer to the manufacturer's information.

Table 3.8 lists the minimum recommended safety equipment to be worn when operating enclosed switches, and Fig. 3.14 illustrates the proper position for the operation. Notice that a backup operator is not required for this procedure; however, secondary assistance is always a good practice. The procedure can be summarized as follows:

1. The operator stands to the side of the switch and/or panel, facing the panel. The operator may stand to either side depending on the physical layout of the area.

2. The operator grasps the handle with the hand closest to the switch.

3. The operator turns his or her head away from the switch and then firmly moves the operating handle to the desired position.

4. If locks or tags are required, they are placed on the switch using the types of equipment described in Chap. 2.

TABLE 3.8 Recommended Minimum Safety Equipment For Operating Enclosed Switches and Disconnects

- Hard hat—ANSI type A, B, E, or G (as required by voltage level)
- Safety glasses with side shields
- Flame-resistant work clothing (select using arc-flash calculations)
- Rubber gloves with leather protectors (class according to voltage level)

(a)

(b)

(c)

(d)

FIGURE 3.15 Various types of low- and medium-voltage, enclosed disconnect switches. (*Courtesy General Electric, Crouse-Hinds Co., and Eaton Corporation, Cutler Hammer Products.*)

Operating Open-Air Disconnects

General Description. Open-air disconnects may be manually operated or mechanism-operated. Mechanism types (Fig. 3.16) are normally installed as overhead devices in medium-voltage substations or on pole lines. The switch has an operating handle close to the ground which is used to open or close the contacts. At ground level, a metal platform is often provided for the operator to stand on. This platform is bonded to the switch mechanism and to the ground grid or ground rod. The operator stands on the grid; thus the operator's hands and feet are at the same electric potential. Note that the operation of some switches is accomplished by moving the handle in the horizontal plane, while others are moved in a vertical direction.

FIGURE 3.16 Mechanism-operated, three-phase, open-air switch. (*Courtesy Alan Mark Franks.*)

Manually operated switches are operated by physically pulling on the blade mechanism. In some cases, such as the one shown in Fig. 3.17, the switch blade is a fuse. The manual operation is accomplished by using a hot stick. Manually operated switches may be located overhead in outdoor installations, or they may be mounted indoors inside metal-clad switchgear.

Operation. Recommended protective clothing depends upon the type and location of the switch. Table 3.9 lists the minimum recommended safety equipment for operating overhead, mechanism-operated switches such as those shown in Fig. 3.16. Figure 3.18 shows the correct operating position for operating such a switch. *Caution:* Not all open-air switches are designed to interrupt load current. Do not use a non–load interrupting switch to interrupt load current.

The basic operating procedure can be summarized as follows:

1. The operator stands on the metal platform (if available).
2. He or she grasps the operating handle firmly with both hands and moves it rapidly and firmly in the open or close direction as required.
3. If locks or tags are required, they are placed on the mechanism using the types of equipment described in Chap. 2.

Table 3.10 lists the recommended equipment for operation of manually operated open-air switches. Note that the use of flash suits is dependent on the location of the switch. If the switch is on an overhead construction that is outdoors, the worker may opt to wear flame-resistant clothing instead of a flash suit. Because the operator must use a hot stick to operate this type of switch, he or she must stand directly in front of the switch.

The general operating procedure is as follows:

1. Stand in front of the switch.
2. Carefully insert the hot stick probe into the switch ring.
3. Look away from the switch and pull it open with a swift, firm motion.
4. Since one side of the switch may be hot, locks and tags are not always applied directly to the switch. If the switch is in an indoor, metal-clad enclosure, the lock and tag may be applied to the door of the gear.

FIGURE 3.17   Open-air, fused disconnect switch.
(*Courtesy General Electric.*)

Operating Motor Starters

General Description.   With some exceptions, the operation of motor starters is very similar to the operation of low- and medium-voltage gear. Figure 3.19 shows a single motor starter in a cabinet suitable for mounting on a wall.

In the motor control center, the starter mounts on a mechanism specially designed for the purpose. In either type of construction, the motor is stopped and started by depressing the appropriate button. The starter also has a fused disconnect or a molded-case circuit breaker that is used to disconnect the motor and its circuitry from the power supply.

TABLE 3.9   Recommended Minimum Safety Equipment for Operating Overhead, Mechanism-Operated Switches

---

- Hard hat—ANSI type A, B, E, or G (as required by voltage level)
- Safety glasses with side shields
- Flame-resistant work clothing (select using arc-flash calculations)
- Rubber gloves with leather protectors (class according to voltage level)

FIGURE 3.18   Correct position for operating an overhead, mechanism-operated switch.

Motor starters used in motor control centers connect to the bus and the line via a set of disconnects. When the starter is open, it can be moved toward the front of the cubicle so that it disconnects from the bus and line. This action is referred to as *racking*. Racking starters is usually accomplished manually. The starter may be completely removed from the motor control center. Note that large medium-voltage contactors may be operated like medium-voltage circuit breakers.

For many types of motor control centers, the front panel provides worker protection from shock, arc, and blast. This means the motor control center is designed to contain arc and blast as long as the door is properly closed and latched.

Closed-Door Operation.   The closed-door operation of motor starters is virtually identical to the closed-door operation of low-voltage circuit breakers. Table 3.6 lists the recommended

TABLE 3.10   Recommended Minimum Safety Equipment for Operating Manually-Operated, Open-Air Disconnect Switches

- Hard hat—ANSI type A, B, E, or G (as required by voltage level)
- Safety glasses with side shields
- Flame-resistant work clothing (select using arc-flash calculations)
- Rubber gloves with leather protectors (class according to voltage level)
- Hot stick of proper length with proper fittings

Motor-
circuit
protector

Operating
mechanism

Magnetic
starter

Overload
relay reset
button

FIGURE 3.19   Combination motor starter in individual metal-clad cabinet.

safety equipment to be used by operators when performing both closed-door and open-door switching on motor starters. Note that both the primary operator and backup operator should be wearing the recommended clothing (Fig. 3.7). The primary operator is the worker who actually manipulates the handle that opens and/or closes the motor starter. The backup operator's responsibility is to back up the primary operator in the event there is a problem. The backup operator may be optional in some facilities.

To operate the starter with a closed door the following steps apply:

1. Depress the stop button to stop the motor.
2. After the motor is stopped, the primary operator stands to the side of the cubicle containing the starter to be operated. The side to which he or she stands should be determined by which side the operating handle is on. If the handle is in the middle, the operator should stand on the hinge side or the handle side of the door depending on which side is stronger. (Refer to the manufacturer.)
3. The primary operator faces away from the gear. *Note:* If the operating handle of the disconnect has a very tight operating mechanism, the primary operator may face the motor starter to obtain the necessary leverage on the cranking handle.
4. The backup operator stands even farther from the cubicle, facing the primary operator.
5. The primary operator reaches across to the operating handle and operates it to open or close the starter disconnect. Note that the primary operator continues to keep his or her face turned away from the gear. Some operators prefer to use a hot stick or a rope for this operation. This keeps the arms as far as possible from any hazard.

6. If the starter can be racked with the door closed (an unusual configuration), and if the starter is to be racked in or out, the primary operator inserts the racking handle and turns it. Note that starters which are racked manually cannot be racked with the door closed. In this operation, the primary operator may have to face the breaker cubicle.

7. If lockout-tagout procedures are required, the primary operator places the necessary tags and/or locks.

Open-Door Operation. Refer to Table 3.6 for a minimum listing of safety equipment for this operation. If the door must be open for racking the starter, the following steps should be observed:

1. The motor is stopped as described earlier under closed-door operation.

2. The primary operator opens the cubicle door and racks the starter to the desired position.

3. If lockout-tagout procedures are required, the primary operator places the necessary tags and/or locks.

ENERGY CONTROL PROGRAMS

An energy control program is a procedure for the proper control of hazardous energy sources. It should include a listing of company-approved steps for the proper and safe energizing and de-energizing of energy isolation devices as well as general company policy statements on policies with respect to preferred methods of operation. Energy control programs fall into two categories—general and specific.

General Energy Control Programs

Overview. A general energy program is one that is inherently generic in nature. Its steps are broad-based and designed in such a way that the program can be used as a procedure for a wide variety of equipment types. General energy control programs should be used only when the equipment being isolated meets all the following criteria:

- The equipment can be disabled, so it has no potential for the release of stored or residual energy after shutdown.
- The equipment is supplied from a single energy source which can be readily identified and isolated.
- The equipment is completely de-energized and deactivated by the isolation and locking out procedure.
- No employees are allowed to work on or near the equipment until it has been tagged and locked. (See the tagout-lockout section later in this chapter.)
- A single lockout device will achieve a locked-out condition.
- The isolating circuit breakers or switches are under the exclusive control of the employee(s) who placed the lock and tag.
- De-energizing and working on the equipment does not create a hazard for other employees.
- There have been no accidents involving unexpected activation or reenergization of the equipment during previous servicing.

Specific Energy Control Programs

When a part of the system or piece of equipment does not meet all the criteria laid out in the overview to this section, a specific energy control program should be written. Although the procedures will vary depending on the specifics of the installation, at a minimum the program should include the following information:

- The description of the system and/or equipment that will be de-energized.
- Any controls, such as motor starter pushbuttons, that exist on the equipment.
- The voltages and short circuit capacities of the parts of the system which will be de-energized.
- The circuit breakers, switches, or contactors which are used to de-energize the system.
- The steps that must be used to de-energize the system. The steps should include:
  1. The methods and order of operation of the circuit breakers, switches, and so on.
  2. Any special requirements for the lockout-tagout procedure.
  3. Special notifications and safety requirements.
- Reenergizing requirements and procedures.

Basic Energy Control Rules

- The safest and securest method to protect personnel from the electrical hazard is to de-energize the conductors which they must work on or near. De-energization is the preferred method.
- If conductors cannot be de-energized, safety equipment and safety-related work practices must be used to protect personnel exposed to the energized conductors.
- Before personnel are allowed to work on or near any exposed, de-energized conductors, the circuit breakers and/or disconnect switches must be locked and tagged to prevent their inadvertent operation.
- All personnel should be instructed to never operate or attempt to operate any circuit breaker and/or disconnect switch when it is tagged and/or locked.
- Only authorized, qualified, and trained personnel should be allowed to operate electric equipment.
- Locks and tags should be removed only by the personnel that placed them. Two exceptions may apply under the following situations:
  1. If the worker who placed the lock and tag is not available, his or her lock and tag may be removed by a qualified person who is authorized to perform such an action. This procedure is often called *bypassing control* as the person who removes the lock and tag is, in fact, bypassing the authority (control) of the person whose tag is being removed.
  2. Some facilities may authorize the concept of a group lock. A group lock is placed by an authorized shift worker, such as the shift operator, and may be removed by another authorized shift worker. This activity should not be used to prevent any employee from placing his or her tag and lock on energy-isolating devices that may feed conductors which they must work on or near.

De-energizing Equipment.   The general energy program for de-energizing equipment should include the following steps:

1. Before beginning the process, carefully identify the voltage levels and short circuit capabilities of the portion of the system which will be de-energized. This information serves to establish the level of the hazard to all personnel.

2. Notify all employees who will be affected by the de-energization that the system is to be de-energized.

3. Perform necessary checks and inspections to ensure that de-energizing the equipment will not introduce additional safety hazards, for example, de-energizing safety-related ventilation systems.

4. Shut down all processes being fed by the electric system which is to be de-energized.

5. Open the appropriate circuit breaker and/or switch.

6. Rack the circuit breaker away from the bus if it is of the type that can be manipulated in this manner.

7. Release stored energy from springs, hydraulic systems, or pneumatics.

8. Discharge and ground any capacitors located in the de-energized portions of the system.

9. Apply tags and/or locks.

10. Attempt to operate the breaker and/or switch to make certain that the locks are preventing operation. If a motor starter is involved, press the start button to make certain the motor will not start.

11. Measure the voltage on the conductors to which employees at the point where they will be exposed.

12. Notify personnel that the system is safely de-energized, locked, and tagged.

Reenergizing Equipment. Reenergization of some systems is more hazardous than de-energization. While the equipment has been out of service, personnel have grown used to its de-energized voltage status. In addition, tools and/or other equipment may have been inadvertently left on or near exposed conductors.

Because of these factors, the same type of rigorous steps should be followed during reenergization.

1. All personnel should be notified that the system is to be reenergized and warned to stay clear of circuits and equipment

2. A trained, qualified person should conduct all tests and visual inspections necessary to verify that all tools, electric jumpers, shorts, grounds, and other such devices have been removed and that the circuits are ready to be reenergized.

3. Close and secure all cabinet doors and other safety-related enclosures.

4. Because the tests and inspections may have required some time, the personnel warnings should be repeated.

5. Locks and tags should be removed by the personnel that placed them.

6. If breakers were racked into disconnected positions, they should be racked in the connected position.

7. Make final checks and tests, and issue final warnings to all personnel.

8. Reenergize the system by closing and reconnecting breakers and switches. These operations are normally carried out in the reverse order of how they were opened.

Procedures Involving More than One Person. When more than one person is required to lock and tag equipment, each person will place his or her own personal lock and tag on the circuit breakers and/or switches. The placement of multiple locks and tags on equipment is often called *ganging*. Since few circuit breakers or switches have the ability to accept multiple locks and tags, this procedure can take one of two common approaches.

1. A multiple-lock hasp may be applied to the breaker or switch. Such hasps will accept up to six locks. If more than six locks are required, multiple-lock hasps may be cascaded. See Fig. 2.55 for examples of such hasps.

2. A lockbox may be used. In such an operation, the lock is applied to the breaker or switch and the key is then placed inside the lockbox. The lockbox is then secured by the use of a multiple-lock hasp. This approach is used when the presence of many locks on the switch or breaker might cause operational problems.

After the work has been completed, each employee removes his or her lock from the lockbox. The key for the lock is retrieved and the lock can them be removed.

LOCKOUT-TAGOUT

Definition and Description

Tags are used to identify equipment that has been removed from service for maintenance or other purposes. They are uniquely designed and have clear warnings printed on them instructing personnel not to operate the equipment. Locks are applied to de-energized equipment to prevent accidental or unauthorized operation. Locks and tags are normally applied together; however, some special circumstances may require the use of a tag without a lock and/or a lock without a tag. See Chap. 2 for a detailed description of the construction of safety locks and tags.

Employers should develop a written specification which clearly defines the lockout-tagout rules for the facility. This specification should be kept on file and reviewed periodically to ensure that it is kept up to date. The following sections define the key elements that should be included in the specification.

When to Use Locks and Tags

Locks and tags should be applied to open circuit breakers, switches, or contactors whenever personnel will be exposed to the conductors which are normally fed by those devices. The application of the tags will warn and inform other employees that the equipment is not available for service, who applied the tag, and why the tag was applied. The lock will prevent the operation of the breaker, switch, or contactor so that the circuit cannot be accidently reenergized.

Minor inspections, adjustments, measurements, and other such servicing activities which are routine, repetitive, and integral to the use of the equipment do not require the placement of locks and tags unless one of the following conditions exist:

1. Guard, insulation, or other safety devices are required to be removed or bypassed.

2. An employee is required to place his or her body into close proximity with an exposed, energized electric conductor.

Locks and tags do not need to be used on plug and cord connected equipment as long as the cord and plug stay under the exclusive control of the employee who is exposed to the electrical hazard.

Locks without Tags or Tags without Locks

Tags may be used without locks under both of the following conditions:

1. The interrupting device is not designed to accept a lock.

2. An extra means of isolation is employed to provide one additional level of protection. Such an extra procedure might take the form of an additional open point such as removing a fuse or disconnecting a wire or the placement of safety grounds to provide an equipotential work area.

Locks may be used without tags under both of the following conditions:

1. The de-energization is limited to only one circuit or piece of equipment.

2. The lockout lasts only as long as the employee who places the lock is on site and in the immediate area.

Rules for Using Locks and Tags

All electric equipment with the capability to be reenergized and harm employees shall be safely isolated by means of a lock and tag during maintenance, repair, or modification of the equipment.

When two or more crafts must both have access to the equipment, authorized employees from both crafts shall place locks and tags on behalf of the members of their craft. This practice is referred to as *ganging*. This should not be construed to limit the right of any employee to place his or her individual lock and tag on the equipment.

Responsibilities of Employees

Employees who are authorized to place locks and tags have certain responsibilities which they must exercise when placing those tags.

- The system must be surveyed to ensure that all sources of power to the system have been identified.
- All the isolating equipment (circuit breakers, switches, etc.) must be identified and correlated with the portions of the system to which they apply.
- The voltage level and short-circuit magnitude for each part of the system to be de-energized must be determined. This step helps to assess the hazard to individuals who will be exposed to the de-energized system parts.
- All personnel who will be affected by the outage must be notified. This includes employees who may be served by the electric power or who may work on or around the equipment which will be affected by the outage.
- The employee(s) who place the locks and tags must maintain knowledge and control of the equipment to which they have affixed their locks and tags.

When locks and tags are removed, authorized, qualified employees must perform certain tasks, including the following:

1. Notify all affected personnel the system is going to be reenergized.

2. If appropriate, in a gang lock situation, make certain that all authorized employees are prepared to remove their locks and tags.

3. Inspect and/or test all parts of the system to make certain that they are ready to be reenergized.

Note: See the Energy Control Program section discussed earlier in this chapter for more information about these requirements.

Sequence

The following steps should be followed when shutting down an energized electric system:

1. Motors and other operational equipment should be shut down using normal or emergency procedures as required. During the shutdown process personnel safety must be the prime consideration.
2. All isolating equipment (circuit breakers, switches, and/or contactors) should be opened.
3. Isolating equipment that is capable of being racked out should be racked to the disconnected position.
4. Stored energy, such as closing/tripping springs, hydraulic pressure, pneumatic pressure, or other such mechanisms should be discharged and released.
5. Discharge and ground capacitors.

Lock and Tag Application

Isolating equipment is capable of being locked out if either of the following conditions is met:

- The equipment has an attachment which is an integral part of the equipment, through which the locking device may be passed in such a way as to prevent operation of the isolating equipment.
- The lock can be attached in some other way to prevent operation without dismantling, rebuilding, or replacing the energy-isolating equipment. This might apply to a breaker which cannot be locked open but can be locked in the racked or disconnected position.

Locks and tags should be applied to all isolating equipment which is capable of being locked. Locks and tags should never be applied to selector switches, pushbuttons, or other such control devices as the sole means of control.

Isolation Verification

After the locks and tags have been applied, the following steps should be employed to verify that the de-energization was successful:

1. Make certain that all employees are clear of the de-energized conductors.
2. Attempt to reenergize the system by operating the circuit breaker control handles, pushing the switch to close it, or whatever other procedure is appropriate.
3. Using proper procedures and test equipment, measure the voltage on conductors at the point(s) where employee exposure will take place. The voltage should be zero.

Removal of Locks and Tags

Normal Removal.   When the work is finished and an employee is ready to remove his or her locks and tags, the following general approaches should be used:

- The work area should be inspected to ensure that nonessential items have been removed and all components are operationally intact.

- The work area should be inspected to ensure that all employees have been safely removed or positioned.
- Remove any specialized equipment such as safety grounds or spring tension blocks.
- Notify all affected employees that locks and tags are to be removed.
- Locks and tags should be removed by the personnel that placed them.

Control Bypass. If the employee who placed the locks and tags is absent and not available to remove them, and if the locks and tags absolutely must be removed, another authorized employee should assume control of the equipment and remove the absent employee's locks and tags. The following steps must be employed:

1. The employee who is assuming control must make an assessment of the situation and determine that a genuine, critical need exists to remove the absent employee's locks and tags. Examples of such critical needs include:
   a. Immediate operation and/or production requires the equipment to resume operation and avoid general shutdown.
   b. Personnel safety requires restoration of power to the de-energized systems.
   c. Equipment must be temporarily returned to service to allow testing or other such evaluation.
2. The employee assuming control must make every reasonable attempt to contact the absent employee to obtain his or her assessment of the equipment.
3. The employee assuming control must contact other employees who may have knowledge of the availability of the equipment and the advisability of removing the absent employee's locks and tags.
4. The employee assuming control must develop and document a formal conclusion with regard to the decision to remove the absent employee's locks and tags.
5. The employee assuming control must follow the normal removal steps to remove the locks and tags and either reinstall his or her own locks and tags or reenergize the system.
6. The area where the absent employee's locks and tags were removed should be posted with large, easy-to-read signs indicating that the locks and tags have been removed.
7. The employee assuming control must contact the absent employee immediately upon his or her return. The returning employee must be told that his or her locks and tags have been removed and completely updated as to the status of the equipment. The absent employees shall remove the signs posted in step 6.

Figure 3.20 shows a form which may be used to document the control bypass procedure. In this figure the employee who is assuming control is referred to as the *authorized controller*. The absent employee is referred to as the *authorized employee*. Notice that each one of the major steps is documented on this form.

Temporary Removal. Locks and tags may be temporarily removed by the employee who placed them. When doing so, the same steps used in normal removal should be observed.

Safety Ground Application

Safety grounds should be applied as an additional safety measure when equipment is removed from service. The only exception to this is when the nature of the work precludes the use of safety grounds. Procedures like insulation measurement cannot be

LOCKOUT/TAGOUT CONTROLLER'S BYPASS FORM

DATE _____ TIME _____

LOCATION _____

AUTHORIZED CONTROLLER _____

DEPARTMENT _____

AUTHORIZED EMPLOYEE _____

TO BE BY-PASSED DEPARTMENT _____

DESCRIPTION OF EQUIPMENT OR DEVICE(S) TO BE BY-PASSED _____
_____

NEED FOR BY PASS PROCEDURE _____
_____
_____
_____
_____
_____

EFFORTS TO CONTACT AUTHORIZED EMPLOYEE _____
_____
_____
_____
_____
_____

EFFORTS TO CONTACT OTHER PERSONNEL THAT MAY BE AWARE OF THE
APPLIED LOCKOUT/TAGOUT DETAILS _____
_____
_____
_____
_____
_____

CONTROLLER EVALUATION OF THE SITUATION _____
_____
_____
_____
_____
_____

ACTION TAKEN BY THE CONTROLLER TO RELEASE EQUIPMENT FOR USE
_____
_____
_____
_____
_____

ACTIONS TAKEN TO PROPERLY MARK AND POST EQUIPMENT AND AREA AS
BEING RE-ENERGIZED _____
_____
_____
_____
_____
_____

FIGURE 3.20 Lockout-tagout control bypass form.

ACTIONS TAKEN TO NOTIFY EFFECTED EMPLOYEES   _____

_____

_____

_____

_____

_____

AUTHORIZED CONTROLLER _____

DEPARTMENT _____ DATE _____

*AUTHORIZED EMPLOYEE _____

DEPARTMENT _____ DATE _____

* To be use after notification of By-Pass

FIGURE 3.20   (*Continued*)

performed when safety grounds are applied; therefore, they may not be used, or must be temporarily disconnected for the measurement process.

Control Transfer

Lockout-tagout control may be transferred from one employee to another as long as both employees are present. The following steps should be used:

1. The employee relinquishing control follows all steps involved in the normal removal of locks and tags.
2. The employee assuming control follows all steps involved in the normal application of locks and tags.

Nonemployees and Contractors

Nonemployees such as contractors should be required to use a lockout-tagout program which provides the same or greater protection as that afforded by the facility's procedure. Contractors should be required to submit their procedure for review and approval. No work should be allowed until the contractor's lockout-tagout program has been reviewed and approved.

All employees should be familiarized with the contractor's procedure and required to comply with it. Copies of the respective standards should be made available for both contractors and employees. Likewise all contractors must be familiar with the facility's procedure and must comply with it.

If the contractor is working on machines or equipment that have not yet been turned over to the facility, the contractor shall be required to lock and tag that equipment if it might operate and endanger employees. The contractor must guarantee the safety of the locked and tagged installation. As an alternative to this procedure, the contractor should be required to allow gang lockout and tagout with employees and contractor personnel.

Lockout-Tagout Training

All employees should be trained in the use, application, and removal of locks and tags. See Chap. 13 for more information on general employee safety training.

Procedural Reviews

The entire lockout-tagout procedure should be reviewed at intervals no longer than 1 year. A review report should be issued which identifies the portions of the procedure which were reviewed, changes which were considered, and changes which were implemented.

VOLTAGE-MEASUREMENT TECHNIQUES

Purpose

No circuit should ever be presumed dead until it has been measured using reliable, prechecked test equipment. Good safety practice and current regulatory standards require that circuits be certified de-energized by measurement as the last definitive step in the lockout-tagout procedure.

Instrument Selection

Voltage-measuring instruments should be selected for the category, voltage level, the application location, short-circuit capacity, sensitivity requirements, and the circuit loading requirements of the circuit which is to be measured.

During voltage measurement, the instrument may be subjected to voltage transients. This is especially critical in low-voltage (below 1000 V) applications. When selecting a low-voltage-measuring instrument, be certain to select the correct category of instrument as defined in IEC 61010. See Chap. 2 for a description of the categories.

Voltage Level. The instrument must be capable of withstanding the voltage of the circuit which is to be measured. Use of underrated instruments, even though the circuit is dead, is a violation of good safety practice.

Application Location. Some instruments are designed for outdoor use only and should not be used in metal-clad switchgear. Always check the manufacturer's information and verify that the instrument is designed for the location in which it is being used.

Internal Short-Circuit Protection. Industrial-grade, safety-rated voltage-measuring instruments are equipped with internal fusing and/or high-resistance elements which will limit the short-circuit current in the event the instrument fails internally. Be certain the instrument being used has internal protection with an interrupting/limiting rating that is at least equal to the short-circuit capacity of the circuit that is being measured.

Sensitivity Requirements. The instrument must be capable of measuring the lowest normal voltage that might be present in the circuit which is being measured. If the instrument has two ranges (high and low, for example), be certain to set it on the correct lowest range which applies to the voltage being measured.

Circuit Loading. A real voltmeter can be represented by an ideal meter in parallel with a resistance. The resistance represents the amount of loading the meter places on the circuit (Fig. 3.21). In the diagram R_S represents that resistance of the system being measured and

Circuit being measured   Voltage measuring instrument

FIGURE 3.21 Equivalent circuit of voltage measurement.

R_M is the internal resistance of the voltmeter. The voltmeter will read a voltage given by the formula:

$$V_M = V \times \frac{R_M}{R_M + R_S} \tag{3.1}$$

If the meter internal resistance is high compared to system resistance, the voltage across the meter will be very close to the actual system voltage. If the meter resistance is very low compared to the system resistance, the meter will read a lower voltage than actually exists because most of the voltage drop will be across the system resistance instead of the meter resistance. In this second situation, the meter is "loading" the circuit.

If a very low resistance meter is used, the meter may not read a potentially lethal static or inductively coupled voltage. This situation is unacceptable since an inductively coupled or static voltage can be lethal. The meter that is selected for a safety-related voltage measurement must have an internal resistance that is high enough to avoid this problem.

Instrument Condition

Before each use, an instrument must be closely inspected to ensure it is in proper working order and that insulation systems have not been damaged. The case physical condition, probe exposure, lead insulation, fusing, and the operability of the instrument should be verified.

Case Physical Condition. The case must be free of breaks, cracks, or other damage that could create a safety hazard or misoperation. Broken instruments should be taken out of service until they can be repaired or replaced.

Probe Exposure. Modern instrument probes have spring-loaded sleeves which cover the probe until forced back. Check to make certain that only the minimum amount of probe required to do the job is exposed.

Lead Insulation Quality. Carefully inspect the lead insulation to make certain it is not damaged in any way.

Fusing. Accessible fuses should be checked to be certain that they are correctly installed and have not been replaced by incorrect units.

Operability. Before each usage (at the beginning of each shift, for example), the instrument should be checked to make certain that it is operable. Do not substitute this check for the instrument checks required in the three-step process.

Three-Step Measurement Process

Step 1—Test the Instrument. Immediately before each measurement, the instrument should be checked on a source which is known to be hot. This step confirms that the instrument is working before the actual circuit verification is made.

   The preferred method is to use an actual power system conductor of the same voltage as that being verified. Finding such a circuit is easier when low-voltage measurements are being made; however, even then a hot circuit may not be available. Some manufacturers supply a device which creates a voltage that is sufficient for testing the instrument.

   Some instruments intended for measurement of medium voltages have a low-voltage switch setting on them which allows them to be checked on low voltage. This is not a preferred method since a problem with the switch could give errorheous, dangerous false readings.

Step 2—Measure the Circuit Being Verified. After the instrument is tested, the worker then measures the circuit being verified to make certain it is de-energized. The actual wires that should be measured and the techniques to be used are discussed later in this chapter.

Step 3—Retest the Instrument. After the circuit has been verified, the instrument should be rechecked as it was in step 1. This ensures that the instrument was operable both before and after the measurement, thus affirming that a zero measurement is zero and not caused by an inoperable instrument.

What to Measure

As a general rule of thumb, all normally energized conductors should be measured to ground and to each other. The readings to ground should be made whether the system is grounded or not. Note that all readings should be made as close to the point of exposure as possible.

   The procedures given in the following paragraphs will apply only to contact type instruments. Proximity instruments measure the electrostatic field set up around an energized conductor rather than the actual voltage between two conductors. If the proximity indicator shows an unexpected voltage, the circuit should be rechecked, possibly with a contact tester.

Single-Phase Systems. The hot wire of single-phase systems should be measured both again neutral and ground. Figure 3.22 shows the points which should be measured.

Two-Phase Systems. A voltage measurement should be taken between the two hot phases, between each hot phase and neutral (one at a time), and between each hot phase and ground. Figure 3.23 shows the measurement locations for a typical two-phase system.

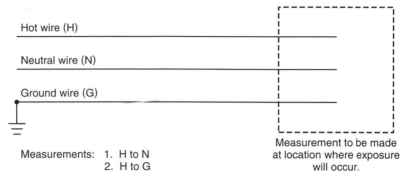

Measurements: 1. H to N
                2. H to G

Measurement to be made
at location where exposure
will occur.

FIGURE 3.22   Measurement points for a single-phase system (one hot wire).

Three-Phase Systems.   Measure between each of the hot wires, two at a time, between each hot wire and neutral, and between each hot wire and ground. Figure 3.24 shows the measurement locations for a typical three-phase system.

Using Voltage Transformer Secondaries for Voltage Measurements.   Many medium-voltage power systems are equipped with step-down voltage transformers which are used for metering or telemetering purposes. Because of the lower voltage on the secondary, some workers wish to measure the secondary winding voltage to verify that the primaries are de-energized. Such transformers may be used for safety-related voltage measurements under the following conditions:

1. The transformers can be visually traced and are known to be connected to the system where exposure will occur.

2. The transformers are located close to the part of the system where exposure will occur.

Measurements: 1. H1 – H2
                2. H1 – N
                3. H2 – N
                4. H1 – G
                5. H2 – G

Measurement to be made
at location where exposure
will occur.

FIGURE 3.23   Measurement points for a single-phase system (two hot wires).

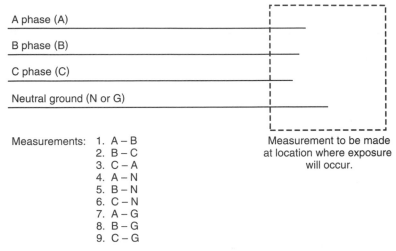

A phase (A)

B phase (B)

C phase (C)

Neutral ground (N or G)

Measurements: 1. A – B
              2. B – C
              3. C – A
              4. A – N
              5. B – N
              6. C – N
              7. A – G
              8. B – G
              9. C – G

Measurement to be made
at location where exposure
will occur.

FIGURE 3.24 Measurement points for a three-phase system.

3. The transformer secondaries must be measured both before and after the circuit is de-energized. This verifies that the transformer fuses are not blown, resulting in an erroneous zero voltage reading.

4. A safety ground is applied to the primary circuit of the transformer after the measurement but before personnel contact occurs.

5. All other safety-related techniques should be used as described in the other parts of this chapter.

Caution: This procedure is considered a secondary, nonpreferred method. Do not use this approach unless absolutely necessary.

Using Panel Voltmeters for Voltage Measurements.   Panel voltmeters may be used as a general indication of system energization; however, they **should not be used** for safety-related voltage measurements.

How to Measure

Preparation.   Table 3.11 identifies the steps which should precede the actual measurement procedure. The area should be cleared of unnecessary personnel. This will prevent their exposure to arc or blast in the event that a problem occurs. The person making the

TABLE 3.11 Voltage Measurement Preparations

- Clear area of unnecessary personnel.
- Wear appropriate safety equipment.
- Expose conductors to be measured.
- Position measuring instrument.

TABLE 3.12   Recommended Minimum Safety Equipment for Making Low-Voltage Safety Measurements

---

- Hard hat—ANSI type A, B, E, or G
- Safety glasses with side shields
- Flame-resistant work clothing (select using arc-flash calculations)
- Class 00 or higher rubber gloves with leather protectors
- Class 00 or higher rubber sleeves (if the measurement requires proximity to energized conductors)

---

measurement should wear and use appropriate safety equipment. (Safety equipment information will be covered later in this chapter.)

After the safety equipment is on, the panels, doors, or other access means should be opened to expose the conductors that are to be measured. The measuring instrument is then carefully positioned. When making medium-voltage measurements with hot sticks, be certain that the hot stick is not contacting your body. The instrument should be securely positioned so that it does not fall to the floor if the leads are inadvertently overextended.

Safety Equipment.   Table 3.12 lists the minimum recommended safety equipment to be worn when making low-voltage measurements. Table 3.13 lists equipment for medium-voltage measurements. The flash suit is identified as optional for outdoor, open-air measurements. Although many line personnel do not wear flash suits when performing overhead work, the use of these uniforms is strongly recommended.

Measurement.   After all preparations are made and the safety equipment is put on, the measuring instrument should be applied to the conductors. If a measurement to ground is being made, one lead should be connected to the ground first and then the phase connection made. When measuring between two hot wires, the order of connection is unimportant. If a contact instrument is being used, each lead should be carefully placed on the appropriate conductor. The meter or readout is then observed to see if the circuit is hot. If a proximity instrument is being used, it should be moved gradually toward the conductor until it indicates or until the conductor is touched.

PLACEMENT OF SAFETY GROUNDS

Safety Grounding Principles

Safety grounds are conductors which are temporarily applied to de-energized system conductors. They are used to provide a safe zone for personnel working on or around

TABLE 3.13   Recommended Minimum Safety Equipment for Making Medium- and High-Voltage Safety Measurements

---

- Hard hat—ANSI type B, E
- Safety glasses with side shields
- Flame-resistant work clothing (select using arc-flash calculations)
- Class 1 or higher rubber gloves with leather protectors (class according to voltage level)
- Class 00 or higher rubber sleeves, if the measurement requires proximity to energized conductors (class according to voltage level)

---

de-energized conductors and to ensure that an accidental reenergization will not cause injury. Power system components should be considered to be energized until safety grounds are in place. Safety grounds should never be placed until the conductors where they are to be installed have been measured and verified to be de-energized.

See Chap. 2 for a detailed description of the design and description of safety grounds. *Caution:* Safety grounds must be properly designed and sized for the conductors and short circuit capacity to which they may be subjected. Refer to Chap. 2 and/or manufacturer's information for specifics.

In the event that the system is accidently reenergized, enormous magnetic forces are exerted on the safety ground wires. The forces can cause the grounds to whip violently and cause injury to personnel. To minimize this effect the safety grounds should be as short as possible, and the conductors should be restrained.

Safety Grounding Location

Equipotential Grounding. Safety grounds should be applied in such a way that a zone of equal potential is formed in the work area. This equipotential zone is formed when fault current is bypassed around the work area by metallic conductors. Figure 3.25 shows the proper location of safety grounds for three different work situations. In each of these situations, the worker is bypassed by the low-resistance metallic conductors of the safety ground.

Assume the worker contacts the center phase in Fig. 3.25*b*. With a fault current capacity of 10,000 A, safety ground resistance (R_j) of 0.001 Ω, and worker resistance (R_w) of 500 Ω, the worker will receive only about 20 mA of current flow.

Figure 3.26*a* through *d* shows two nonpreferred or incorrect locations for the application of safety grounds. In these diagrams, the worker's body is in parallel with the series combination of the jumper resistance (R_j) and the ground resistance (R_g). The placement of grounds in this fashion greatly increases the voltage drop across the worker's body in the event the circuit is reenergized. Such placements should not be used.

Single-Point versus Two-Point Grounding. Single-point grounding is the placement of only one safety ground set. In this procedure, the safety grounds are placed as close to the point of work as is possible. When possible, the grounds are placed between the worker and the source of electric energy.

Two-point grounding is the placement of two safety ground sets. They are usually placed on opposite sides of the work area; that is, one set is placed "upstream" and one set is placed "downstream" from the workers.

In general, more safety ground sets are better than fewer; however, the safety ground system must provide a zone of equal potential as described in the previous section.

Application of Safety Grounds

Safety Equipment. Table 3.14 lists the minimum recommended safety equipment that should be worn and used when applying safety grounds. Full shock, arc, and blast protective clothing is required. No matter how carefully the work area is prepared, the possibility still exists for error and/or inadvertent reenergization during the application process.

Notice also that hot sticks are recommended for the actual application process. The use of hot sticks distances the worker from the point of contact and, therefore, minimizes the possibility of injury caused by arc or blast.

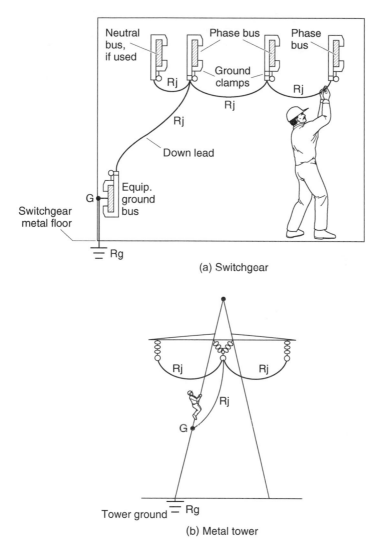

FIGURE 3.25 Application of safety grounds to provide a zone of equal potential.

Procedure. The actual application procedure will be different for each specific application. The following general steps are recommended.

1. Thoroughly inspect the safety ground set which is to be used. Points to check include
   a. Insulation quality
   b. Condition of conductors
   c. Condition of clamps
   d. Condition of ferrule
   e. Condition of cable-to-ferrule connection

(c) Wood pole

(d) Electrical equivalent

FIGURE 3.25   (*Continued*)

TABLE 3.14 Recommended Minimum Safety Equipment for Application of Safety Grounds

- Hard hat—ANSI type B, E
- Safety glasses with side shields
- Flame-resistant work clothing with full head protection (select using arc-flash calculations)
- Class 0 or higher rubber gloves with leather protectors (class according to voltage level)
- Hot stick with suitable fittings for application of safety ground clamps*

*Hot sticks may not be usable in cramped switchgear locations. Workers should exercise best safety judgement in such situations.

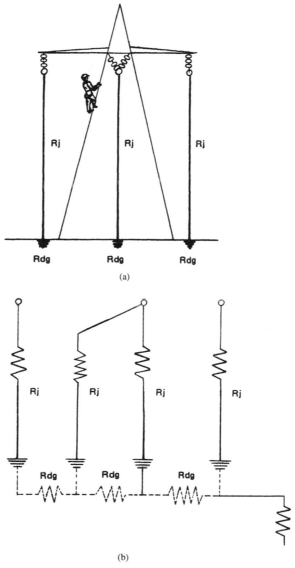

FIGURE 3.26 Two incorrect, nonpreferred safety ground configurations.

3.41

(c)

(d)

FIGURE 3.26 (*Continued*)

2. Identify the point at which each ground clamp will be connected to the system. Be certain to select points which minimize the amount of slack in the safety grounds. This will minimize the whipping action in the event the system is reenergized.
3. Put on required safety equipment.
4. Measure the system voltage to make certain that the system to be grounded is de-energized. (See the previous sections on voltage measurement.)
5. Make certain that all unnecessary personnel have been cleared from the area.
6. Apply the ground end of the safety ground sets first. (These points are labeled with the letter "G" in Fig. 3.25.)
7. Connect the phase-end safety ground clamp to the hot stick.
8. Firmly contact the de-energized conductor with the phase end of the safety ground.
9. Tighten the grounding clamp firmly. Remember the amount of resistance in the clamp connection can make the difference between a safe connection and a hazardous one.
10. Repeat steps 6 through 8 for each of the phases to be grounded.
11. Record the placement of each safety ground by identification number. (See Control of Safety Grounds in the next section.)

Figure 3.27 shows a worker applying safety grounds to a de-energized system.

The Equipotential Zone

When a proper equipotential zone is established, there will be no lethal potential differences which the worker can reach in the work area. Figure 3.25 illustrates configurations that can be used to accomplish this end.

FIGURE 3.27 Worker applying safety grounds. (*Courtesy McGraw-Hill, Inc.*)

Note however, that some situations make the establishment of such a zone difficult or impossible. Consider, for example, the employee who must stand on the earth when he or she is working. Since the earth has a relatively high resistivity, the worker's feet will be at a different potential from that of the metallic elements that he must contact.

There are two approaches that can be used in such circumstances:

1. A metal platform can be laid on the ground and bonded to the grounded metal of the electrical system. Figure 3.18 is an example of this approach. In this situation, as long as the worker stays on the metal pad, he or she will remain in an equipotential zone.
2. The worker can be insulated from the earth or other high resistivity conductors. Rubber mats, gloves, or blankets can be placed so that the worker is completely insulated from electrical contact.

Of course, the best approach is always to establish the equipotential zone; however, these two "work-around" approaches may be used when absolutely necessary.

Removal of Safety Grounds

Safety Equipment.   The removal of safety grounds is no less hazardous than applying them. All the safety equipment listed in Table 3.14 should be worn. Safety grounds should be removed using hot sticks.

Procedure.   Safety grounds should be removed as follows:

1. Put on all required safety equipment.
2. Remove each of the phase connections one at a time.
3. Remove the ground connection.
4. Check the ground off as being removed.

Remember that when safety grounds are not present, the system should be considered to be energized.

Control of Safety Grounds

Safety grounds must be removed before the power system is reenergized. They must also be inspected periodically and before each use. The following sections describe two methods of controlling the safety ground sets.

Inventory Method.   Each safety ground set should be identified with its own unique serial number. The serial number should be etched or impressed on a metal tag which is permanently attached to the safety ground set.

A safety officer should be appointed to control the inventory of safety ground sets. This person will control the use of the safety grounds and is responsible for keeping lists of where the grounds are applied during an outage.

As each safety ground is applied, the safety officer notes its placement on a placement control sheet. As each ground is removed, the officer notes its removal on the sheet. No reenergization is allowed until the safety officer is satisfied that all safety grounds have been removed. Figure 3.28 is a typical safety ground placement control sheet.

The sheet shown in Fig. 3.28 has the minimum required information. Other columns—such as where the ground set is installed, who installed it, and who removed it—may be added as needed.

SAFETY GROUND PLACEMENT CONTROL SHEET			
Safety Officer:		Outage Date:	
Ground Set ID Number	Date Applied:	Date Removed:	Safety Officer's Initials

FIGURE 3.28 Safety ground placement control sheet.

Visual Method.   Some small facilities may not be able to justify the complexity of a control system such as that described in the previous section. A visual control system may suffice for facilities which have only one or two safety grounding sets. In such a system, a brightly colored rope is permanently attached to each grounding set. Nylon ski rope is ideal for this application. The length of the rope should be determined by the applications; however, 3 meters (m) (10 ft) is a good starting point.

At the remote end of the rope, attach a brightly colored warning sign stating "Grounds are Applied." After the safety grounds are attached to the system, the rope should be arranged on the switchgear so that it is easily seen. If it can be arranged, the rope should be placed so the gear cannot be closed up until the rope is removed. The green sign should be placed on the breaker control handle, start pushbutton, or other such device so that it is obviously visible.

When this procedure is used, it is very difficult to reenergize the system with the grounds in place. However, the inventory method is the preferred and recommended method.

FLASH HAZARD CALCULATIONS AND APPROACH DISTANCES

Introduction

How close to an electrical hazard can I get? Obviously, distance provides a safety barrier between a worker and an electrical hazard. As long as workers stay far enough away from any electrical energy sources, there is little or no chance of an electrical injury. This section describes methods that can be used to determine the so-called approach distances. Generally, if work can be carried on outside the approach distances, no personal protective equipment is required. If the worker must cross the approach distances, appropriate personal protective equipment must be worn and appropriate safety procedures must be used.

Approach Distance Definitions

Approach distances generally take two forms—*shock hazard distance* and *flash hazard distance*. Note that the flash or arc hazard distance also includes the *blast hazard distance*. Figure 3.29 illustrates the approach distances that are defined by the NFPA in their publication *NFPA 70E Standard for Electrical Safety in the Workplaces 2000 Edition*. Note that the space inside any given boundary is named for that boundary. Thus the space inside the restricted approach boundary is referred to as the *restricted space*.

No approach boundary may be crossed without meeting the following general requirements:

1. The employee must be qualified to cross the boundary.
2. The employee must be wearing appropriate personal protective equipment.
3. Proper planning must be carried out to prepare the employee for the hazards he/she may face.

Detailed requirements for each boundary will be given in later sections.

Determining Shock Hazard Approach Distances

Table 3.15 may be used to determine the minimum approach distances for employees for shock hazard purposes. Note that the shock approach distances are all based on voltage levels. In general, the higher the voltage level, the greater the approach distance.

Unqualified Persons.   Unqualified persons may not approach exposed energized conductors any closer than the values specified in columns (2) and (3) of Table 3.15. For unqualified persons the "limited" approach boundary is inviolable; that is, no unqualified person

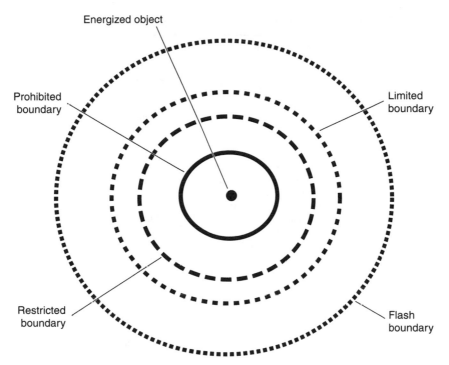

FIGURE 3.29 Minimum approach distance definitions.

may ever enter into the limited approach space under any circumstances. Personal protective equipment is irrelevant to this requirement.

Note that "movable" conductors are those that are not restrained by the installation. Overhead and other types of suspended conductors qualify as movable conductors. Fixed-circuit parts include buses, secured cables, and other such conductors.

Qualified Persons. Qualified personnel approach distances are determined from columns (4) and (5) of Table 3.15. No one may cross the restricted or the prohibited approach boundaries without meeting the requirements which are defined later in this section. Note that the minimum requirements for crossing either of these boundaries include:

1. The worker must be qualified and fully trained for the work and the hazards that will be encountered inside the boundary.
2. The worker must wear all appropriate personnel protective clothing.

Specific Requirements for Crossing the Restricted Approach Boundary. In order to cross the restricted approach boundary, the following criteria must be met:

- The worker must be qualified to do the work.
- There must be a plan in place that is documented and approved by the employer.
- The worker must be certain that no part of the body crosses the prohibited approach boundary.

TABLE 3.15   Approach Boundaries to Live Parts for Shock Protection

(All dimensions are distance from live part to employee.)				
(1)	(2)	(3)	(4)	(5)
	Limited approach boundary		Restricted approach boundary; includes	
Nominal system voltage range, phase to phase	Exposed movable conductor	Exposed fixed circuit part	inadvertent movement adder	Prohibited approach boundary
0 to 50	Not specified	Not specified	Not specified	Not specified
51 to 300	10 ft 0 in.	3 ft 6 in.	Avoid contact	Avoid contact
301 to 750	10 ft 0 in.	3 ft 6 in.	1 ft 0 in.	0 ft 1 in.
751 to 15 kV	10 ft 0 in.	5 ft 0 in.	2 ft 2 in.	0 ft 7 in.
15.1 kV to 36 kV	10 ft 0 in.	6 ft 0 in.	2 ft 7 in.	0 ft 10 in.
36.1 kV to 46 kV	10 ft 0 in.	8 ft 0 in.	2 ft 9 in.	1 ft 5 in.
46.1 kV to 72.5 kV	10 ft 0 in.	8 ft 0 in.	3 ft 3 in.	2 ft 1 in.
72.6 kV to 121 kV	10 ft 8 in.	8 ft 0 in.	3 ft 2 in.	2 ft 8 in.
138 kV to 145 kV	11 ft 0 in.	10 ft 0 in.	3 ft 7 in.	3 ft 1 in.
161 kV to 169 kV	11 ft 8 in.	11 ft 8 in.	4 ft 0 in.	3 ft 6 in.
230 kV to 242 kV	13 ft 0 in.	13 ft 0 in.	5 ft 3 in.	4 ft 9 in.
345 kV to 362 kV	15 ft 4 in.	15 ft 4 in.	8 ft 6 in.	8 ft 0 in.
500 kV to 550 kV	19 ft 0 in.	19 ft 0 in.	11 ft 3 in.	10 ft 9 in.
765 kV to 800 kV	23 ft 9 in.	23 ft 9 in.	14 ft 11 in.	14 ft 5 in.

Notes:
For SI units: 1 in. = 25.4 mm; 1 ft = 0.3048 m.
Source: Courtesy *National Fire Protection Association.*

- The worker must work to minimize the risk that may be caused by inadvertent movement by keeping as much of the body out of the restricted space as possible. Allow only protected body parts to enter the restricted space as necessary to complete the work.
- Personal protective equipment must be used appropriate for the hazards of the exposed energized conductor.

Specific Requirements for Crossing the Prohibited Approach Boundary.   NFPA 70E considers crossing the prohibited approach boundary to be the same as working on or contacting an energized conductor. To cross into the prohibited space, the following requirements must be met:

- The worker must have specified training required to work on energized conductors or circuit parts.
- There must be a plan in place that is documented and approved by the employer.
- A complete risk analysis must be performed.
- Authorized management must review and approve the plan and the risk analysis.
- Personal protective equipment appropriate for the hazards of the exposed energized conductor must be used.

Calculating the Flash Hazard Minimum Approach Distance (Flash Protection Boundary)

Note that the methods given in this section are based solely on the radiant portion of the heat energy from an electric arc. Most of the energy transfer is via radiation, since the electric arc is usually interrupted before convection can contribute a significant amount.

Note, however, that other sources of thermal injury can occur and are not accounted for in these calculations. Flying molten material and/or superheated plasma can burn flesh severely. The clothing suggested in the handbook will help to provide some protection against such events.

Introduction. The concept of a minimum approach boundary for flash protection is based on the amount of tolerance that human tissue has to heat. The current industry standards use the so-called "Stoll Curve" developed by Stoll and Chianta in the 1960s.[1] The flash boundary then represents the closest distance that an unprotected (meaning not wearing thermal protective clothing beyond normal cotton garments) worker may approach an electrical arcing source. Theoretically, at the flash boundary distance, if an electrical arc occurs, the unprotected worker should receive no more than a curable, second-degree burn. Table 3.16 is a reproduction of the Stoll Curve table.

Low-Voltage Calculations (Below 600 V). Table 3.17 illustrates five methods that may be used for calculating the minimum approach distances for electrical flash when the circuit voltage is below 600 V.

$$D_C = \left(\frac{5271 \times t \times (0.0016 I_{SC}^2 - 0.0076 I_{SC} + 0.8938)}{1.2} \right)^{1/1.9593} \quad (3.2)$$

$$D_C = \left(\frac{1038.7 \times t \times (0.0093 I_{SC}^2 - 0.3453 I_{SC} + 5.9675)}{1.2} \right)^{1/1.4738} \quad (3.3)$$

Note that the two bottom methods, represented by Eqs. (3.2) and (3.3), are actually solutions to the Doughty, Neal, and Floyd equations for incident energy = 1.2 cal/cm².[2] This means that methods 3 and 4 must be used only within the constraints under which the empirical equations were developed—circuits with fault currents between 16 and 50 kA, 600 V or less, 18 in or more from the arc source. The "arc-in-a-box" method is based on a test method using a box 20 in on a side.

If we assume a 480 V circuit with a maximum fault current of 20,000 A and a tripping time of 0.1 s, the four methods generate the following results:

Method 1:   4 ft
Method 2:   2.1 ft (top formula only)
Method 3:   2.19 ft
Method 4:   3.44 ft
Method 5:   2.28 ft

Note that the NFPA method (2), the Doughty, Neal, and Floyd equations for open air arcs (3) and method 5 the IEEE calculator agree very closely. Method 4 gives a larger flash boundary, since the "arc-in-a-box" method assumes that the energy is focused toward the worker. That is, most of the energy goes in one direction.

Method 5 employs the IEEE Flash Hazard calculator provided with the IEEE Standard Std 1584 - 2002. This calculator is a Microsoft Excel spreadsheet application that calculates both

TABLE 3.16   Human Tissue Tolerance to Heat, Second-Degree Burn

Exposure time	Heat flux		Total heat		Calorimeter* equivalent		
s	kW/m²	cal/cm²s	kWs/m²	cal/cm²	$\Delta T°C$	$\Delta T°F$	ΔmV
1	50	1.2	50	1.20	8.9	16.0	0.46
2	31	0.73	61	1.46	10.8	19.5	0.57
3	23	0.55	69	1.65	12.2	22.0	0.63
4	19	0.45	75	1.80	13.3	24.0	0.69
5	16	0.38	80	1.90	14.1	25.3	0.72
6	14	0.34	85	2.04	15.1	27.2	0.78
7	13	0.30	88	2.10	15.5	28.0	0.80
8	11.5	0.274	92	2.19	16.2	29.2	0.83
9	10.6	0.252	95	2.27	16.8	30.2	0.86
10	9.8	0.233	98	2.33	17.3	31.1	0.89
11	9.2	0.219	101	2.41	17.8	32.1	0.92
12	8.6	0.205	103	2.46	18.2	32.8	0.94
13	8.1	0.194	106	2.52	18.7	33.6	0.97
14	7.7	0.184	108	2.58	19.1	34.3	0.99
15	7.4	0.177	111	2.66	19.7	35.4	1.02
16	7.0	0.168	113	2.69	19.8	35.8	1.03
17	6.7	0.160	114	2.72	20.2	36.3	1.04
18	6.4	0.154	116	2.77	20.6	37.0	1.06
19	6.2	0.148	118	2.81	20.8	37.5	1.08
20	6.0	0.143	120	2.86	21.2	38.1	1.10
25	5.1	0.122	128	3.05	22.6	40.7	1.17
30	4.5	0.107	134	3.21	23.8	42.8	1.23

*Iron/constantan thermocouple.
Source: Stoll, A. M. and Chianta, M. A., "Method and Rating System for Evaluation of Thermal Protection," Aerospace Medicine, Vol 40, 1968, pp. 1232–1238.

incident energy and flash boundary. It is a relatively "easy to use" application; however, as with all of the calculation methods given in this chapter, it should be employed only by qualified, experienced engineers with sufficient background and experience to use the tool correctly.

Medium- and High-Voltage Calculations.   Much of the research that has been done on flash boundary calculations has concentrated on the low-voltage spectrum (600 V and less); consequently, the availability of empirical support applied to medium voltage systems and higher for the various derived formulas is sparse.

Method 5 is the best choice for calculating the flash boundaries for medium voltage systems up to 15 kV. However, in the absence of the IEEE standard, the engineer may use Method 2 with certain cautions.

Calculating medium- and high-voltage flash boundaries using Method 2 (Table 3.17) appears to give conservatively high results when compared to software programs derived using industry standard heat transfer calculations.

To calculate the flash boundary for medium-voltage systems, use the top formula given in Method 2. Then enter the result into one of the commercially available (or freeware) software programs and verify that the received heat flux density is on the order of 1.2 cal/cm².

TABLE 3.17 Methods for Calculating Flash-Boundary Approach Distances Below 600 Volts

#	Conditions	Method/Formulas	Notes
1	$I_{SC} \times t \leq 5000$ Ampere-seconds	$D_c = 4$ feet (1.22 meters)	I_{SC} = Maximum fault current (Amps) t = Duration of fault in seconds D_c = Flash boundary distance
2	$I_{SC} \times t > 5000$ Ampere-seconds	$D_c = \sqrt{2.65 \times MVA_{bf} \times t}$ $D_c = \sqrt{53 \times MVA \times t}$	I_{SC} = Maximum fault current (Amps) MVA_{bf} = Max fault MVA* MVA = Supply Transformer MVA[†¶] t = Duration of fault in seconds D_c = Flash boundary distance (in feet)
3	$16{,}000\,A < I_{SC} < 50{,}000\,A$	See Eq. 3.2[‡]	I_{SC} = Maximum fault current (kA) t = Duration of fault in seconds D_c = Flash boundary distance
4	$16{,}000\,A < I_{SC} < 50{,}000\,A$	See Eq. 3.3[§]	I_{SC} = Maximum Fault Current (kA) t = Duration of fault in seconds D_c = Flash boundary distance
5	$700\,A \leq I_{SC} \leq 106\,kA$ $208\,V \leq V \leq 15\,kV$	IEEE Std 1584 guide for performing arc-flash hazard calculations	I_{SC} = Maximum fault current V = System phse to phase voltage

*The bolted three-phase fault MVA at the point of exposure. Note that for a three-phase system $MVA_{bf} = kV_{LL} \times I_{SC} \times \sqrt{3} \times 1000$.
[†]This is the maximum self-cooled, full load MVA of the supply transformer.
[‡]This formula applies to open air short circuits.
[§]This formula applies to enclosed short circuits—the so-called "arc-in-a-box."
[¶]For transformers with ratings less than 0.75 MVA, multiply the MVA rating times 1.25.

CALCULATING THE REQUIRED LEVEL OF ARC PROTECTION (FLASH HAZARD CALCULATIONS)

Introduction

Flash protection beyond normal cotton or wool work clothing is not required as long as all parts of the worker's body stay outside the flash boundary as calculated above. Outside the flash boundary the minimum recommendation is either:

1. Flame-resistant clothing with an ATPV of 4.5 cal/cm^2 or higher
2. Natural fabric (cotton or wool) work clothing of 7 oz/yd^2 or more

If, however, the worker must cross the flash boundary and is qualified to do so, he or she must wear flame-resistant clothing to protect against the arc incident energy levels that may be encountered.

The following sections provide suggestions and direction for determining the arc energy levels to which the worker may be exposed. Note that the research into such methods is an on-going effort. The reader should always refer to the most current version of such research as found in IEEE Standard 1584 and NFPA 70E.

To provide arc protection, the worker must select and wear flame-resistant clothing with an ATPV or E_{BT} equal to or greater than the incident energy level. Note that the incident energy level is calculated based on a given distance between the arc source and the worker, for example 24 inches. The worker will not be adequately protected if he or she gets any closer than the distance used in the calculation for the specific situation.

Flash boundaries can be calculated from the formulas in this section by setting the incident energy level to 1.2 cal/cm^2 (5.02 J/cm^2) and then solving the appropriate equation for the distance.

Caution: The following sections present several complex calculation procedures using a variety of mathematical and engineering applications. These calculations should be performed or evaluated only by qualified, experienced engineers or other such technical personnel with the requisite knowledge and skills.

The Lee Method

This method is named after its developer Ralph Lee, PE (deceased). Mr. Lee was one of the early, and most successful researchers into the theoretical models for the amount of energy created by an electrical arc and the amount of incident energy received by workers in the vicinity of the arc.

The Lee method is based on two major assumptions:

1. The maximum energy transfer from the power system to an electric arc is equal to one-half of the bolted short-circuit power available at the point where the arc occurs.

2. All of the arc electrical energy will be converted to the incident heat energy.

Starting with these two assumptions, Mr. Lee and subsequent researchers have refined the Lee method to the form shown in equation 3.4

$$E = K \times VI_{bf}\left(\frac{t}{D^2}\right) \times 10^6 \tag{3.4}$$

Where E = the incident energy (J/cm^2 or cal/cm^2 depending on the value of K)
K = a constant. If K = 2.142 then E is in J/cm^2. If K = 0.512 then E is in cal/cm^2.
V = the system phase-to-phase voltage (kV)
I_{bf} = the bolted fault current (kA)
t = the arcing time (seconds)
D = the distance from the arcing point to the worker (mm)

The Lee method is a very conservative method and was developed without the aid of the empirical results available today from the extensive research in this field. While it can be used for all situations, its principle application is in those cases that have not been modeled based on actual experimental measurements. It is most often used in systems where the system voltage is greater than 600 V and/or where fault currents, arc lengths, or other such parameters are beyond the ranges that have been tested using field measurements.

Although the Lee method does not take the focusing effect of equipment into account, the so-called *arc-in-a-box*, it is very conservative. The use of multipliers should be required only in the most extreme applications.

Methods Outlined in NFPA 70E

In addition to the LEE method, the 2004 Edition of NFPA 70E, *Standard for Electrical Safety in the Workplace* gives two different methods for calculating incident energy levels.

Method #1. This method is based on a research by Richard L. Doughty, Thomas E. Neal, and H. Landis Floyd.[2] They performed research using the thermal manikin approach described in Chap. 1. Their research was bounded by the following conditions:

1. Systems with voltage levels of 600 V and below
2. Systems with maximum available short circuit currents between 16 kA and 50 kA.
3. Working distances of equal to or greater than 18 inches.

After running numerous tests by creating electric arcs in open air, they took their results and performed a curve fit to predict the incident energy. Equation 3.5 is the formula that was developed and which most closely models the energies they measured.

$$E_{MA} = 5271 D_A^{-1.9593} \times t_A (0.0016 I_{SC}^2 - 0.0076 I_{SC} + 0.8938) \qquad (3.5)$$

Where E_{MA} = the maximum open air arc incident energy (cal/cm^2)
  D_A = the distance from the electrodes (in)—note that $DA \geq 18$ in
  t_A = the duration of the arc (seconds)
  I_{SC} = the short circuit current (kA)—note that I_{SC} is the actual arc current, not the maximum bolted fault current.

In the second part of their research they created electrical arcs in a cubic box with one open side. The box was 20 inches on a side and the measurements were taken at various distances from the open end. They then performed a curve fit to model the results. Equation 3.6 is the resulting formula.

$$E_{MB} = 1038.7 D_B^{-1.4738} \times t_A (0.0093 I_{SC}^2 - 0.3453 I_{SC} + 5.9675) \qquad (3.6)$$

Where E_{MB} = the maximum *arc-in-a-box* incident energy (cal/cm^2)
  D_B = the distance from the electrodes (in)—note that $DB \geq 18$ in

All other variables are as described for the open air formula (Eq. 3.5)
  Using either Eqs. 3.5 or 3.6, incident energies can be calculated, then protective clothing can be purchased with the necessary ratings.

Method #2. Method number two is actually the method taken from the IEEE Standard 1584-2002.[3] This method is described in the next section.

IEEE Standard Std 1584-2002

System Model Limitations. Institute of Electrical & Electronics Engineers (IEEE) working group P1584 has performed research and theory development over several years to expand the work done by Doughty, et al.[2] As shown in Table 3.18, the working group has significantly expanded the applicable range for the application of their methods.

Calculation of Arcing Current. One of the principal drawbacks of the Lee method is that the bolted fault current, which is used in that method, is always somewhat greater than the actual current flow when an electrical arc is formed. At voltages below 1000 V, the arcing current is often substantially less than the bolted fault current.
  Using both theoretical and empirical data, Standard 1584, develops a formula to calculate the arcing current that will present in a system when the bolted fault current is known. Equations 3.7, and 3.8 are used to calculate the logarithm to the base ten of arcing current

TABLE 3.18 System Limits for Calculation of Incident
Energy Using IEEE Standard 1584–2002

System phase-to-phase voltage	0.208 to 15 kV
System frequency	50 to 60 Hz
Short-circuit current range	700 to 106 kA
Conductor gap (Arc length)	13 to 152 mm

$(\log_{10}I_a)$. Equation 3.7 is used for systems below 1 kV, and Eq. 3.8 is used for systems that are equal to or greater than 1 kV.

$$\log_{10}I_a = K + 0.662(\log_{10}I_{bf}) + 0.0966V + 0.000526G + 0.5588V(\log_{10}I_{bf}) \\ -0.00304G(\log_{10}I_{bf}) \tag{3.7}$$

$$\log_{10}I_a = 0.00402 + 0.983(\log_{10}I_{bf}) \tag{3.8}$$

Where I_a = the arcing current (kA)
  K = a constant. $K = -0.153$ for open air arcs and $K = -0.097$ for *arc-in-a-box*.
  I_{bf} = the maximum, symmetrical rms, three-phase bolted short circuit current (kA)
  V = the system three phase voltage (kV)
  G = the conductor gap (arc length) (mm)

After $\log_{10}I_a$ has been calculated from Eqs. 3.7 or 3.8, equation 3.9 is used to calculate I_a.

$$I_a = 10^{(\log_{10}I_a)} \tag{3.9}$$

Calculating incident energy. The calculation of actual incident energy is done in two steps using the empirically derived arc current (I_a). In step one, normalized incident energy (E_n) is calculated. E_n is normalized for an arcing time of 0.2 seconds and a distance from the arc of 610 mm. Equations 3.10 and 3.11 are used for this calculation.

$$\log_{10}E_n = k_1 + k_2 + 1.081(\log_{10}I_a) + 0.0011G \tag{3.10}$$

$$E_n = 10^{(\log_{10}E_n)} \tag{3.11}$$

Where E_n = the normalized incident energy (cal/cm²)
  $k_1 = -0.792$ for open air arcs and -0.555 for *arcs-in-a-box*
  $k_2 = 0$ for ungrounded and high-resistance ground systems and -0.113 for grounded systems.
  G = the conductor gap (arc length) (mm)

The actual incident energy at any distance D is then calculated using Eq. 3.12.

$$E = k_{CJ}C_fE_n\left(\frac{t}{0.2}\right)\left(\frac{610}{D}\right)^{\lambda} \tag{3.12}$$

Where E = the incident arc energy at distance D in cal/cm² (k_{CJ} = 1) or J/cm² (k_{CJ} = 4.184)
  k_{CJ} = 1 for E_n is cal/cm² and 4.184 for E_n in J/cm²
  C_f = 1 for systems above 1 kV or 1.5 for systems equal to or less than 1 kV
  E_n = the normalized incident energy calculated from equation 3.11.
  t = the arcing time in seconds
  D = the distance from the arc (mm)
  X = a distance exponent taken from Table 3.19

TABLE 3.19 Distance Factors (X) Used in Eq. 3.19

System voltage (kV)	Type of equipment	Typical conductor gap (mm)	
0.208 to 1	Open-air	10–40	2.000
	switchgear	32	1.473
	MCCs and panelboards	25	1.641
	cables	13	2.000
>1 to 5	Open-air	102	2.000
	switchgear	13–102	0.973
	cables	13	2.000
>5 to 15	Open-air	13–153	2.000
	switchgear	153	0.973
	cables	13	2.000

In Eq. 3.12 the arcing time (t) will be determined by how long the protective devices require to completely interrupt the short circuit. If the arcing current decreases, the arcing time will generally increase. Because of this, NFPA 70E suggests recalculating Eqs. 3.7 through 3.12 using 0.85 I_a and the resulting arcing time. The standard then suggests using the larger incident energy from the two calculations.

Software Solutions

There are several software products on the market that allow the calculation of incident energy and/or flash boundaries. For example, IEEE Standard 1584-2002 comes with a complete set of Microsoft Excel spreadsheet applications. The user only needs to enter the values for his/her particular system to determine the incident energy and the flash boundary using the IEEE method.

At least one freeware software product (MSDOS based) is available for calculating the incident energy for single phase short-circuits. This product was developed by Alan Privette, PE and is available on the internet at a variety of locations. The Cadick Corporation website (http://www.cadickcorp.com) and the Oberon Company website (http://www.oberoncompany.com) are two locations where this product may be found.

In addition to these virtually all of the commercially available engineering software packages such as SKM Systems Analysis, Inc.—PowerTools for Windows, have added arc-flash calculation packages to their short-circuit analysis and coordination study packages.

Required PPE for Crossing the Flash Hazard Boundary

To select the level of flame-resistant protection required, the following procedure may be used:

1. Calculate the incident arc energy value as shown previously:
2. Select clothing that provides an ATPV[*] or E_{BT}[†] that is less than the incident energy value previously calculated. Double layers of protective clothing may be required at the higher energy levels. Refer to the manufacturers for recommendations.

---

[*]Arc Thermal Performance Value—material rating provided by the manufacturer. This is the amount of heat energy that will just cause the onset of a second-degree burn (see Chap. 1).
[†]The average of the five highest incident energy values that did not cause the fabric to break open. Test value supplied by the manufacturer and developed per ASTM F 1959/F 1959M-99.

A Simplified Approach to the Selection of Protective Clothing

The National Fire Protection Association provides a simplified approach to the selection of protective clothing. While this method is convenient and economical, it should be used with extreme caution. The author has found specific situations when the simplified method does not yield the most conservative values, thus putting workers at risk.

Additionally, this method assumes certain short circuit and operating time values. The user must be absolutely certain that his or her power system falls within the assumed values.

1. Identify the Hazard/Risk category from Table 3.20. Note that this is based on the type of work that will be performed. Note also that Table 3.20 identifies whether insulating gloves and/or tools are required.

2. Use Table 3.21 to select the various types of PPE required for the hazard determined in step 1.

3. Use Table 3.22 to select the weight of flame-resistant clothing required for the task.

Although this procedure is quite simple and straightforward, it should be used carefully for at least two reasons:

- Because it is conservative, it tends to result in substantial amounts of clothing for the employee. Workers may tend to disregard necessary equipment out of frustration.

- The standard is task-based as opposed to location-based. Using the quantitative methods described previously will provide, ultimately, a more easily applied set of rules.

- In some rare cases, the simplified approach may give values that do not provide adequate protection. This is especially critical if the limits imposed by the method are ignored.

BARRIERS AND WARNING SIGNS

Barriers and warning signs should be placed to control entrance into a work area where there are exposed energized conductors. The installation of such equipment will vary depending on the layout of the work area. The following general criteria should be applied:

- The signs should be distinctive, easy to read, and posted at all entrances to the work area. The signs should clearly warn personnel of the hazardous or energized condition.

- Barriers and barrier tape should be placed at a height that is easy to see. Three feet or so is a good starting point. Adjust the height as dictated by the specific installation.

- Barriers and barrier tape should be placed so that equipment is not reachable from outside of the barrier. This will prevent the accidental or intentional operation of equipment by personnel not authorized to do so.

- If sufficient work room is not available when barriers are placed, attendants should be used to warn employees of the exposed hazards.

Figure 3.30a shows a sample arrangement for a work area in a switchgear room. The layout given here is for example only; however, the general principles may be used. Key points for this installation are

1. Stanchions are used to provide a firm structure for stringing the barrier tape.

2. Warning signs are placed at both entrances warning personnel that hazards exist inside.

3. Warning signs are also posted on the switchgear itself.

4. Five to ten feet of work clearance is allowed between the tape and the switchgear. This number may vary depending on the space available and the work to be performed.

TABLE 3.20 Hazard Risk Category Classifications

Task (Assumes Equipment Is Energized, and Work Is Done Within the Flash Protection Boundary)	Hazard/Risk Category	V-rated Gloves	V-rated Tools
Panelboards rated 240 V and below—Notes 1 and 3	—	—	—
Circuit breaker (CB) or fused switch operation with covers on	0	N	N
CB or fused switch operation with covers off	0	N	N
Work on energized parts, including voltage testing	1	Y	Y
Remove/install CBs or fused switches	1	Y	Y
Removal of bolted covers (to expose bare, energized parts)	1	N	N
Opening hinged covers (to expose bare, energized parts)	0	N	N
Panelboards or Switchboards rated > 240 V and up to	—	—	—
600 V (with molded case or insulated case circuit breakers)—			
Notes 1 and 3			
CB or fused switch operation with covers on	0	N	N
CB or fused switch operation with covers off	1	N	N
Work on energized parts, including voltage testing	2*	Y	Y
600 V Class Motor Control Centers (MCCs)—Notes 2			
(except as indicated) and 3	—	—	—
CB or fused switch or starter operation with enclosure doors closed	0	N	N
Reading a panel meter while operating a meter switch	0	N	N
CB or fused switch or starter operation with enclosure doors open	1	N	N
Work on energized parts, including voltage testing	2*	Y	Y
Work on control circuits with energized parts 120 V or below, exposed	0	Y	Y
Work on control circuits with energized parts >120 V exposed	2*	Y	Y
Insertion or removal of individual starter "buckets" from MCC—Note 4	3	Y	N
Application of safety grounds, after voltage test	2*	Y	N
Removal of bolted covers (to expose bare, energized parts)	2*	N	N
Opening hinged covers (to expose bare, energized parts)	1	N	N
600 V Class Switchgear (with power circuit breakers or fused			
switches)—Notes 5 and 6	—		—
CB or fused switch operation with enclosure doors closed	0	N	N
Reading a panel meter while operating a meter switch	0	N	N
CB or fused switch operation with enclosure doors open	1	N	N
Work on energized parts, including voltage testing	2*	Y	Y
Work on control circuits with energized parts 120 V or below, exposed	0	Y	Y
Work on control circuits with energized parts >120 V exposed	2*	Y	Y
Insertion or removal (racking) of CBs from cubicles, doors open	3	N	N
Insertion or removal (racking) of CBs from cubicles, doors closed	2	N	N
Application of safety grounds, after voltage test	2*	Y	N
Removal of bolted covers (to expose bare, energized parts)	3	N	N
Opening hinged covers (to expose bare, energized parts)	2	N	N
Other 600 V Class (277 V through 600 V, nominal)			
Equipment—Note 3	—	—	—
Lighting or small power transformers (600 V, maximum)	—	—	—
Removal of bolted covers (to expose bare, energized parts)	2*	N	N
Opening hinged covers (to expose bare, energized parts)	1	N	N
Work on energized parts, including voltage testing	2*	Y	Y
Application of safety grounds, after voltage test	2*	Y	N
Revenue meters (kW-hour, at primary voltage and current)	—	—	—
Insertion or removal	2*	Y	N
Cable trough or tray cover removal or installation	1	N	N
Miscellaneous equipment cover removal or installation	1	N	N
Work on energized parts, including voltage testing	2*	Y	Y
Application of safety grounds, after voltage test	2*	Y	N

(Continued)

TABLE 3.20   Hazard Risk Category Classifications (*Continued*)

Task (Assumes Equipment Is Energized, and Work Is Done Within the Flash Protection Boundary)	Hazard/Risk Category	V-rated Gloves	V-rated Tools
NEMA E2 (fused contactor) Motor Starters, 2.3 kV through 7.2 kV	—	—	—
Contactor operation with enclosure doors closed	0	N	N
Reading a panel meter while operating a meter switch	0	N	N
Contactor operation with enclosure doors open	2*	N	N
Work on energized parts, including voltage testing	3	Y	Y
Work on control circuits with energized parts 120 V or below, exposed	0	Y	Y
Work on control circuits with energized parts >120 V, exposed	3	Y	Y
Insertion or removal (racking) of starters from cubicles, doors open	3	N	N
Insertion or removal (racking) of starters from cubicles, doors closed	2	N	N
Application of safety grounds, after voltage test	3	Y	N
Removal of bolted covers (to expose bare, energized parts)	4	N	N
Opening hinged covers (to expose bare, energized parts)	3	N	N
Metal Clad Switchgear, 1 kV and above	—	—	—
CB or fused switch operation with enclosure doors closed	2	N	N
Reading a panel meter while operating a meter switch	0	N	N
CB or fused switch operation with enclosure doors open	4	N	N
Work on energized parts, including voltage testing	4	Y	Y
Work on control circuits with energized parts 120 V or below, exposed	2	Y	Y
Work on control circuits with energized parts >120 V, exposed	4	Y	Y
Insertion or removal (racking) of CBs from cubicles, doors open	4	N	N
Insertion or removal (racking) of CBs from cubicles, doors closed	2	N	N
Application of safety grounds, after voltage test	4	Y	N
Removal of bolted covers (to expose bare, energized parts)	4	N	N
Opening hinged covers (to expose bare, energized parts)	3	N	N
Opening voltage transformer or control power transformer compartments	4	N	N
Other Equipment 1 kV and above	—	—	—
Metal clad load interrupter switches, fused or unfused	—	—	—
Switch operation, doors closed	2	N	N
Work on energized parts, including voltage testing	4	Y	Y
Removal of bolted covers (to expose bare, energized parts)	4	N	N
Opening hinged covers (to expose bare, energized parts)	3	N	N
Outdoor disconnect switch operation (hookstick operated)	3	Y	Y
Outdoor disconnect switch operation (gang-operated, from grade)	2	N	N
Insulated cable examination, in manhole or other confined space	4	Y	N
Insulated cable examination, in open area	2	Y	N

Legend:
V-rated Gloves are gloves rated and tested for the maximum line-to-line voltage upon which work will be done.
V-rated Tools are tools rated and tested for the maximum line-to-line voltage upon which work will be done.
  2* means that a double-layer switching hood and hearing protection are required for this task in addition to the other Hazard/Risk Category 2 requirements of Table 3-3.9.2 of Part II.
  Y = yes (required)
  N = no (not required)
  Notes:
  1. 25 kA short circuit current available. 0.03 second (2 cycle) fault clearing time.
  2. 65 kA short circuit current available. 0.03 second (2 cycle) fault clearing time.
  3. For <10 kA short circuit current available, the Hazard/Risk Category required may be reduced by one Number.
  4. 65 kA short circuit current available, 0.33 second (20 cycle) fault clearing time.
  5. 65 kA short circuit current available, up to 1.0 second (60 cycle) fault clearing time.
  6. For <25 kA short circuit current available, the Hazard/Risk Category required may be reduced by one Number.
  Source: *Courtesy National Fire Protection Association.*

TABLE 3.21 Protective Clothing and Personal Protective Equipment (PPE) Matrix

Protective clothing and equipment	Protective systems for hazard/risk category					
Hazard/Risk category number	−1 (Note 3)	0	1	2	3	4
Untreated natural fiber	—	—	—	—	—	—
a. T-shirt (short-sleeve)	X			X	X	X
b. Shirt (long-sleeve)		X				
c. Pants (long)	X	X	X (Note 4)	X (Note 6)	X	X
FR Clothing (Note 1)	—	—	—	—	—	—
a. Long-sleeve shirt			X	X	X (Note 9)	X
b. Pants			X (Note 4)	X (Note 6)	X (Note 9)	X
c. Coverall			(Note 5)	(Note 7)	X (Note 9)	(Note 5)
d. Jacket, parka, or rainwear			AN	AN	AN	AN
FR protective equipment	—	—	—	—	—	—
a. Flash suit jacket (2-layer)						X
b. Flash suit pants (2-layer)						X
Head protection	—	—	—	—	—	—
a. Hard hat			X	X	X	X
b. FR hard hat liner					X	X
Eye protection	—	—	—	—	—	—
a. Safety glasses	X	X	X	AL	AL	AL
b. Safety goggles				AL	AL	AL
Face protection (double-layer switching hood)				AR (Note 8)	X	X
Hearing protection (ear canal inserts)				AR (Note 8)	X	X
Leather gloves (Note 2)			AN	X	X	X
Leather work shoes			AN	X	X	X

Legend:
AN = As needed
AL = Select one in group
AR = As required
X = Minimum required
Notes:
1. See Table 3-3.9.3. (ATPV is the Arc Thermal Performance Exposure Value for a garment in cal/cm^2.)
2. If voltage-rated gloves are required, the leather protectors worn external to the rubber gloves satisfy this requirement.
3. Class 1 is only defined if determined by Notes 3 or 6 of Table 3-3.9.1 of Part II.
4. Regular weight (minimum 12 oz/yd^2 fabric weight), untreated, denim cotton blue jeans are acceptable in lieu of FR pants. The FR pants used for Hazard/Risk Category 1 shall have a minimum ATPV of 5.
5. Alternate is to use FR coveralls (minimum ATPV of 5) instead of FR shirt and FR pants.
6. If the FR pants have a minimum ATPV of 8, long pants of untreated natural fiber are not required beneath the FR pants.
7. Alternate is to use FR coveralls (minimum ATPV of 5) over untreated natural fiber pants and T-shirt.
8. A double-layer switching hood and hearing protection are required for the tasks designated 2* in Table 3-3.9.1 of Part II.
9. Alternate is to use two sets of FR coveralls (each with a minimum ATPV of 5) over untreated natural fiber clothing, instead of FR coveralls over FR shirt and FR pants over untreated natural fiber clothing.
Source: *Courtesy National Fire Protection Association.*

TABLE 3.22 Protective Clothing Characteristics

Hazard risk category	Typical protective clothing systems Clothing description (Typical number of clothing layers is given in parentheses)	Required minimum arc rating of PPE [J/cm²(cal/cm²)]
0	Nonmelting, flammable materials (i.e., untreated cotton, wool, rayon, or silk, or blends of these materials) with a fabric weight at least 4.5 oz/yd²(1)	N/A
1	FR shirt and FR pants or FR coverall (1)	16.74(4)
2	Cotton underwear—conventional short sleeve and brief/shorts, plus FR shirt and FR pants (1 or 2)	33.47(8)
3	Cotton underwear plus FR shirt and FR pants plus FR coverall, or cotton underwear plus two FR coveralls (2 or 3)	104.6(25)
4	Cotton underwear plus FR shirt and FR pants plus multilayer flash suit (3 or more)	167.36(40)

Source: Courtesy National Fire Protection Association.

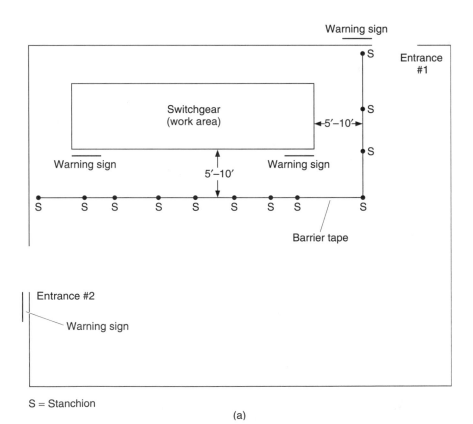

S = Stanchion

(a)

FIGURE 3.30 (*a*) Barriers and warning signs around a work area in a switchgear room with two entrances. (*b*) Barriers and warning signs around a lighting panel which is open for work.

Hallway

S = Stanchion

(b)

FIGURE 3.30 *(Continued)*

Figure 3.30*b* shows a similar barrier system for a hallway lighting panel. The clearance distance here is less because of the need to allow passage down the hallway. Warning signs are used to alert personnel to the hazard.

Illumination

Personnel must not reach into or work in areas which do not have adequate illumination. Although de-energization is the preferred method of eliminating electrical hazards, if de-energization also eliminates illumination, alternative safety measures must be employed.

Conductive Clothing and Materials

When working on or around energized conductors, personnel should remove (preferred) and/or use insulating tape to wrap watches, rings, keys, knives, necklaces, and other such conductive items they have on their bodies.

Conductive materials such as wire, tools, and ladders should be handled in such a way that they do not come into contact with energized conductors. If such material cannot be kept at a safe clearance distance (see section on Approach Distances), they should be wrapped or insulated. Metal ladders should never be used in electrical installations.

Confined Work Spaces

A complete coverage of confined work spaces is beyond the scope of this handbook. Electrical safety hazards in confined work spaces should be avoided. The following steps should be employed:

1. If safe approach distances cannot be maintained, energized conductors should be covered or barricaded so personnel cannot contact them.
2. Doors, hatches, and swinging panels should be secured so that they cannot swing open and push personnel into energized conductors.
3. Confined work spaces should be well ventilated to prevent the concentration of gases that may explode in the presence of an electric arc.
4. Exits should be clearly marked. Employees should be familiarized with the exits before entering the confined work space.
5. Confined work spaces must be well illuminated so that all hazards are clearly visible.

TOOLS AND TEST EQUIPMENT

General

Many electrical accidents involve failures of electric tools or test equipment. Such failures take a number of different forms including insulation failure, open ground return wires, internal shorts to ground, and overheating. These problems can be aggravated or even caused by using tools or test equipment improperly. For example, if extension cords not designed for use in wet areas are used in standing water, severe electric shock can result. The following general procedures should be employed when using electric tools or test equipment:

1. Tools and test equipment should be closely inspected before each use. (See section on Visual Inspection.)
2. Cord-connected tools should never be lifted or handled by their power cord. If tools must be lifted, a rope should be attached.
3. Grounded tools and equipment must have a continuous metallic connection from the tool ground to the supply ground.
4. If a grounded supply system is not available, double-insulated tools should be used.
5. Three-wire connection plugs should never be altered to fit into two-wire sockets.
6. So-called cheater plugs should not be used unless the third wire of the plug can be securely connected to the supply ground.
7. Locking types of plugs should be securely locked before the tool is energized.
8. Cords and tools should not be used in a wet environment unless they are specifically designed for such an application.

Authorized Users

Only authorized, trained persons should be allowed to use electric tools and test equipment. The training should include all the necessary inspection techniques for the tools that will be employed, plus recognition of the common types of safety hazards that are peculiar to the tool being used.

Visual Inspections

Tools and test equipment should be closely inspected before each use. This inspection should occur at the beginning of each work shift and again any time a problem is suspected. The inspector should look for the following types of problems:

- Missing, corroded, or damaged prongs on connecting plugs
- Frayed, worn, or missing insulation on connecting cords and/or test leads
- Improperly exposed conductors
- Bent or damaged prongs or test probes
- Excessive exposure on test prongs
- Loose screws or other poorly made electrical connections
- Missing or mis-sized fuses
- Damaged or cracked cases
- Indication of burning, arcing, or overheating of any type

If any of these problems is noted, the tool or test equipment should be removed from service until the problem can be repaired. If repair is not feasible, the tool or test equipment should be replaced.

Electrical Tests

Tools and extension cords should be electrically tested on a monthly basis. The following tests should be performed:

1. *Ground continuity test.* A high current (25 A minimum) should be applied to the tool's ground circuit. The voltage drop on the ground circuit should be no more than 2.5 V.
2. *Leakage test.* This test determines how much current would flow through the operator in the event that the tool's ground circuit were severed.
3. *Insulation breakdown.* This test applies a high voltage (up to 3000 V) to the tools insulation system and then measures the amount of leakage current.
4. *Operational test.* This test applies rated voltage to the tool and determines how much current the tool draws.

If the tool fails any of the above tests, it should be removed from service until it can be repaired. If repair is not possible, the tool or cord should be replaced. Figure 3.31 is a test set designed especially to check the power circuits for cord-connected tools and extension cords.

Wet and Hazardous Environments

Tools and test equipment should only be used in the environments for which they have been designed. If the work area is wet, only tools which are rated for wet work should be used. Fully insulated, waterproof cords should be used if they will be exposed to water.

If work must be performed in explosive environments, the tools used should be sealed or otherwise designed so that electric arcs will not ignite the explosive materials.

FIGURE 3.31 Tool and appliance tester. (*Courtesy AVO International.*)

⚠	**WARNING**

Arc Flash and Shock Hazard

Appropriate PPE Required

81 inch	Flash Hazard Boundary
14.1	cal/cm^2 Flash Hazard at 18 inches
Class FR	FR Clothing in excess of available incident energy - layer if necessary
480 VAC	Shock Hazard when cover is removed
00	Glove Class
42 inch	Limited Approach (Fixed Circuit)
12 inch	Restricted Approach
1 inch	Prohibited Approach
Bus: Bus #7 Prot: 3A Bus#7	

FIGURE 3.32 Typical warning label to be posted on electrical equipment. (*Label generated by SKM Power Systems Analysis Inc, PowerTools for Windows Software.*)

FIELD MARKING OF POTENTIAL HAZARDS

The type and degree of hazard determined by the various procedures outlined in this chapter should be clearly posted at each piece of electrical equipment in the field. These warning labels should clearly identify the type and degree of the hazard and should include the type and amount of PPE required to work in, on, or around the equipment. Figure 3.32 is an example of such a label.

Warning labels of this type should be placed on switchboards, panelboards, industrial-control panels, metalclad switchgear, meter-socket enclosures, motor-control centers, transformers, motors, generators, and all other such equipment where employees may be exposed to one of the electrical hazards.

Some equipment may have more than one such label. For example, the primary of a transformer may have a different arc-energy level than the secondary. Depending on the mechanical configuration, a label may be required for the primary feeder section of the transformer and the secondary transition cabinet into the switchgear that it feeds.

THE ONE-MINUTE SAFETY AUDIT

Inattention is one of the major root causes of all accidents. Employees whose attention is distracted by work-related or personal problems are prone to miss obvious signs of impending hazards. To make certain that employees are constantly aware of potential hazards, the steps listed in Table 3.23 should be performed by each worker that enters an area where any of the electrical hazards may be present.

The following nine simple steps should not take more than 1 min to perform, even in a very large electrical facility. Of course, not every step will apply in each and every situation. For example, some electrical rooms have no transformers; therefore, step 5 would not be performed.

1. Notify responsible personnel of your presence in the area. In the event that a problem does occur you may not be able to summon the help that you need. If other personnel are aware of your presence in the area, they will know to come to your aid.

2. Listen for any abnormally loud or unusual noises. Sniff for unusual odors. Transformers and other electric equipment do make noise during normal operation. When such noises become too loud or change tone, trouble may be indicated. Burning odors or the smell of ozone can also be signs of incipient faults.

TABLE 3.23 The One-Minute Safety Audit

---

1. Notify responsible personnel of your presence in the area.
2. Listen for any abnormally loud or unusual noises. Sniff for unusual odors.
3. Locate all emergency exits.
4. Locate all fire alarms and telephones.
5. Inspect all transformer liquid level, temperature, and pressure gauges to make certain that they are within acceptable limits.
6. Locate the station one-line diagram and make certain that it is legible and correct.
7. Make certain that the room is neat and tidy. Generally, electrical facilities should not be used for the storage of equipment.
8. Be certain that all required safety equipment is readily available and easily reached.
9. Check to see that all protective relay and other operational flags are properly reset.

---

3. Locate all emergency exits. Electrical accidents are often accompanied by smoke, fire, and noise. Such conditions make it difficult or impossible to find the exits. If exit locations are committed to memory before any problems occur, the worker is more likely to remember them after the problem occurs.

4. Locate all fire alarms and telephones. The same reasoning applies here as locating emergency exits. During an emergency the location of fire alarms and telephones should be a reflex.

5. Inspect all transformer gauges. Transformers will often warn of impending failures by an increase in pressure, temperature, liquid level, or some combination of all these. An inspection of these elements can warn of a problem before it happens.

6. Locate the station one-line diagram. Switching and other such activities must be carried out in a safe manner. To do so requires that the system one-line diagram be accurate. Few things are more hazardous than opening a switch which is believed to be the source to an energized part of the system and then finding out later that, in fact, it was the wrong switch.

7. Make certain that the room is clean and tidy. Storage of some equipment in an electrical facility is a necessary and valuable procedure. For example, what better place to store electrical safety equipment than close to the gear where it may be needed. Ladders, tools, hot sticks, rubber insulating goods, replacement lights, and other such materials are all candidates for storage in the electrical facility. When such storage does take place, however, it should be done in an orderly and safe fashion. Storage cabinets should be procured for small equipment. Ladders should be hung on racks especially mounted for such a purpose. Wire reels and other such materials should not be in the exit paths.

8. Be certain that all safety equipment is readily available. Workers will often use the excuse that "the safety gear is too far away to justify using it for this short procedure." To eliminate such an excuse, the safety equipment should be readily available. Whether it is brought in with the work crew or stored in the electrical facility, it must be readily available.

9. Check all protective relay flags and other indicators. Operation indicators can often be a precursor of a safety problem. A pressure alarm that will not reset, for example, can indicate an impending explosion. Such indicators should be checked. Such an indicator should never be reset without first determining what caused the problem. Also, operations personnel should be contacted before resetting.

These nine simple steps can be quickly accomplished by all personnel . . . and they may save a life.

REFERENCES

1. A.M. Stoll, and M.A. Chianta, "Method and Rating system for Evaluation of Thermal Protection," *Aerospace Medicine*, 40: pp.1232–1238, 1968.
2. Richard L. Doughty, Thomas E. Neal, and H. Landis Floyd, "Predicting Incident Energy to Better Manage the Electric Arc Hazard on 600-V Power Distrbution Systems," IEEE Paper.
3. Institute of Electrical and Electronic Engineers (IEEE), "IEEE Guide for Performing Arc-Flash Calculations," IEEE Std 1584-2002, Piscataway, NJ.

CHAPTER 4
GROUNDING OF ELECTRICAL SYSTEMS AND EQUIPMENT

INTRODUCTION

Electrical grounding seems to be one of the most misunderstood subjects of the National Electrical Code (NEC). This is evidenced by one report which stated that the Occupational Safety and Health Administration (OSHA) documented more than 20,000 violations of grounding in one year. Most of these violations came from loose, damaged, or missing external grounds. Just imagine the increased numbers of violations if the OSHA inspector opened electrical equipment to check equipment ground connections or if they used test equipment to verify proper grounding!

The real problem is that most electrical equipment *will* work without a ground connection. For example, an electric drill *will not* work if the hot or neutral wire is open however, if the ground wire is open, it *will* work properly. Even though the drill works properly without a ground connection, it is not safe to use.

This chapter addresses several issues dealing with general grounding and bonding, system grounding, and equipment grounding. Proper grounding is an issue that must be seriously considered for all electrical installations and equipment. Strict adherence to the OSHA requirements, NEC Article 250, as well as the numerous consensus standards and reference books, reduces the risk of electrical shock from contact with inadvertently energized equipment, enclosures, and structures.

The information in this chapter is not intended to be a substitute for the NEC or OSHA requirements. Nor is it intended to replace or be a substitute for any other standard quoted herein. For proper grounding of electrical systems and equipment, always comply with the requirements contained in the current standards and regulations.

ELECTRIC SHOCK HAZARD

Energized conductors and circuit components, installed within or on electrical equipment, are insulated from the equipment's metal enclosure, to provide protection for personnel who operate the equipment, from being exposed to dangerous voltages. When aging or malfunction causes the insulation to break down, the energized conductors within the equipment can make direct contact with the metal enclosure, thereby energizing it. Anyone making contact with energized equipment could be injured or killed.

Equipment grounding is a way of nullifying this shock hazard. An equipment ground forms a very low impedance path to ground. Personnel making contact with the energized enclosure will be exposed to less voltage than they would otherwise.

Simply stated, proper bonding and grounding of electrical equipment will substantially reduce the risk of electrical shock by effectively eliminating a difference in potential.

GENERAL REQUIREMENTS FOR GROUNDING AND BONDING

Definitions

The following terms are defined here due to their frequent use in this chapter. The definitions contained in this chapter are taken verbatim from OSHA 29 CFR 1910.399. Corresponding definitions can also be found in Article 100 of the current edition of the NEC.

Bonding (Bonded).   The permanent joining of metallic parts to form an electrically conductive path that will ensure electrical continuity and the capacity to conduct safely any current likely to be imposed.

Bonding Jumper.   A reliable conductor to ensure the required electrical conductivity between metal parts required to be electrically connected.

Bonding Jumper, Equipment.   The connection between two or more portions of the equipment grounding conductor.

Bonding Jumper, Main.   The connection between the grounded circuit conductor and the equipment grounding conductor at the service.

Energized.   Electrically connected to a source of potential difference.

Equipment.   A general term including material, fittings, devices, appliances, fixtures, apparatus, and the like used as a part of, or in connection with, an electrical installation.

Ground.   A conducting connection, whether intentional or accidental, between an electrical circuit or equipment and the earth, or to some conducting body that serves in place of the earth.

Grounded.   Connected to earth or to some conducting body that serves in place of the earth.

Grounded, Effectively.   Intentionally connected to earth through a ground connection or connections of sufficiently low impedance and having sufficient current-carrying capacity to prevent the buildup of voltages that may result in undue hazards to connected equipment or to persons.

Grounding Conductor.   A conductor used to connect equipment or the grounded circuit of a wiring system to a grounding electrode or electrodes.

Grounding Conductor, Equipment.   The conductor used to connect the non-current-carrying metal parts of equipment, raceways, and other enclosures to the system grounded conductor, the grounding electrode conductor, or both, at the service equipment or at the source of a separately derived system.

Grounding Electrode Conductor. The conductor used to connect the grounding electrode to the equipment grounding conductor, to the grounded conductor, or to both, of the circuit at the service equipment or at the source of a separately derived system.

Grounding Electrode System. Two or more grounding electrodes that are effectively bonded together shall be considered a single grounding electrode system.

Ground-Fault Circuit Interrupter. A device intended for the protection of personnel that functions to de-energize a circuit or portion thereof within an established period of time when a current to ground exceeds some predetermined value that is less than that required to operate the overcurrent protective device of the supply circuit.

Ground-Fault Protection of Equipment. A system intended to provide protection of equipment from damaging line-to-ground fault currents by operating to cause a disconnecting means to open all ungrounded conductors of the faulted circuit. This protection is provided at current levels less than those required to protect conductors from damage through the operation of a supply circuit overcurrent device.

Grounding of Electrical Systems

Grounded electrical systems are required to be connected to earth in such a way as to limit any voltages imposed by lightning, line surges, or unintentional contact with higher voltage lines. Electrical systems are also grounded to stabilize the voltage to earth during normal operation. If, for example, the neutral of a 120/240 V, wye-connected secondary of a transformer were not grounded, instead of being 120 V to ground, the voltage could reach several hundred volts to ground. A wye-connected electrical system becomes very unstable if it is not properly grounded.

The following requirements are taken from OSHA 29 CFR 1910.304(f) and can also be found in the 2005 National Electrical Code (NEC)® Section 250.20.

(f) **Grounding.** Paragraphs (f)(1) through (f)(7) of this section contain grounding requirements for systems, circuits, and equipment.

(1) **Systems to be grounded.** The following systems, which supply premises wiring, shall be grounded:

(i) All 3-wire DC systems shall have their neutral conductor grounded.

(ii) Two-wire DC systems operating at over 50 volts through 300 volts between conductors shall be grounded unless:

(A) They supply only industrial equipment in limited areas and are equipped with a ground detector; or

(B) They are rectifier-derived from an AC system complying with paragraphs (f)(1)(iii), (f)(1)(iv), and (f)(1)(v) of this section; or

(C) They are fire-protective signaling circuits having a maximum current of 0.030 amperes.

(iii) AC circuits of less than 50 volts shall be grounded if they are installed as overhead conductors outside of buildings or if they are supplied by transformers and the transformer primary supply system is ungrounded or exceeds 150 volts to ground.

(iv) AC systems of 50 volts to 1000 volts shall be grounded under any of the following conditions, unless exempted by paragraph (f)(1)(v) of this section:

(A) If the system can be so grounded that the maximum voltage to ground on the ungrounded conductors does not exceed 150 volts; (see Fig. 4.1)

120V, 1Ø, 2-wire   240/120V, 1Ø, 3-wire   208/120V, 3Ø, 4-wire
250.20(B)(1)   250.20(B)(1)   250.20(B)(2)

The neutrals must be grounded, because the maximum
voltage to ground does not exceed 150 volts from any
other conductor in the system.

95-00543-3

FIGURE 4.1   Voltage-to-ground less than 150 volts. (*Courtesy of AVO Training Institute, Inc.*)

(B) If the system is nominally rated 480Y/277 volt, 3-phase, 4-wire in which the neutral is used as a circuit conductor; (see Fig. 4.2)

(C) If the system is nominally rated 240/120 volt, 3-phase, 4-wire in which the midpoint of one phase is used as a circuit conductor; (see Fig. 4.3)

Or,

(D) If a service conductor is uninsulated.

(v) AC systems of 50 volts to 1000 volts are not required to be grounded under any of the following conditions:

(A) If the system is used exclusively to supply industrial electric furnaces for melting, refining, tempering, and the like.

(B) If the system is separately derived and is used exclusively for rectifiers supplying only adjustable speed industrial drives.

(C) If the system is separately derived and is supplied by a transformer that has a primary voltage rating less than 1000 volts, provided all of the following conditions are met:

{1} The system is used exclusively for control circuits,

All 480-/277V,
3Ø, 4-wire systems,
where the neutral
is used as a circuit
conductor for loads.
250.20(B)(2)

95-00544-3

FIGURE 4.2   480/277 volt systems. (*Courtesy of AVO Training Institute, Inc.*)

240/120V, 3Ø, 4-wire in which the midpoint of one phase is used as a circuit conductor. 250.20(B)(3)

Caution: There exisits a "high leg" voltage from the B phase to ground.
120 x √3 = 208

95-00545-3

FIGURE 4.3 Delta system with neutral. (*Courtesy of AVO Training Institute, Inc.*)

{2} The conditions of maintenance and supervision assure that only qualified persons will service the installation,

{3} Continuity of control power is required, and

{4} Ground detectors are installed on the control system.

(D) If the system is an isolated power system that supplies circuits in health care facilities.

(2) Conductors to be grounded. For AC premises wiring systems the identified conductor shall be grounded.

(4) Grounding path. The path to ground from circuits, equipment, and enclosures shall be permanent and continuous.

(5) Supports, enclosures, and equipment to be grounded—(i) Supports and enclosures for conductors. Metal cable trays, metal raceways, and metal enclosures for conductors shall be grounded, except that:

(7) Grounding of systems and circuits of 1000 volts and over (high voltage.)—(i) General. If high voltage systems are grounded, they shall comply with all applicable provisions of paragraphs (f)(1) through (f)(6) of this section as supplemented and modified by this paragraph (f)(7).

(ii) Grounding of systems supplying portable or mobile equipment. [see 1910.302(b)(3)] Systems supplying portable or mobile high voltage equipment, other than substations installed on a temporary basis, shall comply with the following:

(A) Portable and mobile high voltage equipment shall be supplied from a system having its neutral grounded through an impedance. If a delta-connected high voltage system is used to supply the equipment, a system neutral shall be derived.

(B) Exposed non-current-carrying metal parts of portable and mobile equipment shall be connected by an equipment grounding conductor to the point at which the system neutral impedance is grounded.

(C) Ground-fault detection and relaying shall be provided to automatically de-energize any high voltage system component which has developed a ground fault. The continuity of the equipment grounding conductor shall be continuously monitored so as to de-energize automatically the high voltage feeder to the portable equipment upon loss of continuity of the equipment grounding conductor.

(D) The grounding electrode to which the portable or mobile equipment system neutral impedance is connected shall be isolated from and separated in the ground by at least 20 feet from any other system or equipment grounding electrode, and there shall be no direct connection between the grounding electrodes, such as buried pipe, fence, and so on.

Grounding of Electrical Equipment

OSHA 1910.304(f)(7)(iii) Grounding of equipment. All non-current-carrying metal parts of portable equipment and fixed equipment including their associated fences, housings, enclosures, and supporting structures shall be grounded. However, equipment that is guarded by location and isolated from ground need not be grounded. Additionally, pole-mounted distribution apparatus at a height exceeding 8 feet above ground or grade level need not be grounded.

In 29 CFR 1910.303, "General Requirements," OSHA states under "(b) Examination, installation, and use of equipment (1) Examination" that "Electrical equipment shall be free from recognized hazards that are likely to cause death or serious physical harm to employees." This section continues with "other factors which contribute to the practical safeguarding of employees using or likely to come in contact with the equipment." One of these "other factors" is proper grounding. If the non-current-carrying metal parts of electric equipment are not properly grounded and these parts become energized, then any employee "using or likely to come in contact with the equipment" is at risk of an electrical shock that may or may not be fatal. This is a risk that must not be taken. Proper grounding can effectively eliminate this shock hazard by providing a permanent and continuous low impedance path for ground-fault current to follow in order to clear the circuit protective device(s).

Bonding of Electrically Conductive Materials and Other Equipment

Bonding is the permanent joining of metallic parts of materials and equipment. When different metal parts are not bonded together, a difference in potential could exist between the metal parts. This creates an electrically hazardous condition between the parts. Anyone simultaneously coming into contact with the metal parts would be subject to electrical shock, burns, or electrocution. When all conductive materials and parts of equipment are permanently bonded together, there is only one piece of metal and no potential difference exists between the parts. The metal parts must also be grounded to earth in order to be at earth potential. This minimizes the risk of touch-potential and step-potential hazards when working on or around metal enclosures that could become energized.

Other metallic equipment that is in contact with or adjacent to the electrical equipment should also be grounded to prevent a difference in potential in the event that a ground fault occurs in the electrical equipment. This could include other piping as well as ducts in ventilation systems as shown in Fig. 4.4.

As discussed in the previous section, all non-current-carrying components of electrical equipment must be grounded. Equipment grounding conductors are installed and connected to the required terminal in the equipment to provide the low impedance path for fault current to clear the circuit. All other metallic components of the equipment must be bonded to the grounded portion of the equipment in order to prevent a difference in potential between the components. An example of this would be service enclosures. Figure 4.5 illustrates bonding of service enclosures.

FIGURE 4.4 Bonding of other piping and duct systems. (*Courtesy of AVO Training Institute, Inc.*)

FIGURE 4.5 Bonding service enclosures. (*Courtesy of AVO Training Institute, Inc.*)

Performance of Fault Path

OSHA makes a very simple statement on the performance of the fault path as found in 1910.304(f)(4) which states: *"**Grounding path.** The path to ground from circuits, equipment, and enclosures shall be permanent and continuous."* NEC Section 250.4(A)(5) also requires the fault current path to be able to safely carry the maximum ground-fault current, provide a low impedance path in order to clear the overcurrent devices, and perform in a way that the earth is not used as the sole equipment grounding conductor for fault current.

The fault current path must meet these criteria in order for the overcurrent protection to clear the circuit in the event a ground fault occurs. A low-impedance conductor path must be used because the earth provides an extremely high impedance path and would not allow a sufficient amount of fault current to flow to clear the overcurrent device. Another factor is that a low-impedance conductor path for fault current to flow through minimizes the possibility of step- and touch-potential hazards by limiting the voltage to ground.

Arrangement to Prevent Objectionable Current

Grounding of electrical equipment and systems must be accomplished in a manner that would prevent objectionable current. The main point here is to install an effective grounding system without creating an objectionable current situation. Circulating current is one form of objectionable current that can occur when multiple grounds are utilized.

Alterations to Stop Objectionable Current

Objectionable current flow can occur when using multiple grounds in electrical systems and equipment. If this occurs there are several different options to prevent or at least minimize this current flow:

1. One or more of the grounds may be discontinued. Never discontinue all of the grounds.
2. The grounding connection may be changed to another location.
3. The conductive path for the grounding connections may be interrupted.
4. The authority having jurisdiction may grant other remedial action.

Temporary Currents Not Classified as Objectionable Current

One thing to keep in mind when dealing with this subject is that there are times when current will flow in the grounding system or through the non-current carrying parts of electrical equipment. Ground faults in equipment do occur and that is when the grounding system performs its function. These currents are not classified as objectionable current for the purpose of this discussion.

Connection of Grounding and Bonding Equipment

Conductors for grounding and bonding of electrical equipment must be connected using an approved method such as exothermic welding, listed pressure connections, listed clamps, and other approved means. Solder must not be used as the sole connection. Solder is too soft and has a very low melting point and, therefore, becomes a fuse in the grounding connection. Also, never use sheet metal screws to make connections between the grounding conductor and the enclosure. The use of sheet metal screws would provide limited metal-to-metal contact between the screw and the enclosure and, therefore, may not provide the required low-impedance connection.

Protection of Ground Clamps and Fittings

All grounding connections must be protected from physical damage either by location or by means of an enclosure made of wood, metal, or equivalent. Damaged grounding conductor connections can result in loss of continuity in the ground path, which will create a potential shock hazard.

Clean Surfaces

If the grounding connection point is contaminated with paint or other such coatings, good continuity may not be accomplished. All surfaces must be cleaned as needed to remove any such coatings or other contaminants that could interfere with the continuity of the grounding connection. As was stated earlier, the grounding system must create a sufficiently low impedance path in order for circuit protective devices to clear the circuit in the event of a ground fault.

SYSTEM GROUNDING

Purposes of System Grounding

Power systems are grounded for one or more of the following reasons:

- Control of the voltage to ground to within predictable limits.
- To provide for current flow and allow detection and location of ground faults.
- To reduce shock hazard to personnel.

It is not always possible to accomplish all these goals with a particular method of grounding. In some cases, the chosen method of grounding is a compromise between conflicting goals. Table 4.1 shows the advantages and disadvantages of the various grounding methods.

Grounding Service-Supplied Alternating-Current Systems

The premises wiring system of a grounded electrical service must be grounded. This means that each grounded service must have a grounding electrode conductor connected to the grounding electrode and to the service equipment. The grounded service conductor is also connected to the grounding electrode conductor. OSHA provides this requirement in

TABLE 4.1 Comparison of System Grounding Methods

Method of grounding	Advantages	Disadvantages
Ungrounded—no intentional connection to ground	Little or no ground current. Plant does not need to trip for a single ground fault.	Possiblity of large transient overvoltages. Overvoltages on the unfaulted phases. Hard to detect and locate ground faults. Possibility of ferroresonance.
High Resistance—ground fault < 10 A.	Limit transient overvoltages to 250%. Can usually keep plant running through a single ground fault. Ground fault can be detected.	Overvoltage on the unfaulted phases. Surge arresters need to be rated for the phase-to-phase voltage.
Low Resistance—ground fault > 100 A.	Limit transient overvoltages to 250%. Immediate and selective fault clearing is possible.	Overvoltage on the unfaulted phases. Surge arresters need to be rated phase-to-phase voltage.
Reactance—ground fault at least 25% and preferably 60% of the three-phase fault $X_o < 10 X_1$	Prevent serious transient overvoltages. Limit generator phase to ground fault to the magnitude of the three-phase fault.	High ground fault currents that must be cleared.
Solidly grounded—solid connection between neutral and ground. $R_oX_1X_0 < 3X_1$ Note that Z_oO	Limits system overvoltages. Freedom from ferroresonance. Easier detection, location, and selective tripping of ground faults. Can use phase to ground rated surge arresters.	Extreme ground fault magnitudes. Safety risks (arc blast and flash). Stray voltages due to ground faults (shock hazard). Mechanical and thermal stress and damage during ground faults. Ground fault must be detected and cleared, even if the plant must take an outage.

Source: Courtesy of AVO Training Institue, Inc.

29 CFR 1910.304(3)(i), (ii), and (iii) (NEC Section 250.24 provides a more in-depth requirement) as quoted below:

(3) Grounding connections. (i) For a grounded system, a grounding electrode conductor shall be used to connect both the equipment grounding conductor and the grounded circuit conductor to the grounding electrode (see Fig. 4.6). Both the equipment grounding conductor and the grounding electrode conductor shall be connected to the grounded circuit conductor on the supply side of the service disconnecting means, or on the supply side of the system disconnecting means or overcurrent devices if the system is separately derived.

(ii) For an ungrounded service-supplied system, the equipment grounding conductor shall be connected to the grounding electrode conductor at the service equipment. For an ungrounded

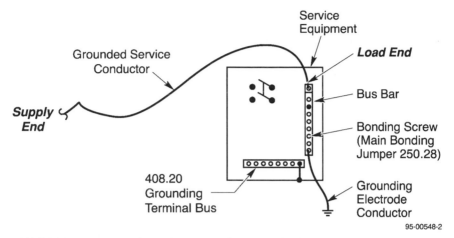

FIGURE 4.6 Grounding service conductor connection to grounding electrode. (*Courtesy of AVO Training Institute, Inc.*)

separately derived system, the equipment grounding conductor shall be connected to the grounding electrode conductor at, or ahead of, the system disconnecting means or overcurrent devices.

(iii) On extensions of existing branch circuits, which do not have an equipment grounding conductor, grounding-type receptacles may be grounded to a grounded cold water pipe near the equipment.

An important point to make here is that the grounding and grounded conductors are only allowed to be connected together on the line-side of the service disconnecting means, they are generally not allowed to be connected on the load-side. This issue will be addressed in more detail in the section titled "Use of Grounded Circuit Conductor for Grounding Equipment."

Conductors to Be Grounded—Alternating-Current Systems

The following conductors of an AC premises wiring system are required to be grounded:

1. One conductor of a single-phase, 2-wire system
2. The neutral conductor of a single-phase, 3-wire system
3. The common conductor of a multiphase system where one wire is common to all phases
4. One phase conductor of a multiphase system requiring a grounded phase
5. The neutral conductor of a multiphase system where one phase is used as the neutral

Main Bonding Jumper

In a grounded electrical system, an unspliced main bonding jumper is required to connect all grounding and grounded conductors to the service equipment enclosure. This connection is made within the service equipment enclosure using either a ground bus, a screw, strap, or wire. Figure 4.7 further illustrates this requirement:

95-00542-1

FIGURE 4.7   Location of the main bonding jumper. (*Courtesy of AVO Training Institute, Inc.*)

Grounding Electrode System

A grounding electrode system is made up of a grounding electrode(s), any required bonding jumpers between electrodes, and a grounding electrode conductor. All conductors and jumpers used in the grounding electrode system must be sized in accordance with NEC Section 250.66 and connected as specified in Section 250.70. There may be several types of grounding electrodes available at each building or structure. All available electrodes must be bonded together and used to form a grounding electrode system. Available electrodes include metal underground water pipe, the metal frame of a building, concrete-encased electrodes, and ground rings. Figure 4.8 illustrates several common types of grounding electrodes. Several of the electrodes mentioned above have specific requirements as noted in Fig. 4.9.

If only a water pipe is available for the grounding electrode, it must be supplemented by one of the other electrodes mentioned above or by a made electrode as illustrated in Fig. 4.10.

Made electrodes, such as a ground rod, pipe, or plate, may also be utilized to supplement existing electrodes. Figure 4.11 illustrates the minimum requirements for a ground rod or pipe.

There are two alternate installation methods for ground rods or pipes that are permitted to be utilized if vertical installation is not possible due to rock or other obstructions. Figure 4.12 illustrates these alternate methods.

Figure 4.13 illustrates the minimum requirements for a ground plate.

Where practicable, these made electrodes are required to be installed below the permanent moisture level and be free from nonconductive coatings. If more than one electrode is used, they must be installed at least 6 ft from each other. A common practice is to place the rods apart at a distance equal to the length of the rod. The best industry practice is to install the rods at least 10 ft apart. This practice will minimize the risk of dissipation overlap in the event of a ground fault.

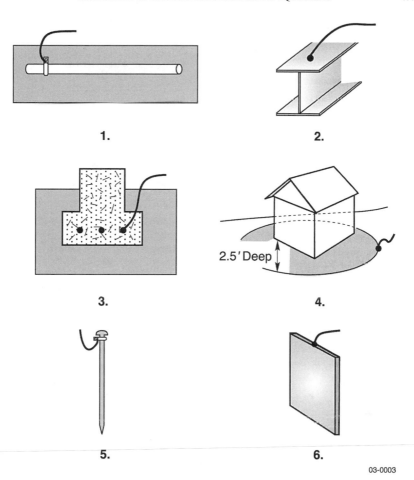

1.

2.

2.5' Deep

3.

4.

5.

6.

03-0003

FIGURE 4.8   Types of Grounding Electrodes. (*Courtesy of AVO Training Institute, Inc.*)

FIGURE 4.9   Electrode connections. (*Courtesy of AVO Training Institute, Inc.*)

FIGURE 4.10 Sole connections. (*Courtesy of AVO Training Institute, Inc.*)

FIGURE 4.11 Minimum requirements for a ground rod or pipe. (*Courtesy of AVO Training Institute, Inc.*)

Grounding Electrode System Resistance

Resistance of all grounding connections must also be addressed briefly. There are three components of resistance to consider: (1) grounding electrode resistance, (2) contact resistance between the electrode and the soil, and (3) the resistance of the soil. These three components are illustrated in Figure 4.14.

The grounding electrode, as a general rule, has very low resistance. The contact of the electrode with the surrounding soil is also generally low. The third component, soil, can very dramatically in resistance from one location to another due to different types and conditions of the soil. For example, the resistivity of inorganic clay can range from 1,000–5,500 ohm/cm whereas gravel can range from 60,000–100,000 ohm/cm. Every type and condition of soil will, as can be seen, vary tremendously. The soil type must be known and grounding electrode testing must be done periodically in order to know whether or not a good grounding system is present. NEC Section 250.56 specifies that the resistance to ground for a made electrode must be 25 ohms or less. If this value cannot be obtained, then multiple electrodes must be used. As was mentioned earlier, make sure the electrodes are spaced far enough apart to prevent dissipation overlap.

Grounding Electrode Conductor

OSHA 29 CFR 1910.399 defines the grounding electrode conductor as: The conductor used to connect the grounding electrode to the equipment grounding conductor and/or to the grounded conductor of the circuit at the service equipment or at the source of a separately derived system.

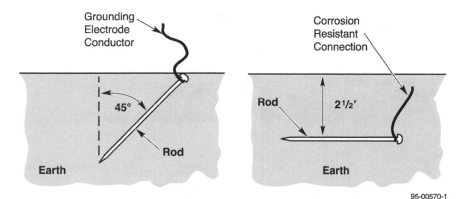

95-00570-1

FIGURE 4.12 Alternate burial methods for ground rods. (*Courtesy of AVO Training Institute, Inc.*)

Although copper is the preferred material for the grounding electrode conductor, aluminum and copper-clad aluminum are also acceptable. There are, however, restrictions with the latter two materials. The aluminum and copper-clad aluminum conductors are not allowed by code to be installed in direct contact with masonry or the earth, or where corrosive conditions exist. If used, these materials must not be installed within 18 in of earth. Copper, on the other hand, is considered a corrosion resistant material and, therefore, is permitted to be installed in most locations.

NEC Section 250.64(C) requires the grounding electrode conductor to be installed continuous without splice or joint. There are, however, two deviations from this rule. A splice can be made using a listed irreversible compression-type connector or the exothermic welding process. More on this subject will be addressed in the next section, "Grounding Conductor Connections to Electrode."

The grounding electrode conductor must be sized by the requirements of NEC Section 250.66. The conductor is sized based on the size of the largest service-entrance conductor

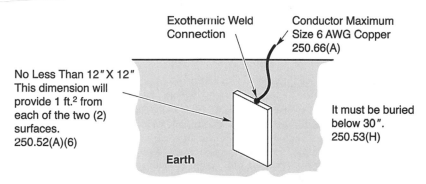

95-00571-2

FIGURE 4.13 Plate electrode minimum requirements. (*Courtesy of AVO Training Institute, Inc.*)

1. Resistance of the electrode itself and connections to it.

2. Contact resistance between the electrode and the soil adjacent to it.

3. Resistance of the surrounding earth.

Current   Current

99-0019

FIGURE 4.14 Components of earth resistance in an earth electrode. (*Courtesy of AVO Training Institute, Inc.*)

or equivalent area of parallel conductors. The size can also vary according to the type of connection used, as will be discussed in the next section. Other deviations to the values presented in NEC Table 250.66 is as follows:

1. *Connections to made electrodes.* The sole connection portion of the conductor is not required to be larger than No. 6 copper wire.

2. *Connections to concrete-encased electrodes.* The sole connection portion of the conductor is not required to be larger than No. 4 copper wire.

3. *Connections to ground rings.* The sole connection portion of the conductor is not required to be larger than the conductor used for the ground ring. NEC Section 250.52(A)(4) states "not smaller than No. 2" copper wire.

Grounding Conductor Connection to Electrodes

NEC Section 250.68(B) states that the connection of the grounding electrode conductor must be made in a manner that will ensure a permanent and effective grounding path. Section 250.70 provides the means of connection of the grounding electrode conductor to the grounding electrode. These methods include exothermic welding, listed lugs, listed pressure connectors, listed clamps, and other listed means of connection. Note the term "listed" for various connection methods. These are designed, manufactured, listed, and labeled for the purpose and are the only ones permitted for use in grounding electrode systems. The code goes on to state that connections made with solder must not be used. Solder has a very low melting point and, therefore, would disconnect the grounding connection in the event of a ground-fault, because the solder would act like a fuse. The National Electrical Safety Code (NESC) states that joints in grounding conductors must be made and maintained so as not to materially increase the resistance of the grounding conductor. The NESC also states that the connection must have appropriate mechanical and corrosion-resistant characteristics.

IEEE Std. 80, "Guide for Safety in AC Substation Grounding" is an excellent resource for determining the minimum size conductor based on the type of connection used. The Onderdonk AC equation is used to calculate fusing current of a conductor based on the type of connection used. Connectors and splice connections must also meet the requirements of IEEE Std. 837 in order to be acceptable. The following examples will further illustrate this point:

For this example, a shortened version of the Onderdonk equation will be used to determine conductor size and to determine fusing current levels based on the type of connection.

<table>
<tr><th>Conductor Size</th><th>Fusing Current</th></tr>
<tr><td>$A = KI\sqrt{S}$</td><td>$I = \dfrac{A}{K\sqrt{S}}$</td></tr>
</table>

where A = Cable size, circular mil
$\quad K$ = Connector factor
$\quad S$ = Maximum fault time, seconds
$\quad I$ = Maximum fault current, amperes

For copper conductors these equations use the following "K" values to represent connection temperature:

- Conductor only—1083°C temperature rating—value of "K" = 6.96
- Welded connections—1083°C temperature rating—value of "K" = 6.96
- Irreversible compression-type connections—1083°C temperature rating— value of "K" = 6.96
- Brazed connections—450°C temperature rating—value of "K" = 9.12
- Pressure-type connections—250°C temperature rating – value of "K" = 11.54

The following three examples, used to determine minimum conductor size, will use a 20,000-A fault current with a protective device clearing time of 5 cycles (.083 second) using $A = KI\sqrt{S}$:

1. Pressure type connection:

$$A = 11.54 \times 20,000 \times \sqrt{.083}$$

$$A = 66,493 \text{ cm (minimum size conductor)}$$

According to NEC Table 8, the size that corresponds to the minimum size conductor would be a No. 1 AWG copper conductor (83,690 cm)

2. Brazed connection:

$$A = 9.12 \times 20,000 \times \sqrt{.083}$$

$$A = 52,549 \text{ cm (minimum size conductor)}$$

According to NEC Table 8, the size that corresponds to the minimum size conductor would be a No. 3 AWG copper conductor (52,620 cm)

3. Welded or irreversible compression-type connection:

$$A = 6.96 \times 20,000 \times \sqrt{.083}$$

$$A = 40,103 \text{ cm (minimum size conductor)}$$

According to NEC Table 8, the size that corresponds to the minimum size conductor would be a No. 4 AWG copper conductor (41,740 cm)

As can be seen by the previous examples, if a pressure type connection were used, it would require a conductor approximately 60 percent larger than a connection made with a welded or irreversible compression-type connection. The brazed connection would require a conductor approximately 31 percent larger than a welded or irreversible compression-type connection. For practical purposes, the welded or irreversible compression-type connection would be the preferred choice.

Another consideration would be the approximate fusing current of the conductor based on the type of connection. This example will use a 4/0 (211,600 cm) conductor with a protective device clearing time of 5 cycles (.083 second) using $I = A/K\sqrt{S}$.

1. Pressure type connection:

$$I = \frac{211600}{11.54\sqrt{.083}}$$

$I = 63{,}646$ amperes fusing current

2. Welded or irreversible compression-type connection:

$$I = \frac{211600}{6.96\sqrt{.083}}$$

$I = 105{,}528$ amperes fussing current

As can be seen in the above examples the conductor utilizing the welded or irreversible compression-type connection will handle much more current before fusing than the pressure-type connection.

When reliability and cost are taken into consideration, it is plain to see that the welded, as well as the irreversible compression-type, connection are far superior to any other type of connection.

Bonding

If only one statement were to be made about grounding and bonding, it would be that all non-current-carrying parts of electrical equipment and nonelectrical equipment, likely to become energized, be effectively grounded and bonded together. By using this general philosophy, there will be a minimum (near zero) risk that the non-current-carrying parts of equipment could become energized. This would greatly reduce the risk of electrical shock or electrocution of any person likely to come into contact with the equipment. As discussed earlier, a general statement in OSHA 29 CFR 1910.303(b)(1)(vii) includes, "other factors which contribute to the practical safeguarding of employees using or likely to come in contact with the equipment." One of these "other factors" refers to grounding and bonding of equipment likely to become energized.

NEC Section 250.90 states that bonding must be provided where necessary to ensure electrical continuity. Bonding is also required to ensure the capacity to safely conduct any fault current that is likely to be imposed on the equipment. This requirement applies to all types of equipment, systems, and structures. NEC Article 250, Part V. Bonding, must be complied with in order to size the bonding jumper correctly. Table 4.2 illustrates further why Part V must be adhered to.

TABLE 4.2   Reference Table for Sizing Bonding Jumpers

Paragraph	Reference for sizing
Supply side of service	Table 250.66
Load side of service	Table 250.122
Interior water pipe	Table 250.66
Multiple occupancies	Table 250.122
Multiple buildings-common service	Section 250.122
Separately derived systems	Table 250.66
Other metal piping	Table 250.122
Structural steel	Table 250.66

The key here is to know where the bonding jumper is and what it is being used for; it makes a significant difference. If the wrong section or table is used, the bonding jumper may not be of adequate current-carrying capacity to safely conduct any fault current likely to be imposed on it.

EQUIPMENT GROUNDING

Equipment to be Grounded

Equipment to be grounded essentially means that all non-current-carrying metal parts of electrical equipment, whether fastened in place or portable, must be grounded. NEC Article 250, Part VI, "Equipment Grounding and Equipment Grounding Conductors" very specifically lays out the requirements for equipment grounding for electrical as well as nonelectrical equipment that could become energized. NEC Section 250.110 lists the requirements for equipment fastened in place (fixed) and has three exceptions. Exception No. 2 itemizes, "distribution apparatus, such as transformer and capacitor cases, mounted on wooden poles, at a height exceeding 8 ft above the ground or grade level." This exception would protect the general public from possible contact with the ungrounded apparatus; however, it does not protect the person working on the pole. With regard to these issues, OSHA 29 CFR 1910.269(l)(9) states:

> *Non-current-carrying metal parts.* Non-current-carrying metal parts of equipment or devices, such as transformer cases and circuit breaker housings, shall be treated as energized at the highest voltage to which they are exposed, unless the employer inspects the installation and determines that these parts are grounded before work is performed.

The best practice is to always ground every case or enclosure that contains electrical equipment or conductors. When in doubt, ground it.

Grounding Cord- and Plug-Connected Equipment

NEC Section 250.114 states the same philosophy for cord- and plug-connected equipment, as does Sections 250.110 through 250.112 for fixed equipment. It directs that all exposed non-current-carrying metal parts that are likely to become energized must be grounded. A key point here is "likely to become energized." Any metal housing or enclosure that contains electrical components is, at some time or another, likely to become energized. Proper equipment grounding techniques will provide the sufficiently low impedance path required to cause the overcurrent device to operate and clear the ground fault condition.

In the material written about electrical safety-related work practices, OSHA has provided specific requirements for the use of portable electrical equipment and extension cords. This requirement is found in OSHA 29 CFR 1910.334 where it states:

Use of equipment. (a) Portable electric equipment. (3) Grounding type equipment. (i) A flexible cord used with grounding type equipment shall contain an equipment grounding conductor.

(ii) Attachment plugs and receptacles may not be connected or altered in a manner which would prevent proper continuity of the equipment grounding conductor at the point where plugs are attached to receptacles. Additionally, these devices may not be altered to allow the grounding pole of a plug to be inserted into slots intended for connection to the current-carrying conductors.

(iii) Adapters which interrupt the continuity of the equipment grounding connection may not be used.

OSHA makes it very clear that this type of equipment must be grounded. Note as well the statement in (iii) concerning adapters. The adapters referred to here are used when a grounded plug is required to be used where an ungrounded receptacle exists. These adapters are UL approved and can be used, but only if used properly, that is, the ground connection must be attached to a return ground path. The problem with adapters is that the ground connection device is generally cut off or otherwise not used, thus defeating the ground continuity.

Equipment grounding conductors are required as stated previously, however, grounding alone does not give complete protection when using portable cord- and plug-connected, hand-held equipment and extension cords. The use of a ground-fault circuit interrupter (GFCI) when using portable equipment will provide additional safety for the user. Grounding does provide a path for ground fault current to flow to cause the overcurrent device to operate, however, it does not provide protection when current leakage occurs due to moisture in a piece of equipment or when there is an undetected cut in the cord or crack in the equipment case. The GFCI is designed to trip at 4 to 6 mA of current. To explain this further, the following example is provided.

Given a 20-A molded case circuit breaker, the maximum load allowed by code is 16 A (80 percent of rating). At 16 A this circuit breaker will, under normal conditions, run indefinitely. Even at 20 A the circuit breaker will remain closed for several minutes to infinity. Taking this into consideration, the amount of current leakage from moisture in the equipment (a hand-held drill for example) will not trip this circuit breaker, however, a solid ground fault will. Since the circuit breaker will not trip, a person contacting this piece of equipment is exposed to a shock hazard. It takes approximately one-tenth of an ampere to cause ventricular fibrillation in most people, which is generally fatal. The point is that the circuit breaker will not protect a person in this situation but a GFCI will.

NEC Section 590.6 requires ground-fault protection for all temporary wiring installations for construction, remodeling, maintenance, repair, demolition of buildings, structures, equipment, or similar activities. In other words, any time an extension cord is used for one of these activities, a GFCI is required. Although not specifically required here, and because the hazard risk is the same, all hand-held cord- and plug-connected portable electrical equipment should utilize a GFCI for the protection of the worker.

Figure 4.15 further illustrates how a GFCI works.

Use a GFCI; it may save your life. Look at it this way—it is cheap life insurance.

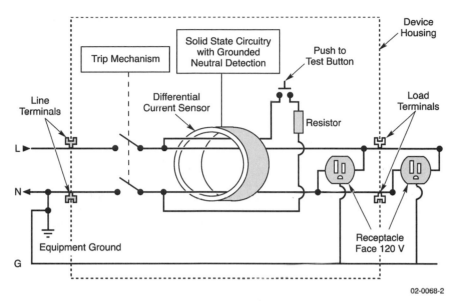

02-0068-2

FIGURE 4.15   Internal diagram of a ground-fault circuit interrupter (GFCI). (*Courtesy of AVO Training Institute, Inc.*)

Equipment Grounding Conductors

NEC Section 250.118 identifies several types of equipment grounding conductors. These types vary from an equipment grounding conductor run with circuit conductors or enclosing them. It can consist of one or more of the following:

1. A copper, aluminum, or copper-clad aluminum conductor.
2. Rigid metal conduit.
3. Intermediate metal conduit.
4. Electrical metallic tubing.
5. Flexible metal conduit and fittings listed for grounding.
6. Liquid-tight flexible metal conduit and fittings listed for grounding.
7. Flexible metallic tubing, under conditions.
8. Type AC armored cable
9. Copper sheath of type MI cable.
10. Type MC cable metal sheath.
11. Cable trays meeting the requirements of NEC Section 392.3(C) and 392.7.
12. Cablebus framework permitted by NEC Section 370.3.
13. Other metal raceways listed for grounding that are electrically continuous.
14. Surface metal raceways listed for grounding.

The equipment grounding conductor can either be bare, covered, or insulated. Where the conductor is insulated it must be identified by a continuous outer finish of green, green with one or more yellow stripes, or bare. Equipment grounding conductors larger than No. 6 and multiconductor cable, where one or more conductors are designated for use as an equipment grounding conductor, must be permanently identified as the equipment grounding conductor by green stripping, coloring, or taping.

Sizing Equipment Grounding Conductors

Equipment grounding conductors are required to be sized not smaller than is stated in NEC Table 250.122 and are not required to be larger than the supply conductors. Section 250.122(B) states that if conductors are adjusted in size to compensate for voltage drop, the equipment grounding conductor must be adjusted proportionately according to circular mil area. The following example further illustrates this:

A circuit utilizing a No. 1 THW copper conductor that is protected by a 100-A circuit breaker would use a No. 8 copper equipment grounding conductor (according to NEC Table 250.122). Due to voltage drop, the ungrounded conductor size must be increased to a 1/0 conductor. In order to now adjust the size of the equipment grounding conductor proportionately by circular mil area, the following calculation must take place (values of circular mil area are found in NEC Table 8):

given
No. 1 conductor = 83,690 circular mil area
No. 8 conductor = 16,510 circular mil area
1/0 conductor = 105,600 circular mil area

$$\frac{\text{No.1 ungrounded conductor}}{\text{No.8 grounded conductor}} = \frac{83,690}{16,510} = 5.07$$

$$\frac{\text{No.1/0 ungrounded conductor}}{\text{answer from previous equation}} = \frac{105,600}{5.07} = 20,828$$

This calculation determines that the minimum size of equipment grounding conductor requires a 20,828 circular mil area. Referring again to NEC Table 8, it is noted that this does not correspond to a standard size conductor; therefore, the next larger conductor must be used. In this case, a No. 6 copper conductor must be used, which has a 26,240 circular mil area. This is what the NEC refers to when it says "adjusted proportionately according to circular mil area."

There is an important note just below NEC Table 250.122 that is very often overlooked. It states, "Where necessary to comply with Section 250.4(A)(5) or (B)(4), the equipment grounding conductor shall be sized larger than this table." Section 250.4(A)(5) states that grounding conductors "shall be capable of safely carrying the maximum ground-fault current likely to be imposed on it." Too many times, equipment grounding conductors are sized according to Table 250.122 without considering these other facts. In this case, the conductor may not be able to safely carry the fault current. As was seen earlier, at a given value of current, the conductor will fuse (melt), the ground is now lost and the equipment case or enclosure would become energized, which would create a shock or electrocution hazard.

Use of Grounded Circuit Conductor for Grounding Equipment

NEC Section 250.24(A)(5) states that the grounded conductor (current-carrying neutral) and the grounding conductor (non-current-carrying) shall not be connected together on the load-side of the service disconnecting means (see Fig. 4.16).

Figure 4.17 further illustrates this point.

The hazard of making this connection on the load-side is that a parallel return path has now been established. The metallic raceway becomes a parallel return path with the grounded conductor. The metallic raceway will generally have higher impedance than the

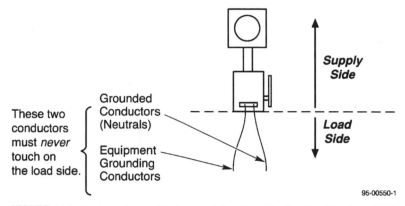

FIGURE 4.16 Neutral and ground to be separated on load side of service. (*Courtesy of AVO Training Institute, Inc.*)

grounding conductor. As a result, the I × R values will be different in the parallel return paths. A different voltage rise will occur and thus a difference in potential will exist between the grounded conductor and the metallic raceway. A difference in potential may result in a shock hazard for anyone coming into contact with the metallic raceway. See Fig. 4.18 for more on this issue.

Another problem with these parallel return paths is what is known as a ground loop. Ground loops are dangerous because of the resonating pattern established with frequency. Lightning is generated along a diminishing sine wave or damper wave. There is a varying frequency along this damper wave. If the frequencies are in resonation with each other, lightning may be drawn to the ground loop. This would help in explaining why lightning strikes some structures. Figure 4.19 further presents this phenomenon.

Another issue that must be addressed is proper grounding for two or more buildings where a common service is used. This issue deals with the grounding electrode, grounded

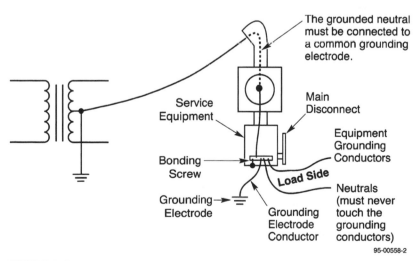

FIGURE 4.17 Grounding requirements at the service (supply side). (*Courtesy of AVO Training Institute, Inc.*)

$E_1 \neq E_2$

Voltage Potential !

$E_1 = I \times R_1$

$E_2 = I \times R_2$

Sub-Panel

Neutral Bus Bar

Bonding Screw
Code Violation

95-00551-1

FIGURE 4.18   Ground loop. (*Courtesy of AVO Training Institute, Inc.*)

Neutral

Ground Loop

Grounding

Sub-Panel

E

(f)

Time

E

Lightning Damper Wave

95-00552-1

FIGURE 4.19   Lightning hazard. (*Courtesy of AVO Training Institute, Inc.*)

(neutral) conductor, equipment grounding conductor, and bonding. NEC Section 250.50 requires the grounding electrodes at each building to be bonded together to form the grounding electrode system. Where no grounding electrode exists, one must be installed. There are, however, exceptions to this rule. Where the second building has only one branch circuit and an equipment grounding conductor is installed with the supply conductors, the grounding electrode at the main building is all that is required. Figure 4.20 illustrates this in simple form.

Section 250.32(B)(1) requires an equipment grounding conductor to be run to the separate building where there is equipment that is required to be grounded. In this case, the grounding and grounded conductors are prohibited from being connected together. The grounded conductor is considered to "float' in this case because it is not bonded to the equipment enclosure and grounding conductor. See Fig. 4.21 for further information on this.

Section 250.32(B)(2) states another situation that can be used. Where there is no equipment grounding conductor run with the supply conductors, the buildings are not bonded together through metal raceways or other piping, and there is no ground-fault equipment installed at the common service, the grounded conductor from the main building must be connected to the grounding electrode and bonded to the disconnecting means at the second building. This type of situation is essentially that same as individual services at each building. Figure 4.22 illustrates this further.

Ferroresonance

Another very important safety issue associated with proper system grounding is ferroresonance. Ferroresonance is a rare phenomenon in power systems, but one that deserves mention. In response to a voltage transient, phase to ground fault, circuit breaker opening, equipment energization or de-energization, lightning induced overvoltages, or any number of other sudden changes, the power system can take a sudden non-linear jump to a sustained

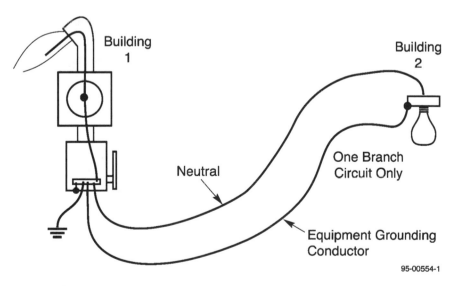

FIGURE 4.20 One branch circuit with equipment grounding conductor run, no electrode is required. (*Courtesy of AVO Training Institute, Inc.*)

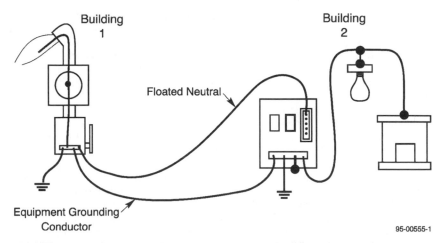

FIGURE 4.21   Equipment grounding conductor run to second building. (*Courtesy of AVO Training Institute, Inc.*)

condition of severe harmonic distortion and high (several per-unit) overvoltages that can severely damage power system equipment. Ferroresonance is an electrical resonance between the non-linear inductance of a transformer and system capacitance. Effective grounding decreases the chances of ferroresonance in three ways:

- Placing a very low impedance in parallel with the system capacitance.
- Fixing the electrical neutral of the system. Ferroresonance seems to require one point in the system whose potential is not fixed, like the floating neutral of an ungrounded system.
- Controlling and limiting system overvoltages that can initiate a ferroresonance.

For more information, consult any power system analysis text. An excellent reference is a technical paper by P. Ferracci entitled *Ferroresonance*, Cahier Technique No. 190(ECT 190), March 1998. This paper can be obtained from the Internet at http://www.Schneiderelectric. com.tr/ftp/literature/Publications/ECT190.PDF.

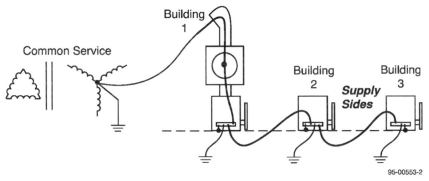

FIGURE 4.22   Common service with no equipment grounding conductor run. (*Courtesy of AVO Training Institute, Inc.*)

SUMMARY

Proper grounding is an essential part of electrical safety. Without proper grounding, the non-current-carrying metal components of electrical equipment have the risk of becoming energized. Design and install a grounding system based on the requirements of the National Electrical Code, the IEEE Std. 142 (Green Book), and IEEE Std. 80, as well as any other national standards that are relevant to the application.

Figure 4.23 provides an overall look at electrical system grounding requirements based on the current edition of the National Electrical Code.

Note: Don't take a shortcut *with grounding*; it may cost someone his or her life. That someone might be you.

250.64(C): The grounding electrode conductor shall be installed in one continuous length without a splice or joint, unless connectors listed for purpose.

95-00573-3

FIGURE 4.23 Service grounding and bonding requirements. (*Courtesy of AVO Training Institute, Inc.*)

CHAPTER 5

ELECTRICAL MAINTENANCE AND ITS RELATIONSHIP TO SAFETY

INTRODUCTION

All maintenance and testing of the electrical protective devices addressed in this chapter must be accomplished in accordance with the manufacturer's instructions. The NETA "Maintenance Testing Specifications for Electrical Power Distribution Equipment and Systems"[1] 2005 Edition is an excellent source of information for performing the required maintenance and testing of these devices. Visit the NETA website (http://www.netaworld.org) for further information.

Because of the electrical hazards to which they may be exposed, all workers involved in electrical testing and maintenance must be qualified personnel as defined in the OSHA and NFPA standards. NFPA 70B, *Recommended Practice for Electrical Equipment Maintenance*, states that equipment deterioration is normal, but equipment failure is not inevitable. Normal deterioration begins as soon as new equipment is installed and if unchecked can eventually cause equipment failure. Deterioration can be accelerated by factors such as environment, overload, or severe duty cycles. A periodic maintenance program can identify these factors and provide measures for coping with them. There are other potential causes of equipment failure that can also be detected and corrected. Among these are load changes or additions, circuit alterations, improperly set or improperly selected protective devices, and changing voltage conditions.[2]

Without an effective periodic maintenance and testing program, management assumes a greatly increased risk of a serious electrical failure and its consequences. An effective electrical preventive maintenance program can reduce accidents, save lives, and minimize costly breakdowns and unplanned shutdowns of production equipment. Failures can be averted by identifying and correcting impending troubles before they become major problems requiring more expensive, time consuming solutions.[2]

THE SAFETY-RELATED CASE FOR ELECTRICAL MAINTENANCE

Overview

The relationship between safety and preventive maintenance is not a difficult one to establish. Properly designed equipment that is properly installed is well capable of doing its job when it is new. As equipment ages however, several factors begin to take their toll on electrical equipment.

- Dust, dirt, and other contaminants collect on equipment causing the equipment to overheat and bearings and other moving parts to bind.
- Vibration causes hardware to loosen. Subsequent operations of equipment can cause joints and equipment to fail explosively.
- Heat and age can cause insulation to fail, resulting in shock hazards to personnel.
- Increased loads, motor starting surges, and power quality issues such as harmonics combine to increase the aging process and set the stage for equipment failure.

Unfortunately, the ultimate failure of unmaintained equipment usually occurs when the equipment is needed the most—during electrical faults. Such failures result in arc and blast events that can and do harm workers in the area. They also result in significant downtime, loss of equipment, and construction cost incurred in rebuilding the equipment.

The only way to ensure that electrical equipment continues to operate in an optimal manner is to maintain it so that it stays in factory-new-operating condition.

Regulatory

As discussed above and in previous chapters, the catastrophic failure of electrical equipment creates severe hazards for personnel working in the area. Recognizing this the *Standard for Electrical Safety in the Workplace* (NFPA 70E)[3] requires that electrical equipment be properly maintained to minimize the possibility of failure.

Relationship of Improperly Maintained Electrical Equipment to the Hazards of Electricity

Improperly maintained equipment may expose workers to any of the three electrical hazards. For example:

1. Improperly maintained tools or flexible cord sets (extension cords) can have frayed insulation which exposes the energized conductors and allows them to contact the worker or the metallic tool the worker is using. The result is an electric shock.
2. Improperly maintained protective devices, such as circuit breakers or fuses, can fail when interrupting an overcurrent. Such a failure is likely to be explosive; consequently, the worker is exposed to electrical arc and electrical blast.

⚠ **WARNING**

Arc Flash Hazard.
Appropriate PPE Required.
Failure To Comply Can Result in Death or Injury.
Refer to NFPA 70 E.

02-0008

FIGURE 5.1 Minimum required warning label for posting on electrical equipment to warn of hazards.

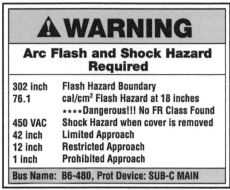

02-0007-2

FIGURE 5.2   Preferred required warning label for posting on electrical equipment to warn of hazards.

3. Improperly maintained connections can overheat resulting in any of the following:
   a. melted insulation, exposed conductors, and the attendant electrical shock hazard
   b. fire
   c. failed connections resulting in electrical arc and blast
4. Improperly maintained switchgear, motor control centers, or panelboards can fail explosively when an arc occurs internally. This exposes workers to the effects of electrical blast and possibly electrical arc.

Hazards Associated with Electrical Maintenance

Anytime workers approach electrical equipment they might face one or more of the electrical hazards. This includes activities associated with electrical maintenance. Both the *National Electrical Code* (NFPA 70)[4] and the *Standard for Electrical Safety in the Workplace* (NFPA 70E)[3] require that

> "Switchboards, panelboards, industrial control panels, and motor control centers that are in other than dwelling occupancies and are likely to require examination, adjustment, servicing, or maintenance while energized shall be field marked to warn qualified persons of potential electric arc flash hazards. The marking shall be located so as to be clearly visible to qualified persons before examination, adjustment, servicing, or maintenance of the equipment."[3,4]

Figure 5.1 illustrates an example of the minimum labeling requirements, and Figure 5.2 shows a preferred label for the application. Refer to Chap. 3 for details on the methods for determining the extent of the hazards.

THE ECONOMIC CASE FOR ELECTRICAL MAINTENANCE

Although a complete discussion of the economics of an electrical maintenance program is beyond the scope of the handbook, Figure 5.3, taken from *Recommended Practice for*

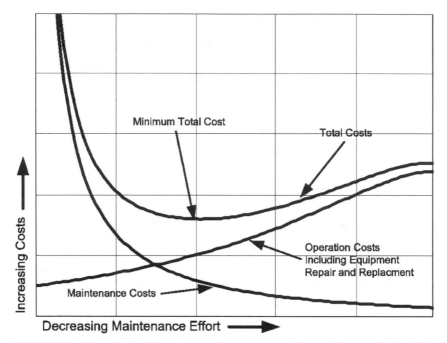

FIGURE 5.3   Effect of electrical maintenance frequency on overall costs. (*Courtesy National Fire Protection Association.*)

Electrical Equipment Maintenance (NFPA 70B)[2] provides a clear definition of the economic importance of such maintenance.

This diagram charts the relationship between costs and maintenance. As can be seen, the cost of electrical maintenance (EPM) decreases as the interval between inspections increase. In other words, the cost of maintenance decreases as less maintenance is done. The cost of equipment repair and replacement, on the other hand, increases as less maintenance is done.

Unfortunately, many managers focus only on the cost of maintenance without taking the costs of equipment repair and replacement into consideration. The correct way to approach the economics of a maintenance program is to look at the total cost as shown in the figure. Little or no maintenance (right-hand side of the figure) results in very high repair and replacement costs. As the electrical maintenance program is intensified (moving the left on the figure), repair and replacement costs decrease.

At some point, the minimum total cost is realized. This is the point at which the maintenance program should be operated to provide for the best overall efficiency of the system.

RELIABILITY CENTERED MAINTENANCE (RCM)

(The following paragraphs are taken from *RCM White Paper* by Al Rose, (c) Cadick Corporation, 2001–2005)

What is Reliability Centered Maintenance?

Reliability-Centered Maintenance (RCM) is the process of determining the most effective maintenance approach. The RCM philosophy employs Preventive Maintenance (PM), Predictive Maintenance (PdM), Real-time Monitoring (RTM), Run-to-Failure (RTF- also called reactive maintenance) and Proactive Maintenance techniques in an integrated manner to increase the probability that a machine or component will function in the required manner over its design life cycle with a minimum of maintenance. The goal of the philosophy is to provide the stated function of the facility, with the required reliability and availability at the lowest cost. RCM requires that maintenance decisions be based on maintenance requirements supported by sound, technical, and economic justification.

A Brief History of RCM

RCM originated in the Airline industry in the 1960s. By the late 1950s, the cost of maintenance activities in this industry had become high enough to warrant a special investigation into the effectiveness of those activities. Accordingly, in 1960, a task force was formed consisting of representatives of both the airlines and the Federal Aviation Administration (FAA) to investigate the capabilities of preventive maintenance. The establishment of this task force subsequently led to the development of a series of guidelines for airlines and aircraft manufacturers to use, when establishing maintenance schedules for their aircraft.

This led to the 747 Maintenance Steering Group (MSG) document MSG-1; Handbook: Maintenance Evaluation and Program Development from the Air Transport Association in 1968. MSG-1 was used to develop the maintenance program for the Boeing 747 aircraft, the first maintenance program to apply RCM concepts. MSG-2, the next revision, was used to develop the maintenance programs for the Lockheed L-1011 and the Douglas DC-10. The success of this program is demonstrated by comparing maintenance requirements of a DC-8 aircraft, maintained using standard maintenance techniques, and the DC-10 aircraft, maintained using MSG-2 guidelines. The DC-8 aircraft has 339 items that require an overhaul, verses only seven items on a DC-10. Using another example, the original Boeing 747 required 66,000 labor hours on major structural inspections before a major heavy inspection at 20,000 operating hours. In comparison, the DC-8 - a smaller and less sophisticated aircraft using standard maintenance programs of the day required more than 4 million labor hours before reaching 20,000 operating hours.

In 1974 the U.S. Department of Defense commissioned United Airlines to write a report on the processes used in the civil aviation industry for the development of maintenance programs for aircraft. This report, written by Stan Nowlan and Howard Heap and published in 1978, was entitled *Reliability Centered Maintenance*,[5] and has become the report upon which all subsequent Reliability Centered Maintenance approaches have been based. What Nowlan and Heap found was that many types of failures could not be prevented no matter how intensive the maintenance activities were. Additionally it was discovered that for many items the probability of failure did not increase with age. Consequently, a maintenance program based on age will have little, if any effect on the failure rate.

RCM in the Industrial and Utility Arena

As with any philosophy, there are many paths, or processes that lead to a final goal. This is especially true for RCM where the consequences of failure can vary dramatically.

Rigorous RCM analysis has been used extensively by the aircraft, space, defense, and nuclear industries where functional failures have the potential to result in large losses of life, national security implications, and/or extreme environmental impact. A rigorous RCM analysis is based on a detailed Failure Modes and Effects Analysis (FMEA) and includes probabilities of failure and system reliability calculations. The analysis is used to determine appropriate maintenance tasks to address each of the identified failure modes and their consequences.

While this process is appropriate for these industries, it is not necessarily the most practical or best approach to use for industrial and utility systems maintenance. For these systems a streamlined or intuitive RCM analysis process may be more appropriate. This is due to the high analysis cost of the rigorous approach, the relative low impact of failure of most industrial systems, the type of systems and components maintained, and the amount of redundant systems in place. The streamlined approach uses the same principles as the rigorous, but recognizes that not all failure modes will be analyzed. For most industrial and utility systems the most economical and efficient approach is to use a combination of rigorous (formal) and intuitive analysis depending on system criticality and failure impact. Failure modes that result in high costs or personnel injury, or where the resultant reliability is still unacceptable in terms of safety or operational impact still receive the rigorous approach, but all other failure modes will use intuitive analysis.

The primary RCM principles

1. **RCM is Function Oriented**—It seeks to preserve system or equipment function, not just operability for operability's sake. Redundancy of function through multiple equipment improves functional reliability, but increases life cycle cost in terms of procurement and operating costs.
2. **RCM is System Focused**—It is more concerned with maintaining system function than individual component function.
3. **RCM is Reliability Centered**—It treats failure statistics in an actuarial manner. The relationship between operating age and the failures experienced is important. RCM is not overly concerned with simple failure rate; it seeks to know the conditional probability of failure at specific ages (the probability that failure will occur in each given operating age bracket).
4. **RCM Acknowledges Design Limitations**—Its objective is to maintain the inherent reliability of the equipment design, recognizing that changes in inherent reliability are the province of design rather than maintenance. Maintenance can, at best, only achieve and maintain the level of reliability for equipment, which is provided for by design. However, RCM recognizes that maintenance feedback can improve on the original design. In addition, RCM recognizes that a difference often exists between the perceived design life and the intrinsic or actual design life, and addresses this through the Age Exploration (AE) process.
5. **RCM is Driven by Safety and Economics**—Safety must be ensured at any cost; thereafter, cost-effectiveness becomes the criterion.
6. **RCM Defines Failure as Any Unsatisfactory Condition**—Therefore, failure may be either a loss of function (operation ceases) or a loss of acceptable quality (operation continues).
7. **RCM Uses a Logic Tree to Screen Maintenance Tasks**—This provides a consistent approach to the maintenance of all kinds of equipment. See Fig. 5.4.
8. **RCM Tasks Must Be Applicable**—The tasks must address the failure mode and consider the failure mode characteristics.
9. **RCM Tasks Must Be Effective**—The tasks must reduce the probability of failure and be cost effective.

FIGURE 5.4   RCM Logic Tree.

10. RCM Acknowledges Three Types of Maintenance Tasks
   a. Time-directed (PM)—Scheduled when appropriate
   b. Condition-directed (PdM and real-time monitoring)—Performed when conditions indicate they are needed
   c. Failure finding (one of several aspects of Proactive Maintenance)—Equipment is run-to-failure. This is acceptable for some situations and some types of equipment.
11. RCM is a Living System—It gathers data from the results achieved and feeds this data back to improve design and future maintenance. This feedback is an important part of the Proactive Maintenance element of the RCM program.

   The RCM analysis carefully considers the following questions:

1. What does the system or equipment do; what is its function?
2. What functional failures are likely to occur?

3. What are the likely consequences of these functional failures?

4. What can be done to reduce the probability of the failure, identify the onset of failure, or reduce the consequences of the failure?

Answers to these four questions determine the necessary actions required to maintain the systems or equipment. Figure 1 shows the RCM logic tree used to answer these questions.

Failure

Failure can be defined in many ways. In a broad sense failure is simply an unsatisfactory condition. RCM, however, requires us to look at failure from not just an equipment standpoint, but from a system standpoint as well. A piece of equipment can be operating (an HVAC unit in a clean room, for example), but if its output is less than required it would be considered failed. On the other hand, a protective relay in a power system can have failed, but if it has not taken the circuit off-line (the circuit is still energized) the function of the system is unchanged. Essentially, defining failure depends on the function of an item or system and the operating context in which the item or system is used.

Functional Failure.   A functional failure is essentially the inability of an item or system to meet its specified performance standard. A complete loss of function is a functional failure; however, in the HVAC example above if the system's output is less than specified, a functional failure has occurred even if the system is still operating.

Consequences of Failure.   The consequences of failure determine the priority of maintenance activities or design improvement required to prevent occurrence. If failure of an item results in little or no consequence, mimimal maintenance activities are generally required. If, however, failure of an item results in a large economic hardship, personnel injury, or environmental damage maximum maintenance activities or a redesign may be called for.

Maintenance Actions in an RCM Program

RCM goals are to identify the most cost-effective and applicable maintenance techniques to minimize the risk and impact of failure in facility and utility equipment and systems. This allows systems and equipment functionality to be maintained in the most economical manner. Specific RCM objectives as stated by Nowlan and Heap[5] are:

- To ensure realization of the inherent safety and reliability levels of the equipment.
- To restore the equipment to these inherent levels when deterioration occurs.
- To obtain the information necessary for design improvement of those items whose inherent reliability proves to be inadequate.
- To accomplish these goals at a minimum total cost, including maintenance costs, support costs, and economic consequences of operational failures.

To this end there are four outcomes from an RCM analysis:

1. Perform no Maintenance—This is referred to as reactive maintenance, repair, fix-on-fail, or Run-To-Fail (RTF). This type of maintenance assumes that failure is equally likely to occur in any part, and that a failure is not detrimental to the operation. When this is the sole type of maintenance practiced high failure rates, large parts inventories, and excessive amounts of overtime become common. A purely RTF maintenance program ignores many of the opportunities to influence equipment survivability.

2. **Perform Preventive Maintenance (PM)**—PM consists of regularly scheduled inspections, adjustments, cleanings, lubrication, and replacement of components and equipment. PM is also referred to as time-driven or interval-based maintenance. It is performed without regard to equipment condition. PM schedules inspection and maintenance at predefined intervals in an attempt to reduce equipment failures. However, as Nowlan and Heap discovered, a PM program can result in a significant increase in inspections and cost without any increase in reliability.

3. **Perform Condition Based Maintenance (CBM)**—CBM consists of Predictive Maintenance (PdM) and real-time monitoring. PdM primarily uses non-intrusive testing techniques to measure and trend equipment performance. Real-time monitoring uses current performance data to asses machinery condition. CBM replaces arbitrarily timed maintenance tasks with maintenance that is scheduled only when warranted by equipment condition. Continuing analysis of equipment condition data allows for planning and scheduling maintenance activities or repairs prior to functional or catastrophic failure.

4. **Redesign**—When failure of a system or piece of equipment is an unacceptable risk and none of the above tasks can help mitigate the failure, a redesign of the equipment or system is in order. In most cases adding redundancy eliminates the risk and adds very little to overall maintenance costs.

An RCM program will include all four of the above analysis outcomes, as no one is more important than another. Figure 2 shows the relationship maintenance activities have to RCM (See Fig. 5.5).

Impact of RCM on a Facilities Life Cycle

A facilities life cycle is often divided into two broad stages: acquisition (planning, design, and construction) and operations. RCM affects all phases of the acquisition and operations stages to some degree.

Decisions made early in the acquisition cycle profoundly affect the life-cycle cost of a facility. Even though expenditures for plant and equipment may occur later during the acquisition process, their cost is committed at an early stage. As shown conceptually in Fig. 5.6, planning (including conceptual design) fixes two-thirds of the facility's overall life-cycle costs. The subsequent design phases determine an additional 29 percent of the life-cycle cost, leaving only about 5 percent of the life-cycle cost that can be impacted by the later phases.

Thus, the decision to include a facility in an RCM program, including condition monitoring which will have a major impact on its life-cycle cost, is best made during the planning

FIGURE 5.5 RCM structure.

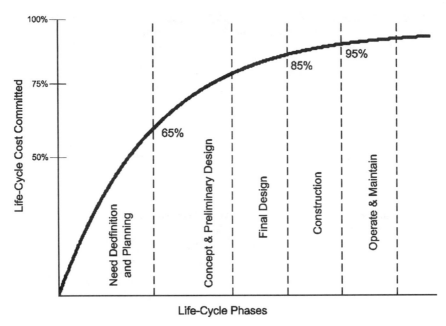

FIGURE 5.6   Source Blanchard, B.S, Design and Manage to Life Cycle Cost, Forest Grove, OR, MA Press, 1978

phase. As RCM decisions are made later in the life cycle, it becomes more difficult to achieve the maximum possible benefit from the RCM program.

Even though maintenance is a relatively small portion of the overall life-cycle cost, typically 3 to 5 percent of a facility's operating cost, RCM is still capable of introducing significant savings during the O&M phase of the facility's life. Savings of 30 to 50 percent in the annual maintenance budget are often obtained through the introduction of a balanced RCM program.

Conclusion

Reliability-Centered Maintenance is the process of determining the most effective maintenance approach. As the airline industry showed 40 years ago, RCM can not only improve the reliability of a system, it can significantly reduce the required maintenance. In today's competitive world economy that translates into monies saved, both from reduced failures and reduced work. And a properly implemented RCM program will continue saving money year after year. To remain competitive most companies can no longer afford to remain on the sidelines doing things like they've always done.

THE EIGHT STEP MAINTENANCE PROGRAM

Some facilities, especially very small ones, may not want to embark on such an ambitious process as a complete RCM program. For such locations, a minimal program such as the one described below may be more desirable. This information in the following paragraphs is taken from *The Eight Step Maintenance Program* by John Cadick.[6]

Introduction

There are eight simple steps that should be at the heart of an electrical preventive maintenance program. These key steps are plan, inspect, clean, tighten, lubricate, test, record, and evaluate. The following paragraphs briefly describe each step and illustrate the part that they play in a good preventive maintenance program.

Step 1—Plan

Before working on anything, take time to think about what you are going to do and how you are going to do it. A few minutes of thought before the job is started results in a safer and more efficient job. This step is especially important when you are performing emergency repair work.

Start by identifying the specific task or tasks that you intend to perform. In this initial stage you do not need to go into too much detail, rather, concentrate on the broad goals that you need to accomplish. List each goal, the personnel that you will use, the equipment that you will need, and the amount of time that will be required.

Be certain that you obtain and use the proper instruction manuals and design drawings before you start. The joke of "getting the manual after smoke is present" has too much basis in fact to be very funny for experienced electricians. Trouble shooting without accurate and correct schematics and wiring diagrams is close to impossible in all but the simplest circuits.

Develop detailed plans and procedures from the preliminary plans discussed above. Steps should be detailed as much as required by the complexity of the job and the experience of your personnel.

Always start any procedure with a job briefing as described in Chap. 3.

Step 2—Inspect

Periodic inspections should be set up on a routine basis. Inspection can be done with the senses that nature gave us, called unaided inspection, or with instruments such as infrared viewers, ultra sound transducers, and so on.

Unaided Inspection. Your eyes are your most important inspection tools. They can detect dirt, note current readings, determine levels, discover discolored, overheated insulation, determine the presence of insects and rodents, and find a whole host of other problems. The more experienced you become, the more information your eyes can provide; however, even the most inexperienced electrician can detect spider webs in a circuit breaker operating mechanism. Always observe what you are working on and look for any abnormalities that can give you an idea of any thing that can cause these problems.

Your ears can also be useful inspection tools. Overloaded transformers are normally noisier than lightly loaded ones, corona has a distinctive hissing sound, and motors with unbalanced voltages or bad bearings vibrate. Any change in sound should be investigated.

Nothing in electrical systems is more distinctive than the odor of overheated or burning insulation. Your nose can detect such problems long before your other senses. Again, the most useful indicator of problems is a change in odor.

Touch may also be employed to inspect electrical equipment. Feeling for excessive vibration or heating is a time-tested technique. Of course, you should be extremely cautious before touching any piece of equipment. If the equipment is not de-energized, contact should be made only when wearing appropriate insulating PPE, such as rubber gloves.

Inspection with Tools or Instruments. Except for your own senses, infrared viewers are the most cost-effective maintenance tool in existence. Annual infrared scans of your power system invariably discover problems which, if allowed to continue, could result in severe problems. Whether this service is done "in house" or contracted to an outside service, you *should* perform an annual (minimum) infrared scan of your system.

Ultrasound transducers are normally used on major overhauls to evaluate the condition of metals which are subjected to constant vibration.

Many other such tools and instruments are available to help in the performance of maintenance tasks. Such tools should be used as required by qualified personnel trained in their proper and safe use.

Step 3—Clean

Dirt and electricity don't mix. After you have completed the inspection of your electrical equipment, the next step is to clean all components.

The first step is general cleanliness. Dust, dirt, and foreign material must be cleaned from motors, switchgear, transformers, and other such electrical equipment. For such cleaning we recommend the use of high suction, insulated hose shop vacuum, clean, lint free wiping cloth, and general purpose solvents. A citrus-based general purpose solvent is effective and biodegradable.

Each insulator or conductor has its own way to be cleaned. Porcelain, for example, can be cleaned with soap, water, and appropriate cleaning brushes or pads. Plastic insulation, on the other hand, cannot take any abrasion. Always refer to the manufacturer's recommendations for specific techniques. Also, always avoid using steel wool or other conductive abrasives in any area where contact with energized conductors is possible.

Conductor connecting surfaces need special attention. Contact surfaces of copper conductors should be cleaned to a bright finish where they are connected together. Aluminium must be cleaned and a protective lubricant must be applied immediately after the cleaning. The lubricant cuts through any aluminum oxide (a good insulator) that may be present, and prevents the formation of additional aluminium oxide. (See the section on lubrication.)

Refer to the manufacturer's instruction books, which sometimes specify the cleaning material to be used for their equipment. Be sure to consult this literature before attempting to clean electrical components.

Step 4—Tighten

Tightening fasteners on electrical equipment takes special care and should be done to the torque recommended by the equipment manufacturer. Electrical connections are especially critical. Recommendations for recommended torque values can be found in the maintenance and acceptance testing specifications published by the InterNational Electrical Testing Association.[1] Be certain to torque connections per this table and/or manufacturer's recommendations.

Step 5—Lubricate

Lubrication of electrical system components is an often overlooked or improperly performed procedure. Too much lubricant can be worse than no lubricant at all. This is especially true of devices such as protective relays and circuit breakers which may go for years without operating. Be especially careful of conductive lubricants to make sure they do not cause a short circuit. Always use the lubricants recommended by the equipment manufacturer or one that has the same characteristics. Lubrication falls into two important areas:

1. Non-conductors

   Non-conductive joints or moving members should be lubricated using an appropriate material. In some cases grease, oil, or some synthetic lubricant may be recommended. Motor bearings, mechanism pivots, and other such equipment fall into this category. Remember that some pieces of equipment, such as protective relays, do not require any lubricant at all.

2. Conductors

   Most manufacturers make and/or recommend a lubricant to be used on conductors, stabs, and other such equipment. The ubiquitous "black grease" has been and continues to be a good lubricant when used properly. The correct approach to lubrication of electrical conductors is to apply a small amount and then wipe the contact surface clean with a clean, lint free cloth. Do not *scrub* the surface however. You wish to leave a fine film of lubricant in place.

Remember: **Use the proper lubricants in the proper amounts.**

Step 6—Test

Electrical equipment should be tested periodically. Insulation resistance should be measured, breaker trip times should be checked, relays should be calibrated, and a variety of other procedures should be performed. The only exercise that many protective devices get is during the test interval.

A detailed review of all procedures is beyond the scope of this course; however, the recommendations made by the InterNational Electrical Testing Association (NETA) in their publication entitled *Maintenance Testing Specifications*[1] should be reviewed and applied as appropriate.

Step 7—Record

Doing work without keeping records is almost useless. Recording test and other data can help set maintenance intervals, isolate troublesome equipment (or manufacturers), and provide a baseline that helps to know when insulation or other components are starting to fail. Besides test results records should always include the date, equipment identification, and all pertinent data on any problems discovered. System conditions like load current, voltage, temperature, and other such information can help determine how close components are to being overloaded.

Test and record forms should be developed and tailored to your company's specific needs. Examples of such forms may be found at the back of the NFPA 70B Electrical Equipment Maintenance. Review that document for ideas and examples.

Step 8—Evaluate

Evaluation of the test results is, possibly, the single most important step in the entire process. Since 1990, significant progress has been made in the statistical analysis of small population data such as that gathered during maintenance intervals. Whether the analysis is done using sophisticated mathematical techniques or simple field rules of thumb, no program is complete without it. Analysis of records will allow the company to determine what, if any, additional maintenance needs to be performed and to pinpoint problems which might otherwise go unobserved.

FREQUENCY OF MAINTENANCE

How often should a power system be maintained? Unfortunately there is no simple answer to this question. The following should help in determining frequency of maintenance in a power system. The steps listed in this procedure are valid for all of the testing and maintenance information given in this text.

Determining Testing Intervals

1. Start initially with a once per year test program. Continue this procedure for the first two intervals.
2. Review the test records from your maintenance intervals.
3. If frequent and/or severe problems are evident, decrease the maintenance intervals.
4. If no major or frequent problems are evident, increase the maintenance intervals.
5. Repeat steps 2 through 5 throughout the life of your maintenance program.

The average maintenance interval in industrial systems is about two years. This means that every two years, most industrial plants completely maintain their electrical system. Electrical utilities have similar intervals although their average is probably closer to $1\frac{1}{2}$ years. Recent advances in statistical analysis have enabled maintenance programs based on technology rather than "estimate." Such *Condition Based Maintenance* programs will be the way maintenance is done in the 21st century.

Summary

The organization of an electrical preventive maintenance program may be greatly enhanced by adhering to the simple, eight step program outlined in this chapter. These steps contain all of the elements required for a comprehensive program. If you *plan, inspect, clean, tighten, lubricate, test, and record,* you will be well on your way to a safe, efficient, and profitable preventive maintenance program.

MAINTENANCE REQUIREMENTS FOR SPECIFIC EQUIPMENT AND LOCATIONS

Although comprehensive maintenance procedures are beyond the scope of this chapter; the following should help to provide a starting point for those new to electrical maintenance or a mid-course correction for those with more experience. For detailed information refer to the documents in the references of this chapter and other standard texts.

General Maintenance Requirements

Table 5.1 shows the general maintenance requirements for electrical systems that are laid out in NFPA 70E.[3] Some of the requirements have been slightly modified from that document to be consistent with the authors' beliefs with regard to proper maintenance techniques.

TABLE 5.1 General Maintenance Requirements for Electrical Systems

Topic	Required maintenance
Qualified persons	Only qualified persons are allowed to perform maintenance on electrical equipment and installations. This means that they must be trained in and familiar with the operation of the equipment, the safety hazards of the equipment, the specific maintenance procedures, and the tests that they will perform.
Single line diagram	A legible, accurate, and up-to-date single-line diagram should be maintained to assist in day-to-day operations including maintenance, testing, and switching activities. (Note that NFPA 70E does not require a single-line diagram, but does require that it be properly maintained if it exists.)
Spaces about electrical equipment	The working spaces that are required by NFPA 70, NFPA 70E, and 29CFR 1910.302 - .308 must be maintained.
Grounding and bonding	All grounding a bonding conductors must be maintained to ensure electrical continuity.
Guarding of live parts	Enclosures around live parts must be maintained to guard against accidental contact.
Safety equipment	Locks, interlocks, and all other types of safety equipment (including personal protective equipment) must be maintained in proper working condition so that they provide the desired protection.
Clear spaces	Access to working space and escape passages must be maintained. (This refers to exits and aisles as opposed to actual working spaces about electrical equipment.)
Identification of components and circuits	All electrical components and circuits should be identified with legible, secure signs, tags, or other such devices. Operating procedures, such as switching and/or lockout/tagout instructions fall under this requirement.
Conductors and cables	All types of conductors and cables, including flexible cord sets (extension cords) should be maintained so that they are not allowed to become damaged, frayed, or strained.

Substations, Switchgear, Panel boards, Motor Control Centers, and Disconnect Switches

Enclosures (Electrical Rooms, Buildings, and Other Such Locations).   Enclosures must be kept clear of extraneous materials that could cause a tripping, fire, or any other type of hazard.

Area enclosures (Fences and Other Such Enclosures).   Fences and other such guards shall be maintained to prevent unauthorized access or accidental exposure to electrical hazards. This requirement is especially important for fences that protect the public from access to electrical facilities.

Conductors.   All current-carrying conductors must be maintained so that they can carry rated current without overheating. They must also be maintained so they can carry available short-circuit current without failing and introducing arc and blast hazards.

Electrical Insulation.   Electrical insulation must be maintained so that it will continue to support the impressed voltage. The maintenance of insulation is very important. Insulation maintenance typically accounts for over 70 percent of all maintenance procedures.

Protective Devices.   Protective devices must be maintained so that they can adequately withstand and/or interrupt the available fault current. They must also be capable of operating as they are designed to do. The next sections provide detail for many of the maintenance procedures for protective devices.

Fuse Maintenance Requirements

Fuses may well be the most widely used electrical protective device. They are certainly the oldest dating back to the late 1800s when it was noticed that the smallest piece of wire in a series of wires was always the first one to open. Although fuses are usually not electrically tested, the following specific steps should be employed during electrical maintenance:

1. Inspect fuses and fuse blocks closely for any signs of overheating that may be caused by loose connections, overload currents, or environmental conditions.
2. Make certain that the correct size fuse is being used in the application.
3. In a three-phase or two-phase fuse block, the fuses should be of the same size and from the same manufacturer.
4. Make certain that fuses have sufficient interrupting rating for the available fault current at their point of application.
5. Check the tightness of fuse mounting clips and hardware.

Although simple in application, these five simple steps can help to ensure that fuses will operate properly when called upon.

Molded-Case Circuit Breakers

The need for inspection of molded-case breakers will vary depending on operating conditions. Suggested inspection and testing is defined in ANSI/NEMA AB 4, *Guidelines for Inspection and Preventive Maintenance of Molded Case Circuit Breakers Used in Commercial and Industrial Applications*. As a part of these guidelines, AB 4 also provides some basic procedures for the inspection and maintenance of molded-case circuit breakers, by qualified persons.

Generally, maintenance on molded-case circuit breakers is limited to proper mechanical mounting, electrical connections, and periodic manual operation. Most lighting, appliance, and power panel circuit breakers have riveted frames and are not designed to be opened for internal inspection or maintenance. All other molded-case circuit breakers, that are Underwriter's Laboratory (UL) approved, are factory-sealed to prevent access to the calibrated elements. An unbroken seal indicates that the mechanism has not been tampered with and that it should function as specified by UL. A broken seal voids the UL listing and the manufacturers' warranty of the device. In this case, the integrity of the device would be questionable. The only exception to this would be a seal being broken by a manufacturer's authorized facility.

Molded-case circuit breakers receive initial testing and calibration at the manufacturers' plants. These tests are performed in accordance with UL 489, *Standard for Safety, Molded-Case Circuit Breakers, Molded-Case Switches and Circuit Breaker Enclosures*. Molded-case circuit breakers, other than the riveted frame types, are permitted to be reconditioned and

FIGURE 5.7   Principle components of a molded-case circuit breaker.

returned to the manufacturer's original condition. In order to conform to the manufacturer's original design, circuit breakers must be reconditioned according to recognized standards. An example of a recognized standard is the Professional Electrical Apparatus Recyclers League (PEARL) Reconditioning Standards. In order to ensure equipment reliability it is highly recommended that only authorized professionals recondition molded-case circuit breakers.

Circuit breakers installed in a system are often forgotten. Even though the breakers have been sitting in place supplying power to a circuit for years, there are several things that can go wrong. The circuit breaker can fail to open due to a burned out trip coil or because the mechanism is frozen due to dirt, dried lubricant, or corrosion. The overcurrent device can fail due to inactivity or a burned out electronic component. Many problems can occur when proper maintenance is not performed and the breaker fails to open under fault conditions. This combination of events can result in fires, damage to equipment, or injuries to personnel.

Very often, a circuit breaker fails because the minimum maintenance (as specified by the manufacturer) was not performed or was performed improperly. Small things, like failing to properly clean and/or lubricate a circuit breaker, can lead to operational failure or complete destruction due to overheating of the internal components. Common sense, as well as manufacturers' literature, must be used when maintaining circuit breakers. Most manufacturers, as well as NFPA 70B,[1] recommend that if a molded-case circuit breaker has not been operated, opened or closed, either manually or by automatic means, within as little as six months time, it should be removed from service and manually exercised several times. This manual exercise helps to keep the contacts clean, due to their wiping action, and ensures that the operating mechanism moves freely. This exercise however does not operate the mechanical linkages in the tripping mechanism, Fig. 5.7. The only way to properly exercise the entire breaker operating and tripping mechanisms is to remove the breaker from service and test the overcurrent and short-circuit tripping capabilities. A stiff or sticky mechanism can cause an unintentional time delay in its operation under fault conditions. This could dramatically increase the arc/flash incident energy level to a value in excess of the rating of personal protective equipment.

Another consideration is addressed by OSHA in 29 CFR 1910.334(b)(2) which states:

> Reclosing circuits after protective device operation. After a circuit is deenergized by a circuit protective device, the circuit may NOT be manually reenergized until it has been determined

that the equipment and circuit can be safely reenergized. The repetitive manual reclosing of circuit breakers or reenergizing circuits through replaced fuses is prohibited.

NOTE: When it can be determined from the design of the circuit and the overcurrent devices involved that the automatic operation of a device was caused by an overload rather than a fault condition, no examination of the circuit or connected equipment is needed before the circuit is reenergized.

The safety of the worker manually operating the circuit breaker is at risk if the short circuit condition still exists when reclosing the breaker as stated above. OSHA no longer allows the past practice of resetting a circuit breaker one, two, or three times before investigating the cause of the trip. This previous practice has caused numerous burn injuries that resulted from the explosion of electrical equipment. Before resetting a circuit breaker, it, along with the circuit and equipment, must be tested and inspected, by a qualified person, to ensure a short- circuit condition does not exist and that it is safe to reset the breaker.

Any time a circuit breaker has operated and the reason is unknown, the breaker, circuit, and equipment must be inspected for a short circuit condition. Melted arc chutes will not interrupt fault currents. If the breaker cannot interrupt a second fault, it will fail and may destroy its enclosure and create a hazard for anyone working near the equipment.

To further emphasize this point the following quote is provided:

> After a high level fault has occurred in equipment that is properly rated and installed, it is not always clear to investigating electricians what damage has occurred inside encased equipment. The circuit breaker may well appear virtually clean while its internal condition is unknown. For such situations, the NEMA AB4 'Guidelines for Inspection and Preventive Maintenance of MCCBs Used in Commercial and Industrial Applications' may be of help. Circuit breakers unsuitable for continued service may be identified by simple inspection under these guidelines. Testing outlined in the document is another and more definite step that will help to identify circuit breakers that are not suitable for continued service.
>
> After the occurrence of a short circuit, it is important that the cause be investigated and repaired and that the condition of the installed equipment be investigated. A circuit breaker may require replacement just as any other switching device, wiring or electrical equipment in the circuit that has been exposed to a short circuit. Questionable circuit breakers must be replaced for continued, dependable circuit protection.[7]

The condition of the circuit breaker must be known to ensure that it functions properly and safely before it is put back into service.

Low-voltage Power circuit Breakers

Low-voltage power circuit breakers are manufactured under a high degree of quality control, of the best materials available, and with a high degree of tooling for operational accuracy. Manufacturer's tests show these circuit breakers to have durability beyond the minimum standards requirements. All of these factors give these circuit breakers a very high reliability rating when proper maintenance is performed per the manufacturer's instructions. However, because of the varying application conditions and the dependence placed upon them for protection of electrical systems and equipment as well as the assurance of service continuity, inspections and maintenance checks must be made on a regular basis. Several studies have shown that low-voltage power circuit breakers, which were not maintained within a 5 year period, have a 50 percent failure rate.

Maintenance of these breakers will generally consist of keeping them clean and properly lubricated. The frequency of maintenance will depend to some extent on the cleanliness and environmental conditions of the surrounding area. If there were very much dust, lint, moisture, or other foreign matter present then more frequent maintenance would be required.

Industry standards for and manufacturers of low-voltage power circuit breakers recommend a general inspection and lubrication after a specified number of operations or at least once per year, whichever comes first. Some manufacturers also recommend this same inspection and maintenance be performed after the first six months of service for a new circuit breaker, regardless of the number of operations. If the breaker remains open or closed for a long period of time, it is recommended that arrangements be made to open and close the breaker several times in succession. Environmental conditions would also play a major role in the scheduling of inspections and maintenance. If the initial inspection indicates that maintenance is not required at that time, the period may be extended to a more economical point. However, more frequent inspections and maintenance may be required if severe load conditions exist or if an inspection reveals heavy accumulations of dirt, moisture, or other foreign matter that might cause mechanical, insulation, or electrical failure. Mechanical failure would include an unintentional time delay in the circuit breakers tripping operation due to dry, dirty, or corroded pivot points, or by hardened or sticky lubricant in the moving parts of the operating mechanism. The manufacturer's instructions must be followed in order to minimize the risk of any unintentional time delay.

1. Shunt Trip Device	10. Ratchet Wheel	19. Reset Spring
2. Trip Shaft	11. Hold Pawl	20. Closing Spring Anchor
3. Roller Constraining Link	12. Drive Plate	21. Pole Shaft
4. Trip Latch	13. Emergency Charge Pawl	22. Motor
5. Close Cam	14. Oscillator	23. Emergency Charge Handle
6. Stop Roller	15. Crank Shaft	24. Motor Crank and Handle
7. Spring Release Latch	16. Emergency Charge Device	25. Moving Contact Assembly
8. Spring Release Device	17. Crank Arm	26. Insulating Link
9. Oscillator Pawl	18. Closing Spring	27. Main Drive Link

96-0077

FIGURE 5.8   Power operated mechanism for a low-voltage power circuit breaker.

Figure 5.8 provides an illustration of the operating mechanism for such a circuit brekaer and the numerous points where lubrication would be required and where dirt, moisture, corrosion, or other foreign matter could accumulate causing a time delay in, or complete failure of, the circuit breaker operation.

Medium voltage circuit breakers

Most of the inspection and maintenance requirements for low-voltage power circuit breakers also apply to medium-voltage power circuit breakers. Manufacturers recommend that these breakers be removed from service and inspected at least once a year. They also state that the number and severity of interruptions may indicate the need for more frequent maintenance checks. Always follow the manufacturer's instructions because every breaker is different. Fig. 5.9 and 5.10 illustrate two types of operating mechanisms for medium-voltage power circuit breakers. These mechanisms are typical of the types used for air, vacuum, oil, and SF-6 (sulfur-hexaflouride) circuit breakers. As can be seen in these figures, there are many points that would require cleaning and lubrication in order to function properly.

1. TRIPPING MAGNET
2. TRIPPING LATCH
3. CENTER POLE UNIT LEVER
4. MAIN CONTACT
   OPERATING ROD
5. MAIN LINK
6. CLOSING CAM
   FOLLOWING ROLLER
7. CLOSING CAM
8. CRANK SHAFT
9. TRIPPING CAM
10. TRIPPING TRIGGER
11. TRIPPING CAM
    CONNECTING LINK
12. FRONT PANEL
13. MECH BACK PLATE
14. BUMPER
15. DOLLY BRACKET
16. TRIPPING CAM
    ADJUSTING SCREW
17. LOCKING NUT
18. TRIP LATCH ROLLER

96-0284a

FIGURE 5.9 Operating mechanism for a medium-voltage air circuit breaker.

FIGURE 5.10   Solenoid operated mechanism for a medium-voltage circuit breaker.

Protective Relays

Relays must continuously monitor complex power circuit conditions, such as current and voltage magnitudes, phase angle relationships, direction of power flow, and frequency. When an intolerable circuit condition, such as a short circuit (or fault) is detected, the relay responds and closes its contacts, and the abnormal portion of the circuit is de-energized via the circuit breaker. The ultimate goal of protective relaying is to disconnect a faulty system element as quickly as possible. Sensitivity and selectivity are essential to ensure that the proper circuit breakers are tripped at the proper speed to clear the fault, minimize damage to equipment, and to reduce the hazards to personnel.

A clear understanding of the possible causes of primary relaying failure is necessary for a better appreciation of the practices involved in backup relaying. One of several things may happen to prevent primary relaying from disconnecting a power system fault:

- Current or voltage supplies to the relays are incorrect.
- DC tripping voltage supply is low or absent.
- Protective relay malfunctions.
- Tripping circuit or breaker mechanism hangs up.

There are two groups of protective relays: primary and backup. Primary relaying is the so-called first line of defense, and backup relaying is sometimes considered to be a subordinate type of protection. Many companies, however, prefer to supply two lines of relaying and do not think of them as primary and backup. Figure 5.11 shows the zones of protection for a primary relay system. Circuit breakers are found in the connections to

Generator Zone

Low Voltage Switchgear Zone

Transformer Zone

High Voltage Switchgear With Tiebreaker Zone

Transmission Line Zone

High Voltage Switchgear Zone

95-00142-1

FIGURE 5.11 Zones of protection for a primary protective relaying system.

each power system element. This provision makes it possible to disconnect only the faulty part of the system. Each element of the system has zones of protection surrounding the element. A fault within the given zone should cause the tripping of all circuit breakers within that zone and no tripping of breakers outside that zone. Adjacent zones of protection can overlap, and in fact, this practice is preferred, because for failures anywhere in the zone, except in the overlap region, the minimum number of circuit breakers are tripped.

In addition, if faults occur in the overlap region, several breakers respond and isolate the sections from the power system. Backup relaying is generally used only for protection against short circuits. Since most power system failures are caused by short circuits, short circuit primary relaying is called on more often than most other types. Therefore, short circuit primary relaying is more likely to fail.

Voltage and current transformers play a vital role in the power protection scheme. These transformers are used to isolate and protect both people and devices from high voltage, and to allow current carrying devices such as relays, meters, and other instruments to have a reasonable amount of insulation. It should be clearly understood that the performance of a relay is only as good as the voltage and current transformers connected to it. A basic understanding of the operating characteristics, application, and function of instrument transformers is essential to the certified relay technician.

Some overcurrent relays are equipped with an instantaneous overcurrent unit, which operates when the current reaches its minimum pickup point, Fig. 5.12. An instantaneous unit is a relay having no intentional time delay. Should an overcurrent of sufficient magnitude be applied to the relay, both the induction disc and the instantaneous unit will operate.

However, the instantaneous unit will trip the circuit breaker, since it has no intentional time delay.

The instantaneous trip unit is a small, AC-operated clapper device. A magnetic armature, to which leaf-spring-mounted contacts are attached, is attracted to the magnetic core upon energization. When the instantaneous unit closes, the moving contacts bridge two stationary contacts and complete the trip circuit. The core screw, accessible from the top of the unit, provides the adjustable pickup range. Newer designs also feature tapped coils to allow even greater ranges of adjustment.

The instantaneous unit, like the ICS unit, is equipped with an indicator target. This indication shows that the relay has operated. It is important to know which relay has operated, and no relay target should be reset without the supervisor's knowledge and permission.

As can be seen, several things can go wrong that would prevent the instantaneous unit from operating properly. These things include an open

FIGURE 5.12 Electromechanical instantaneous unit.

or shunted current transformer, open coil, or dirty contacts. Protective relays, like circuit breakers, require periodic inspection, maintenance, and testing to function properly. Most manufacturers recommend that periodic inspections and maintenance be performed at intervals of one to two years. The intervals between periodic inspection and maintenance will vary depending upon environment, type of relay, and the user's experience with periodic testing.

The periodic inspections, maintenance, and testing are intended to ensure that the protective relays are functioning properly and have not deviated from the design settings. If deviations are found, the relay must be retested and serviced as described in the manufacturer's instructions.

Rotating Equipment

The following maintenance requirements are taken verbatim from NFPA 70E.[3]

Terminal Boxes.   Terminal chambers, enclosures, and terminal boxes shall be maintained to guard against accidental contact with live parts and other electrical hazards.

Guards, Barriers, and Access Plates.   Guards, barriers, and access plates shall be maintained to prevent employees from contacting moving or energized parts.

Portable Electric Tools and Equipment

The maintenance requirements for these types of devices focus primarily on the attachment plugs, receptacles, cover plates, flexible cords, and cord connectors. These should be maintained to ensure that:

1. There are no breaks, damage, or cracks which may expose energized live parts.
2. All cover plates are in place, undamaged, and secure.
3. All terminations are complete, and have no exposed strands or loose terminals.

4. That all pins and connectors in the plugs are intact, in place, and firm.

5. That the polarity is correct, that is, the hot, neutral, and ground leads are all connected to their correct terminations.

Personal Safety and Protective Equipment

The maintenance, inspection, and testing requirements for this type of equipment are all covered in detail in Chap. 3 of this handbook.

CONCLUSION

Electrical maintenance is critical to the safety of all personnel who must work in the vicinity of electrical equipment. Whether justified on a safety or an economic basis, all facilities should have a comprehensive electrical maintenance program.

Although the topic is too detailed for complete coverage in this handbook, there are a number of excellent publications and consultants available to help with the establishment and continuation of a good electrical maintenance program.

REFERENCES

1. International Electrical Testing Association (NETA), *Maintenance Testing Specifications for Electrical Power Distribution Equipment and Systems*, Morrison, CO, 2001.

2. NFPA 70B, *Recommended Practice for Electrical Equipment Maintenance*, National Fire Protection Association, Quincy, MA, 2002.

3. NFPA 70E, *Standard for Electrical Safety Requirements for Employee Workplaces*, National Fire Protection Association, Quincy, MA, 2004.

4. NFPA 70, *National Electrical Code*, National Fire Protection Association, Quincy, MA, 2005.

5. F.S. Nowlan and H.F. Heap, *Reliability-Centered Maintenance*, United Airlines, for Office of Assistant Secretary of Defense, Washington, D.C., December 29, 1978.

6. John Cadick, *The Eight Step Maintenance Program*, Cadick Corporation, Dallas, Texas, 2000–2005.

7. National Equipment Manufacturer's Association (NEMA), Vince A. Baclawski, Power Distribution Products, NEMA, *EC&M magazine*, p. 10, January 1995.

CHAPTER 6

REGULATORY AND LEGAL SAFETY REQUIREMENTS AND STANDARDS

INTRODUCTION

The modern trend in electrical safety is toward more and more individual responsibility as employees are being held increasingly responsible for their actions. For individuals to successfully fulfill their responsibilities, they must be aware of the rules that apply to their performance. This means that company safety rules; national and international standards; and local, state, and federal laws must be part of an employee's knowledge base.

At the same time, the Occupational Safety and Health Administration (OSHA) has become increasingly severe in its enforcement of safety standards. Employers must redouble their efforts to interpret and implement the best possible electrical safety program. To do this, employers must become familiar with all of the various standards—both regulatory and consensus—so that they can comply with them.

Electrical safety standards and/or requirements are produced by a variety of organizations. Some of the standards are voluntary and some are local or federal law. Whether voluntary or mandatory, these standards provide guidance for the proper way to work on or around electric energy. Much of the material covered in other parts of this handbook is drawn from the standards that will be covered in this chapter.

Since electrical standards are periodically reviewed and updated, the information included in this chapter is necessarily general. Always refer to the most recent edition of each of these standards when current information is required. Note that not all standards are covered in this chapter. The reader should refer to industry-specific literature for their own industries.

Because of the very critical role that standards organizations play, coverage of them is also included in this chapter. The organizational structure, the standard-making procedure, and the historical development of each standards organization are covered.

THE REGULATORY BODIES

The American National Standards Institute (ANSI)

Overview. Founded in 1918, the ANSI is a private, nonprofit, membership organization. The ANSI coordinates the United States voluntary consensus standards systems and approves American national standards. The ANSI membership includes over 1000 companies; hundreds of professional, technical, trade, labor, and consumer organizations; and dozens of government agencies. The ANSI's primary purpose is to ensure that a single set of noncon-flicting American national standards is developed by ANSI-accredited standards developers and that all interests concerned have the opportunity to participate in the development process. Since all interested parties are able to participate in the standard-making process, ANSI standards are true consensus standards.

Functions.   The ANSI serves five extremely important functions for industry:

- *Coordinates voluntary standards activities.*   The ANSI assists standards developers and standards users from the private sector and government to reach agreement on the need for standards and to establish priorities. It accredits qualified organizations to develop standards, helps them avoid duplication of effort, and offers a neutral forum for resolving differences. The ANSI's Executive Standards Council and standards boards spearhead this function.

- *Approves American national standards.*   The ANSI administers the only recognized US system for establishing American national standards. The ANSI approval procedures ensure that any interested party can participate in a standard's development or comment on its provisions. These requirements for due process guarantee that American national standards earn high levels of confidence and credibility and achieve broad acceptance.

- *Represents US interests in international standardization.*   The ANSI is the sole US representative to two of the major, nontreaty standards organizations: the International Organization for Standardization (ISO) and the International Electrotechnical Commission (IEC). The ISO, founded in 1946, develops, coordinates, and promotes international standards with national standards organizations from over 70 countries around the world. The IEC, founded in 1906, develops and promotes electrotechnical standards with national committees from over 40 countries. The ANSI participates in almost the entire technical program of both organizations and administers many key committees and subgroups.

- *Provides information on and access to world standards.*   The ANSI maintains copies of approved and draft American national standards, ISO and IEC standards, proposals of regional groups tied to the European Community, and the specifications of 90 national standards organizations that belong to the ISO. These standards are available to interested parties.

Member Companies.   The ANSI is a membership organization and draws its principle revenue from the company members. Company members help to govern the ANSI by determining policies, procedures, and long-range plans for standards development. They identify commercial and industrial needs for standards, both nationally and internationally. Based on these needs, they encourage ANSI to initiate new activities for standards development and accelerate improvement of existing standards.

Standards Developers.   Standards developers are primarily national trade, technical, professional, consumer, and labor organizations. They voluntarily submit standards to the ANSI for recognition as national consensus standards. Many of these organizations are also members of the ANSI and are represented on the boards, councils, and committees that help govern the ANSI and coordinate national and international standardization activities.

Government Agencies.   The ANSI works with governmental agencies in the same way they work with other organizations. That is, government agencies are ANSI members, and government representatives serve on ANSI boards and councils.

Coordinating committees, formed with the OSHA and the Consumer Product Safety Commission, provide forums to discuss activities affecting the voluntary and governmental standards communities.

The Institute of Electrical and Electronic Engineers (IEEE)

Overview.   The IEEE is the world's largest professional technical society, with over 350,000 members from approximately 150 countries around the world. The American Institute of Electrical Engineers (AIEE) was founded in 1884, and the Institute of Radio Engineers (IRE) was founded in 1912. These two organizations merged in 1963 to form the IEEE.

The IEEE is a membership society composed of electrical engineers and technicians. The interest areas of IEEE members covers the gamut of technical areas including computer engineering, biomedical technology, telecommunications, electric power, aerospace, and consumer products.

Functions.   Among the IEEE's primary activities are technical publishing, technical conferences, and development of technical standards. The IEEE is responsible for approximately 30 percent of the world's published literature in electrotechnology. Ranging from published papers, to textbooks, to technical journals, IEEE literature is typically on the cutting edge of technology.

The IEEE is divided into a variety of technical societies with a broad spectrum of technical interests. Primarily through these societies the IEEE holds more than 300 meetings annually which allow engineers and technicians to exchange information. The IEEE also develops and publishes voluntary technical standards which regulate a variety of electrical technical activities.

National Fire Protection Association (NFPA)

Overview.   The NFPA was founded in 1896 for the express purpose of protecting people, property, and the environment from the effects of fire. To fulfill its purposes, the NFPA focuses its efforts into intensive technical activities and extensive educational programs. The NFPA is a nonprofit organization and has a membership of approximately 67,000 from 85 nations.

The NFPA has an extremely well recognized and respected set of voluntary consensus codes and standards. These codes and standards, hundreds of them, are used extensively by the NFPA's network of fire service, industrial, health care, building code, and other professional members. The NFPA technical committees are staffed by thousands of volunteers from a balanced cross section of affected interests.

The NFPA is responsible for the National Electrical Code, a document which is one of the most recognized electrical safety standards in the world.

Functions.   The NFPA is involved in a wide variety of functions and activities.

- The Learn Not to Burn Curriculum is for children from kindergarten through eighth grade. This program, introduced by the NFPA in 1979, has a documented record of saving hundreds of lives because of lessons remembered by children.
- Fire Prevention Week is observed each year on October 9—the anniversary of the great Chicago fire of 1871. The first Fire Prevention Week was proclaimed by President Warren G. Harding in 1922. The NFPA has sponsored the week ever since.
- The NFPA maintains one of the major fire safety literature collections.
- NFPA conducts investigations of major fires. These fires are investigated as a result of their technical or educational interest.

The NFPA Standards-Making System. The NFPA and its standards-making efforts grew from a meeting of a small group in Boston in 1895. That group was meeting to discuss sprinkler installation standards in the Boston area. From that initial standard has grown the NFPA standards-making system of today.

The standards-making procedure starts with the submission of a request for a new standard by an interested party. The request is reviewed by the NFPA Standards Council who publishes an announcement in the membership newsletter, *Fire News.* The announcement acknowledges the request for a standard and asks for comments on the need for the project, information on organizations that may be active in the subject matter of the proposed project, a listing of resource material that is available, and an indication of who would be willing to participate in the project if it is approved.

If the council determines a need for the proposed project, it either assigns the project to an existing technical committee or a new committee whose membership reflects a fair balance of concerned interests. Members of technical committees are appointed by the standards council and include volunteer experts representing the government, educational institutions, business, insurance companies, industry, and consumers. Each committee member is classified according to his or her interest, and each committee is structured so that no more than one-third of the membership is from a single interest.

Once the technical committee has been established, the NFPA issues public notices announcing the committee's next meeting date and calling for specific proposals from interested persons. Anyone may submit a proposal for text to be included in a new document or added to an existing document. The committee then meets to consider all proposals received and drafts a proposed document or amendments to an existing document. Submitters may address the committee on their proposals at the meeting. Letter ballot approval by at least two-thirds of all committee members eligible to vote is required to approve a proposal for inclusion in the document.

After suitable procedures to allow for objections, and a vote of the membership, the document is presented to the standards council for final issuance. If the standards council issues the standard, based on the vote of the membership, it is published in pamphlet form and is also included in the appropriate volume of NFPA's National Fire Codes. If the standard is to be an American national standard, it is submitted to the ANSI.

American Society for Testing and Materials (ASTM)

Overview. The ASTM was organized in 1898 and has grown to an extremely large, voluntary standards development organization. The organization develops standards in a variety of areas including materials, products, systems, and services. The ASTM has over 129 standards-writing committees which write standards in such diverse areas as metals, paints, plastics, textiles, petroleum, construction, energy, the environment, consumer products, medical services and devices, computerized systems, and electronics.

The ASTM headquarters has no technical research or testing facilities because such work is done voluntarily by over 32,000 technically qualified ASTM members located throughout the world. The ASTM publishes approximately 10,000 standards each year in the 73 volumes of the *Annual Book of ASTM Standards.*

Function. The ASTM develops six principal types of full consensus standards including

- *Standard test method.* This is a definitive procedure for the identification, measurement, and evaluation of one or more qualities, characteristics, or properties of a material, product, system, or service that produces a test result.

- *Standard specification.*   This is a precise statement of a set of requirements to be satisfied by a material, product, system, or service that also indicates the procedures for determining whether each of the requirements is satisfied.

- *Standard practice.*   A definitive procedure for performing one or more specific operations or functions that does not produce a test result.

- *Standard terminology.*   A document comprising terms, definitions, descriptions of terms, and explanations of symbols, abbreviations, or acronyms.

- *Standard guide.*   A series of options or instructions that do not recommend a specific course of action.

- *Standard classification.*   A systematic arrangement or division of materials, products, systems, or services into groups based on similar characteristics such as origin, composition, properties, or use.

The ASTM Standards-Making System.   Like most consensus standards organizations, the ASTM begins its standards-making activities when a need is identified. Task group members prepare a draft standard, which is reviewed by its parent subcommittee through a letter ballot. After the subcommittee approves the document, it is submitted to a main committee letter ballot. Once approved at the main committee level, the document is submitted for balloting to the ASTM. All negative votes cast during the balloting process, which must include a written explanation of the voter's objections, must be fully considered before the document can be submitted to the next level in the process. Final approval of a standard depends on concurrence by the ASTM Committee on Standards that proper procedures were followed and due process was achieved.

American Society of Safety Engineers (ASSE)

Overview.   The ASSE was founded in 1911. It is a membership organization composed of safety professionals throughout the world. The 33,000 members manage, supervise, and consult on safety, health, and environmental issues in industry, government, and education.

The ASSE's mission is to enhance the status and promote the advancement of the safety profession and to foster the technical, scientific, managerial, and ethical knowledge, skills, and competency of safety professionals.

The ASSE has 32,600 members who manage, supervise, and consult on safety, health, and environmental issues in industry, insurance, government, and education. Safety professionals help prevent accidents, injuries, and occupational diseases; create safe environments for work and leisure; and develop safe products for use in all areas of human activity.

ASSE is guided by a 15-member board of directors, which consists of 8 regional vice presidents; 3 council vice presidents; and the Society president, president-elect, senior vice president, vice president of finance, and executive director. The board is guided in decision making by five standing committees and four councils.

Function.   The ASSE meets its responsibilities by providing continuing education, membership services, standards-making activities, peer recognition awards, marketing, and safety public relations.

   Continuing Education.   The ASSE sponsors professional development seminars, publications, conference proceedings, training packages, computer resources, and audiovisual aids.

   Membership Services.   The ASSE provides a variety of related membership services including professional development seminars, technical publications, and leadership conferences.

ASSE Standards. The ASSE serves as the secretariat for a number of American national standards. Those of special interest to electrical safety include Eye and Face Protection (Z87.1), Safety Requirements for Confined Spaces (Z117.1), and Safety Requirements for Ladders (A14.1 to A14.8).

The Occupational Safety and Health Administration (OSHA)

Standards issued by the OSHA differ from other standards in one very significant aspect— they are enforceable under federal law. The following sections outline some of the more important facts about the OSHA and its standards. You should contact your local OSHA office if more detailed information is required.

The Occupational Safety and Health Act of 1970. Part of the material in the following sections was drawn from the government publication OSHA 2056 entitled *All About OSHA.*

In 1970, the United States was faced with the following statistics:

• Job-related accidents accounted for more than 14,000 worker deaths per year.

• Nearly 2.5 million workers were disabled.

• Ten times as many person-days were lost from job-related disabilities than from strikes.

• Estimated new cases of occupational diseases totaled 300,000.

To address these problems, Congress passed the Occupational Safety and Health Act of 1970 (the Act). The execution of the Act is the responsibility of the Department of Labor under the Secretary of Labor. The Act created the OSHA to continually create, review, and redefine specific standards and practices.

Coverage under the Act extends to all employers and their employees in the 50 states, the District of Columbia, Puerto Rico, and all other territories under federal government jurisdiction. The Act exempts three important categories:

• Self-employed persons

• Farms at which only immediate members of the farm employer's family are employed

• Working conditions regulated by other federal agencies under other federal statutes

The Act has special provisions for some employees and employers. State and local governments and the federal government have special consideration under the Act. Employers and employees should contact their local OSHA office to determine what their obligations are.

The OSHA's Purpose. The OSHA is the administration that is given the principle responsibilities for implementation of the Act. A few of the OSHA's responsibilities include

• Encourage employers and employees to reduce workplace hazards and to implement new or improved existing safety and health programs.

• Establish separate but dependent responsibilities and rights for employers and employees with respect to achieving safe and healthful working conditions.

• Set and enforce mandatory occupational safety and health standards applicable to businesses affecting interstate commerce.

• Maintain a reporting and record-keeping system to monitor job-related illnesses and injuries.

- Establish training programs to increase the number and competence of occupational safety and health personnel.
- Provide for the development, analysis, evaluation, and approval of state occupational safety and health programs.

The OSHA is a branch of the Department of Labor and is responsible to the Secretary of Labor.

Responsibilities and Rights of Employers.   Employers have the principal responsibility to guarantee that all OSHA rules and regulations are properly and promptly administered in the workplace. Employers who do not carry out their responsibilities can be heavily fined. The following list identifies some of the more significant employer responsibilities. Employers must

- Meet their general responsibility to provide a workplace free from recognized hazards that are causing or are likely to cause death or serious physical harm to employees.
- Be familiar with mandatory OSHA standards and make copies available to employees for review upon request.
- Inform all employees about the OSHA.
- Examine workplace conditions to make sure they conform to applicable standards.
- Minimize or reduce hazards.
- Make sure that employees have, use, and maintain safe tools and equipment. These tools and equipment include appropriate personal protective equipment such as flash suits, rubber gloves, and hard hats.
- Use color codes, posters, labels, or signs when needed to warn employees of potential hazards.
- Establish or update operating procedures and communicate them so that employees follow safety and health requirements.
- Provide medical examinations when required by OSHA standards.
- Provide training required by OSHA standards. For example, the Safety Related Work Practices Rule and the Control of Hazardous Energy Source Rule both require a certain amount of employee training for complete implementation.
- Report to the nearest OSHA office within 48 h any fatal accident or one that results in the hospitalization of five or more employees.
- Keep OSHA-required records of work-related injuries and illnesses, and post a copy of the totals from the last page of OSHA No. 200 during the entire month of February of each year. *Note:* This applies only to employers with 11 or more employees and employers who are chosen by the Bureau of Labor Statistics.
- Post the OSHA poster (OSHA 2203) informing employees of their rights and responsibilities at a prominent location in the workplace. In states with OSHA-approved state job safety and health programs, the state's equivalent poster and/or the OSHA poster may be required.
- Provide employees, former employees, and their representatives access to the Log and Summary of Occupational Injuries and Illnesses (OSHA No. 200) at a reasonable time and in a reasonable manner.
- Provide access to employee medical records and exposure records to employees or their authorized representatives.
- Cooperate with the OSHA compliance officer by furnishing names of authorized employee representatives who may be asked to accompany the compliance officer during an inspection.

(If none, the compliance officer will consult with a reasonable number of employees concerning safety and health in the workplace.)

- Not discriminate against employees who properly exercise their rights under the Act.
- Put OSHA citations at or near the work site involved.
- Abate cited violations within the prescribed period.

Employers also have rights under the OSHA act. The following items list some of the more important employer's rights. Employers have the right to

- Seek advice and off-site consultation as needed by writing, calling, or visiting the nearest OSHA office. The OSHA will not inspect merely because an employer requests assistance.
- Be active in their industry association's involvement in safety and health.
- Request and receive proper identification of the OSHA compliance officer prior to inspection.
- Be advised by the compliance officer of the reason for an inspection.
- Have an opening and closing conference with the compliance officer.
- Accompany the compliance officer on the inspection.
- File a Notice of Contest with the OSHA area director within 15 working days or receipt of a notice of citation and proposed penalty.
- Apply to the OSHA for a temporary variance from a standard if unable to comply because of the unavailability of materials, equipment, or personnel needed to make necessary changes within the required time.
- Apply to the OSHA for a permanent variance from a standard if they can furnish proof that their facilities or method of operation provide employee protection at least as effective as that required by the standard.
- Take an active role in developing safety and health standards through participation in OSHA standards and advisory committees, through nationally recognized standards-setting organizations, and through evidence and views presented in writing or at hearings.
- Be assured of the confidentiality of any trade secrets observed by an OSHA compliance officer during inspection.
- Submit a written request to the National Institute of Occupational Safety and Health (NIOSH) for information on whether any substance in your workplace has potentially toxic effects in the concentrations being used.

Responsibilities and Rights of Employees. The Act was designed to establish workplace rules and regulations which will enhance the health and safety of employees. Because of this, the employers are considered to have principal responsibility for the implementation of OSHA rules. Employees, therefore, are not cited for violations. In spite of this, employees still have responsibilities under the Act. Employees must

- Read the OSHA poster at the job site.
- Comply with all applicable OSHA standards.
- Follow all employer safety and health rules and regulations, and wear or use prescribed protective equipment while engaged in work.
- Report hazardous conditions to the supervisor.
- Report any job-related injury or illness to the employer.
- Promptly seek treatment from job-related injuries or illnesses.

- Cooperate with the OSHA compliance officer conducting an inspection if he or she inquires about safety and health conditions in your workplace.
- Exercise their rights under the Act in a responsible manner.

  Employees also have significant rights under the Act. Employees have a right to

- Seek safety and health on the job without fear of punishment.
- Complain to an employer, union, the OSHA, or any other government agency about job safety and health hazards.
- File safety or health grievances.
- Participate on a workplace safety and health committee or in union activities concerning job safety and health.
- Participate in OSHA inspections, conferences, hearings, or other OSHA-related activities.

OSHA'S Electrical Safety Standards.   OSHA safety standards are published in the Code of Federal Regulations, Title 29, Subtitle B, Chapter XVII. The OSHA has issued or will issue electrical safety standards in at least four categories:

1. Design and installation safety
2. Safety-related work practices
3. Safety-related maintenance requirements
4. Special equipment

In addition, the OSHA has issued or will issue standards regulating a variety of related areas such as Control of Hazardous Energy. Electrical safety standards for general industry are found in Part 1910, Subpart S, which includes Paragraphs 1910-301 through 1910-399. Electrical safety standards for employees in the construction industry are found in Part 1926, Subpart K, which includes Paragraphs 1926-400 through 1926-449. Table 6.1 identifies the various locations for the standards.

Note that by its very nature, Table 6.1 is time-sensitive. Be certain to check the current edition of the OSHA standard for complete information. These standards can be procured from your local OSHA office. The various OSHA standards will be covered in more detail later in this chapter.

OSHA Technical Consultation.   The OSHA will provide technical support and consultation at no charge. The consultation services are provided by state government agencies or universities employing professional safety and health consultants. Although intended for smaller employers with more hazardous operations, the consultation service is available to companies of all sizes.

The technical consultation starts when the employer requests the service and makes a commitment to correct any possible violations discovered by the inspector. The complete consultation involves six basic steps: the request, the opening conference, the walk-through inspection, the closing conference, the written report, and the corrective actions. Table 6.2 lists and describes each of these six steps.

Employers are exempted from inspections or fines by OSHA compliance personnel during the consultation process. In some cases, if employers can demonstrate that an effective safety and health program is in operation, they may be exempted from OSHA general schedule enforcement inspections. This exemption does not apply to complaint or accident investigations.

Voluntary Protection Programs.   Voluntary protection programs (VPP) (Table 6.3) are designed to extend worker protection beyond the minimums required by the OSHA standards. These three programs are designed to

TABLE 6.1 Partial Listing of Applicable OSHA Safety Standards and the Industries to Which They Apply

Title	Industry	Paragraphs	Comments
Design Safety Standards for Electrical Equipment	General Industry & Maritime	1910.302–1910.308	
Safety-Related Work Practices	General Industry & Maritime	1910.331–1910.335	Not implemented
Safety-Related Maintenance Requirements	General Industry & Maritime	1910.361–1910.380	Not implemented
Safety Requirements for Special Equipment	General Industry & Maritime	1910.381–1910.398	
Installation Safety Requirements	Construction Industry	1926.402–1926.408	
Safety-Related Work Practices	Construction Industry	1926.416–1926.417	
Safety-Related Maintenance and Environmental Considerations	Construction Industry	1926.431–1926.432	
Safety Requirements for Special Equipment	Construction Industry	1926.441–1926.448	Only batteries implemented
Power Generation, Transmission, and Distribution	Construction Industry	1926.950–1926.960	
Power Transmission and Distribution	Power Generation, Transmission, and Distribution	1910.269	
Communications Safety	Telecommunications	1910.268	

TABLE 6.2   Six-Step OSHA Technical Consultation Process

Employer requests assistance	The employer contacts the state agency responsible for OSHA consultation services. Each state has its own agency. Phone numbers may be obtained by contacting the local OSHA office or requesting OSHA bulletins #2056 and #3047. Employer must make initial commitment to correct any observed discrepancies.
Consultant holds on-site opening conference	In this conference the ground rules for the consultation are thoroughly explained to the employer, the plan for the walk-through visit is determined, and any preliminary questions are answered.
Walk-through inspection	The employer representative and the consultant perform a walk-through inspection of the workplace. Specific hazards are noted and an examination of the aspects of the employer's safety and health program which relate to the scope of the visit are observed.
Consultant holds an on-site closing conference	The closing conference is used to apprise the employer of any extremely serious violations which were observed. Generally only those hazards which represent an immediate and serious hazard to employees are itemized.
Written report	A written report is issued by the consultant in which the hazards and/or violations of OSHA regulations are delineated.
Corrective actions	The employer corrects those problems which were outlined by the consultant.

1. Recognize outstanding achievements of those who have successfully incorporated comprehensive safety and health programs into their total management system.
2. Motivate others to achieve excellent safety and health results in the same outstanding way.
3. Establish a relationship between employers, employees, and the OSHA that is based on cooperation rather than coercion or confrontation.

Voluntary protection programs are available in all states under federal jurisdiction, and some state OSHA agencies have similar programs. Employers should contact their local OSHA offices for details on the VPPs. The OSHA reevaluates VPPs on a regular basis—annually for Merit and Demonstration Programs and triennially for Star Programs.

Training and Education.   The OSHA can provide training and education in one of three basic ways. First, OSHA's area offices are full-service centers that provide a wealth of information in the form of pamphlets, brochures, publications, audiovisual aids, technical advice, and availability for speaking engagements.

Second, the OSHA Training Institute in Des Plaines, Illinois, provides both basic and advanced training in areas such as electrical hazards, machine guarding, ventilation, and ergonomics. The institute is a full-service training facility with all necessary classrooms, laboratories, library, and an audiovisual unit. Training courses at the institute are open to federal and state compliance officers; state consultants; federal agency personnel; and private-sector employers, employees, and their representatives.

Finally, the OSHA provides funds in the form of grants to nonprofit organizations to conduct workplace training and education in subjects where the OSHA believes there is a lack of current workplace training.

State OSHA Organizations.   The Occupational Safety and Health Act encourages states to develop and operate state job safety and health plans. These plans are operated under OSHA guidance. The OSHA funds up to 50 percent of the state program's operating costs.

TABLE 6.3   OSHA Voluntary Protection Programs

Star program	This is the most demanding and prestigious of the VPPs. It is open to any employer in any industry who has successfully managed a comprehensive safety and health program to reduce injury rates below the national average for the industry. Specific requirements for the program include: management commitment and employee participation; a high-quality work site analysis program; hazard prevention and control programs; and comprehensive safety and health training for all employees. These requirements must all be in place and operating effectively.
Merit program	This is a stepping stone to the Star program. An employer with a basic safety and health program built around the Star requirements who is committed to improving the company's program and who has the resources to do so within a specified period of time may work with the OSHA to meet Star requirements.
Demonstration program	This VPP is for companies that provide Star-quality worker protection in industries where certain Star requirements may not be appropriate or effective. It allows the OSHA both the opportunity to recognize outstanding safety and health programs that would otherwise be unreached by VPP and to determine if general Star requirements can be changed to include these companies as Star participants.

The requirements for development of a state program are beyond the scope of this book. The entire setup process is monitored by the OSHA with milestones and performance requirements established from the outset. At the end of the process, if the state program is operating at least as effectively as the federal OSHA program and other administrative requirements are met, the OSHA issues final approval and federal authority ends in the areas over which the state has jurisdiction.

Other Electrical Safety Organizations

There are many other organizations that are directly or indirectly involved with electrical safety issues. Table 6.4 identifies a number of groups, including those discussed in this chapter. Note that the World Wide Web is a rapidly changing structure, and that URLs change on a daily basis.

THE NATIONAL ELECTRICAL SAFETY CODE (NESC)—ANSI C-2

General Description

The NESC was first developed in 1913 by the National Bureau of Standards as a consensus standard and remains so in the present. The NESC is an American national standard published by the IEEE, the secretariat for the NESC. It is intended to provide practical rules for safeguarding personnel during the installation, operation, or maintenance of electric supply and communications lines and associated equipment.

The code has three general rules as follows:

1. All electric supply and communication lines and equipment shall be designed, constructed, operated, and maintained to meet its requirements.

TABLE 6.4 Electrical Safety-Related Organizations and Their URLs

Organization	URL
American Burn Association	www.ameriburn.org
American National Standards Institute (ANSI)	www.ansi.org
American Society of Safety Engineers (ASSE)	www.asse.org
American Society for Testing and Materials (ASTM)	www.astm.org
International Electrotechnical Commission (IEC)	www.iec.ch
International Organization for Standardization (ISO)	www.iso.ch
Institute of Electrical and Electronic Engineers (IEEE)	www.ieee.org
National Electrical Safety Foundations (NESF)	www.nesf.org
International Electrical Testing Association (NETA)	www.netaworld.org
National Fire Protection Association (NFPA)	www.nfpa.org
Occupational Safety and Health Administration (OSHA)	www.osha.gov
Petroleum and Chemical Industry Committee (PCIC)	www.ieee-pcic.org
IAS Electrical Safety Workshop	www.ieee-pcic.org/safety1/esw.htm
National Electrical Manufacturer's Association	www.nema.org
Professional Electrical Apparatus Recyclers League (PEARL)	www.pearl1.org
American Academy of Forensic Sciences (AAFS)	www.aafs.org
The Canadian Society of Forensic Science (CSFS)	www.csfs.ca
American Society of Training and Development	www.astd.org

2. The utilities, authorized contractors, or other entities, as applicable, performing design, construction, operation, or maintenance tasks for electric supply or communication lines or equipment covered by the NESC are held responsible for meeting the applicable requirements.

3. For all particulars not specified in the NESC, construction and maintenance should be done in accordance with accepted good practice for the given local conditions.

Industries and Facilities Covered

The NESC rules cover supply and communication lines, equipment, and associated work practices used by both public and private electric supply, communications, railway, or similar utilities. The rules also cover similar systems which are under the control of qualified persons. This means that facilities such as cogeneration plants, industrial complexes, or utility interactive systems are covered by the NESC.

Note that the NESC specifically states it does not cover installations in mines, ships, railway rolling equipment, aircraft, automotive equipment, or utilization wiring except in certain cases. It also specifies that building utilization requirements are covered by the National Electrical Code, ANSI/NFPA 70.

Technical/Safety Items Covered

The NESC covers five major areas in numerous parts. The following sections describe each of the major areas of coverage and list the type of information included in each. The information provided in the following sections is intended to be general information only. Always refer to the most recent edition of the NESC for specific information.

Grounding Methods for Electric Supply and Communications Facilities.   This section covers the methods to implement protective grounding for electric supply and communications facilities. The information is technical in nature and includes topics such as the point of connection, the composition of the grounding conductor, the means of connection, grounding electrodes, and many other grounding topics.

Rules for Installation and Maintenance of Electric Supply Stations and Equipment. This section covers those parts of the electrical supply system and associated structural arrangements which are accessible only to qualified personnel. It also covers the conductors and equipment employed primarily for the utilization of electric power when such conductors and equipment are used by the utility in the exercise of its function as a utility. Protective arrangements in electric supply stations, illumination, installation and maintenance of equipment, rotating equipment, storage batteries, transformers and regulators, conductors, circuit breakers, switchgear and enclosed metal bus, and surge arrestors are included in this section.

Safety Rules for the Installation and Maintenance of Overhead Electric Supply and Communications Lines. This section applies to overhead lines, associated structural assemblies, and extension of such lines into buildings. Typical examples of topics included in this part are inspection and tests of lines and equipment, grounding of circuits, arrangement of switches, relations between various classes of equipment, joint use of structures, clearances, grades of construction, line insulation, and other such topics.

Safety Rules for the Installation and Maintenance of Underground Electric Supply and Communication Lines. This part is similar to the previous part except the equipment it includes is underground equipment. Topics include general requirements applying to underground lines, underground conduit systems, supply cable, underground structures, direct buried cable, risers, supply cable terminations, and installations in tunnels.

Rules for the Operation of Electric Supply and Communications Lines and Equipment. This final part of the NESC discusses rules for the operation of supply and communications lines and equipment. The first major section provides operational rules for both communications and supply employees. Personal general precautions, general operating routines, overhead line operating procedures, and underground line operating procedures are also included.

The second major section has additional rules which pertain to communications workers. Approaching energized conductors, joint-use structures, attendant on surface at joint-use manhole, and sheath continuity rules are included.

The final major part lists additional rules which pertain strictly to electrical supply employees. This section has six significant sections including energized conductors or parts, switching control procedures, working on energized lines, de-energizing equipment or lines, protective grounds, and live-line work.

THE NATIONAL ELECTRICAL CODE (NEC)— ANSI/NFPA 70

General Description

The NEC is arguably the oldest of all the various standards. It is an installation design standard that covers industrial, commercial, and residential electric utilization systems.

The NEC was first developed in 1897 by a variety of insurance, engineering, architectural, electrical, and other such interest groups. In 1911, the sponsorship of the Code was assumed by the NFPA. Later, the NEC was adopted as an American national standard by the ANSI. The NEC is only one of many fire safety codes that are published by the NFPA.

The NEC is unique among the other codes and standards in a very significant way—all or portions of it have been adopted as local law in many states, municipalities, cities, and other such areas. Many countries outside of the United States have adopted the NEC in its entirety, or in part, for their national installation codes. Note that both the NFPA and ANSI consider the NEC to be purely an advisory document.

The NEC has become a large document with a wealth of design, installation, and general technical information included. The principal objective of the NEC is to provide standardized, technically sound electrical installation standards to help minimize the possibility of electric fires.

Industries and Facilities Covered

The NEC covers four basic types of installations. Note that all of these installations are so-called utilization installations as opposed to transmission and distribution.

1. Installations of electric conductors and equipment within or on public and private buildings or other structures, including mobile homes, recreational vehicles, and floating buildings. This section also includes premises wiring for such facilities as yards, carnival lots, parking lots, and industrial substations.
2. Installations of conductors and equipment that connect to the supply of electricity.
3. Installations of other outside conductors and equipment on the premises.
4. Installations of optical fiber cable.

The NEC also lists several installations to which it does not apply including ships, railway rolling stock, underground mine installations, railway generation, communications equipment under the control of communications utilities, and certain facilities under the control of electric utilities. The utility facilities that are excluded are the communications, metering, generation, transformation, transmission, and distribution in buildings or substations used specifically for that purpose. The NEC does include utility office buildings and other such public and commercial structures.

Technical and Safety Items Included

The NEC includes technical specifications and requirements for virtually all electric lines and equipment. The Code has multiple chapters which include rules for wiring and protection, wiring methods and materials, equipment for general use, special occupancies, special equipment, special conditions, and communications systems.

Tables for conductor ampacities are included along with formulas for calculating the required size and number of conductors for special applications, as well as sizing requirements for protective devices. Grounding requirements including what, when, and how are included. Rules for interrupting capacities and other such information are also included. The NEC has become an engineering reference, an installation guide, a design standard, and a fire safety code all in one volume.

ELECTRICAL EQUIPMENT MAINTENANCE— ANSI/NFPA 70B

General Description

The Electrical Equipment Maintenance document (70B) is a companion publication to the NEC. Many interested parties had been requesting that safety-related maintenance requirements be added to the NEC. A special committee determined that the NEC, which is basically a design and installation standard, is not the correct vehicle for maintenance information.

In studying the need for this document, the committee determined that electrical safety concerns logically fall into four basic categories: product design standards, installation standards (covered by the NEC and the NESC), safety-related maintenance information, and usage instructions. Because of the high number of injuries and deaths related to improper or nonexistent equipment maintenance practices, the committee decided that an electric equipment maintenance document should be developed. To give the document as much prestige as possible, the decision was reached to make this new document a companion piece to the NEC. The final result, ANSI/NFPA 70B, contains recommended practice information for proper maintenance of electric equipment.

Industries and Facilities Covered

Document 70B covers maintenance of industrial-type electric systems and equipment. The covered systems and equipment are typical of those installed in commercial buildings, industrial plants, and large multifamily dwellings. Document 70B specifically states that it does not cover consumer appliances or home-type equipment. It also states that it is not intended to supersede manufacturer-recommended maintenance procedures.

Technical and Safety Items Covered

Document 70B includes 22 chapters and 10 appendices. One of the most important chapters develops a concise case for the implementation of proper preventive maintenance procedures. Subsequent chapters cover maintenance fundamentals and how to plan a preventive-maintenance program. Then 12 chapters provide specific, detailed maintenance techniques and expected results for switchgear, circuit breakers, cables, motor control centers, rotating equipment, wiring devices, and other such electric equipment.

One chapter details the specifics of electrical testing in areas such as insulation resistance, protective device testing, infrared, fault-gas analysis, and many other such modern testing procedures. Still other chapters cover the maintenance of equipment which is subject to long intervals between shutdowns and methods to be used for de-energizing equipment in such a way that maintenance personnel are protected.

The appendices include information on walk-through inspections, instruction techniques, symbols, diagrams, and a variety of recommended test sheets and forms to be used for maintenance programs.

STANDARD FOR ELECTRICAL SAFETY IN THE WORKPLACE—ANSI/NFPA 70E

General Description

Consensus standards are developed by personnel who are intimately familiar with the safety hazards of any given discipline; therefore, they will be accurate and extremely useful to the purpose for which they were intended. Because of this, when OSHA develops a safety regulation, they prefer to work with existing consensus standards. In 1976, the NFPA formed a committee to assist OSHA in the preparation of their electrical safety regulations.

The decision was made to develop a consensus standard which would embody four parts: Part I, Installation Safety Requirements; Part II, Safety-Related Work Practices; Part III,

Safety-Related Maintenance Requirements; and Part IV, Safety Requirements for Special Equipment. These four parts are identical to the four major rules which OSHA has developed or is developing. These four parts existed in this order through the 2000 edition of NFPA 70E. It was determined, by NFPA and the technical committee, that a more logical order and numbering system would be more user friendly throughout industry, therefore the 2004 edition was revised to be in accordance with the NEC Style Guide. It was also determined that the former Part I, based on the National Electrical Code, did not represent the most important part of the standard, therefore it was relocated. The NFPA 70E-2004 is now divided into four chapters: Chap. 1, Safety-Related Work Practices; Chap. 2, Safety-Related Maintenance Requirements; Chap. 3, Safety Requirements for Special Equipment; and Chap. 4, Installation Safety Requirements. The 2004 edition also includes thirteen Annexes: Annex A, Referenced Publications; Annex B, Informational Publications; Annex C, Limits of Approach; Annex D, Sample Calculations of Flash Protection Boundary; Annex E, Electrical Safety Program; Annex F, Hazard/Risk Evaluation Procedure; Annex G, Sample Lockout/Tagout Procedure; Annex H, Simplified, Two-Category, Flame-Resistant (FR) Clothing System; Annex I, Job Briefing and Planning Checklist; Annex J, Energized Electrical Work Permit; Annex K, General Categories of Electrical Hazards; Annex L, Typical Application of Safeguards in the Cell Line Working Zone; and Annex M, Cross-Reference Tables.

Many industries are using the NFPA 70E as the key element in the development of their electrical safety programs and procedures for at least three reasons:

1. OSHA has identified NFPA 70E to be part of and integral to its electrical safety rule-making effort.

2. OSHA maintains representation on the NFPA 70E committee, thereby helping to ensure that the NFPA 70E is consistent with the OSHA requirements and intent for electrical safety.

3. NFPA 70E is more detailed and performance-oriented than the equivalent OSHA regulations; consequently 70E is easier to employ as the basis for a good electrical safety program.

Industries and Facilities Covered

NFPA 70E is a companion document to the NEC; therefore, it covers exactly the same industries as the NEC. Specifically NFPA 70E addresses electrical safety requirements that are necessary for safeguarding employees in the workplace. It covers three major installations:

1. Building electrical utilization installations including carnival and parking lots, mobile homes and recreational vehicles, and industrial substations.

2. Conductors which connect installations to a supply of electricity.

3. Other outside premises wiring.

   Document 70E specifically excludes five basic categories of installation:

1. Ships, watercraft, railway rolling stock, aircraft, and automotive vehicles other than mobile homes and recreational vehicles.

2. Installations underground in mines.

3. Installations of railways for generation, transformation, transmission, or distribution of power used exclusively for operation of rolling stock or installations used exclusively for signaling and communication purposes.

4. Installation of communication equipment under the exclusive control of communications utilities, located outdoors or in building spaces used exclusively for such installations.

5. Installations under the exclusive control of electric utilities used for communication, metering, or for the generation, control transformation, transmission, and distribution of

electric energy located in buildings used exclusively by utilities for such purposes or located outdoors on property owned or leased by the utility or on public highways, streets, roads, and so on, or outdoors by established rights on private property.

Technical Safety Items Covered

Chapter 1, Safety-Related Work Practices.   Chapter 1 of 70E is intended to give directions to employees in how to perform their work in such a manner that they can avoid electrical hazards. It includes information on working on or near energized parts, working on or near de-energized parts, safe de-energizing of systems, lockout-tagout procedures, safe use of tools and equipment, personal and other protective equipment, safe approach distances for overhead lines, switching of protective devices after operation, safety training, and safe voltage-measurement practices.

The majority of the material in Chap. 1 was adopted by OSHA and is included in it's safety-related work practices regulation in 29 CFR 1910.331-.335.

Chapter 2, Safety-Related Maintenance Requirements.   Chapter 2 includes a variety of information on maintenance; for the most part it addresses the basic, common sense procedures for many common types of electric equipment. Included are substation and switchgear assemblies, premises wiring, controller equipment, fuses and molded-case circuit breakers, rotating electric equipment, portable electric tools and equipment, safety and protective equipment, hazardous (classified) locations, and batteries and battery rooms.

Chapter 2 has only minimal broad recommendations. Specific maintenance techniques needed to meet the provisions of Chap. 2 are found in NFPA 70B, *Recommended Practice for Electrical Equipment Maintenance.*

Chapter 3, Safety Requirements for Special Equipment.   Chapter 3 covers special safety requirements for employees working on or near electrolytic cells, batteries and battery rooms, lasers, and power electronics equipment. These sections provide critical information for personnel working on or near such equipment.

Chapter 4, Installation Safety Requirements.   In developing document NFPA 70E, seven basic criteria were used:

1. The provisions should give employees protection from electrical hazards.
2. It's provisions should be excerpted from the NEC in such a way that they will maintain their intent as they apply to electrical safety.
3. The material should be selected so that frequent revision will not be required.
4. Compliance should be determined by means of an inspection during the normal state of employee inspection during normal work hours, without removal of parts requiring shutdown of the electric installation or by damaging the building structure.
5. The provisions should not have unnecessary details.
6. The provisions should be written so as to enhance their understanding by employers and employees.
7. The provisions must not add any requirements not found in the NEC.

NFPA 70E, Chap. 4 includes information on installing equipment, providing working clearances, guarding live parts, wiring design and protection, grounding connections, wiring methods, damp and wet locations, specific-purpose equipment, electric signs, cranes and hoists, electric welders, portable electric equipment, hazardous locations, and special systems. Chapter 4 is very similar to OSHA 29 CFR 1910.302-.308, which will be covered in more detail later in this chapter.

THE AMERICAN SOCIETY FOR TESTING AND MATERIALS (ASTM) STANDARDS

The ASTM publishes a variety of standards primarily associated with safety equipment design, usage, and testing. Table 6.5 summarizes some of the ASTM standards and briefly describes their coverage. Refer to Chapter 2 for many specific references and uses of the ASTM standards.

OCCUPATIONAL SAFETY AND HEALTH ADMINISTRATION (OSHA) STANDARDS

Overview

OSHA standards represent federal law and must be implemented by all covered industries, except in those states which have adopted state safety programs which are approved by OSHA. OSHA has been developing many electrical and related safety standards since its creation in 1980.

Because the OSHA standards apply so universally, several of the more important ones are reproduced in their entirety, along with explanations from the Federal Register Preamble as well as the Directorate of Compliance, in this section. *Note:* The standards reproduced in this section are the current standards at the date of publication of this handbook. The user should contact OSHA for the current, up-to-date version of each standard. All OSHA standards are kept updated at their website http://www.osha.gov.

General Industry

At the time of this publication, OSHA had not released proposed or final rules for maintenance requirements or for special equipment. Contact OSHA for the status of these two standards.

Scope. The general industry OSHA standards apply to electric utilization systems installed or used within or on buildings, structures, and other premises. Other premises include:

- Yards
- Carnivals
- Parking and other lots
- Mobile homes
- Recreational vehicles
- Industrial substations
- Conductors that connect the installations to a supply of electricity
- Other outside conductors on the premises

TABLE 6.5 Summary of ASTM Standards for Electrical Protective Equipment

Standard number/name	Description
D 120/Specification for Rubber Insulating Gloves	Acceptance testing of rubber insulating gloves used for worker protection.
D 178/Specification for Rubber Insulating Matting	Acceptance testing of rubber insulating matting used as floor covering.
D 1048/Specification for Rubber Insulating Blankets	Acceptance testing of rubber insulating blankets.
D 1049/Specification for Rubber Insulating Covers	Acceptance testing of rubber insulating covers. Insulator hoods, line hose connectors, and so on.
D 1050/Specification for Rubber Insulating Line Hose	Acceptance testing of rubber line hose.
D 1051/Specification for Rubber Insulating Sleeves	Acceptance testing of rubber insulating sleeves.
F 478/In-Service Care of Insulating Line Hose & Covers	Care, inspection, testing, and use voltage of insulating line hose and covers.
F 479/In-Service Care of Insulating Blankets	Care, inspection, testing, and use voltage of insulating blankets for shock protection.
F 496/In-Service Care of Insulating Gloves and Sleeves	Care, inspection, testing, and use voltage of insulating gloves and sleeves for shock protection.
F 696/Specification for Leather Protectors for Rubber Insulating Gloves and Mittens	Specification for the design and manufacture of leather protectors for rubber insulating goods. Note that the OSHA requires leather protectors over rubber insulating goods if they may be damaged.
F 711/Specification for Fiberglass Reinforced Plastic Rod and Tube Used in Live Line Tools	Specification for the design and manufacture of insulating foam-filled tubes and rods made from fiberglass- reinforced plastic used for live line tools.
F 712/Test Methods for Electrical Insulating Plastic Guard Equipment for Protection of Workers	Withstand voltage, flashover voltage, and leakage current test methods are defined for line guards, guard connectors, deadend covers, pole guards, and other such equipment.
F 819 of Terms Relating to Electrical Protective Equipment	A concise definition list of key terms.
F 855/Specifications for Temporary Grounding Systems to Be Used on De-Energized Electric Power Lines and Equipment	Construction, testing, and materials used for temporary grounding systems.
F 887/Specifications for Personal Climbing Equipment	Acceptance testing specifications for climbers, climber straps, body belts, and pole straps.
F 914/Test Methods for Acoustic Emission for Insulated Aerial Personnel Devices	Acoustic emission testing methods for insulated aerial personnel devices.

F 968/Specification for Electrically Insulating Plastic Guard Equipment for Protection of Workers	Construction and materials specification for line guards, guard connectors, deadend covers, and other such equipment.
F 1116/Test Method for Determining Dielectric Strength of Overshoe Footwear	Recommended methods for determining the dielectric strength of insulating overshoe footwear.
F 1117/Specification for Dielectric Overshoe Footwear	Acceptance testing of overshoe footwear.
F 1236/Guide for Visual Inspection of Electrical Protective Rubber Products	Recommended methods and techniques for proper in-service inspection of electrical protective rubber products. Chapter 3 describes many of these methods.
F 1505/Specification for Insulated and Insulating Hand Tools	Design, construction, testing, and materials to be used in insulated hand tools below 600 V.
F 1506/Performance Specification for Textile Materials for Wearing Apparel for Use by Electrical Workers Exposed to Momentary Electric Arc and Related Thermal Hazards	Covers the flame-resistance requirements for materials used in electrical workers' clothing. Covers only classic flame testing. Predates the APTV requirement.
F 1564/Specification for Structure-Mounted Insulating Work Platforms for Electrical Workers	Design testing standard for mechanical and electrical characteristics of structure-mounted work platforms. Covers only single-worker platforms.
F 1742/Specification for PVC Insulating Sheeting	Design specs for PVC sheeting material.
F 1796/Specification for High Voltage Detectors—Part 1	Design and construction requirements for capacitive voltage detectors 600V to 80 kV.
F 1825/Specification for Fixed Length Clampstick T Live Line Tools	Establishes the technical specifications for the design and manufacture of clampsticks.
F 1826/Specification for Telescoping Live Line Tools	Establishes the technical specifications for the design and manufacture of telescoping live line tools.
F 1891/Specification for Arc and Flame-Resistant Rainwear	Similar to F 1506, but covers rainwear materials.
F 1959/F 19S9MJStandard Test Method for Determining Arc Thermal Performance Value of Materials for Clothing	Establishes test values and techniques for APTV of flame-resistant clothing.

OSHA has issued or will issue standards on design safety; electrical safety-related work practices; lockout/tagout; maintenance-related safety; electric power generation transmission, and distribution; and special equipment safety. All general industry electrical safety standards are published in Subpart S of Part 1910. Some related standards are published in Subpart J and others in Subpart R. Always refer to the most recent edition. They are reproduced here for reference only.

Design Safety Standards.   The design safety standards cover all utilization systems which have been installed or undergone major modification since March 15, 1972. These standards provide design and installation requirements. The design safety standards are found in the United States Code of Federal Regulations, Title 29, Part 1910, Subpart S, Paragraphs 1910.303 through 1910.308. They are reproduced in Fig. 6.1. *Note:* explanations for these paragraphs are found in the National Electrical Code Handbook.

Safety-Related Work Practices.   This OSHA standard covers the procedures required to ensure that employees use optimum safety-related procedures when working around energized or potentially energized equipment. It covers a variety of topics including lockout-tagout procedures. Electrical lockout-tagout must follow the methods given in Paragraph 1910.333; however, with certain additions, 1910.333 does allow the use of the Control of Hazardous Energy Rule (1910.147).

The Electrical Safety-Related Work Practices Rule is found in the United States Code of Federal Regulations, Title 29, Part 1910, Subpart S, Paragraphs 1910.331 through 1910.335. It is reproduced in this text, along with explanations extracted from the preamble of the Federal Register as well as the Directorate of Compliance, as Fig. 6.2.

Power Generation, Transmission, and Distribution.   Employees involved in the operation or maintenance of equipment used for power generation, transmission, or distribution and under the control of an electric utility are covered by this rule. The rule is published in the United States Code of Federal Regulations, Title 29, Part 1910, Subpart R, Paragraph 1910.269. Figure 6.3 is the rule, including explanations extracted from the preamble of the Federal Register.

Control of Hazardous Energy Source (Lockout/Tagout).   This rule is not an electrical standard; in fact, section (a)(1)(ii)(C) specifically excludes electrical hazards. However, the Safety-Related Work Practices Rule, Paragraph 1910.333(b)(2), Note 2 allows the use of the Lockout-Tagout Rule with certain restrictions. Because of its importance, it is reproduced here, along with explanations from the preamble of the Federal Register, as Fig. 6.4. This rule is published in the United States Code of Federal Regulations, Title 29, Part 1910, Subpart J, Paragraph 1910.147.

Definitions.   Figure 6.5 lists the definitions that the OSHA uses in Subpart S of Part 1910. It is published in the United States Code of Federal Regulations, Part 1910, Subpart S, Paragraph 1910.399.

Construction Industry

Scope.   The OSHA publishes electrical safety standards for the construction industry in two subparts. Subpart K applies to construction sites and facilities exclusive of those intended exclusively for power generation, transmission, and distribution which is covered in Subpart V.

Installation Safety Requirements. Figure 6.6 is a reproduction of the Installation Safety Requirements for construction sites. This standard is published in the United States Code of Federal Regulations, Title 29, Part 1926, Subpart K, Paragraphs 1926.402 through 1926.408.

Safety-Related Work Practices. Because of the nature of electrical exposure of construction personnel, this rule is much shorter and less comprehensive than the safety-related work practices rule which applies to the general industry. This standard is published in the United States Code of Federal Regulations, Title 29, Part 1926, Sub part K, Paragraphs 1926.416 and 1926.417. Figure 6.7 is a reproduction of the rule.

Safety-Related Maintenance and Environmental Considerations (Construction). This standard is published in the United States Code of Federal Regulations, Title 29, Part 1926, Subpart K, Paragraphs 431-432. Figure 6.8 is a reproduction of this standard.

Safety Requirements for Special Equipment (Construction). Batteries are the only apparatus included in this rule to date. It is published in the United States Code of Federal Regulations, Part 1926, Subpart K, Paragraph 1926.441. Figure 6.9 reproduces this standard.

Definitions. Figure 6.10 lists the definitions that the OSHA uses in Subpart K of Part 1926. It is published in the United States Code of Federal Regulations, Part 1926, Subpart K, Paragraph 1910.449.

DESIGN SAFETY STANDARDS FOR
ELECTRICAL SYSTEMS

§ 1910.302 Electric utilization systems.

Sections 1910.302 through 1910.308 contain design safety standards for electric utilization systems.

(a) *Scope*–

(a)(1) *Covered.* The provisions of 1910.302 through 1910.308 of this subpart cover electrical installations and utilization equipment installed or used within or on buildings, structures, and other premises including:

(a)(1)(i) Yards,

(a)(1)(ii) Carnivals,

(a)(1)(iii) Parking and other lots,

(a)(1)(iv) Mobile homes,

(a)(1)(v) Recreational vehicles,

(a)(1)(vi) Industrial substations,

(a)(1)(vii) Conductors that connect the installations to a supply of electricity, and

(a)(1)(viii) Other outside conductors on the premises.

§ 1910.302(a)(2)

(a)(2) Not covered. The provisions of 1910.302 through 1910.308 of this subpart do not cover:

(a)(2)(i) Installations in ships, watercraft, railway rolling stock, aircraft, or automotive vehicles other than mobile homes and recreational vehicles.

(a)(2)(ii) Installations underground in mines.

(a)(2)(iii) Installations of railways for generation, transformation, transmission, or distribution of power used exclusively for operation of rolling stock or installations used exclusively for signaling and communication purposes.

(a)(2)(iv) Installations of communication equipment under the exclusive control of communication utilities, located outdoors or in building spaces used exclusively for such installations.

(a)(2)(v) Installations under the exclusive control of electric utilities for the purpose of communication or metering; or for the generation, control, transformation, transmission, and distribution of electric energy located in buildings used exclusively by utilities for such purposes or located outdoors on property owned or leased by the utility or on public highways, streets, roads, etc., or outdoors by established rights on private property.

§ 1910.302(b)

(b) Extent of application.

(b)(1) The requirements contained in the sections listed below shall apply to all electrical installations and utilization equipment, regardless of when they were designed or installed.

Sections:

1910.303(b)	Examination, installation, and use of equipment.
1910.303(c)	Splices.
1910.303(d)	Arcing parts.
1910.303(e)	Marking.
1910.303(f)	Identification of disconnecting means.
1910.303(g)(2)	Guarding of live parts.
1910.304(e)(1)(i)	Protection of conductors and equipment.
1910.304(e)(1)(iv)	Location in or on premises.
1910.304(e)(1)(v)	Arcing or suddenly moving parts.
1910.304(f)(1)(ii)	2-Wire DC systems to be grounded:
1910.304(f)(1)(iii) and 1910.304(f)(1)(iv)	AC Systems to be grounded.
1910.304(f)(1)(v)	AC Systems 50 to 1000 volts not required to be grounded.
1910.304(f)(3)	Grounding connections.
1910.304(f)(4)	Grounding path.
1910.304(f)(5)(iv)(a) through 1910.304(f)(5)(iv)(d)	Fixed equipment required to be grounded.
1910.304(f)(5)(v)	Grounding of equipment connected by cord and plug.
1910.304(f)(5)(vi)	Grounding of nonelectrical equipment.
1910.304(f)(6)(i)	Methods of grounding fixed equipment.
1910.305(g)(1)(i) and 1910.305(g)(1)(ii)	Flexible cords and cables, uses.
1910.305(g)(1)(iii)	Flexible cords and cables prohibited.
1910.305(g)(2)(ii)	Flexible cords and cables, splices.
1910.305(g)(2)(iii)	Pull at joints and terminals of flexible cords and cables.
1910.307	Hazardous (classified) locations.

(b)(2) Every electric utilization system and all utilization equipment installed after March

FIGURE 6.1 Design Safety Standards for Electrical Systems (OSHA-CFR, Title 29, Part 1910, Paragraphs 302–308). (*Courtesy OSHA Web site.*)

15, 1972, and every major replacement, modification, repair, or rehabilitation, after March 15, 1972, of any part of any electric utilization system or utilization equipment installed before March 15, 1972, shall comply with the provisions of 1910.302 through 1910.308.

NOTE: "Major replacements, modifications, repairs, or rehabilitations" include work similar to that involved when a new building or facility is built, a new wing is added, or an entire floor is renovated.

(b)(3) The following provisions apply to electric utilization systems and utilization equipment installed after April 16, 1981:

1910.303(h)(4) (i) and (ii)	Entrance and access to workspace (over 600 volts).
1910.304(e)(1)(vi)(b)	Circuit breakers operated vertically.
1910.304(e)(1)(vi)(c)	Circuit breakers used as switches.
1910.304(f)(7)(ii)	Grounding of systems of 1000 volts or more supplying portable or mobile equipment.
1910.305(j)(6)(ii)(b)	Switching series capacitors over 600 volts.
1910.306(c)(2)	Warning signs for elevators and escalators.
1910.306(i)	Electrically controlled irrigation machines.
1910.306(j)(5)	Ground-fault circuit interrupters for fountains.
1910.308(a)(1)(ii)	Physical protection of conductors over 600 volts.
1910.308(c)(2)	Marking of Class 2 and Class 3 power supplies.
1910.308(d)	Fire protective signaling circuits.

[46 FR 4056, Jan. 16, 1981; 46 FR 40185, Aug. 7, 1981]

§ 1910.303 General requirements.

(a) Approval. The conductors and equipment required or permitted by this subpart shall be acceptable only if approved.

(b) Examination, installation, and use of equipment—

(b)(1) Examination. Electrical equipment shall be free from recognized hazards that are likely to cause death or serious physical harm to employees. Safety of equipment shall be determined using the following considerations:

(b)(1)(i) Suitability for installation and use in conformity with the provisions of this subpart. Suitability of equipment for an identified purpose may be evidenced by listing or labeling for that identified purpose.

(b)(1)(ii) Mechanical strength and durability, including, for parts designed to enclose and protect other equipment, the adequacy of the protection thus provided.

(b)(1)(iii) Electrical insulation.

(b)(1)(iv) Heating effects under conditions of use.

§ 1910.303(b)(1)(v)

(b)(1)(v) Arcing effects.

(b)(1)(vi) Classification by type, size, voltage, current capacity, specific use.

(b)(1)(vii) Other factors which contribute to the practical safeguarding of employees using or likely to come in contact with the equipment.

(b)(2) Installation and use. Listed or labeled equipment shall be used or installed in accordance with any instructions included in the listing or labeling.

(c) Splices. Conductors shall be spliced or joined with splicing devices suitable for the use or by brazing, welding, or soldering with a fusible metal or alloy. Soldered splices shall first be so spliced or joined as to be mechanically and electrically secure without solder and then soldered. All splices and joints and the free ends of conductors shall be covered with an insulation equivalent to that of the conductors or with an insulating device suitable for the purpose.

(d) Arcing parts. Parts of electric equipment which in ordinary operation produce arcs, sparks, flames, or molten metal shall be enclosed or separated and isolated from all combustible material.

§ 1910.303(e)

(e) Marking. Electrical equipment may not be used unless the manufacturer's name, trademark, or other descriptive marking by which the organization responsible for the product may be identified is placed on the equipment. Other markings shall be provided giving voltage, current, wattage, or other ratings as necessary. The marking shall be of sufficient durability to withstand the environment involved.

(f) Identification of disconnecting means and circuits. Each disconnecting means required by this subpart for motors and appliances shall be

FIGURE 6.1 (*Continued*)

legibly marked to indicate its purpose, unless located and arranged so the purpose is evident. Each service, feeder, and branch circuit, at its disconnecting means or overcurrent device, shall be legibly marked to indicate its purpose, unless located and arranged so the purpose is evident. These markings shall be of sufficient durability to withstand the environment involved.

(g) 600 Volts, nominal, or less—

(g)(1) Working space about electric equipment. Sufficient access and working space shall be provided and maintained about all electric equipment to permit ready and safe operation and maintenance of such equipment.

§ 1910.303(g)(1)(i)

(g)(1)(i) Working clearances. Except as required or permitted elsewhere in this subpart, the dimension of the working space in the direction of access to live parts operating at 600 volts or less and likely to require examination, adjustment, servicing, or maintenance while alive may not be less than indicated in Table S-1. In addition to the dimensions shown in Table S-1, workspace may not be less than 30 inches wide in front of the electric equipment. Distances shall be measured from the live parts if they are exposed, or from the enclosure front or opening if the live parts are enclosed. Concrete, brick, or tile walls are considered to be grounded. Working space is not required in back of assemblies such as dead-front switchboards or motor control centers where there are no renewable or adjustable parts such as fuses or switches on the back and where all connections are accessible from locations other than the back.

(g)(1)(ii) Clear spaces. Working space required by this subpart may not be used for storage. When normally enclosed live parts are exposed for inspection or servicing, the working space, if in a passageway or general open space, shall be suitably guarded.

(g)(1)(iii) Access and entrance to working space. At least one entrance of sufficient area shall be provided to give access to the working space about electric equipment.

§ 1910.303(g)(1)(iv)

(g)(1)(iv) Front working space. Where there are live parts normally exposed on the front of switchboards or motor control centers, the working space in front of such equipment may not be less than 3 feet.

(g)(1)(v) Illumination. Illumination shall be provided for all working spaces about service equipment, switchboards, panelboards, and motor control centers installed indoors.

(g)(1)(vi) Headroom. The minimum headroom of working spaces about service equipment, switchboards, panel-boards, or motor control centers shall be 6 feet 3 inches.

Note: As used in this section a motor control center is an assembly of one or more enclosed sections having a common power bus and principally containing motor control units.

(g)(2) Guarding of live parts.

(g)(2)(i) Except as required or permitted elsewhere in this subpart, live parts of electric equipment operating at 50 volts or more shall be guarded against accidental contact by approved cabinets or other forms of approved enclosures, or by any of the following means:

(g)(2)(i)(A) By location in a room, vault, or similar enclosure that is accessible only to qualified persons.

TABLE S-1 Working Clearances

	Minimum clear distance for condition[2] (ft)		
Nominal voltage to ground	(a)	(b)	(c)
0–150	[1]3	[1]3	3
151–600	[1]3	3½	4

(1) Minimum clear distances may be 2 feet 6 inches for installations built prior to April 16, 1981.
(2) Conditions (a), (b), and (c), are as follows: (a) Exposed live parts on one side and no live or grounded parts on the other side of the working space, or exposed live parts on both sides effectively guarded by suitable wood or other insulating material. Insulated wire or insulated busbars operating at not over 300 volts are not considered live parts. (b) Exposed live parts on one side and grounded parts on the other side. (c) Exposed live parts on both sides of the workspace [not guarded as provided in Condition (a)] with the operator between.

FIGURE 6.1 *(Continued)*

§ 1910.303(g)(2)(i)(B)

(g)(2)(i)(B) By suitable permanent, substantial partitions or screens so arranged that only qualified persons will have access to the space within reach of the live parts. Any openings in such partitions or screens shall be so sized and located that persons are not likely to come into accidental contact with the live parts or to bring conducting objects into contact with them.

(g)(2)(i)(C) By location on a suitable balcony, gallery, or platform so elevated and arranged as to exclude unqualified persons.

(g)(2)(i)(D) By elevation of 8 feet or more above the floor or other working surface.

(g)(2)(ii) In locations where electric equipment would be exposed to physical damage, enclosures or guards shall be so arranged and of such strength as to prevent such damage.

(g)(2)(iii) Entrances to rooms and other guarded locations containing exposed live parts shall be marked with conspicuous warning signs forbidding unqualified persons to enter.

(h) Over 600 volts, nominal—

(h)(1) General. Conductors and equipment used on circuits exceeding 600 volts, nominal, shall comply with all applicable provisions of paragraphs (a) through (g) of this section and with the following provisions which supplement or modify those requirements. The provisions of paragraphs (h)(2), (h)(3), and (h)(4) of this section do not apply to equipment on the supply side of the service conductors.

§ 1910.303(h)(2)

(h)(2) Enclosure for electrical installations. Electrical installations in a vault, room, closet or in an area surrounded by a wall, screen, or fence, access to which is controlled by lock and key or other approved means, are considered to be accessible to qualified persons only. A wall, screen, or fence less than 8 feet in height is not considered to prevent access unless it has other features that provide a degree of isolation equivalent to an 8 foot fence. The entrances to all buildings, rooms, or enclosures containing exposed live parts or exposed conductors operating at over 600 volts, nominal, shall be kept locked or shall be under the observation of a qualified person at all times.

(h)(2)(i) Installations accessible to qualified persons only. Electrical installations having exposed live parts shall be accessible to qualified persons only and shall comply with the applicable provisions of paragraph (h)(3) of this section.

(h)(2)(ii) Installations accessible to unqualified persons. Electrical installations that are open to unqualified persons shall be made with metal-enclosed equipment or shall be enclosed in a vault or in an area, access to which is controlled by a lock. If metal-enclosed equipment is installed so that the bottom of the enclosure is less than 8 feet above the floor, the door or cover shall be kept locked. Metal-enclosed switchgear, unit substations, transformers, pull boxes, connection boxes, and other similar associated equipment shall be marked with appropriate caution signs. If equipment is exposed to physical damage from vehicular traffic, suitable guards shall be provided to prevent such damage. Ventilating or similar openings in metal-enclosed equipment shall be designed so that foreign objects inserted through these openings will be deflected from energized parts.

§ 1910.303(h)(3)

(h)(3) Workspace about equipment. Sufficient space shall be provided and maintained about electric equipment to permit ready and safe operation and maintenance of such equipment. Where energized parts are exposed, the minimum clear workspace may not be less than 6 feet 6 inches high (measured vertically from the floor or platform), or less than 3 feet wide (measured parallel to the equipment). The depth shall be as required in Table S-2. The workspace shall be adequate to permit at least a 90 degree opening of doors or hinged panels.

(h)(3)(i) Working space. The minimum clear working space in front of electric equipment such as switchboards, control panels, switches, circuit breakers, motor controllers, relays, and similar equipment may not be less than specified in Table S-2 unless otherwise specified in this subpart. Distances shall be measured from the live parts if they are exposed, or from the enclosure front or opening if the live parts are enclosed. However, working space is not required in back of equipment such as deadfront switchboards or control assemblies where there are no renewable or adjustable parts (such as fuses or switches) on the back and where all connections are accessible from locations other than the back. Where rear access is required to work on de-energized parts on the back of enclosed equipment, a minimum working space of 30 inches horizontally shall be provided.

FIGURE 6.1 (*Continued*)

TABLE S-2   Minimum Depth of Clear Working Space in Front of Electric Equipment

Nominal voltage to ground	Conditions[2a] (ft)		
	(a)	(b)	(c)
601 to 2,500	3	4	5
2,501 to 9,000	4	5	6
9,001 to 25,000	5	6	9
25,001 to 75kV[1a]	6	8	10
Above 75kV[1a]	8	10	12

(1a) Minimum depth of clear working space in front of electric equipment with a nominal voltage to ground above 25,000 volts may be the same as for 25,000 volts under Conditions (a), (b), and (c) for installations built prior to April 16, 1981.

(2a) Conditions (a), (b), and (c) are as follows: (a) Exposed live parts on one side and no live or grounded parts on the other side of the working space, or exposed live parts on both sides effectively guarded by suitable wood or other insulating materials. Insulated wire or insulated busbars operating at not over 300 volts are not considered live parts. (b) Exposed live parts on one side and grounded parts on the other side. Concrete, brick, or tile walls will be considered as grounded surfaces. (c) Exposed live parts on both sides of the workspace not guarded as provided in Condition (a) with the operator between.

§ 1910.303(h)(3)(ii)

(h)(3)(ii) Illumination. Adequate illumination shall be provided for all working spaces about electric equipment. The lighting outlets shall be so arranged that persons changing lamps or making repairs on the lighting system will not be endangered by live parts or other equipment. The points of control shall be so located that persons are not likely to come in contact with any live part or moving part of the equipment while turning on the lights.

(h)(3)(iii) Elevation of unguarded live parts. Unguarded live parts above working space shall be maintained at elevations not less than specified in Table S-3.

(h)(4) Entrance and access to workspace. (See § 1910.302(b)(3).)

(h)(4)(i) At least one entrance not less than 24 inches wide and 6 feet 6 inches high shall be provided to give access to the working space about electric equipment. On switchboard and control panels exceeding 48 inches in width, there shall be one entrance at each end of such board where practicable. Where bare energized parts at any voltage or insulated energized parts above 600 volts are located adjacent to such entrance, they shall be suitably guarded.

(h)(4)(ii) Permanent ladders or stairways shall be provided to give safe access to the working space around electric equipment installed on platforms, balconies, mezzanine floors, or in attic or roof rooms or spaces.

[46 FR 4056, Jan. 16, 1981; 46 FR 40185, Aug. 7, 1981]

§ 1910.304 Wiring design and protection.

(a) Use and identification of grounded and grounding conductors.

TABLE S-3   Elevation of Unguarded Energized Parts Above Working Space

Nominal voltage between phases	Minimum elevation
601 to 7,500	8 feet 6 inches.*
7,501 to 35,000	9 feet.
Over 35kV	9 feet + 0.37 inches kV above 35kV.

*Note.—Minimum elevation may be 8 feet 0 inches for installations built prior to April 16, 1981 if the nominal voltage between phases is in the range of 601–6600 volts.

FIGURE 6.1   (*Continued*)

(a)(1) Identification of conductors. A conductor used as a grounded conductor shall be identifiable and distinguishable from all other conductors. A conductor used as an equipment grounding conductor shall be identifiable and distinguishable from all other conductors.

(a)(2) Polarity of connections. No grounded conductor may be attached to any terminal or lead so as to reverse designated polarity.

(a)(3) Use of grounding terminals and devices. A grounding terminal or grounding-type device on a receptacle, cord connector, or attachment plug may not be used for purposes other than grounding.

(b) Branch circuits—

(b)(1) [Reserved]

(b)(2) Outlet devices. Outlet devices shall have an ampere rating not less than the load to be served.

§ 1910.304(c)

(c) Outside conductors, 600 volts, nominal, or less. Paragraphs (c)(1), (c)(2), (c)(3), and (c)(4) of this section apply to branch circuit, feeder, and service conductors rated 600 volts, nominal, or less and run outdoors as open conductors. Paragraph (c)(5) applies to lamps installed under such conductors.

(c)(1) Conductors on poles. Conductors supported on poles shall provide a horizontal climbing space not less than the following:

(c)(1)(i) Power conductors below communication conductors—30 inches.

(c)(1)(ii) Power conductors alone or above communication conductors: 300 volts or less—24 inches; more than 300 volts—30 inches.

(c)(1)(iii) Communication conductors below power conductors with power conductors 300 volts or less—24 inches; more than 300 volts—30 inches.

(c)(2) Clearance from ground. Open conductors shall conform to the following minimum clearances:

(c)(2)(i) 10 feet—above finished grade, sidewalks, or from any platform or projection from which they might be reached.

(c)(2)(ii) 12 feet—over areas subject to vehicular traffic other than truck traffic.

(c)(2)(iii) 15 feet—over areas other than those specified in paragraph (c)(2)(iv) of this section that are subject to truck traffic.

§ 1910.304(c)(2)(iv)

(c)(2)(iv) 18 feet—over public streets, alleys, roads, and driveways.

(c)(3) Clearance from building openings. Conductors shall have a clearance of at least 3 feet from windows, doors, porches, fire escapes, or similar locations. Conductors run above the top level of a window are considered to be out of reach from that window and, therefore, do not have to be 3 feet away.

(c)(4) Clearance over roofs. Conductors shall have a clearance of not less than 8 feet from the highest point of roofs over which they pass, except that:

(c)(4)(i) Where the voltage between conductors is 300 volts or less and the roof has a slope of not less than 4 inches in 12, the clearance from roofs shall be at least 3 feet, or

(c)(4)(ii) Where the voltage between conductors is 300 volts or less and the conductors do not pass over more than 4 feet of the overhang portion of the roof and they are terminated at a through-the-roof raceway or approved support, the clearance from roofs shall be at least 18 inches.

§ 1910.304(c)(5)

(c)(5) Location of outdoor lamps. Lamps for outdoor lighting shall be located below all live conductors, transformers, or other electric equipment, unless such equipment is controlled by a disconnecting means that can be locked in the open position or unless adequate clearances or other safeguards are provided for relamping operations.

(d) Services —

(d)(1) Disconnecting means—

(d)(1)(i) General. Means shall be provided to disconnect all conductors in a building or other structure from the service-entrance conductors. The disconnecting means shall plainly indicate whether it is in the open or closed position and shall be installed at a readily accessible location nearest the point of entrance of the service-entrance conductors.

(d)(1)(ii) Simultaneous opening of poles. Each service disconnecting means shall simultaneously disconnect all ungrounded conductors.

(d)(2) Services over 600 volts, nominal. The following additional requirements apply to services over 600 volts, nominal.

(d)(2)(i) Guarding. Service-entrance conductors installed as open wires shall be guarded to make them accessible only to qualified persons.

(d)(2)(ii) Warning signs. Signs warning of high voltage shall be posted where other than qualified employees might come in contact with live parts.

FIGURE 6.1 (*Continued*)

(e) Overcurrent protection.

(e)(1) 600 volts, nominal, or less. The following requirements apply to overcurrent protection of circuits rated 600 volts, nominal, or less.

§ 1910.304(e)(1)(i)

(e)(1)(i) Protection of conductors and equipment. Conductors and equipment shall be protected from overcurrent in accordance with their ability to safely conduct current.

(e)(1)(ii) Grounded conductors. Except for motor running overload protection, overcurrent devices may not interrupt the continuity of the grounded conductor unless all conductors of the circuit are opened simultaneously.

(e)(1)(iii) Disconnection of fuses and thermal cutouts. Except for service fuses, all cartridge fuses which are accessible to other than qualified persons and all fuses and thermal cutouts on circuits over 150 volts to ground shall be provided with disconnecting means. This disconnecting means shall be installed so that the fuse or thermal cutout can be disconnected from its supply without disrupting service to equipment and circuits unrelated to those protected by the overcurrent device.

(e)(1)(iv) Location in or on premises. Overcurrent devices shall be readily accessible to each employee or authorized building management personnel. These overcurrent devices may not be located where they will be exposed to physical damage nor in the vicinity of easily ignitable material.

(e)(1)(v) Arcing or suddenly moving parts. Fuses and circuit breakers shall be so located or shielded that employees will not be burned or otherwise injured by their operation.

§ 1910.304(e)(1)(vi)

(e)(1)(vi) Circuit breakers.

(e)(1)(vi)(A) Circuit breakers shall clearly indicate whether they are in the open (off) or closed (on) position.

(e)(1)(vi)(B) Where circuit breaker handles on switchboards are operated vertically rather than horizontally or rotationally, the up position of the handle shall be the closed (on) position. (See 1910.302(b)(3).)

(e)(1)(vi)(C) If used as switches in 120-volt, fluorescent lighting circuits, circuit breakers shall be approved for the purpose and marked "SWD." (See 1910.302(b)(3).)

(e)(2) Over 600 volts, nominal. Feeders and branch circuits over 600 volts, nominal, shall have short-circuit protection.

(f) Grounding. Paragraphs (f)(1) through (f)(7) of this section contain grounding requirements for systems, circuits, and equipment.

(f)(1) Systems to be grounded. The following systems which supply premises wiring shall be grounded:

(f)(1)(i) All 3-wire DC systems shall have their neutral conductor grounded.

(f)(1)(ii) Two-wire DC systems operating at over 50 volts through 300 volts between conductors shall be grounded unless:

§ 1910.304(f)(1)(ii)(A)

(f)(1)(ii)(A) They supply only industrial equipment in limited areas and are equipped with a ground detector; or

(f)(1)(ii)(B) They are rectifier-derived from an AC system complying with paragraphs (f)(1)(iii), (f)(1)(iv), and (f)(1)(v) of this section; or

(f)(1)(ii)(C) They are fire-protective signaling circuits having a maximum current of 0.030 amperes.

(f)(1)(iii) AC circuits of less than 50 volts shall be grounded if they are installed as overhead conductors outside of buildings or if they are supplied by transformers and the transformer primary supply system is ungrounded or exceeds 150 volts to ground.

(f)(1)(iv) AC systems of 50 volts to 1000 volts shall be grounded under any of the following conditions, unless exempted by paragraph (f)(1)(v) of this section:

(f)(1)(iv)(A) If the system can be so grounded that the maximum voltage to ground on the ungrounded conductors does not exceed 150 volts;

(f)(1)(iv)(B) If the system is nominally rated 480Y/277 volt, 3-phase, 4-wire in which the neutral is used as a circuit conductor;

(f)(1)(iv)(C) If the system is nominally rated 240/120 volt, 3-phase, 4-wire in which the midpoint of one phase is used as a circuit conductor; or

§ 1910.304(f)(1)(iv)(D)

(f)(1)(iv)(D) If a service conductor is uninsulated.

(f)(1)(v) AC systems of 50 volts to 1000 volts are not required to be grounded under any of the following conditions:

(f)(1)(v)(A) If the system is used exclusively to supply industrial electric furnaces for melting, refining, tempering, and the like.

(f)(1)(v)(B) If the system is separately derived and is used exclusively for rectifiers supplying only adjustable speed industrial drives.

FIGURE 6.1   *(Continued)*

(f)(1)(v)(C) If the system is separately derived and is supplied by a transformer that has a primary voltage rating less than 1000 volts, provided all of the following conditions are met:

(f)(1)(v)(C)(1) The system is used exclusively for control circuits,

(f)(1)(v)(C)(2) The conditions of maintenance and supervision assure that only qualified persons will service the installation,

(f)(1)(v)(C)(3) Continuity of control power is required, and

(f)(1)(v)(C)(4) Ground detectors are installed on the control system.

§ 1910.304(f)(1)(v)(D)

(f)(1)(v)(D) If the system is an isolated power system that supplies circuits in health care facilities.

(f)(2) Conductors to be grounded. For AC premises wiring systems the identified conductor shall be grounded.

(f)(3) Grounding connections.

(f)(3)(i) For a grounded system, a grounding electrode conductor shall be used to connect both the equipment grounding conductor and the grounded circuit conductor to the grounding electrode. Both the equipment grounding conductor and the grounding electrode conductor shall be connected to the grounded circuit conductor on the supply side of the service disconnecting means, or on the supply side of the system disconnecting means or overcurrent devices if the system is separately derived.

(f)(3)(ii) For an ungrounded service-supplied system, the equipment grounding conductor shall be connected to the grounding electrode conductor at the service equipment. For an ungrounded separately derived system, the equipment grounding conductor shall be connected to the grounding electrode conductor at, or ahead of, the system disconnecting means or overcurrent devices.

(f)(3)(iii) On extensions of existing branch circuits which do not have an equipment grounding conductor, grounding-type receptacles may be grounded to a grounded cold water pipe near the equipment.

§ 1910.304(f)(4)

(f)(4) Grounding path. The path to ground from circuits, equipment, and enclosures shall be permanent and continuous.

(f)(5) Supports, enclosures, and equipment to be grounded—

(f)(5)(i) Supports and enclosures for conductors. Metal cable trays, metal raceways, and metal enclosures for conductors shall be grounded, except that:

(f)(5)(i)(A) Metal enclosures such as sleeves that are used to protect cable assemblies from physical damage need not be grounded; or

(f)(5)(i)(B) Metal enclosures for conductors added to existing installations of open wire, knob-and-tube wiring, and nonmetallic-sheathed cable need not be grounded if all of the following conditions are met:

(f)(5)(i)(B)(1) Runs are less than 25 feet;

(f)(5)(i)(B)(2) enclosures are free from probable contact with ground, grounded metal, metal laths, or other conductive materials; and

(f)(5)(i)(B)(3) enclosures are guarded against employee contact.

(f)(5)(ii) Service equipment enclosures. Metal enclosures for service equipment shall be grounded.

§ 1910.304(f)(5)(iii)

(f)(5)(iii) Frames of ranges and clothes dryers. Frames of electric ranges, wall-mounted ovens, counter-mounted cooking units, clothes dryers, and metal outlet or junction boxes which are part of the circuit for these appliances shall be grounded.

(f)(5)(iv) Fixed equipment. Exposed non-current-carrying metal parts of fixed equipment which may become energized shall be grounded under any of the following conditions:

(f)(5)(iv)(A) If within 8 feet vertically or 5 feet horizontally of ground or grounded metal objects and subject to employee contact.

(f)(5)(iv)(B) If located in a wet or damp location and not isolated.

(f)(5)(iv)(C) If in electrical contact with metal.

(f)(5)(iv)(D) If in a hazardous (classified) location.

(f)(5)(iv)(E) If supplied by a metal-clad, metal-sheathed, or grounded metal raceway wiring method.

(f)(5)(iv)(F) If equipment operates with any terminal at over 150 volts to ground; however, the following need not be grounded:

(f)(5)(iv)(F)(1) Enclosures for switches or circuit breakers used for other than service equipment and accessible to qualified persons only;

§ 1910.304(f)(5)(iv)(F)(2)

(f)(5)(iv)(F)(2) Metal frames of electrically heated appliances which are permanently and effectively insulated from ground; and

FIGURE 6.1 (*Continued*)

(f)(5)(iv)(F)(*3*) The cases of distribution apparatus such as transformers and capacitors mounted on wooden poles at a height exceeding 8 feet above ground or grade level.

(f)(5)(v) Equipment connected by cord and plug. Under any of the conditions described in paragraphs (f)(5)(v)(A) through (f)(5)(v)(C) of this section, exposed non-current-carrying metal parts of cord- and plug-connected equipment which may become energized shall be grounded.

(f)(5)(v)(A) If in hazardous (classified) locations (see 1910.307).

(f)(5)(v)(B) If operated at over 150 volts to ground, except for guarded motors and metal frames of electrically heated appliances if the appliance frames are permanently and effectively insulated from ground.

(f)(5)(v)(C) If the equipment is of the following types:

(f)(5)(v)(C)(*1*) Refrigerators, freezers, and air conditioners;

(f)(5)(v)(C)(*2*) Clothes-washing, clothes-drying and dishwashing machines, sump pumps, and electrical aquarium equipment;

(f)(5)(v)(C)(*3*) Hand-held motor-operated tools;

§ 1910.304(f)(5)(v)(C)(4)

(f)(5)(v)(C)(*4*) Motor-operated appliances of the following types: hedge clippers, lawn mowers, snow blowers, and wet scrubbers;

(f)(5)(v)(C)(*5*) Cord- and plug-connected appliances used in damp or wet locations or by employees standing on the ground or on metal floors or working inside of metal tanks or boilers;

(f)(5)(v)(C)(*6*) Portable and mobile X-ray and associated equipment;

(f)(5)(v)(C)(*7*) Tools likely to be used in wet and conductive locations; and

(f)(5)(v)(C)(*8*) Portable hand lamps.

Tools likely to be used in wet and conductive locations need not be grounded if supplied through an isolating transformer with an ungrounded secondary of not over 50 volts. Listed or labeled portable tools and appliances protected by an approved system of double insulation, or its equivalent, need not be grounded. If such a system is employed, the equipment shall be distinctively marked to indicate that the tool or appliance utilizes an approved system of double insulation.

(f)(5)(vi) Nonelectrical equipment. The metal parts of the following nonelectrical equipment shall be grounded: frames and tracks of electrically operated cranes; frames of nonelectrically driven elevator cars to which electric conductors are attached; hand operated metal shifting ropes or cables of electric elevators, and metal partitions, grill work, and similar metal enclosures around equipment of over 750 volts between conductors.

§ 1910.304(f)(6)

(f)(6) Methods of grounding fixed equipment.

(f)(6)(i) Non-current-carrying metal parts of fixed equipment, if required to be grounded by this subpart, shall be grounded by an equipment grounding conductor which is contained within the same raceway, cable, or cord, or runs with or encloses the circuit conductors. For DC circuits only, the equipment grounding conductor may be run separately from the circuit conductors.

(f)(6)(ii) Electric equipment is considered to be effectively grounded if it is secured to, and in electrical contact with, a metal rack or structure that is provided for its support and the metal rack or structure is grounded by the method specified for the non-current-carrying metal parts of fixed equipment in paragraph (f)(6)(i) of this section. For installations made before April 16, 1981, only, electric equipment is also considered to be effectively grounded if it is secured to, and in metallic contact with, the grounded structural metal frame of a building. Metal car frames supported by metal hoisting cables attached to or running over metal sheaves or drums of grounded elevator machines are also considered to be effectively grounded.

(f)(7) Grounding of systems and circuits of 1000 volts and over (high voltage.)—

(f)(7)(i) General. If high voltage systems are grounded, they shall comply with all applicable provisions of paragraphs (f)(1) through (f)(6) of this section as supplemented and modified by this paragraph (f)(7).

(f)(7)(ii) Grounding of systems supplying portable or mobile equipment. (See 1910.302(b)(3).) Systems supplying portable or mobile high voltage equipment, other than substations installed on a temporary basis, shall comply with the following:

§ 1910.304(f)(7)(ii)(A)

(f)(7)(ii)(A) Portable and mobile high voltage equipment shall be supplied from a system

FIGURE 6.1 (*Continued*)

having its neutral grounded through an impedance. If a delta-connected high voltage system is used to supply the equipment, a system neutral shall be derived.

(f)(7)(ii)(B) Exposed non-current-carrying metal parts of portable and mobile equipment shall be connected by an equipment grounding conductor to the point at which the system neutral impedance is grounded.

(f)(7)(ii)(C) Ground-fault detection and relaying shall be provided to automatically deenergize any high voltage system component which has developed a ground fault. The continuity of the equipment grounding conductor shall be continuously monitored so as to deenergize automatically the high voltage feeder to the portable equipment upon loss of continuity of the equipment grounding conductor.

(f)(7)(ii)(D) The grounding electrode to which the portable or mobile equipment system neutral impedance is connected shall be isolated from and separated in the ground by at least 20 feet from any other system or equipment grounding electrode, and there shall be no direct connection between the grounding electrodes, such as buried pipe, fence, etc.

(f)(7)(iii) Grounding of equipment. All noncurrent-carrying metal parts of portable equipment and fixed equipment including their associated fences, housings, enclosures, and supporting structures shall be grounded. However, equipment which is guarded by location and isolated from ground need not be grounded. Additionally, pole-mounted distribution apparatus at a height exceeding 8 feet above ground or grade level need not be grounded.

[46 FR 4056, Jan. 16, 1981; 46 FR 40185, Aug. 7, 1981, as amended at 55 FR 32015, Aug. 6, 1990]

§ 1910.305 Wiring methods, components, and equipment for general use.

(a) Wiring methods. The provisions of this section do not apply to the conductors that are an integral part of factory-assembled equipment.

(a)(1) General requirements—
(a)(1)(i) Electrical continuity of metal raceways and enclosures. Metal raceways, cable armor, and other metal enclosures for conductors shall be metallically joined together into a continuous electric conductor and shall be so connected to all boxes, fittings, and cabinets as to provide effective electrical continuity.

(a)(1)(ii) Wiring in ducts. No wiring systems of any type shall be installed in ducts used to transport dust, loose stock or flammable vapors. No wiring system of any type may be installed in any duct used for vapor removal or for ventilation of commercial-type cooking equipment, or in any shaft containing only such ducts.

(a)(2) Temporary wiring. Temporary electrical power and lighting wiring methods may be of a class less than would be required for a permanent installation. Except as specifically modified in this paragraph, all other requirements of this subpart for permanent wiring shall apply to temporary wiring installations.

§ 1910.305(a)(2)(i)

(a)(2)(i) Uses permitted, 600 volts, nominal, or less. Temporary electrical power and lighting installations 600 volts, nominal, or less may be used only:
(a)(2)(i)(A) During and for remodeling, maintenance, repair, or demolition of buildings, structures, or equipment, and similar activities;
(a)(2)(i)(B) For experimental or development work, and
(a)(2)(i)(C) For a period not to exceed 90 days for Christmas decorative lighting, carnivals, and similar purposes.
(a)(2)(ii) Uses permitted, over 600 volts, nominal. Temporary wiring over 600 volts, nominal, may be used only during periods of tests, experiments, or emergencies.
(a)(2)(iii) General requirements for temporary wiring.
(a)(2)(iii)(A) Feeders shall originate in an approved distribution center. The conductors shall be run as multiconductor cord or cable assemblies, or, where not subject to physical damage, they may be run as open conductors on insulators not more than 10 feet apart.

§ 1910.305(a)(2)(iii)(B)

(a)(2)(iii)(B) Branch circuits shall originate in an approved power outlet or panelboard. Conductors shall be multiconductor cord or cable assemblies or open conductors. If run as open conductors they shall be fastened at ceiling height every 10 feet. No branch-circuit conductor may be laid on the floor. Each branch circuit that supplies receptacles or fixed equipment

FIGURE 6.1 (*Continued*)

shall contain a separate equipment grounding conductor if run as open conductors.

(a)(2)(iii)(C) Receptacles shall be of the grounding type. Unless installed in a complete metallic raceway, each branch circuit shall contain a separate equipment grounding conductor and all receptacles shall be electrically connected to the grounding conductor.

(a)(2)(iii)(D) No bare conductors nor earth returns may be used for the wiring of any temporary circuit.

(a)(2)(iii)(E) Suitable disconnecting switches or plug connectors shall be installed to permit the disconnection of all ungrounded conductors of each temporary circuit.

(a)(2)(iii)(F) Lamps for general illumination shall be protected from accidental contact or breakage. Protection shall be provided by elevation of at least 7 feet from normal working surface or by a suitable fixture or lampholder with a guard.

(a)(2)(iii)(G) Flexible cords and cables shall be protected from accidental damage. Sharp corners and projections shall be avoided. Where passing through doorways or other pinch points, flexible cords and cables shall be provided with protection to avoid damage.

§ 1910.305(a)(3)

(a)(3) Cable trays.

(a)(3)(i) Uses permitted.

(a)(3)(i)(A) Only the following may be installed in cable tray systems:

(a)(3)(i)(A)(*1*) Mineral-insulated metal-sheathed cable (Type MI);

(a)(3)(i)(A)(*2*) Armored cable (Type AC);

(a)(3)(i)(A)(*3*) Metal-clad cable (Type MC);

(a)(3)(i)(A)(*4*) Power-limited tray cable (Type PLTC);

(a)(3)(i)(A)(*5*) Nonmetallic-sheathed cable (Type NM or NMC);

(a)(3)(i)(A)(*6*) Shielded nonmetallic-sheathed cable (Type SNM);

(a)(3)(i)(A)(*7*) Multiconductor service-entrance cable (Type SE or USE);

(a)(3)(i)(A)(*8*) Multiconductor underground feeder and branch-circuit cable (Type UF);

(a)(3)(i)(A)(*9*) Power and control tray cable (Type TC);

(a)(3)(i)(A)(*10*) Other factory-assembled, multiconductor control, signal, or power cables which are specifically approved for installation in cable trays; or

§ 1910.305(a)(3)(i)(A)(*11*)

(a)(3)(i)(A)(*11*) Any approved conduit or raceway with its contained conductors.

(a)(3)(i)(B) In industrial establishments only, where conditions of maintenance and supervision assure that only qualified persons will service the installed cable tray system, the following cables may also be installed in ladder, ventilated trough, or 4 inch ventilated channel-type cable trays:

(a)(3)(i)(B)(*1*) Single conductor cables which are 250 MCM or larger and are Types RHH, RHW, MV, USE, or THW, and other 250 MCM or larger single conductor cables if specifically approved for installation in cable trays. Where exposed to direct rays of the sun, cables shall be sunlight-resistant.

(a)(3)(i)(B)(*2*) Type MV cables, where exposed to direct rays of the sun, shall be sunlight-resistant.

(a)(3)(i)(C) Cable trays in hazardous (classified) locations shall contain only the cable types permitted in such locations.

(a)(3)(ii) Uses not permitted. Cable tray systems may not be used in hoistways or where subjected to severe physical damage.

(a)(4) Open wiring on insulators—

(a)(4)(i) Uses permitted. Open wiring on insulators is only permitted on systems of 600 volts, nominal, or less for industrial or agricultural establishments and for services.

§ 1910.305(a)(4)(ii)

(a)(4)(ii) Conductor supports. Conductors shall be rigidly supported on noncombustible, nonabsorbent insulating materials and may not contact any other objects.

(a)(4)(iii) Flexible nonmetallic tubing. In dry locations where not exposed to severe physical damage, conductors may be separately enclosed in flexible nonmetallic tubing. The tubing shall be in continuous lengths not exceeding 15 feet and secured to the surface by straps at intervals not exceeding 4 feet 6 inches.

(a)(4)(iv) Through walls, floors, wood cross members, etc. Open conductors shall be separated from contact with walls, floors, wood cross members, or partitions through which they pass by tubes or bushings of noncombustible, nonabsorbent insulating material. If the bushing is shorter than the hole, a waterproof sleeve of nonconductive material shall be inserted in the hole and an insulating bushing slipped into the sleeve at each end in such a

FIGURE 6.1 (*Continued*)

manner as to keep the conductors absolutely out of contact with the sleeve. Each conductor shall be carried through a separate tube or sleeve.

(a)(4)(v) Protection from physical damage. Conductors within 7 feet from the floor are considered exposed to physical damage. Where open conductors cross ceiling joints and wall studs and are exposed to physical damage, they shall be protected.

§ 1910.305(b)

(b) Cabinets, boxes, and fittings—

(b)(1) Conductors entering boxes, cabinets, or fittings. Conductors entering boxes, cabinets, or fittings shall also be protected from abrasion, and openings through which conductors enter shall be effectively closed. Unused openings in cabinets, boxes, and fittings shall be effectively closed.

(b)(2) Covers and canopies. All pull boxes, junction boxes, and fittings shall be provided with covers approved for the purpose. If metal covers are used they shall be grounded. In completed installations each outlet box shall have a cover, faceplate, or fixture canopy. Covers of outlet boxes having holes through which flexible cord pendants pass shall be provided with bushings designed for the purpose or shall have smooth, well-rounded surfaces on which the cords may bear.

(b)(3) Pull and junction boxes for systems over 600 volts, nominal. In addition to other requirements in this section for pull and junction boxes, the following shall apply to these boxes for systems over 600 volts, nominal:

(b)(3)(i) Boxes shall provide a complete enclosure for the contained conductors or cables.

(b)(3)(ii) Boxes shall be closed by suitable covers securely fastened in place. Underground box covers that weigh over 100 pounds meet this requirement. Covers for boxes shall be permanently marked "HIGH VOLTAGE." The marking shall be on the outside of the box cover and shall be readily visible and legible.

§ 1910.305(c)

(c) Switches—

(c)(1) Knife switches. Single-throw knife switches shall be so connected that the blades are dead when the switch is in the open position. Single-throw knife switches shall be so placed that gravity will not tend to close them. Single-throw knife switches approved for use in the inverted position shall be provided with

a locking device that will ensure that the blades remain in the open position when so set. Double-throw knife switches may be mounted so that the throw will be either vertical or horizontal. However, if the throw is vertical a locking device shall be provided to ensure that the blades remain in the open position when so set.

(c)(2) Faceplates for flush-mounted snap switches. Flush snap switches that are mounted in ungrounded metal boxes and located within reach of conducting floors or other conducting surfaces shall be provided with faceplates of nonconducting, noncombustible material.

(d) Switchboards and panelboards. Switchboards that have any exposed live parts shall be located in permanently dry locations and accessible only to qualified persons. Panelboards shall be mounted in cabinets, cutout boxes, or enclosures approved for the purpose and shall be dead front. However, panelboards other than the dead front externally-operable type are permitted where accessible only to qualified persons. Exposed blades of knife switches shall be dead when open.

(e) Enclosures for damp or wet locations.

(e)(1) Cabinets, cutout boxes, fittings, boxes, and panelboard enclosures in damp or wet locations shall be installed so as to prevent moisture or water from entering and accumulating within the enclosures. In wet locations the enclosures shall be weatherproof.

(e)(2) Switches, circuit breakers, and switchboards installed in wet locations shall be enclosed in weatherproof enclosures.

§ 1910.305(f)

(f) Conductors for general wiring. All conductors used for general wiring shall be insulated unless otherwise permitted in this Subpart. The conductor insulation shall be of a type that is approved for the voltage, operating temperature, and location of use. Insulated conductors shall be distinguishable by appropriate color or other suitable means as being grounded conductors, ungrounded conductors, or equipment grounding conductors.

(g) Flexible cords and cables—

(g)(1) Use of flexible cords and cables.

(g)(1)(i) Flexible cords and cables shall be approved and suitable for conditions of use and location. Flexible cords and cables shall be used only for:

(g)(1)(i)(A) Pendants;

(g)(1)(i)(B) Wiring of fixtures;

FIGURE 6.1 (*Continued*)

(g)(1)(i)(C) Connection of portable lamps or appliances;

(g)(1)(i)(D) Elevator cables;

(g)(1)(i)(E) Wiring of cranes and hoists;

(g)(1)(i)(F) Connection of stationary equipment to facilitate their frequent interchange;

(g)(1)(i)(G) Prevention of the transmission of noise or vibration;

(g)(1)(i)(H) Appliances where the fastening means and mechanical connections are designed to permit removal for maintenance and repair; or

§ **1910.305(g)(1)(i)(i)**

(g)(1)(i)(I) Data processing cables approved as a part of the data processing system.

(g)(1)(ii) If used as permitted in paragraphs (g)(1)(i)(C), (g)(1)(i)(F), or (g)(1)(i)(H) of this section, the flexible cord shall be equipped with an attachment plug and shall be energized from an approved receptacle outlet.

(g)(1)(iii) Unless specifically permitted in paragraph (g)(1)(i) of this section, flexible cords and cables may not be used:

(g)(1)(iii)(A) As a substitute for the fixed wiring of a structure;

(g)(1)(iii)(B) Where run through holes in walls, ceilings, or floors;

(g)(1)(iii)(C) Where run through doorways, windows, or similar openings;

(g)(1)(iii)(D) Where attached to building surfaces; or

(g)(1)(iii)(E) Where concealed behind building walls, ceilings, or floors.

(g)(1)(iv) Flexible cords used in show windows and showcases shall be Type S, SO, SJ, SJO, ST, STO, SJT, SJTO, or AFS except for the wiring of chain-supported lighting fixtures and supply cords for portable lamps and other merchandise being displayed or exhibited.

§ **1910.305(g)(2)**

(g)(2) Identification, splices, and terminations.

(g)(2)(i) A conductor of a flexible cord or cable that is used as a grounded conductor or an equipment grounding conductor shall be distinguishable from other conductors. Types SJ, SJO, SJT, SJTO, S, SO, ST, and STO shall be durably marked on the surface with the type designation, size, and number of conductors.

(g)(2)(ii) Flexible cords shall be used only in continuous lengths without splice or tap. Hard service flexible cords No. 12 or larger may be repaired if spliced so that the splice retains the insulation, outer sheath properties, and usage characteristics of the cord being spliced.

(g)(2)(iii) Flexible cords shall be connected to devices and fittings so that strain relief is provided which will prevent pull from being directly transmitted to joints or terminal screws.

(h) Portable cables over 600 volts, nominal. Multiconductor portable cable for use in supplying power to portable or mobile equipment at over 600 volts, nominal, shall consist of No. 8 or larger conductors employing flexible stranding. Cables operated at over 2,000 volts shall be shielded for the purpose of confining the voltage stresses to the insulation. Grounding conductors shall be provided. Connectors for these cables shall be of a locking type with provisions to prevent their opening or closing while energized. Strain relief shall be provided at connections and terminations. Portable cables may not be operated with splices unless the splices are of the permanent molded, vulcanized, or other approved type. Termination enclosures shall be suitably marked with a high voltage hazard warning, and terminations shall be accessible only to authorized and qualified personnel.

§ **1910.305(i)**

(i) Fixture wires—

(i)(1) General. Fixture wires shall be approved for the voltage, temperature, and location of use. A fixture wire which is used as a grounded conductor shall be identified.

(i)(2) Uses permitted. Fixture wires may be used:

(i)(2)(i) For installation in lighting fixtures and in similar equipment where enclosed or protected and not subject to bending or twisting in use; or

(i)(2)(ii) For connecting lighting fixtures to the branch-circuit conductors supplying the fixtures.

(i)(3) Uses not permitted. Fixture wires may not be used as branch-circuit conductors except as permitted for Class 1 power limited circuits.

(j) Equipment for general use—

(j)(1) Lighting fixtures, lampholders, lamps, and receptacles.

(j)(1)(i) Fixtures, lampholders, lamps, rosettes, and receptacles may have no live parts normally exposed to employee contact. However, rosettes and cleat-type lampholders and receptacles located at least 8 feet above the floor may have exposed parts.

(j)(1)(ii) Handlamps of the portable type supplied through flexible cords shall be

FIGURE 6.1   *(Continued)*

equipped with a handle of molded composition or other material approved for the purpose, and a substantial guard shall be attached to the lampholder or the handle.

§ 1910.305(j)(1)(iii)

(j)(1)(iii) Lampholders of the screw-shell type shall be installed for use as lampholders only. Lampholders installed in wet or damp locations shall be of the weatherproof type.

(j)(1)(iv) Fixtures installed in wet or damp locations shall be approved for the purpose and shall be so constructed or installed that water cannot enter or accumulate in wireways, lampholders, or other electrical parts.

(j)(2) Receptacles, cord connectors, and attachment plugs (caps).

(j)(2)(i) Receptacles, cord connectors, and attachment plugs shall be constructed so that no receptacle or cord connector will accept an attachment plug with a different voltage or current rating than that for which the device is intended. However, a 20-ampere T-slot receptacle or cord connector may accept a 15-ampere attachment plug of the same voltage rating.

(j)(2)(ii) A receptacle installed in a wet or damp location shall be suitable for the location.

(j)(3) Appliances.

(j)(3)(i) Appliances, other than those in which the current-carrying parts at high temperatures are necessarily exposed, may have no live parts normally exposed to employee contact.

(j)(3)(ii) A means shall be provided to disconnect each appliance.

§ 1910.305(j)(3)(iii)

(j)(3)(iii) Each appliance shall be marked with its rating in volts and amperes or volts and watts.

(j)(4) Motors. This paragraph applies to motors, motor circuits, and controllers.

(j)(4)(i) In sight from. If specified that one piece of equipment shall be "in sight from" another piece of equipment, one shall be visible and not more than 50 feet from the other.

(j)(4)(ii) Disconnecting means.

(j)(4)(ii)(A) A disconnecting means shall be located in sight from the controller location. However, a single disconnecting means may be located adjacent to a group of coordinated controllers mounted adjacent to each other on a multi-motor continuous process machine. The controller disconnecting means for motor branch circuits over 600 volts, nominal, may be out of sight of the controller, if the controller is

marked with a warning label giving the location and identification of the disconnecting means which is to be locked in the open position.

(j)(4)(ii)(B) The disconnecting means shall disconnect the motor and the controller from all ungrounded supply conductors and shall be so designed that no pole can be operated independently.

§ 1910.305(j)(4)(ii)(C)

(j)(4)(ii)(C) If a motor and the driven machinery are not in sight from the controller location, the installation shall comply with one of the following conditions:

(j)(4)(ii)(C)(*1*) The controller disconnecting means shall be capable of being locked in the open position.

(j)(4)(ii)(C)(*2*) A manually operable switch that will disconnect the motor from its source of supply shall be placed in sight from the motor location.

(j)(4)(ii)(D) The disconnecting means shall plainly indicate whether it is in the open (off) or closed (on) position.

(j)(4)(ii)(E) The disconnecting means shall be readily accessible. If more than one disconnect is provided for the same equipment, only one need be readily accessible.

(j)(4)(ii)(F) An individual disconnecting means shall be provided for each motor, but a single disconnecting means may be used for a group of motors under any one of the following conditions:

(j)(4)(ii)(F)(*1*) If a number of motors drive special parts of a single machine or piece of apparatus, such as a metal or woodworking machine, crane, or hoist;

(j)(4)(ii)(F)(*2*) If a group of motors is under the protection of one set of branch-circuit protective devices; or

§ 1910.305(j)(4)(ii)(F)(*3*)

(j)(4)(ii)(F)(*3*) If a group of motors is in a single room in sight from the location of the disconnecting means.

(j)(4)(iii) Motor overload, short-circuit, and ground-fault protection. Motors, motor-control apparatus, and motor branch-circuit conductors shall be protected against overheating due to motor overloads or failure to start, and against short-circuits or ground faults. These provisions shall not require overload protection that will stop a motor where a shutdown is likely to introduce additional or increased hazards, as in the case of fire pumps, or where continued operation

FIGURE 6.1   (*Continued*)

of a motor is necessary for a safe shutdown of equipment or process and motor overload sensing devices are connected to a supervised alarm.

(j)(4)(iv) Protection of live parts—all voltages.

(j)(4)(iv)(A) Stationary motors having commutators, collectors, and brush rigging located inside of motor end brackets and not conductively connected to supply circuits operating at more than 150 volts to ground need not have such parts guarded. Exposed live parts of motors and controllers operating at 50 volts or more between terminals shall be guarded against accidental contact by any of the following:

(j)(4)(iv)(A)(*1*) By installation in a room or enclosure that is accessible only to qualified persons;

§ 1910.305(j)(4)(iv)(A)(2)

(j)(4)(iv)(A)(*2*) By installation on a suitable balcony, gallery, or platform, so elevated and arranged as to exclude unqualified persons; or

(j)(4)(iv)(A)(*3*) By elevation 8 feet or more above the floor.

(j)(4)(iv)(B) Where live parts of motors or controllers operating at over 150 volts to ground are guarded against accidental contact only by location, and where adjustment or other attendance may be necessary during the operation of the apparatus, suitable insulating mats or platforms shall be provided so that the attendant cannot readily touch live parts unless standing on the mats or platforms.

(j)(5) Transformers.

(j)(5)(i) The following paragraphs cover the installation of all transformers except the following:

(j)(5)(i)(A) Current transformers;

(j)(5)(i)(B) Dry-type transformers installed as a component part of other apparatus;

(j)(5)(i)(C) Transformers which are an integral part of an X-ray, high frequency, or electrostatic-coating apparatus;

§ 1910.305(j)(5)(i)(D)

(j)(5)(i)(D) Transformers used with Class 2 and Class 3 circuits, sign and outline lighting, electric discharge lighting, and power-limited fire-protective signaling circuits; and

(j)(5)(i)(E) Liquid-filled or dry-type transformers used for research, development, or testing, where effective safeguard arrangements are provided.

(j)(5)(ii) The operating voltage of exposed live parts of transformer installations shall be indicated by warning signs or visible markings on the equipment or structure.

(j)(5)(iii) Dry-type, high fire point liquid-insulated, and askarel-insulated transformers installed indoors and rated over 35kV shall be in a vault.

(j)(5)(iv) If they present a fire hazard to employees, oil-insulated transformers installed indoors shall be in a vault.

(j)(5)(v) Combustible material, combustible buildings and parts of buildings, fire escapes, and door and window openings shall be safeguarded from fires which may originate in oil-insulated transformers attached to or adjacent to a building or combustible material.

(j)(5)(vi) Transformer vaults shall be constructed so as to contain fire and combustible liquids within the vault and to prevent unauthorized access. Locks and latches shall be so arranged that a vault door can be readily opened from the inside.

§ 1910.305(j)(5)(vii)

(j)(5)(vii) Any pipe or duct system foreign to the vault installation may not enter or pass through a transformer vault.

(j)(5)(viii) Materials may not be stored in transformer vaults.

(j)(6) Capacitors.

(j)(6)(i) All capacitors, except surge capacitors or capacitors included as a component part of other apparatus, shall be provided with an automatic means of draining the stored charge after the capacitor is disconnected from its source of supply.

(j)(6)(ii) Capacitors rated over 600 volts, nominal, shall comply with the following additional requirements:

(j)(6)(ii)(A) Isolating or disconnecting switches (with no interrupting rating) shall be interlocked with the load interrupting device or shall be provided with prominently displayed caution signs to prevent switching load current.

(j)(6)(ii)(B) For series capacitors (see 1910.302(b)(3)), the proper switching shall be assured by use of at least one of the following:

(j)(6)(ii)(B)(*1*) Mechanically sequenced isolating and bypass switches,

(j)(6)(ii)(B)(*2*) Interlocks, or

§ 1910.305(j)(6)(ii)(B)(3)

(j)(6)(ii)(B)(*3*) Switching procedure prominently displayed at the switching location.

(j)(7) Storage batteries. Provisions shall be made for sufficient diffusion and ventilation of

FIGURE 6.1 (*Continued*)

gases from storage batteries to prevent the accumulation of explosive mixtures.

[46 FR 4056, Jan. 16, 1981; 46 FR 40185, Aug. 7, 1981]

§ 1910.306 Specific purpose equipment and installations.

(a) Electric signs and outline lighting—
(a)(1) Disconnecting means. Signs operated by electronic or electromechanical controllers located outside the sign shall have a disconnecting means located inside the controller enclosure or within sight of the controller location, and it shall be capable of being locked in the open position. Such disconnecting means shall have no pole that can be operated independently, and it shall open all ungrounded conductors that supply the controller and sign. All other signs, except the portable type, and all outline lighting installations shall have an externally operable disconnecting means which can open all ungrounded conductors and is within the sight of the sign or outline lighting it controls.

(a)(2) Doors or covers giving access to uninsulated parts of indoor signs or outline lighting exceeding 600 volts and accessible to other than qualified persons shall either be provided with interlock switches to disconnect the primary circuit or shall be so fastened that the use of other than ordinary tools will be necessary to open them.

§ 1910.306(b)

(b) Cranes and hoists. This paragraph applies to the installation of electric equipment and wiring used in connection with cranes, monorail hoists, hoists, and all runways.

(b)(1) Disconnecting means. A readily accessible disconnecting means—
(b)(1)(i) shall be provided between the runway contact conductors and the power supply.
(b)(1)(ii) Another disconnecting means, capable of being locked in the open position, shall be provided in the leads from the runway contact conductors or other power supply on any crane or monorail hoist.
(b)(1)(ii)(A) If this additional disconnecting means is not readily accessible from the crane or monorail hoist operating station, means shall be provided at the operating station to open the power circuit to all motors of the crane or monorail hoist.
(b)(1)(ii)(B) The additional disconnect may be omitted if a monorail hoist or hand-

propelled crane bridge installation meets all of the following:
(b)(1)(ii)(B)(1) The unit is floor controlled;
(b)(1)(ii)(B)(2) The unit is within view of the power supply disconnecting means; and
(b)(1)(ii)(B)(3) No fixed work platform has been provided for servicing the unit.

§ 1910.306(b)(2)

(b)(2) Control. A limit switch or other device shall be provided to prevent the load block from passing the safe upper limit of travel of any hoisting mechanism.
(b)(3) Clearance. The dimension of the working space in the direction of access to live parts which may require examination, adjustment, servicing, or maintenance while alive shall be a minimum of 2 feet 6 inches. Where controls are enclosed in cabinets, the door(s) shall either open at least 90 degrees or be removable.
(c) Elevators, dumbwaiters, escalators, and moving walks—
(c)(1) Disconnecting means. Elevators, dumbwaiters, escalators, and moving walks shall have a single means for disconnecting all ungrounded main power supply conductors for each unit.
(c)(2) Warning signs. If interconnections between control panels are necessary for operation of the system on a multicar installation that remains energized from a source other than the disconnecting means, a warning sign shall be mounted on or adjacent to the disconnecting means. The sign shall be clearly legible and shall read "Warning—Parts of the control panel are not de-energized by this switch." (See 1910.302(b)(3).)
(c)(3) Control panels. If control panels are not located in the same space as the drive machine, they shall be located in cabinets with doors or panels capable of being locked closed.

§ 1910.306(d)

(d) Electric welders—disconnecting means.
(d)(1) A disconnecting means shall be provided in the supply circuit for each motor-generator arc welder, and for each AC transformer and DC rectifier arc welder which is not equipped with a disconnect mounted as an integral part of the welder.
(d)(2) A switch or circuit breaker shall be provided by which each resistance welder and its control equipment can be isolated from the supply circuit. The ampere rating of this

FIGURE 6.1 (*Continued*)

disconnecting means may not be less than the supply conductor ampacity.

(e) Data processing systems—disconnecting means. A disconnecting means shall be provided to disconnect the power to all electronic equipment in data processing or computer rooms. This disconnecting means shall be controlled from locations readily accessible to the operator at the principal exit doors. There shall also be a similar disconnecting means to disconnect the air conditioning system serving this area.

(f) X-Ray equipment. This paragraph applies to X-ray equipment for other than medical or dental use.

(f)(1) Disconnecting means.

(f)(1)(i) A disconnecting means shall be provided in the supply circuit. The disconnecting means shall be operable from a location readily accessible from the X-ray control. For equipment connected to a 120-volt branch circuit of 30 amperes or less, a grounding-type attachment plug cap and receptacle of proper rating may serve as a disconnecting means.

§ **1910.306(f)(1)(ii)**

(f)(1)(ii) If more than one piece of equipment is operated from the same high-voltage circuit, each piece or each group of equipment as a unit shall be provided with a high-voltage switch or equivalent disconnecting means. This disconnecting means shall be constructed, enclosed, or located so as to avoid contact by employees with its live parts.

(f)(2) Control—

(f)(2)(i) Radiographic and fluoroscopic types. Radiographic and fluoroscopic-type equipment shall be effectively enclosed or shall have interlocks that de-energize the equipment automatically to prevent ready access to live current-carrying parts.

(f)(2)(ii) Diffraction and irradiation types. Diffraction- and irradiation-type equipment shall be provided with a means to indicate when it is energized unless the equipment or installation is effectively enclosed or is provided with interlocks to prevent access to live current-carrying parts during operation.

(g) Induction and dielectric heating equipment—

(g)(1) Scope. Paragraphs (g)(2) and (g)(3) of this section cover induction and dielectric heating equipment and accessories for industrial and scientific applications, but not for medical or dental applications or for appliances.

(g)(2) Guarding and grounding.

(g)(2)(i) Enclosures. The converting apparatus (including the DC line) and high-frequency electric circuits (excluding the output circuits and remote-control circuits) shall be completely contained within enclosures of noncombustible material.

§ **1910.306(g)(2)(ii)**

(g)(2)(ii) Panel controls. All panel controls shall be of dead-front construction.

(g)(2)(iii) Access to internal equipment. Where doors are used for access to voltages from 500 to 1000 volts AC or DC, either door locks or interlocks shall be provided. Where doors are used for access to voltages of over 1000 volts AC or DC, either mechanical lockouts with a disconnecting means to prevent access until voltage is removed from the cubicle, or both door interlocking and mechanical door locks, shall be provided.

(g)(2)(iv) Warning labels. "Danger" labels shall be attached on the equipment and shall be plainly visible even when doors are open or panels are removed from compartments containing voltages of over 250 volts AC or DC.

(g)(2)(v) Work applicator shielding. Protective cages or adequate shielding shall be used to guard work applicators other than induction heating coils. Induction heating coils shall be protected by insulation and/or refractory materials. Interlock switches shall be used on all hinged access doors, sliding panels, or other such means of access to the applicator. Interlock switches shall be connected in such a manner as to remove all power from the applicator when any one of the access doors or panels is open. Interlocks on access doors or panels are not required if the applicator is an induction heating coil at DC ground potential or operating at less than 150 volts AC.

§ **1910.306(g)(2)(vi)**

(g)(2)(vi) Disconnecting means. A readily accessible disconnecting means shall be provided by which each unit of heating equipment can be isolated from its supply circuit.

(g)(3) Remote control. If remote controls are used for applying power, a selector switch shall be provided and interlocked to provide power from only one control point at a time. Switches operated by foot pressure shall be provided with a shield over the contact button to avoid accidental closing of the switch.

FIGURE 6.1 *(Continued)*

(h) Electrolytic cells.

(h)(1) Scope. These provisions for electrolytic cells apply to the installation of the electrical components and accessory equipment of electrolytic cells, electrolytic cell lines, and process power supply for the production of aluminum, cadmium, chlorine, copper, fluorine, hydrogen peroxide, magnesium, sodium, sodium chlorate, and zinc. Cells used as a source of electric energy and for electroplating processes and cells used for production of hydrogen are not covered by these provisions.

(h)(2) Definitions applicable to this paragraph.

Cell line: An assembly of electrically interconnected electrolytic cells supplied by a source of direct-current power.

Cell line attachments and auxiliary equipment: Cell line attachments and auxiliary equipment include, but are not limited to: auxiliary tanks; process piping; duct work; structural supports; exposed cell line conductors; conduits and other raceways; pumps; positioning equipment and cell cutout or by-pass electrical devices. Auxiliary equipment also includes tools, welding machines, crucibles, and other portable equipment used for operation and maintenance within the electrolytic cell line working zone. In the cell line working zone, auxiliary equipment includes the exposed conductive surfaces of ungrounded cranes and crane-mounted cell-servicing equipment.

Cell line working zone: The cell line working zone is the space envelope wherein operation or maintenance is normally performed on or in the vicinity of exposed energized surfaces of cell lines or their attachments.

Electrolytic cells: A receptacle or vessel in which electrochemical reactions are caused by applying energy for the purpose of refining or producing usable materials.

(h)(3) Application. Installations covered by paragraph (h) of this section shall comply with all applicable provisions of this subpart, except as follows:

§ **1910.306(h)(3)(i)**

(h)(3)(i) Overcurrent protection of electrolytic cell DC process power circuits need not comply with the requirements of 1910.304(e).

(h)(3)(ii) Equipment located or used within the cell line working zone or associated with the cell line DC power circuits need not comply with the provisions of 1910.304(f).

(h)(3)(iii) Electrolytic cells, cell line conductors, cell line attachments, and the wiring of auxiliary equipment and devices within the cell line working zone need not comply with the provisions of 1910.303, and 1910.304 (b) and (c).

(h)(4) Disconnecting means.

(h)(4)(i) If more than one DC cell line process power supply serves the same cell line, a disconnecting means shall be provided on the cell line circuit side of each power supply to disconnect it from the cell line circuit.

(h)(4)(ii) Removable links or removable conductors may be used as the disconnecting means.

§ **1910.306(h)(5)**

(h)(5) Portable electric equipment.

(h)(5)(i) The frames and enclosures of portable electric equipment used within the cell line working zone may not be grounded. However, these frames and enclosures may be grounded if the cell line circuit voltage does not exceed 200 volts DC or if the frames are guarded.

(h)(5)(ii) Ungrounded portable electric equipment shall be distinctively marked and may not be interchangeable with grounded portable electric equipment.

(h)(6) Power supply circuits and receptacles for portable electric equipment.

(h)(6)(i) Circuits supplying power to ungrounded receptacles for hand-held, cord- and plug-connected equipment shall be electrically isolated from any distribution system supplying areas other than the cell line working zone and shall be ungrounded. Power for these circuits shall be supplied through isolating transformers.

(h)(6)(ii) Receptacles and their mating plugs for ungrounded equipment may not have provision for a grounding conductor and shall be of a configuration which prevents their use for equipment required to be grounded.

(h)(6)(iii) Receptacles on circuits supplied by an isolating transformer with an ungrounded secondary shall have a distinctive configuration, shall be distinctively marked, and may not be used in any other location in the plant.

§ **1910.306(h)(7)**

(h)(7) Fixed and portable electric equipment.

(h)(7)(i) AC systems supplying fixed and portable electric equipment within the cell line working zone need not be grounded.

(h)(7)(ii) Exposed conductive surfaces, such as electric equipment housings, cabinets, boxes, motors, raceways and the like that are within

FIGURE 6.1 (*Continued*)

the cell line working zone need not be grounded.

(h)(7)(iii) Auxiliary electrical devices, such as motors, transducers, sensors, control devices, and alarms, mounted on an electrolytic cell or other energized surface, shall be connected by any of the following means:

(h)(7)(iii)(A) Multiconductor hard usage or extra hard usage flexible cord;

(h)(7)(iii)(B) Wire or cable in suitable raceways; or

(h)(7)(iii)(C) Exposed metal conduit, cable tray, armored cable, or similar metallic systems installed with insulating breaks such that they will not cause a potentially hazardous electrical condition.

(h)(7)(iv) Fixed electric equipment may be bonded to the energized conductive surfaces of the cell line, its attachments, or auxiliaries. If fixed electric equipment is mounted on an energized conductive surface, it shall be bonded to that surface.

§ 1910.306(h)(8)

(h)(8) Auxiliary nonelectric connections. Auxiliary nonelectric connections, such as air hoses, water hoses, and the like, to an electrolytic cell, its attachments, or auxiliary equipment may not have continuous conductive reinforcing wire, armor, braids, and the like. Hoses shall be of a nonconductive material.

(h)(9) Cranes and hoists.

(h)(9)(i) The conductive surfaces of cranes and hoists that enter the cell line working zone need not be grounded. The portion of an overhead crane or hoist which contacts an energized electrolytic cell or energized attachments shall be insulated from ground.

(h)(9)(ii) Remote crane or hoist controls which may introduce hazardous electrical conditions into the cell line working zone shall employ one or more of the following systems:

(h)(9)(ii)(A) Insulated and ungrounded control circuit;

(h)(9)(ii)(B) Nonconductive rope operator;

(h)(9)(ii)(C) Pendant pushbutton with nonconductive supporting means and having nonconductive surfaces or ungrounded exposed conductive surfaces; or

(h)(9)(ii)(D) Radio.

(i) Electrically driven or controlled irrigation machines. (See 1910.302(b)(3).)

§ 1910.306(i)(1)

(i)(1) Lightning protection. If an electrically driven or controlled irrigation machine has a

stationary point, a driven ground rod shall be connected to the machine at the stationary point for lightning protection.

(i)(2) Disconnecting means. The main disconnecting means for a center pivot irrigation machine shall be located at the point of connection of electrical power to the machine and shall be readily accessible and capable of being locked in the open position. A disconnecting means shall be provided for each motor and controller.

(j) Swimming pools, fountains, and similar installations—

(j)(1) Scope. Paragraphs (j)(2) through (j)(5) of this section apply to electric wiring for and equipment in or adjacent to all swimming, wading, therapeutic, and decorative pools and fountains, whether permanently installed or storable, and to metallic auxiliary equipment, such as pumps, filters, and similar equipment. Therapeutic pools in health care facilities are exempt from these provisions.

(j)(2) Lighting and receptacles—

(j)(2)(i) Receptacles. A single receptacle of the locking and grounding type that provides power for a permanently installed swimming pool recirculating pump motor may be located not less than 5 feet from the inside walls of a pool. All other receptacles on the property shall be located at least 10 feet from the inside walls of a pool. Receptacles which are located within 15 feet of the inside walls of the pool shall be protected by ground-fault circuit interrupters.

NOTE: In determining these dimensions, the distance to be measured is the shortest path the supply cord of an appliance connected to the receptacle would follow without piercing a floor, wall, or ceiling of a building or other effective permanent barrier.

§ 1910.306(j)(2)(ii)

(j)(2)(ii) Lighting fixtures and lighting outlets.

(j)(2)(ii)(A) Unless they are 12 feet above the maximum water level, lighting fixtures and lighting outlets may not be installed over a pool or over the area extending 5 feet horizontally from the inside walls of a pool. However, a lighting fixture or lighting outlet which has been installed before April 16, 1981, may be located less than 5 feet measured horizontally from the inside walls of a pool if it is at least 5 feet above the surface of the maximum water level and shall be rigidly attached to the existing structure. It shall also be protected by a

FIGURE 6.1 *(Continued)*

ground-fault circuit interrupter installed in the branch circuit supplying the fixture.

(j)(2)(ii)(B) Unless installed 5 feet above the maximum water level and rigidly attached to the structure adjacent to or enclosing the pool, lighting fixtures and lighting outlets installed in the area extending between 5 feet and 10 feet horizontally from the inside walls of a pool shall be protected by a ground-fault circuit interrupter.

(j)(3) Cord- and plug-connected equipment. Flexible cords used with the following equipment may not exceed 3 feet in length and shall have a copper equipment grounding conductor with a grounding type attachment plug.

(j)(3)(i) Cord- and plug-connected lighting fixtures installed within 16 feet of the water surface of permanently installed pools.

(j)(3)(ii) Other cord- and plug-connected, fixed or stationary equipment used with permanently installed pools.

(j)(4) Underwater equipment.

(j)(4)(i) A ground-fault circuit interrupter shall be installed in the branch circuit supplying underwater fixtures operating at more than 15 volts. Equipment installed underwater shall be approved for the purpose.

§ 1910.306(j)(4)(ii)

(j)(4)(ii) No underwater lighting fixtures may be installed for operation at over 150 volts between conductors.

(j)(5) Fountains. All electric equipment operating at more than 15 volts, including power supply cords, used with fountains shall be protected by ground-fault circuit interrupters. (See 1910.302(b)(3).)

[46 FR 4056, Jan. 16, 1981; 46 FR 40185, Aug. 7, 1981]

§ 1910.307 Hazardous (classified) locations.

(a) Scope. This section covers the requirements for electric equipment and wiring in locations which are classified depending on the properties of the flammable vapors, liquids or gases, or combustible dusts or fibers which may be present therein and the likelihood that a flammable or combustible concentration or quantity is present. Hazardous (classified) locations may be found in occupancies such as, but not limited to, the following: aircraft hangars, gasoline dispensing and service stations, bulk storage plants for gasoline or other volatile flammable liquids, paint-finishing process plants, health care facilities, agricultural or other facilities where excessive combustible

dusts may be present, marinas, boat yards, and petroleum and chemical processing plants. Each room, section or area shall be considered individually in determining its classification. These hazardous (classified) locations are assigned six designations as follows:

Class I, Division 1 Class I, Division 2 Class II, Division 1 Class II, Division 2 Class III, Division 1 Class III, Division 2

For definitions of these locations see 1910.399(a). All applicable requirements in this subpart shall apply to hazardous (classified) locations, unless modified by provisions of this section.

(b) Electrical installations. Equipment, wiring methods, and installations of equipment in hazardous (classified) locations shall be intrinsically safe, approved for the hazardous (classified) location, or safe or for the hazardous (classified) location. Requirements for each of these options are as follows:

§ 1910.307(b)(1)

(b)(1) Intrinsically safe. Equipment and associated wiring approved as intrinsically safe shall be permitted in any hazardous (classified) location for which it is approved.

(b)(2) Approved for the hazardous (classified) location.

(b)(2)(i) Equipment shall be approved not only for the class of location but also for the ignitable or combustible properties of the specific gas, vapor, dust, or fiber that will be present.

NOTE: NFPA 70, the National Electrical Code, lists or defines hazardous gases, vapors, and dusts by "Groups" characterized by their ignitable or combustible properties.

(b)(2)(ii) Equipment shall be marked to show the class, group, and operating temperature or temperature range, based on operation in a 40 degrees C ambient, for which it is approved. The temperature marking may not exceed the ignition temperature of the specific gas or vapor to be encountered. However, the following provisions modify this marking requirement for specific equipment:

(b)(2)(ii)(A) Equipment of the non-heat-producing type, such as junction boxes, conduit, and fittings, and equipment of the heat-producing type having a maximum temperature not more than 100 degrees C (212 degrees F) need not have a marked operating temperature or temperature range.

FIGURE 6.1 (*Continued*)

(b)(2)(ii)(B) Fixed lighting fixtures marked for use in Class I, Division 2 locations only, need not be marked to indicate the group.

§ 1910.307(b)(2)(ii)(C)

(b)(2)(ii)(C) Fixed general-purpose equipment in Class I locations, other than lighting fixtures, which is acceptable for use in Class I, Division 2 locations need not be marked with the class, group, division, or operating temperature.

(b)(2)(ii)(D) Fixed dust-tight equipment, other than lighting fixtures, which is acceptable for use in Class II, Division 2 and Class III locations need not be marked with the class, group, division, or operating temperature.

(b)(3) Safe for the hazardous (classified) location. Equipment which is safe for the location shall be of a type and design which the employer demonstrates will provide protection from the hazards arising from the combustibility and flammability of vapors, liquids, gases, dusts, or fibers.

NOTE: The National Electrical Code, NFPA 70, contains guidelines for determining the type and design of equipment and installations which will meet this requirement. The guidelines of this document address electric wiring, equipment, and systems installed in hazardous (classified) locations and contain specific provisions for the following: wiring methods, wiring connections; conductor insulation, flexible cords, sealing and drainage, transformers, capacitors, switches, circuit breakers, fuses, motor controllers, receptacles, attachment plugs, meters, relays, instruments, resistors, generators, motors, lighting fixtures, storage battery charging equipment, electric cranes, electric hoists and similar equipment, utilization equipment, signaling systems, alarm systems, remote control systems, local loud speaker and communication systems, ventilation piping, live parts, lightning surge protection, and grounding. Compliance with these guidelines will constitute one means, but not the only means, of compliance with this paragraph.

(c) Conduits. All conduits shall be threaded and shall be made wrench-tight. Where it is impractical to make a threaded joint tight, a bonding jumper shall be utilized.

(d) Equipment in Division 2 locations. Equipment that has been approved for a Division 1 location may be installed in a Division 2 location of the same class and group. General-purpose equipment or equipment in general-purpose enclosures may be installed in Division 2 locations if the equipment does not constitute a source of ignition under normal operating conditions.

[46 FR 4056, Jan. 16, 1981; 46 FR 40185, Aug. 7, 1981]

§ 1910.308 Special systems.

(a) Systems over 600 volts, nominal. Paragraphs (a)(1) through (4) of this section cover the general requirements for all circuits and equipment operated at over 600 volts.

(a)(1) Wiring methods for fixed installations.

(a)(1)(i) Above-ground conductors shall be installed in rigid metal conduit, in intermediate metal conduit, in cable trays, in cablebus, in other suitable raceways, or as open runs of metal-clad cable suitable for the use and purpose. However, open runs of non-metallic-sheathed cable or of bare conductors or busbars may be installed in locations accessible only to qualified persons. Metallic shielding components, such as tapes, wires, or braids for conductors, shall be grounded. Open runs of insulated wires and cables having a bare lead sheath or a braided outer covering shall be supported in a manner designed to prevent physical damage to the braid or sheath.

(a)(1)(ii) Conductors emerging from the ground shall be enclosed in approved raceways. (See 1910.302(b)(3).)

§ 1910.308(a)(2)

(a)(2) Interrupting and isolating devices.

(a)(2)(i) Circuit breaker installations located indoors shall consist of metal-enclosed units or fire-resistant cell-mounted units. In locations accessible only to qualified personnel, open mounting of circuit breakers is permitted. A means of indicating the open and closed position of circuit breakers shall be provided.

(a)(2)(ii) Fused cutouts installed in buildings or transformer vaults shall be of a type approved for the purpose. They shall be readily accessible for fuse replacement.

(a)(2)(iii) A means shall be provided to completely isolate equipment for inspection and repairs. Isolating means which are not designed to interrupt the load current of the circuit shall be either interlocked with an approved circuit interrupter or provided with a sign warning against opening them under load.

(a)(3) Mobile and portable equipment.

(a)(3)(i) Power cable connections to mobile machines. A metallic enclosure shall be provided on the mobile machine for enclosing the terminals of the power cable. The enclosure

FIGURE 6.1 *(Continued)*

shall include provisions for a solid connection for the ground wire(s) terminal to effectively ground the machine frame. The method of cable termination used shall prevent any strain or pull on the cable from stressing the electrical connections. The enclosure shall have provision for locking so only authorized qualified persons may open it and shall be marked with a sign warning of the presence of energized parts.

§ 1910.308(a)(3)(ii)

(a)(3)(ii) Guarding live parts. All energized switching and control parts shall be enclosed in effectively grounded metal cabinets or enclosures. Circuit breakers and protective equipment shall have the operating means projecting through the metal cabinet or enclosure so these units can be reset without locked doors being opened. Enclosures and metal cabinets shall be locked so that only authorized qualified persons have access and shall be marked with a sign warning of the presence of energized parts. Collector ring assemblies on revolving-type machines (shovels, draglines, etc.) shall be guarded.

(a)(4) Tunnel installation—

(a)(4)(i) Application. The provisions of this paragraph apply to installation and use of high-voltage power distribution and utilization equipment which is portable and/or mobile, such as substations, trailers, cars, mobile shovels, draglines, hoists, drills, dredges, compressors, pumps, conveyors, and underground excavators.

(a)(4)(ii) Conductors. Conductors in tunnels shall be installed in one or more of the following:

(a)(4)(ii)(A) Metal conduit or other metal raceway,

(a)(4)(ii)(B) Type MC cable, or

(a)(4)(ii)(C) Other approved multiconductor cable.

Conductors shall also be so located or guarded as to protect them from physical damage. Multiconductor portable cable may supply mobile equipment. An equipment grounding conductor shall be run with circuit conductors inside the metal raceway or inside the multiconductor cable jacket. The equipment grounding conductor may be insulated or bare.

(a)(4)(iii) Guarding live parts. Bare terminals of transformers, switches, motor controllers, and other equipment shall be enclosed to prevent accidental contact with energized parts. Enclosures for use in tunnels shall be drip-proof, weatherproof, or submersible as required by the environmental conditions.

§ 1910.308(a)(4)(iv)

(a)(4)(iv) Disconnecting means. A disconnecting means that simultaneously opens all ungrounded conductors shall be installed at each transformer or motor location.

(a)(4)(v) Grounding and bonding. All nonenergized metal parts of electric equipment and metal raceways and cable sheaths shall be effectively grounded and bonded to all metal pipes and rails at the portal and at intervals not exceeding 1000 feet throughout the tunnel.

(b) Emergency power systems—

(b)(1) Scope. The provisions for emergency systems apply to circuits, systems, and equipment intended to supply power for illumination and special loads, in the event of failure of the normal supply.

(b)(2) Wiring methods. Emergency circuit wiring shall be kept entirely independent of all other wiring and equipment and may not enter the same raceway, cable, box, or cabinet or other wiring except either where common circuit elements suitable for the purpose are required, or for transferring power from the normal to the emergency source.

(b)(3) Emergency illumination. Where emergency lighting is necessary, the system shall be so arranged that the failure of any individual lighting element, such as the burning out of a light bulb, cannot leave any space in total darkness.

§ 1910.308(c)

(c) Class 1, Class 2, and Class 3 remote control, signaling, and power-limited circuits—

(c)(1) Classification. Class 1, Class 2, or Class 3 remote control, signaling, or power-limited circuits are characterized by their usage and electrical power limitation which differentiates them from light and power circuits. These circuits are classified in accordance with their respective voltage and power limitations as summarized in paragraphs (c)(1)(i) through (c)(1)(iii) of this section.

(c)(1)(i) Class 1 circuits.

(c)(1)(i)(A) A Class 1 power-limited circuit is supplied from a source having a rated output of not more than 30 volts and 1000 volt-amperes.

(c)(1)(i)(B) A Class 1 remote control circuit or a Class 1 signaling circuit has a voltage which does not exceed 600 volts; however, the power output of the source need not be limited.

(c)(1)(ii) Class 2 and Class 3 circuits.

(c)(1)(ii)(A) Power for Class 2 and Class 3 circuits is limited either inherently (in which no

FIGURE 6.1   (Continued)

overcurrent protection is required) or by a combination of a power source and overcurrent protection.

(c)(1)(ii)(B) The maximum circuit voltage is 150 volts AC or DC for a Class 2 inherently limited power source, and 100 volts AC or DC for a Class 3 inherently limited power source.

(c)(1)(ii)(C) The maximum circuit voltage is 30 volts AC and 60 volts DC for a Class 2 power source limited by overcurrent protection, and 150 volts AC or DC for a Class 3 power source limited by overcurrent protection.

§ 1910.308(c)(1)(iii)

(c)(1)(iii) The maximum circuit voltages in paragraphs (c)(1)(i) and (c)(1)(ii) of this section apply to sinusoidal AC or continuous DC power sources, and where wet contact occurrence is not likely.

(c)(2) Marking. A Class 2 or Class 3 power supply unit shall be durably marked where plainly visible to indicate the class of supply and its electrical rating. (See 1910.302(b)(3).)

(d) Fire protective signaling systems. (See 1910.302(b)(3).)

(d)(1) Classifications. Fire protective signaling circuits shall be classified either as non-power limited or power limited.

(d)(2) Power sources. The power sources for use with fire protective signaling circuits shall be either power limited or nonlimited as follows:

(d)(2)(i) The power supply of non-power-limited fire protective signaling circuits shall have an output voltage not in excess of 600 volts.

(d)(2)(ii) The power for power-limited fire protective signaling circuits shall be either inherently limited, in which no overcurrent protection is required, or limited by a combination of a power source and overcurrent protection.

§ 1910.308(d)(3)

(d)(3) Non-power-limited conductor location. Non-power-limited fire protective signaling circuits and Class 1 circuits may occupy the same enclosure, cable, or raceway provided all conductors are insulated for maximum voltage of any conductor within the enclosure, cable, or raceway. Power supply and fire protective signaling circuit conductors are permitted in the same enclosure, cable, or raceway only if connected to the same equipment.

(d)(4) Power-limited conductor location. Where open conductors are installed, power-limited fire protective signaling circuits shall be separated at least 2 inches from conductors of any light, power, Class 1, and non-power-limited fire protective signaling circuits unless a special and equally protective method of conductor separation is employed. Cables and conductors of two or more power-limited fire protective signaling circuits or Class 3 circuits are permitted in the same cable, enclosure, or raceway. Conductors of one or more Class 2 circuits are permitted within the same cable, enclosure, or raceway with conductors of power-limited fire protective signaling circuits provided that the insulation of Class 2 circuit conductors in the cable, enclosure, or raceway is at least that needed for the power-limited fire protective signaling circuits.

(d)(5) Identification. Fire protective signaling circuits shall be identified at terminal and junction locations in a manner which will prevent unintentional interference with the signaling circuit during testing and servicing. Power-limited fire protective signaling circuits shall be durably marked as such where plainly visible at terminations.

§ 1910.308(e)

(e) Communications systems—

(e)(1) Scope. These provisions for communication systems apply to such systems as central-station-connected and non-central-station-connected telephone circuits, radio and television receiving and transmitting equipment, including community antenna television and radio distribution systems, telegraph, district messenger, and outside wiring for fire and burglar alarm, and similar central station systems. These installations need not comply with the provisions of 1910.303 through 1910.308(d), except 1910.304(c)(1) and 1910.307(b).

(e)(2) Protective devices.

(e)(2)(i) Communication circuits so located as to be exposed to accidental contact with light or power conductors operating at over 300 volts shall have each circuit so exposed provided with a protector approved for the purpose.

(e)(2)(ii) Each conductor of a lead-in from an outdoor antenna shall be provided with an antenna discharge unit or other suitable means that will drain static charges from the antenna system.

(e)(3) Conductor location—

(e)(3)(i) Outside of buildings.

(e)(3)(i)(a) Receiving distribution lead-in or aerial-drop cables attached to buildings and

FIGURE 6.1 (*Continued*)

lead-in conductors to radio transmitters shall be so installed as to avoid the possibility of accidental contact with electric light or power conductors.

(e)(3)(i)(b) The clearance between lead-in conductors and any lightning protection conductors may not be less than 6 feet.

§ 1910.308(e)(3)(ii)

(e)(3)(ii) On poles. Where practicable, communication conductors on poles shall be located below the light or power conductors. Communications conductors may not be attached to a crossarm that carries light or power conductors.

(e)(3)(iii) Inside of buildings. Indoor antennas, lead-ins, and other communication conductors attached as open conductors to the inside of buildings shall be located at least 2 inches from conductors of any light or power or Class 1 circuits unless a special and equally protective method of conductor separation, approved for the purpose, is employed.

(c)(4) Equipment location. Outdoor metal structures supporting antennas, as well as self-supporting antennas such as vertical rods or dipole structures, shall be located as far away from overhead conductors of electric light and power circuits of over 150 volts to ground as necessary to avoid the possibility of the antenna or structure falling into or making accidental contact with such circuits.

(e)(5) Grounding—

(e)(5)(i) Lead-in conductors. If exposed to contact with electric light and power conductors, the metal sheath of aerial cables entering buildings shall be grounded or shall be interrupted close to the entrance to the building by an insulating joint or equivalent device. Where protective devices are used, they shall be grounded in an approved manner.

§ 1910.308(e)(5)(ii)

(e)(5)(ii) Antenna structures. Masts and metal structures supporting antennas shall be permanently and effectively grounded without splice or connection in the grounding conductor.

(e)(5)(iii) Equipment enclosures. Transmitters shall be enclosed in a metal frame or grill or separated from the operating space by a barrier, all metallic parts of which are effectively connected to ground. All external metal handles and controls accessible to the operating personnel shall be effectively grounded. Unpowered equipment and enclosures shall be considered grounded where connected to an attached coaxial cable with an effectively grounded metallic shield.

[46 FR 4056, Jan. 16, 1981; 46 FR 40185, Aug. 7, 1981]

FIGURE 6.1 (*Continued*)

Prior to August 6, 1990 the electrical standards in Subpart S of the General Industry Standards covered electrical equipment and installations rather than work practices. The electrical safety-related work practice standards that did exist were distributed in other subparts of 29 CFR 1910. Although unsafe work practices appear to be involved in most workplace electrocutions, OSHA has had very few regulations addressing work practices necessary for electrical safety. Because of this, OSHA determined that standards were needed to minimize these hazards and the risk to employees.

This rule addresses practices and procedures that are necessary to protect employees working on or near exposed energized and deenergized parts of electric equipment. This rule also promotes uniformity and reduces redundancy among the general industry standards. 29 CFR 1910.331-.335 "Electrical Safety-Related Work Practices" is based largely on NFPA 70E "Electrical Safety in the workplace," Chapter I "Safety-Related Work Practices."

On September 1, 1989, OSHA promulgated a generic standard on the control of hazardous energy, 29 CFR 1910.147 (lockout/tagout).

This standard addresses practices and procedures that are necessary to deenergize machinery or equipment and to prevent the release of potentially hazardous energy while maintenance and servicing activities are being performed.

Although this rule is related to electrical energy, it specifically excludes "exposure to electrical hazards from work on, near, or with conductors or equipment in electric utilization installations, which is covered by Subpart S of 29 CFR 1910." Therefore, the lockout/tagout standard does not cover electrical hazards.

The Electrical Safety-Related Work Practice standard has provisions to achieve maximum safety by deenergizing energized parts and, secondly, when lockout/tagout is used, it is done to ensure that the deenergized state is maintained.

The effective date for this standard (29 CFR 1910.331-.335) was December 4, 1990, except for 1910.332 (training), which became effective August 6, 1991.

As of August 6, 1991, the training practices of the employer for qualified and unqualified employees will be evaluated by the OSHA inspector to assess whether the training provided is appropriate to the tasks being performed or to be performed.

Summary (from Preamble):

These performance-oriented regulations complement the existing electrical installation standards. The new standard includes requirements for work performed on or near exposed energized and deenergized parts of electric equipment; use of electrical protective equipment; and the safe use of electric equipment. Compliance with these safe work practices will reduce the number of electrical accidents resulting from unsafe work practices by employees.

The following is a table of electrical fatalities that was compiled by the National Institute of Occupational Safety and Health (NIOSH). This is an annual average.

TABLE 4 Electrical Fatalities by Occupation

Job title	Number of fatalities
Laborer	28
Lineman	20
Electrician	16
Painter	13
Truck driver	7
Machine operator	6
Groundman	6
Maintenance man	5
Technician	5
Crew leader/foreman	4
Construction worker	4
Carpenter	3
Iron worker	3
Brick layer	3
Bill poster	3
Welders	2
Restaurant worker	2
Fireman	2

1910.331 Scope

(a) *Covered work by both qualified and unqualified persons.* The provisions of 1910.331 through 1910.335 cover electrical safety-related work practices for both qualified persons

FIGURE 6.2 Electrical Safety-Related Work Practices (OSHA-CFR, Title 29, Part 1910, Paragraphs 331–335). (*Courtesy OSHA Web site.*)

(those who have training in avoiding the electrical hazards of working on or near exposed energized parts)...

"On" or "near" is mentioned here and will be mentioned throughout this subpart. The definition of "on" is actual contact with the circuit or part (involving either direct contact or contact by means of tools or material); "near" however needs further explanation. "Near" is when a person is near enough to the energized circuits or parts to be exposed to any hazards they present.

More plainly stated; If a person is within reaching, stumbling, or falling distance of an energized circuit or part or close enough to the energized parts that he/she could contact these parts with any part of their body or with any conductive object he/she is handling, that person is doing energized work.

...and unqualified persons (those with little or no such training) working on, near, or with the following installations:

(1) *Premises wiring.* Installations of electric conductors and equipment within or on buildings or other structures, and on other premises such as yards, carnival, parking, and other lots, and industrial substations;

(2) *Wiring for connection to supply.* Installations of conductors that connect to the supply of electricity; and

(3) *Other wiring.* Installations of other outside conductors on the premises.

Any wiring system that has been installed in accordance with the NEC and OSHA Subpart S "Electrical Standard").

(4) *Optical fiber cable.* Installations of optical fiber cable where such installations are made along with electric conductors.

There are several types of optical fiber cable that contain noncurrent-carrying conductive members such as metallic strength members and metallic vapor barriers.

Note: See 1910.399 for the definition of "qualified person." See 1910.332 for training requirements that apply to qualified and unqualified persons.

According to the National Electrical Code (NEC) and OSHA 29 CFR 1910.399, a qualified person is "one that is familiar with the construction and operation of the equipment and the hazards involved."

OSHA states that qualified persons are intended to be only those who are well acquainted with and thoroughly conversant in the electric equipment and electrical hazards involved with the work being performed.

Definitions applicable to this subpart for a qualified person;

"**Note 1:** Whether an employee is considered to be a qualified person will depend on various circumstances in the workplace. It is possible and, in fact, likely for an individual to be considered qualified with regard to certain equipment in the workplace, but unqualified as to other equipment."

In other words just because a person is a journeyman, master or 1st class electrician does not mean he is qualified on all systems and equipment.

Example: If a new piece of equipment is installed and the electrician is not familiar with the construction, operation and hazards involved, then he/she is unqualified as to that equipment.

Note 2: An employee who is undergoing on-the-job training and who, in the course of such training, has demonstrated an ability to perform duties safely at his or her level of training and who is under the direct supervision of a qualified person is considered to be a qualified person for the performance of those duties."

There are numerous references throughout this subpart that state "qualified persons only". For example: Only qualified persons are allowed to work on or near energized circuits or parts; a qualified person shall operate equipment to verify it cannot restart; a qualified person shall use test equipment to test the circuit to verify it is de-energized; only qualified persons may perform testing work on electric circuits, etc. Therefore, if a person were not qualified he/she could never become qualified because they could never do anything. This is the reason for Note 2, quoted above. A person can become qualified if working under the direct supervision of a qualified person.

(b) *Other covered work by unqualified persons.* The provisions of 1910.331 through 1910.335 also cover work performed by unqualified persons on, near, or with the installations listed in paragraphs (c)(1) through (c)(4) of this section.

(c) *Excluded work by qualified persons.* The provisions of 1910.331 through 1910.335

FIGURE 6.2 (*Continued*)

do not apply to work performed by qualified persons on or directly associated with the following installations:

(1) *Generation, transmission, and distribution installations.* Installations for the generation, control, transformation, transmission, and distribution of electric energy (including communication and metering) located in buildings used for such purposes or located outdoors.

OSHA (D of C) states that "Work on the specified electrical installations is excluded, but work on the other electric equipment in the buildings is not excluded."

Note 1: Work on or directly associated with installations of utilization equipment used for purposes other than generating, transmitting, or distributing electric energy (such as installations which are in office buildings, warehouses, garages, machine shops, or recreational buildings, or other utilization installations which are not an integral part of a generating installation, substation, or control center) is covered under paragraph (a)(1) of this section.

Note 2: Work on or directly associated with generation, transmission, or distribution installations includes:

1) Work performed directly on such installations, such as repairing overhead or underground distribution lines or repairing a feed-water pump for the boiler in a generating plant.

2) Work directly associated with such installations, such as line-clearance tree trimming and replacing utility poles.

3) Work on electric utilization circuits in a generating plant provided that:

A) Such circuits are commingled with installations of power generation equipment or circuits, and

B) The generation equipment or circuits present greater electrical hazards than those posed by the utilization equipment or circuits (such as exposure to higher voltages or lack of overcurrent protection).

(2) *Communications installations.* Installations of communication equipment to the extent that the work is covered under 1910.268.

(3) *Installations in vehicles.* Installations in ships, watercraft, railway rolling stock, aircraft, or automotive vehicles other than mobile homes and recreational vehicles.

The OSHA Federal Register for ships and watercraft (marine) is 29 CFR 1915.

(4) *Railway installations.* Installations of railways for generation, transformation, transmission, or distribution of power used exclusively for operation of rolling stock or installations of railways used exclusively for signaling and communication purposes.

1910.332 Training

(a) *Scope.* The training requirements contained in this section apply to employees who face a risk of electric shock that is not reduced to a safe level by the electrical installation requirements of 1910.303 through 1910.308.

(NFPA 70E, Section 1.A) "Safety Training. Employees shall be trained in the safety-related work practices, safety procedures, and other personnel safety requirements in this standard that pertain to their respective job assignments.

Employees shall not be permitted to work in an area where they are likely to encounter electrical hazards unless they have been trained to recognize and avoid the electrical hazards to which they will be exposed.

For the purpose of this standard, an electrical hazard is a recognizable dangerous electrical condition. Electrical conditions such as exposed energized parts and unguarded electric equipment which may become energized unexpectedly are examples of electrical hazards."

(Preamble) "Employees are required to be trained in the safety-related work practices of this standard, as well as any other practices necessary for safety from electrical hazards."

This would include systems and equipment that are enclosed or otherwise rendered inaccessible to employees or are at voltages less than 50 volts to ground. However, when an enclosure cover has been removed or opened for inspection or maintenance of the energized components, the employee faces a risk of electric shock and the requirements of 1910.331 through 1910.335 would apply.

(D of C) "All employees who face a risk of electric shock, burns, or other related injuries, not reduced to a safe level by the installation safety requirements of Subpart S, must be trained in safety-related work practices required by 29 CFR 1910.331-.335."

(Preamble) "The final rule does not contain provisions addressing the qualifications of trainers or the demonstration of effectiveness of the training."

FIGURE 6.2 *(Continued)*

TABLE S-4 Typical Occupational Categories of Employees Facing A Higher Than Normal Risk of Electrical Accident

Occupation
Blue collar supervisors.*
Electrical and electronic engineers.*
Electrical and electronic equipment assemblers.*
Electrical and electronic technicians.*
Electricians
Industrial machine operators.*
Material handling equipment operators.*
Mechanics and repairers.*
Painters.*
Riggers and roustabouts.*
Stationary engineers.*
Welders.

*Workers in these groups do not need to be trained if their work or the work of those they supervise does not bring them or the employees they supervise close enough to exposed parts of electric circuits operating at 50 volts or more to ground for a hazard to exist.

Note: Employees in occupations listed in Table S-4 face such a risk and are required to be trained …

…Other employees who also may reasonably be expected to face comparable risk of injury due to electric shock or other electrical hazards must also be trained.

This last statement places an additional responsibility on the employer to evaluate the work situations of every employee, not just the ones listed in Table S-4, to determine if they face a risk of electrical shock or other electrical hazards and to provide training for them.

(b) Content of training. (1) Practices addressed in this standard. Employees shall be trained in and familiar with the safety-related work practices required by 1910.331 through 1910.335 that pertain to their respective job assignments.

This statement makes it clear that if an employee is not exposed to all of the hazards or equipment addressed by this requirement, he/she is not required to be trained on it but must be trained on those that do present a hazard.

Example: A painter is not a qualified electrical worker and therefore would not

require training in the use of electrical test equipment, the need to defeat an interlock, use electrical protective equipment, or deenergize an electrical system and install a lockout/tagout. However he/she would require training on clearance distances from overhead lines, the purpose of lockout/tagout, requirements for use of portable cord- and plug-connected equipment and the use of flammable liquids near possible spark producing equipment.

(2) Additional requirements for unqualified persons. Employees who are covered by paragraph (a) of this section but who are not qualified persons shall also be trained in and familiar with any electrically related safety practices not specifically addressed by 1910.331 through 1910.335 but which are necessary for their safety.

(OSHA's D of C) "In addition to being trained in and familiar with safety-related work practices, unqualified employees must be trained in the inherent hazards of electricity, such as high voltages, electric current, arcing, grounding, and lack of guarding. Any electrically related safety practices not specifically addressed by Sections 1910.331 through 1910.335 but necessary for safety in specific workplace conditions shall be included."

This requirement will cause an employer to evaluate the work hazards of every employee to determine if there are any other hazards that were not addressed in this standard.

(3) Additional requirements for qualified persons. Qualified persons (i.e. those permitted to work on or near exposed energized parts)...

(D of C) "The standard defines a qualified person as one familiar with the construction and operation of the equipment and the hazards involved.

Qualified Persons are intended to be only those who are well acquainted with and thoroughly conversant in the electric equipment and electrical hazards involved with the work being performed."

...shall, at a minimum, be trained in and familiar with the following:
(i) The skills and techniques necessary to distinguish exposed live parts from other parts of electric equipment.
(ii) The skills and techniques necessary to determine the nominal voltage of exposed live parts, and

FIGURE 6.2 *(Continued)*

(iii) The clearance distances specified in 1910.333(c) and the corresponding voltages to which the qualified person will be exposed.

Note 1: For the purposes of 1910.331 through 1910.335, a person must have the training required by paragraph (b)(3) of this section in order to be considered a qualified person.

Note 2: Qualified persons whose work on energized equipment involves either direct contact or contact by means of tools or materials must also have the training needed to meet 1910.333(C)(2).

(c) *Type of training.* The training required by this section shall be of the classroom or on-the-job type. The degree of training provided shall be determined by the risk to the employee.

In most cases, both classroom and on-the-job type training would be needed.

(D of C) "The failure to train 'qualified' and 'unqualified' employees, as required for their respective classifications, shall normally be cited as a serious violation."

1910.333 Selection and Use of Work Practices

(Preamble) "The basic intent of 1910.333 is to require employers to take one of three options to protect employees working on electric circuits and equipment: (1) Deenergize the equipment involved and lockout its disconnecting means; or (2) deenergize the equipment and tag the disconnecting means, if the employer can demonstrate that tagging is as safe as locking; or (3) work the equipment energized if the employer can demonstrate that it is not feasible to deenergize it.

Paragraph (a) of 1910.333 sets forth general requirements on the selection and use of work practices. The requirements of this paragraph mainly address accidents which involve the hazards of exposure to live parts of electric equipment. A deenergized part is obviously safer than an energized one. Because the next best method of protecting an employee working on exposed parts of electric equipment (the use of personal protective equipment) would continue to expose that employee to a risk of injury from electric shock, 1910.333(a) makes equipment deenergizing the primary method of protecting employees.

Under certain conditions, however, deenergizing need not be employed. Employees may be allowed to work on or near exposed energized parts as allowed under 1910.333(a)(1)."

(a) *General.* Safety-related work practices shall be employed to prevent electric shock or other injuries resulting from either direct or indirect electrical contacts, when work is performed near or on equipment or circuits which are or may be energized. The specific safety-related work practices shall be consistent with the nature and extent of the associated electrical hazards.

(D of C) "1910.333(a)(1) requires that live parts be deenergized before a potentially exposed employee works on or near them. OSHA believes that this is the preferred method for protecting employees from electrical hazards. The employer is permitted to allow employees to work on or near exposed live parts only:

(1) If the employer can demonstrate that deenergizing introduces additional or increased hazards (see Note 1 below),

(2) If the employer can demonstrate that deenergizing is infeasible due to equipment design or operational limitations (see Note 2 below)."

(1) *Deenergized parts.* Live parts to which an employee may be exposed shall be deenergized before the employee works on or near them, unless the employer can demonstrate that deenergizing introduces additional or increased hazards, or is infeasible due to equipment design or operational limitations. Live parts that operate at less than 50 volts to ground need not be deenergized if there will be no increased exposure to electrical burns or to explosion due to electric arcs.

Note 1: Examples of increased or additional hazards include interruption of life support equipment, deactivation of emergency alarm systems, shutdown of hazardous location ventilation equipment, or removal of illumination for an area.

Note 2: Examples of work that may be performed on or near energized circuit parts because of infeasibility due to equipment design or operational limitations include testing of electric circuits that can only be performed with the circuit energized and work on circuits that form an integral part of a continuous industrial process in a chemical plant that would otherwise need to be completely shut down in order to permit work on one circuit or piece of equipment.

FIGURE 6.2 *(Continued)*

Note 3: Work on or near deenergized parts is covered by paragraph (b) of this section.

(D of C) "Under 1910.333(a)(2) if the employer does not deenergize (under the conditions permitted in 1910.333(a)(1)), then suitable safe work practices for the conditions under which the work is to be performed shall be included in the **written procedures** and **strictly enforced**. These work practices are given in 1910.333(c) and 1910.335.

Only qualified persons shall be allowed to work on energized parts or equipment."

(2) Energized parts. If the exposed live parts are not deenergized (i.e., for reasons of increased or additional hazards or infeasibility), other safety-related work practices shall be used to protect employees who may be exposed to the electrical hazards involved. Such work practices shall protect employees against contact with energized circuit parts directly with any part of their body or indirectly through some other conductive object. The work practices that are used shall be suitable for the conditions under which the work is to be performed and for the voltage level of the exposed electric conductors or circuit parts. Specific work practice requirements are detailed in paragraph (c) of this section.

(D of C) "If the employer does not deenergize (under the conditions permitted in 1910.333(a)(1)), then suitable safe work practices for the conditions under which the work is to be performed shall be included in the written procedure and strictly enforced."

(b) *Working on or near exposed deenergized parts.* **(1)** *Application.* This paragraph applies to work on exposed deenergized parts or near enough to them to expose the employee to any electrical hazard they present. Conductors and parts of electric equipment that have been deenergized but have not been locked out or tagged in accordance with paragraph (b) of this section shall be treated as energized parts, and paragraph (c) of this section applies to work on or near them.

(D of C) "Circuits parts that cannot be deenergized using the procedures outlined in 1910.333(b)(2) must be treated as energized, regardless of whether the parts are turned off and assumed to be deenergized."

The lockout/tagout requirements of this regulation were derived from the existing ANSI and NFPA standards.

(Preamble) "The NFPA 70E requirements were intended to apply any time work is performed on or near deenergized circuit parts or equipment in any situation which presents a danger that the circuit parts or equipment might become unexpectedly energized. Thus, the NFPA provisions not only address the hazard of contact with energized parts, but also cover other hazards which are presented by start-up of equipment during maintenance operations. This is expressed in NFPA 70E, Part II, Section 1.B, second paragraph, which states:

'Where the work to be performed requires employees to work on or near exposed circuit parts or equipment, and there is danger of injury due to electric shock, unexpected movement of equipment, or other electrical hazards, the circuit parts and equipment that endanger the employees shall be deenergized and locked out or tagged out in accordance with the policies and procedures specified.'

In contrast to this, OSHA Subpart S standard is intended to cover employee exposure to electrical hazards which might occur from the unexpected energizing of the circuit parts and does not cover other equipment-related hazards which do not involve exposed live parts. Thus, the Subpart S standard will protect the electrician working on a circuit but does not address a mechanic working on the mechanical parts of an electrically powered machine. The generic lockout standard, 1910.147, covers the hazards of unexpected activation or energization of machinery or equipment during servicing or maintenance activities."

(2) Lockout and Tagging. While any employee is exposed to contact with parts of fixed electric equipment or circuits which have been deenergized, the circuits energizing the parts shall be locked out or tagged or both in accordance with the requirements of this paragraph. The requirements shall be followed in the order in which they are presented (i.e., paragraph (b)(2)(i) first, then paragraph (b)(2)(ii), etc.).

Note 1: As used in this section, fixed equipment refers to equipment fastened in place or connected by permanent wiring methods.

Note 2: Lockout and tagging procedures that comply with paragraphs (c) through (f) of 1910.147 will also be deemed to comply with paragraph (b)(2) of this section provided that:

(1) The procedures address the electrical safety hazards covered by this Subpart; and

FIGURE 6.2 *(Continued)*

(2) The procedures also incorporate the requirements of paragraphs (b)(2)(iii)(D) and (b)(2)(iv)(B) of this section.

(Preamble) "In comparing 1910.333(b)(2) with final 1910.147, OSHA determined that the generic lockout standard encompassed all the lockout and tagging requirements contained in the electrical work practices with two exceptions. First, the electrical standard more tightly restricted the use of tags without locks and called for additional protection when tags were permitted. Secondly, the electrical work practices contain specific requirements for testing circuit parts for voltage before they could be considered as deenergized.

OSHA believes that a lockout and tagging program which meets 1910 .147 will, with these two exceptions, provide protection for servicing and maintenance involving electrical work and live parts. Accordingly, the final rule on electrical safety-related work practices incorporates this finding."

(D of C) "Only qualified persons may place and remove locks and tags."

(Preamble) "OSHA has decided to accept any lockout and tagging program that conforms to 1910.147 if it also meets paragraphs 1910.333 (b)(2)(iii)(D) and (b)(2)(iv) (B). This will enable employers to use a single lockout and tagout program to cover all hazards addressed by these two standards, as long as that program includes procedures that meet the two additional paragraphs when exposure to electric shock is involved.

OSHA has limited the application of the lockout and tagging provisions to fixed equipment. Employees can safely work on cord-and plug-connected, portable and stationary equipment which is disconnected from the circuit. Generally, such equipment is returned to a maintenance shop for repair, and the danger of accidental energizing of equipment parts is eliminated in most cases. However, in 1910.147 (the generic lockout/ tagout standard) it states: "This standard does not apply to the following:

(A) Work on cord- and plug-connected equipment for which exposure to the hazards of unexpected energization or start-up of the equipment is controlled by the unplugging of the equipment from the energy source and by the plug being under the exclusive control of the employee performing the servicing or maintenance."

This requirement clearly states that if the plug is not under the "exclusive control" of the employee, it must be locked out or tagged out.

"The generic lockout standard (OSHA) and NFPA 70E's lockout provisions contain requirements for lockout procedures to be in writing.

OSHA believes that written procedures are an important part of a successful lockout and tagging program and that they should, therefore, be required. However, the final electrical standard lockout provisions are themselves a step-by-step procedure for safely deenergizing electrical circuits. For this reason, the final electrical work practices standard incorporates a requirement for employers to maintain a copy of the lockout procedure outlined in 1910.333(b)(2) and make it available for inspection by employees and OSHA."

(1910.147) The following is quoted from 1910.147(a)(3) Purpose: "(ii) When other standards in this part (meaning 1910) require the use of lockout or tagout, they shall be used and supplemented by the procedural and training requirements of this section." Therefore, the requirements of the generic standard (1910.147) also apply, along with the additional lockout and tagout requirements of this standard, for all electrical lockout or tagout situations.

(i) *Procedures.* The employer shall maintain a written copy of the procedures outlined in paragraph (b)(2) and shall make it available for inspection by employees and by the Assistant Secretary of Labor and his or her authorized representatives.

(D of C) "During walk around inspections, compliance officers shall evaluate any electrical related work being performed to ascertain conformance with the employer's written procedures as required by 1910.333 (b)(2)(i) and all safety-related work practices in Sections 1910.333 through 1910. 335.

A deficiency in the employer's program that could contribute to a potential exposure capable of producing serious physical harm or death shall be cited as a serious violation."

Note: The written procedures may be in the form of a copy of paragraph (b) of this section.

(ii) *Deenergizing equipment.* **(A)** Safe procedures for deenergizing circuits and equipment shall be determined before circuits or equipment are deenergized.

FIGURE 6.2 *(Continued)*

1910.147(d)(1) states this requirement much more clearly, as follows:

"Preparation for shutdown. Before an authorized or affected employee turns off a machine or equipment, the authorized employee shall have knowledge of the type and magnitude of the energy, the hazards of the energy to be controlled, and the method or means to control the energy."

(B) The circuits and equipment to be worked on shall be disconnected from all electric energy sources....

There can be several electric energy sources such as the power circuit, control circuit energized from a separate source, and interlocks from other devices. In the case of a power circuit breaker there could be AC and/or DC sources for the trip coil, closing coil, spring charging motor, and any number of circuits that operate through the auxiliary contacts.

...Control circuit devices, such as push buttons, selector switches, and interlocks, may not be used as the sole means for deenergizing circuits or equipment...

It is a good practice to turn a control device off so the load will be removed prior to opening the disconnecting means. This practice will reduce the risk of arcing in the device and a possible flash-over from phase-to-phase and/or phase-to-ground. These control devices should also be locked and/or tagged before the main disconnecting means is opened.

If this is not done, the disconnecting means could be opened at the same time a control is turned to the on position resulting in a current level far beyond its rating. The following example is used to further illustrate:

If the disconnect switch was rated for a load of 30 amperes, and the full load current was 24 amperes, and if the disconnect switch were to be opened at the same time the motor was being started, the disconnect switch contacts would experience approximately locked motor current which is about six times the full load current or, in this case, 144 amperes. In this case a flash-over from phase-to-phase and/or phase-to-ground would be very likely. This could cause the enclosure to explode and injure the person who is switching the device.

...Interlocks for electric equipment may not be used as a substitute for lockout and tagging procedures.

(C) Stored electric energy which might endanger personnel shall be released. Capacitors shall be discharged and high capacitance elements shall be short-circuited and grounded, if the stored electric energy might endanger personnel.

Note: If the capacitors or associated equipment are handled in meeting this requirement, they shall be treated as energized.

(D) Stored non-electrical energy in devices that could reenergize electric circuit parts shall be blocked or relieved to the extent that the circuit parts could not be accidentally energized by the device.

A good example of this would be charged closing springs on a power circuit breaker, or hydraulic or pneumatic pressure in an oil circuit breaker operating mechanism.

(iii) *Application of locks and tags.* **(A)** A lock and a tag shall be placed on each disconnecting means used to deenergize circuits and equipment on which work is to be performed, except as provided in paragraphs (b)(2)(iii)(C) and (b)(2)(iii)(E) of this section. The lock shall be attached so as to prevent persons from operating the disconnecting means unless they resort to undue force or the use of tools.

1910.147 states, "(C) Substantial, (1) Lockout devices. Lockout devices shall be substantial enough to prevent removal without the use of excessive force or unusual techniques, such as with the use of bolt cutters or other metal cutting tools."

(B) Each tag shall contain a statement prohibiting unauthorized operation of the disconnecting means and removal of the tag.

1910.147 states, "Tagout devices shall warn against hazardous conditions if the machine or equipment is energized and shall include a legend such as the following: Do Not Start, Do Not Open, Do Not Close, Do Not Energize, Do Not Operate."

(C) If a lock cannot be applied, or if the employer can demonstrate that tagging procedures will provide a level of safety equivalent to that obtained by the use of a lock, a tag may be used without a lock.

(D) A tag used without a lock, as permitted by paragraph (b)(2)(iii)(C) of this section, shall be supplemented by at least one additional

FIGURE 6.2 (*Continued*)

safety measure that provides a level of safety equivalent to that obtained by use of a lock. Examples of additional safety measures include the removal of an isolating circuit element, blocking of a controlling switch, or opening of an extra disconnecting device.

(Preamble) "Because a person could operate the disconnecting means before reading or recognizing the tag, the standard also requires that, where tags are used, one or more additional safety measures be taken to provide added safety. The additional measures used must either: (1) Ensure that the closing of the tagged single switch would not reenergize the circuit on which employees are working or (2) virtually prevent the accidental closing of the disconnecting means. The following examples from the record illustrate protective techniques that are commonly used to supplement tags and protect workers:

(1) The removal of a fuse or fuses for a circuit.
(2) The removal of a draw-out circuit breaker from a switchboard (i.e., racking out the breaker).
(3) The placement of a blocking mechanism over the operating handle of a disconnecting means so that the handle is blocked from being placed in the closed position.
(4) The opening of a switch (other than the disconnecting means) which also opens the circuit between the source of power and the exposed parts on which work is to be performed.
(5) The opening of a switch for a control circuit that operates a disconnect that is itself open and disconnected from the control circuit or is otherwise disabled.
(6) Grounding of the circuit upon which work is to be performed.

The additional safety measure is necessary because, at least for electrical disconnecting means, tagging alone is significantly less safe than locking out. A disconnecting means could be closed by an employee who has failed to recognize the purpose of the tag. The disconnect could also be closed accidentally."

It should be noted that, in addition to the measures noted in the preamble, there are other measures that can be taken, such as:

(1) Removing the control power fuses.
(2) Lifting the wires from the circuit breaker, disconnect, or contactor terminals or from a terminal.

It should also be noted that these additional measures could also be used in the case of a lockout, as an additional safety measure.

(Preamble) "Any measure or combination of measures, that will protect employees as well as the application of a lock, can be used to supplement the tagging of the disconnecting means. In determining whether an individual employer's 'tagging-only' program provides safety equivalent to a lock, OSHA will consider factors such as:

(1) The safety record under the employer's tagging program;
(2) Whether the tagged disconnecting means are accessible to qualified persons only;
(3) Whether the tag and its attachment mechanism clearly identify the disconnecting means that is open and effectively inhibit reenergizing the electric circuit (for example, when a tag is used on an energy isolating device that is capable of being locked out, the tag must be attached at the same location that the lock would have been attached); and
(4) Whether employees are thoroughly familiar with the tagging procedures, especially their duties and responsibilities with respect to the procedures and the meaning and significance of the tags used with the procedure."

It is also a requirement of 1910.147 that if the tag is used in the place of a lock the tag attachment means shall be non-reusable, attachable by hand, self-locking, and non-releasable and have a pull strength of no less than 50 pounds and be at least equivalent to an all environment tolerant nylon cable tie.

It is also a requirement that if the energy isolating device is not capable of being locked out, the tag must be attached to or immediately adjacent to the device in such a manner that it would be obvious to anyone attempting to operate the device, that it is tagged out.

(E) A lock may be placed without a tag only under the following conditions:
{1} Only one circuit or piece of equipment is deenergized, and
{2} The lockout period does not extend beyond the work shift, and
{3} Employees exposed to the hazards associated with reenergizing the circuit or equipment are familiar with this procedure.

FIGURE 6.2 (*Continued*)

(iv) *Verification of deenergized condition.* The requirements of this paragraph shall be met before any circuits or equipment can be considered and worked as deenergized.

(D of C) "Verification of deenergization is mandatory. This verification must be done by a "qualified person." This is done by following the requirements of paragraph (A) and (B) of this section."

(A) A qualified person shall operate the equipment operating controls or otherwise verify that the equipment cannot be restarted.

(B) A qualified person shall use test equipment to test the circuit elements and electrical parts of equipment to which employees will be exposed and shall verify that the circuit elements and equipment parts are deenergized. The test shall also determine if any energized condition exists as a result of inadvertently induced voltage or unrelated voltage backfeed even though specific parts of the circuit have been deenergized and presumed to be safe....

A current one-line diagram can be a very useful tool to verify that the right circuit was deenergized, that all electrical sources were deenergized, and will identify all possible backfeed sources.

The key to this is that the one-line diagram must be current or correct.

...If the circuit to be tested is over 600 volts, nominal, the test equipment shall be checked for proper operation immediately before and immediately after this test.

(Preamble) "Voltages over 600 volts are more likely than lower voltages to cause test equipment itself to fail, leading to false indications of no-voltage conditions. To prevent accidents resulting from such failure of test equipment, the final standard requires checking operation of the test equipment immediately before and after use, if voltages over 600 volts are involved."

(D of C) "Testing instruments and equipment shall be visually inspected for external defects or damage before being used to determine deenergization according to the requirements of 29 CFR 1910.334(c)(2)."

There is evidence of electrical injuries and fatalities with voltages as low as 50 volts. Because of this hazard, it is a good practice to verify the proper operation of the test equipment on these lower voltages as well as the requirement for over 600 volts.

(v) *Reenergizing equipment.* These requirements shall be met, in the order given, before circuits or equipment are reenergized, even temporarily.

(A) A qualified person shall conduct tests and visual inspections, as necessary, to verify that all tools, electrical jumpers, shorts, grounds, and other such devices have been removed, so that the circuits and equipment can be safely energized.

(B) Employees exposed to the hazards associated with reenergizing the circuit or equipment shall be warned to stay clear of circuits and equipment.

(C) Each lock and tag shall be removed by the employee who applied it or under his or her direct supervision. However, if this employee is absent from the workplace, then the lock or tag may be removed by a qualified person designated to perform this task provided that:

{1} The employer ensures that the employee who applied the lock or tag is not available at the workplace, and

An additional requirement, found in 1910.147 for energy control, is that every reasonable effort be made to contact the employee before locks or tags are removed.

{2} The employer ensures that the employee is aware that the lock or tag has been removed before he or she resumes work at that workplace.

(Preamble) This requirement ensures that locks and tags are removed with the full knowledge of the employee who applied them. Without such a requirement, it is likely that a lock or a tag could be removed by persons who are not responsible for the lockout, endangering employees working on the 'deenergized parts.' This provision also allows for removal of locks and tags if the employee who applied them is absent from the workplace, under conditions like those in the exception to 1910.147(e)(3)."

(D) There shall be a visual determination that all employees are clear of the circuits and equipment.

Often this would require more than one person, i.e., one at the equipment and one at the disconnecting means, both using two-way radios or other effective communications methods.

(c) *Working on or near exposed energized parts.* **(1)** *Application.* This paragraph applies

FIGURE 6.2 *(Continued)*

to work performed on exposed live parts (involving either direct contact or by means of tools or materials) or near enough to them for employees to be exposed to any hazard they present.

All of 1910.333(c) addresses the required energized work procedures and employee qualification requirements.

(2) *Work on energized equipment.* Only qualified persons may work on electric circuit parts or equipment that have not been deenergized under the procedures of paragraph (b) of this section. Such persons shall be capable of working safely on energized circuits and shall be familiar with the proper use of special precautionary techniques, personal protective equipment, insulating and shielding materials, and insulated tools.

(3) *Overhead lines.* If work is to be performed near overhead lines, the lines shall be deenergized and grounded, or other protective measures shall be provided before work is started. If the lines are to be deenergized, arrangements shall be made with the person or organization that operates or controls the electric circuits involved to deenergize and ground them. If protective measures, such as guarding, isolating, or insulating, are provided, these precautions shall prevent employees from contacting such lines directly with any part of their body or indirectly through conductive materials, tools, or equipment.

(D of C) "This standard for Electrical Safety-Related Work Practices can be applied with respect to electrical hazards related to any size, utilization, or configuration of overhead power lines in general industry; e.g., residential power lines, remotely located overhead power lines, temporarily rigged overhead power lines, and overhead power lines along streets and alleys."

It should also be noted that overhead power lines and bus structures in substations and switchyards present the same hazard as all other overhead power lines. Therefore the same work practices should be used when work is being performed near these overhead systems.

(D of C) "Working On or Near Overhead Power Lines;

(a) OSHA believes that the preferred method of protecting employees working near overhead power lines is to deenergize and ground the lines when work is to be performed near them.

(b) In addition to other operations, this standard also applies to tree trimming operations performed by tree workers who are not qualified persons. In this respect the exclusion in 1910.331(c)(1) applies only to qualified persons performing line-clearance tree trimming (trimming trees that are closer than 10 feet to overhead power lines).

(c) This standard does not prohibit workers who are not qualified persons from working in a tree that is closer than 10 feet to power lines so long as that person or any object he or she may be using, does not come within 10 feet of a power line. However, it would require qualified persons to perform the work if the worker or any object he or she may be using will come within 10 feet of an exposed energized part or if a branch being cut may be expected to come within 10 feet of an exposed energized part while falling from the tree.

(d) The purpose for the approach distance requirements is to prevent contact with, and/or arcing, from energized overhead power lines. The approach distance applies to tools used by employees as well as the employees them-selves. Table S-5 calls for the approach distances for qualified persons only. Unqualified persons are required to adhere to the 10 foot minimum."

Note: The work practices used by qualified persons installing insulating devices on overhead power transmission or distribution lines are not covered by 1910.332 through 1910.335. Under paragraph (c)(2) of this section, unqualified persons are prohibited from performing this type of work.

Qualified persons installing insulating equipment on or deenergizing and grounding the lines and line-clearance tree trimming are covered under 1910.269, "Electric Power Generation, Transmission, and Distribution" Federal Register.

(i) *Unqualified persons.* (A) When an unqualified person is working in an elevated position near overhead lines, the location shall be such that the person and the longest conductive object he or she may contact cannot come

FIGURE 6.2   *(Continued)*

TABLE S-5 Approach Distances For Qualified
Employees—Alternating Current

Voltage range (phase to phase)	Minimum approach distance
300V and less	Avoid Contact
Over 300V, not over 750V	1 ft 0 in (30.5 cm).
Over 750V, not over 2kV	1 ft 6 in (46 cm).
Over 2kV, not over 15kV	2 ft 0 in (61 cm).
Over 15kV, not over 37kV	3 ft 0 in (91 cm).
Over 37kV, not over 87.5kV	3 ft 6 in (107 cm).
Over 87.5kV, not over 121kV	4 ft 0 in (122 cm).
Over 121kV, not over 140kV	4 ft 6 in (137 cm).

closer to any unguarded, energized overhead line than the following distances:

{1} For voltages to ground 50 kV or below - 10 feet (305 cm);

{2} For voltages to ground over 50 kV - 10 feet (305 cm) plus 4 inches (10 cm) for every 10 kV over 50 kV.

(B) When an unqualified person is working on the ground in the vicinity of overhead lines, the person may not bring any conductive object closer to unguarded, energized overhead lines than the distances given in paragraph (c)(3)(i)(A) of this section.

Note: For voltages normally encountered with overhead power line, objects which do not have an insulating rating for the voltage involved are considered to be conductive.

(ii) *Qualified persons.* When a qualified person is working in the vicinity of overhead lines, whether in an elevated position or on the ground, the person may not approach or take any conductive object without an approved insulating handle closer to exposed energized parts than shown in Table S-5 unless:

(A) The person is insulated from the energized part (gloves, with sleeves if necessary, rated for the voltage involved are considered to be insulation of the person from the energized part on which work is performed), or

(B) The energized part is insulated both from all other conductive objects at a different potential and from the person, or

(C) The person is insulated from all conductive objects at a potential different from that of the energized part.

(iii) *Vehicular and mechanical equipment.* (A) Any vehicle or mechanical equipment capable of having parts of its structure elevated near energized overhead lines...

(D of C) "Employees working on or around vehicles and mechanical equipment, such as gin-pole trucks, forklifts, cherry pickers, garbage trucks, cranes, dump trucks, and elevating platforms, who are potentially exposed to hazards related to equipment component contact with overhead lines, shall have been trained by their employers in the inherent hazards of electricity and means of avoiding exposure to such hazards."

...shall be operated so that a clearance of 10 ft (305 cm) is maintained. If the voltage is higher than 50 kV, the clearance shall be increased 4 in (10 cm) for every 10 kV over that voltage. However, under any of the following conditions, the clearance may be reduced:

{1} If the vehicle is in transit with its structure lowered, the clearance may be reduced to 4 ft (122 cm). If the voltage is higher than 50 kV, the clearance shall be increased 4 in (10 cm) for every 10 kV over that voltage.

{2} If insulating barriers are installed to prevent contact with the lines, and if the barriers are rated for the voltage of the line being guarded and are not a part of or an attachment to the vehicle or its raised structure, the clearance may be reduced to a distance within the designed working dimensions of the insulating barrier.

{3} If the equipment is an aerial lift insulated for the voltage involved, and if the work is performed by a qualified person, the clearance (between the uninsulated portion of the aerial lift and the power line) may be reduced to the distance given in Table S-5.

(B) Employees standing on the ground may not contact the vehicle or mechanical equipment or any of its attachments, unless:

(1) The employee is using protective equipment rated for the voltage; or

FIGURE 6.2 (*Continued*)

(2) The equipment is located so that no uninsulated part of its structure (that portion of the structure that provides a conductive path to employees on the ground) can come closer to the line than permitted in paragraph (c)(3)(iii) of this section.

(C) If any vehicle or mechanical equipment capable of having parts of its structure elevated near energized overhead lines is intentionally grounded, employees working on the ground near the point of grounding may not stand at the grounding location whenever there is a possibility of overhead line contact. Additional precautions, such as the use of barricades or insulation, shall be taken to protect employees from hazardous ground potentials, depending on earth resistivity and fault currents, which can develop within the first few feet or more outward from the grounding point.

(4) *Illumination.* **(i)** Employees may not enter spaces containing exposed energized parts, unless illumination is provided that enables the employees to perform the work safely.

(ii) Where lack of illumination or an obstruction precludes observation of the work to be performed, employees may not perform tasks near exposed energized parts. Employees may not reach blindly into areas which may contain energized parts.

(5) *Confined or enclosed work spaces.* When an employee works in a confined or enclosed space (such as a manhole or vault) that contains exposed energized parts, the employer shall provide, and the employee shall use, protective shields, protective barriers, or insulating materials as necessary to avoid inadvertent contact with these parts.

Doors, hinged panels, and the like shall be secured to prevent their swinging into an employee and causing the employee to contact exposed energized parts.

Confined or enclosed work spaces can present even greater hazards than exposed energized parts. On January 14, 1993, OSHA signed into Final Rule the "Permit Required Confined Spaces for General Industry" Federal Regulation 29 CFR 1910.146. This regulation identifies the major hazards associated with confined spaces. It also requires specific training for anyone entering a confined space. The major hazards of a permit required confined space are:

(1) Oxygen–Less than 19.5% or greater than 23.5%;

(2) Carbon Monoxide–Greater than 50 parts per million (PPM) for an 8-hour exposure;

(3) Hydrogen Sulfide–Greater than 20 PPM "acceptable ceiling concentration" and 50 PPM "acceptable maximum peak above the acceptable ceiling concentration for an 8-hour shift" for a maximum duration of "10 minutes once, only if no other measurable exposure occurs"; and

(4) Greater than 10% by volume of any flammable or explosive substance.

(6) *Conductive materials and equipment.* Conductive materials and equipment that are in contact with any part of an employee's body shall be handled in a manner that will prevent them from contacting exposed energized conductors or circuit parts. If an employee must handle long dimensional conductive objects (such as ducts and pipes) in areas with exposed live parts, the employer shall institute work practices (such as the use of insulation, guarding, and material handling techniques) which will minimize the hazard.

(7) *Portable ladders.* Portable ladders shall have non-conductive siderails if they are used where the employee or the ladder could contact exposed energized parts.

Do not get over confident with wooden or fiberglass ladders. They can become contaminated from moisture, petroleum products, dirt or paint and thus become electrically conductive. A ladder must be treated the same as any other piece of electrical protective equipment, in that it should be inspected and cleaned prior to each use and stored properly when not in use to prevent damage or contamination from occurring.

(8) *Conductive apparel.* Conductive articles of jewelry and clothing (such as watch bands, bracelets, rings, key chains, necklaces, metalized aprons, cloth with conductive thread, or metal headgear) may not be worn if they might contact exposed energized parts. However, such articles may be worn if they are rendered nonconductive by covering, wrapping, or other insulating means.

Additional items have also been noted as presenting a hazard. They are: earrings, metal framed glasses, belt buckles, tool belts and knife scabbards.

It is permissible to render these articles nonconductive so that a person could still wear them. However, it is advisable that these articles be removed. There are numerous accounts of rings being smashed while being on the finger or a person jumping down from a piece of equipment and the ring

FIGURE 6.2   *(Continued)*

snagging on the equipment and the persons finger being pulled off. There are other hazards besides the electrical hazard. Be alert to all potentially hazardous situations.

(9) *Housekeeping duties.* Where live parts present an electrical contact hazard, employees may not perform housekeeping duties at such close distances to the parts that there is a possibility of contact, unless adequate safeguards (such as insulating equipment or barriers) are provided. Electrically conductive cleaning materials (including conductive solids such as steel wool, metalized cloth, and silicon carbide, as well as conductive liquid solutions) may not be used in proximity to energized parts unless procedures are followed which will prevent electrical contact.

(10) *Interlocks.* Only a qualified person following the requirements of paragraph (c) of this section may defeat an electrical safety interlock, and then only temporarily while he or she is working on the equipment. The interlock system shall be returned to its operable condition when this work is completed.

(D of C) "Interlocks found on panels, covers and guards are designed to deenergize circuits to prevent electric shock to persons using the equipment or performing minor maintenance or adjustments and shall not be defeated or bypassed by an unqualified person."

A qualified person must be allowed to defeat or bypass a safety interlock in order to perform such tasks as troubleshooting while the circuit is energized. However, as it is stated above, the interlock must be returned to its operable condition when the work is completed so as to prevent access to the energized parts by unqualified persons.

1910.334 Use of Equipment

(a) *Portable electric equipment.* This paragraph applies to the use of cord- and plug-connected equipment, including flexible cord sets (extension cords).

(1) *Handling.* Portable equipment shall be handled in a manner which will not cause damage. Flexible electric cords connected to equipment may not be used for raising or lowering the equipment. Flexible cords may not be fastened with staples or otherwise hung in such a fashion as could damage the outer jacket or insulation.

(2) *Visual inspection.* **(i)** Portable cord and plug connected equipment and flexible cord sets (extension cords) shall be visually inspected before use on any shift for external defects (such as loose parts, deformed and missing pins, or damage to outer jacket or insulation) and for evidence of possible internal damage (such as pinched or crushed outer jacket). Cord and plug connected equipment and flexible cord sets (extension cords) which remain connected once they are put in place and are not exposed to damage need not be visually inspected until they are relocated.

(ii) If there is a defect or evidence of damage that might expose an employee to injury, the defective or damaged item shall be removed from service, and no employee may use it until repairs and tests necessary to render the equipment safe have been made.

(iii) When an attachment plug is to be connected to a receptacle (including an on a cord set), the relationship of the plug and receptacle contacts shall first be checked to ensure that they are of proper mating configurations.

(3) *Grounding type equipment.* **(i)** A flexible cord used with grounding type equipment shall contain an equipment grounding conductor.

(ii) Attachment plugs and receptacles may not be connected or altered in a manner which would prevent proper continuity of the equipment grounding conductor at the point where plugs are attached to receptacles. Additionally, these devices may not be altered to allow the grounding pole of a plug to be inserted into slots intended for connection to the current-carrying conductors.

(iii) Adapters which interrupt the continuity of the equipment grounding connection may not be used.

Its not that adapters are not safe, they are UL approved. They may not be used because they are used incorrectly most of the time, in that the ground connection is not connected or it is connected to a device that is not grounded. This interrupts the equipment grounding connection and creates a possible shock hazard.

(4) *Conductive work locations.* Portable electric equipment and flexible cords used in highly conductive work locations (such a those inundated with water or other conductive liquids), or in job locations where employees are likely to contact water or conductive liquids, shall be approved for those locations.

(5) *Connecting attachment plugs.* **(i)** Employees' hands may not be wet when plugging and unplugging flexible cords and cord and plug

FIGURE 6.2 (*Continued*)

connected equipment, if energized equipment is involved.

(D of C) "Cord- and plug-connected equipment. Energized equipment here means either the equipment being plugged or the receptacle into which it is being plugged, or both."

(ii) Energized plug and receptacle connections may be handled only with insulating protective equipment if the condition of the connection could provide a conducting path to the employee's hand (if, for example, a cord connector is wet from being immersed in water).

(iii) Locking type connectors shall be properly secured after connection.

Additional safety can be achieved by the use of a ground- fault circuit-interrupter (GFCI) whenever cord- and plug-connected equipment or flexible cord sets are used. This would be for both the grounded type and the double-insulated type of equipment.

(b) *Electric power and lighting circuits.* **(1)** Routine opening and closing of circuits. Load rated switches, circuit breakers, or other devices specifically designed as disconnecting means shall be used for the opening, reversing, or closing of circuits under load conditions. Cable connectors not of the load break type, fuses, terminal plugs, and cable splice connections may not be used for such purposes, except in an emergency.

It is a good practice to remove the load, if at all possible, before opening any device whether it is load rated or not. This practice will extend the life of the device and reduce the risk of arcing which could cause a flashover from phase-to-phase and/or from phase-to-ground.

(2) Reclosing circuits after protective device operation. After a circuit is deenergized by a circuit protective device, the circuit may not be manually reenergized until it has been determined that the equipment and circuit can be safely energized. The repetitive manual reclosing of circuit breakers or reenergizing circuits through replaced fuses is prohibited.

Note: When it can be determined from the design of the circuit and the overcurrent devices involved that the automatic operation of a device was caused by an overload rather than a fault condition, no examination of the circuit or connected equipment is needed before the circuit is reenergized.

It has been proven on many occasions that a circuit breaker will explode if it is reclosed on a fault. Sometimes it can be closed several times and on other occasions only once before it explodes. The determining factor being the amount of fault current.

The standard practice in the general industry has been to reset the breaker once or twice before troubleshooting to find the cause. This practice is very dangerous and must be avoided.

(3) *Overcurrent protection modification.* Overcurrent protection of circuits and conductors may not be modified, even on a temporary basis, beyond that allowed by 1910.304(e), the installation safety requirements for overcurrent protection.

(c) *Test instruments and equipment.* **(1)** *Use.* Only qualified persons may perform testing work on electric circuits or equipment.

(2) *Visual inspection.* Test instruments and equipment and all associated test leads, cables, power cords, probes, and connectors shall be visually inspected for external defects and damage before the equipment is used. If there is a defect or evidence of damage that might expose an employee to injury, the defective or damaged item shall be removed from service, and no employee may use it until repairs and tests necessary to render the equipment safe have been made.

Care should be taken to verify that the setting, scale, and so on are correct.

(3) *Rating of equipment.* Test instruments and equipment and their accessories shall be rated for the circuits and equipment to which they will be connected and shall be designed for the environment in which they will be used.

(d) *Occasional use of flammable or ignitable materials.* Where flammable materials are present only occasionally, electric equipment capable of igniting them shall not be used, unless measures are taken to prevent hazardous conditions from developing. Such materials include, but are not limited to: flammable gases, vapors, or liquids; combustible dust and ignitable fibers or flyings.

Note: Electrical installation requirements for locations where flammable materials are present on a regular basis are contained in 1910.307.

It is very possible to produce a hazardous (classified) location when using solvents and

FIGURE 6.2 *(Continued)*

cleaning fluids. This is especially hazardous when it is around electrical equipment that can produce arcing such as circuit breakers, contactors, and relays. If these types of solvents and cleaning fluids are going to be used, adequate ventilation must be utilized to prevent the buildup of an explosive mixture.

1910.335 Safeguards for Personnel Protection

(a) *Use of protective equipment.* **(1)** *Personal protective equipment.* **(i)** Employees working in areas where there are potential electrical hazards shall be provided with, and shall use, electrical protective equipment that is appropriate for the specific parts of the body to be protected and for the work to be performed.

Note: Personal protective equipment requirements are contained in subpart I of this part.

(ii) Protective equipment shall be maintained in a safe reliable condition and shall be periodically inspected or tested, as required by 1910.137.

The following is a summary of 1910.137:

OSHA's 1910.137 "Electrical Protective Equipment" provides the requirements for the design, manufacture, and in-service care and use of rubber temporary insulation equipment. (Based on ASTM standards)

The following table outlines six different voltage classifications for rubber protective equipment per ASTM.

OSHA states in 1910.137 that "Electrical protective equipment shall be maintained in a safe, reliable condition." The following specific requirements apply to insulating equipment made of rubber:

1. Maximum use voltage shall conform to those values listed in the above table.

2. Insulating equipment shall be inspected for damage before each day's use and anytime damage is suspected.
3. Gloves shall be given an air test.

Insulating equipment with the following defects may not be used:

1. Holes, tears, punctures, or cuts;
2. Ozone cutting or checking (fine interlacing cracks);
3. Embedded foreign objects;
4. Texture changes such as: swelling, softening, hardening, or becoming sticky or inelastic (swelling, softening and becoming sticky are results of petroleum deterioration and hardening and becoming inelastic are the results of ageing, which can be accelerated by heat); and
5. Any other defects that could damage the insulating properties of the rubber.

Protector (leather) gloves should also be inspected for any damage, contamination from petroleum products, any imbedded objects, or anything that could damage the rubber gloves.

Insulating equipment shall be cleaned as needed to remove foreign substances (wash with a mild detergent, no abrasives or bleaching agents, and water, rinse thoroughly and allow the equipment to dry completely before use).

Insulating equipment shall be stored to protect it from the following:

1. Light (primarily sunlight and any light that produces UV; UV produces ozone);
2. Temperature extremes (high temperatures accelerate the ageing process; examples of heat sources are hot forced air, steam lines, near radiators, and so on.);
3. Excessive humidity (because of condensation);

Voltage Classifications for Rubber Protective Equipment

ID tag	Class	AC proof test voltage	AC max. use voltage	DC proof test voltage
Beige	00	2,500	500	10,000
Red	0	5,000	1,000	20,000
White	1	10,000	7,500	40,000
Yellow	2	20,000	17,000	50,000
Green	3	30,000	26,500	60,000
Orange	4	40,000	36,000	70,000

Notes: Class 00 does not currently appear in OSHA, 1910.137, but was added by the Author based on ASTM F496

FIGURE 6.2 (*Continued*)

4. Ozone (produced by corona, arcing, and ultra-violet rays); and
5. Other injurious substances and conditions (beware of chemicals and petroleum products of any kind, such as oil, gasoline, fuels, hydraulic fluids, solvents, hand lotions, waterless hand cleaners, ointments, and so on).

Protector gloves shall be worn over insulating gloves, except as follows:

1. Class 0, under limited-use and where high finger dexterity is needed;
2. Other classes, provided there is little chance of physical damage, one class higher is used, and the gloves are tested prior to use on a higher voltage.

Electrical protective equipment shall be subjected to periodic electrical tests, as stated in the following table:

The rubber insulating equipment that fails to pass inspections or electrical tests may not be used, except as follows:

1. Rubber insulating line hose may be used in shorter lengths after the defective portion has been cut off;
2. Rubber insulating blankets may be repaired by using a compatible patch purchased from the manufacture of the blanket (not a tire patch), or the defective portion may be cut off, provided the undamaged portion is not smaller than 22 inches by 22 inches for Class 1, 2, 3, and 4 blankets (cutting off the defective portion cannot include that part of the blanket that contains the ID tag); and
3. Rubber insulating gloves and sleeves may be repaired if the defect is minor such as small cuts, tears, or punctures. The repairs on gloves are permitted only on the gauntlet area not the hand area. A compatible patch may be used or, if the defect is only a

minor surface blemish, a compatible liquid compound may be used.

If any rubber insulating equipment is repaired, it must be retested before it may be used.

The employer shall certify that the equipment has been tested in accordance with 1910.137. This certification must identify the equipment tested and the date it was tested.

Detailed information on rubber protective equipment is available from the American Society for Testing and Materials (ASTM), 1916 Race Street, Philadelphia, Pennsylvania 19103.

ASTM Standards for the in-service care of electrical protective equipment are listed in the following table:

(iii) If the insulating capability of protective equipment may be subject to damage during use, the insulating material shall be protected. (For example, an outer covering of leather is sometimes used for the protection of rubber insulating material.)

(iv) Employees shall wear non-conductive head protection wherever there is a danger of head injury from electric shock or burns due to contact with exposed energized parts.

Head protection, for electrical used, must be labeled ANSI Z89.2. All other hard hats are labeled ANSI Z89.1.

(v) Employees shall wear protective equipment for the eyes or face wherever there is danger of injury to the eyes or face from electric arcs or flashes or from flying objects resulting from electrical explosion.

(2) *General protective equipment and tools.* (i) When working near exposed energized conductors or circuit parts, each employee shall use insulated tools or handling equipment if the tools or handling equipment might make contact with such conductors or parts. If the insulating

Rubber Insulating Equipment Test Intervals

Type of equipment	When to test
Rubber insulating line hose	Upon indication that insulating value is suspect
Rubber insulating covers	Upon indication that insulating value is suspect
Rubber insulating blankets	Before first issue and every 12 months thereafter*
Rubber insulating gloves	Before first issue and every 6 months thereafter*
Rubber insulating sleeves	Before first issue and every 12 months thereafter*

*If the insulating equipment has been electrically tested, but not issued for service, it may not be placed into service unless it has been electrically tested within the previous 12 months.

FIGURE 6.2   (*Continued*)

Standards on Electrical Protective Equipment

Subject	Number and title
Gloves/Sleeves	ASTM F 496, Specification for In-Service Care of Insulating Gloves and Sleeves
Blankets	ASTM F 479, Specification for In-Service Care of Insulating Blankets
Line Hose/Covers	ASTM F 478, Specification for In-Service Care of Insulating Line Hose and Covers

The above listed standards, along with other standards for electrical protective equipment are found in ASTM, Section 10, Volume 10.03.

capability of insulated tools or handling equipment is subject to damage, the insulating material shall be protected.

(OSHA's D of C) "This means a tool incased within material of composition and thickness that is recognized as electrical insulation."

OSHA has stated verbally that tape of any type, heat shrink tubing, or any of the dipping processes do not meet the "recognized as electrical insulation" requirement and that they must not be used.

Insulated tools that are manufactured in accordance with the IEC 900 International Standard as well as the soon to be published ASTM standard will meet this requirement. Additional information is found in IEEE Standard 935.

(A) Fuse handling equipment, insulated for the circuit voltage, shall be used to remove or install fuses when the fuse terminals are energized.

(B) Ropes and handlines used near exposed energized parts shall be nonconductive.

(ii) Protective shields, protective barriers, or insulating materials shall be used to protect each employee from shock, burns, or other electrically related injuries while that employee is working near exposed energized parts which might be accidentally contacted or where dangerous electric heating or arcing might occur.

When normally enclosed live parts are exposed for maintenance or repair, they shall be guarded to protect unqualified persons from contact with the live parts.

(b) *Alerting techniques.* The following alerting techniques shall be used to warn and protect employees from hazards which could cause injury due to electric shock, burns, or failure of electric equipment parts:

(1) *Safety signs and tags.* Safety signs, safety symbols, or accident prevention tags shall be used where necessary to warn employees about electrical hazards which may endanger them, as required by 1910.145.

(2) *Barricades.* Barricades shall be used in conjunction with safety signs where it is necessary to prevent or limit employee access to work areas exposing employees to uninsulated energized conductors or circuit parts. Conductive barricades may not be used where they might cause an electrical contact hazard.

(3) *Attendants.* If signs and barricades do not provide sufficient warning and protection from electrical hazards, an attendant shall be stationed to warn and protect employees.

These alerting techniques are being used for a reason, a hazard exists. Never attempt to go through or bypass a safety sign or tag, barricade, or an attendant.

FIGURE 6.2   (*Continued*)

ELECTRIC POWER GENERATION, TRANSMISSION, AND DISTRIBUTION (OSHA 29 CFR 1910.269) (FINAL RULE JANUARY 31, 1994)

1910.269 Electric power generation, transmission, and distribution

1910.269 contains requirements for the prevention of injuries to employees performing operation or maintenance work on electric power generation, transmission, or distribution installations.

The standard consists largely of work practice requirements that are necessary for employee safety. The Agency believes that it is important to apply these work practices in full to existing installations, as well as to conductors and equipment that are installed in the future.

The Agency has determined that employees are presently facing significant risk. The risk that an electric utility employee will be seriously injured or die from a fall or an electric shock is significant. OSHA has determined that risk can be reduced by adopting a standard that requires the industry to change existing protective measures in certain cases. The areas for which this holds true are explained in the discussion of individual provisions.

(a) General. (1) Application. (i) This section covers the operation and maintenance of electric power generation, control, transformation, transmission, and distribution lines and equipment. These provisions apply to:

Paragraph (a)(1) sets forth the scope of the standard. Under the terms of paragraph (a)(1)(i), the provisions of 1910.269 apply to the operation and maintenance of electric power generation, transmission, and distribution systems, to electrical testing of such systems, and to line-clearance tree trimming.

The standard primarily covers the following types of work operations:

(1) Inspection,
(2) Switching (connection and disconnection of facilities),
(3) Maintenance of lines and equipment,
(4) Line-clearance tree trimming,
(5) Testing and fault locating,
(6) Streetlight relamping,
(7) Chemical cleaning of boilers, and
(8) Other operation and maintenance activities.

According to proposed 1910.269(a)(1)(ii)(B), OSHA would only have applied the regulation to installations for the generation, transmission, or distribution of electric energy that are owned or operated by electric utilities and to work performed on such installations owned by a utility. Many industrial generation, transmission, and distribution systems are essentially the same as those of a utility, and the work performed on these systems is nearly identical to that performed on electric utility installations. Existing OSHA and national consensus standards, Subpart V of Part 1926 and ANSI C2, respectively, do extend their coverage to anyone doing electric-utility-type work.

OSHA received many comments on this issue, from utilities, from electrical contractors, from other industries, and from unions. In general, the utilities supported extending coverage to all generation, transmission, and distribution installations. For example, the New York State Electric and Gas Corporation stated that their personnel perform work on transmission and distribution interconnect facilities as well as inspect, oversee, and approve protection system design, installation, testing, and maintenance on non-utility protection systems. Their employees also provide assistance to industrial customers under emergency conditions.

Unions also supported extending the scope of 1910.269. The International Brotherhood of Electrical Workers (IBEW) stated that the hazards, training, and work practices are the same for electric power generation, transmission, and distribution facilities regardless of who owns or operates them. Therefore, they argued, the safety and health requirements should be the same.

The National Electrical Contractors Association (NECA) represents the contractors who perform work on utility and on industrial power generation, transmission, and distribution installations. NECA agreed with IBEW that these installations were the same, no matter who owned or operated them, and that the accident prevention measures should be the same.

FIGURE 6.3 Electrical Power Generation, Transmission, and Distribution Rule (OSHA-CFR, Title 29, Part 1910, Paragraph 269).

Note: Reading the following paragraphs is a must in order to fully understand the application of 1910.269.

The installation safety requirements in Subpart S of Part 1910 (1910.302 through 1910.308) do not cover "installations under the exclusive control of electric utilities for the generation, control, transformation, transmission, and distribution of electric energy" (1910.302(a)(2)(v)). Additionally, OSHA has interpreted the Subpart S installation requirements to exempt industrial power generation and distribution systems that are similar to electric utility installations. This exclusion reflects the unique hazards and work practices involved in generation, transmission, and distribution of electric energy. The work practice requirements in Subpart S of Part 1910 (1910.332 through 1910.335) are designed to complement the installation safety provisions in Subpart S and do not cover work practices for qualified persons who work on or near electric generation, transmission, or distribution installations. Also, because electric power generation, transmission, and distribution installations involve similar hazards and work practices whether or not they are controlled by electric utilities, the Subpart S work practices standard does not apply to qualified persons who work on or near any such installation, regardless of who owns or controls the installation.

OSHA believes that there are hazards related to electric power generation, transmission, and distribution work that are not adequately addressed elsewhere in the General Industry Standards. The hazards related to transmission systems are the same whether the system is owned by a steel plant, a chemical plant, or an electric utility. There are currently no OSHA standards governing the design or installation of these systems, and the electrical standards in Subpart S of Part 1910 do not apply.

Coverage of electric power generation and distribution systems is slightly different from the coverage of transmission systems. Utility-type generation and distribution installations are not covered by the provisions of 1910.303 through 1910.308 or (if the work is performed by a qualified employee) by 1910.332 through 1910.335. Commercial-type systems, however, are covered by the Subpart S requirements. Additionally, some employers voluntarily comply with OSHA's electrical standards in Subpart S for their large-scale generation and distribution installations.

From an electrical viewpoint, the hazards faced by employees working on an installation that conforms to the design requirements of 1910.303 through 1910.308 are different from those faced by employees working on an installation that was designed to conform to the National Electrical Safety Code. OSHA believes that whether an employer should comply with the Subpart S work practice requirements or with the provisions of 1910.269 depends on the hazards faced by an employee. The hazards posed by an installation are related to the type of installation involved and to whether or not it conforms to the design standards in Subpart S. The risk faced by an employee working on the installation depends on what the hazards are and on whether or not the employee is trained to recognize and avoid the hazards. Therefore, the Agency has made application of most of the electrical requirements in the new standard dependent on whether or not the installation conforms to 1910.303 through 1910.308 and on whether or not the employee is qualified to perform the work, not on whether or not the work is performed by an employee of an electric utility.

OSHA has determined which provisions of final 1910.269 address electrical hazards that are already addressed in 1910.332 through 1910.335 of Subpart S for electrical installations that meet the design requirements in 1910.302 through 1910.308 of Subpart S. In short, when qualified employees work on such installations, the Agency will consider these installations and work practices conforming to 1910.332 through 1910.335 to be in compliance with the provisions of 1910.269 that are identified in of Appendix A-2.

OSHA has also identified requirements in 1910.269 that are not adequately addressed in Subpart S, and these requirements must be followed at all times. These provisions are listed in of Appendix A-2 as well. It should be noted that, if unqualified employees are working on, near, or with electric power generation, transmission, and distribution installations, 1910.332 through 1910.335 apply in any event. Appendices A-1 and A-2 illustrate the application of 1910.269 and Subpart S to the various types of electrical installations.

FIGURE 6.3 *(Continued)*

The non-electrical provisions in 1910.269 (for example, paragraph (g)(2) on fall protection and paragraph (p)(1) on mechanical equipment) address only unique aspects of electric power generation, transmission, and distribution work. As noted in paragraph (a)(1)(iii), the requirements of 1910.269 supplement those elsewhere in Part 1910, unless an exception is specifically mentioned. The non-electrical requirements in this section have been handled individually throughout the standard to allow alternative methods of compliance already recognized in the General Industry Standards. For example, the lockout and tagging provisions of paragraph (d) recognize compliance with the generic standard on control of hazardous energy sources in 1910.147.

(A) Power generation, transmission, and distribution installations, including related equipment for the purpose of communication or metering, which are accessible only to qualified employees;

Note: The types of installations covered by this paragraph include the generation, transmission, and distribution installations of electric utilities, as well as equivalent installations of industrial establishments. Supplementary electric generating equipment that is used to supply a workplace for emergency, standby, or similar purposes only is covered under Subpart S of this Part. (See paragraph (a)(1)(ii)(B) of this section.)

Paragraph (a)(1)(i)(A) sets forth the scope of 1910.269 as it relates to industrial and utility power generation, transmission, and distribution.

OSHA believes that this language will effectively extend the scope of the standard to the types of installations that the standard is intended to cover, namely, electric power generation, transmission, and distribution systems of electric utilities and equivalent industrial systems. It also makes it clear that supplementary generating equipment, such as emergency and standby generators used to provide temporary power at a workplace, is not covered. These installations are considered to be part of the utilization system rather than separate generation installations and are addressed by the existing Subpart S regulations.

Section 1910.269 applies to the parts of a facility that are directly involved with the generation, transmission, or distribution of electric power. Installations not used for one of these purposes are not covered by the standard. For example, office buildings, warehouses, machine shops, and other installations which are not integral parts of generating plants, substations, or control centers are not covered by final 1910.269. Work performed on these installations is not of a type addressed by the standard.

(B) Other installations at an electric power generating station, as follows:

(1) Fuel and ash handling and processing installations, such as coal conveyors,

(2) Water and steam installations, such as penstocks, pipelines, and tanks providing a source of energy for electric generators, and

(3) Chlorine and hydrogen systems;

(C) Test sites where electrical testing involving temporary measurements associated with electric power generation, transmission, and distribution is performed in laboratories, in the field, in substations, and on lines, as opposed to metering, relaying, and routine line work;

(D) Work on or directly associated with the installations covered in paragraphs (a)(1)(i)(A) through (a)(1)(i)(C) of this section; and

(E) Line-clearance tree-trimming operations, as follows:

(1) Entire 1910.269 of this Part, except paragraph (r)(1), applies to line-clearance tree-trimming operations performed by qualified employees (those who are knowledgeable in the construction and operation of electric power generation, transmission, or distribution equipment involved, along with the associated hazards).

(2) Paragraphs (a)(2), (b), (c), (g), (k), (p), and (r) of this section apply to line-clearance tree-trimming operations performed by line-clearance tree trimmers who are not qualified employees.

Standards on the construction of transmission and distribution lines and equipment are contained in 29 CFR Part 1926, Subpart V. So as not to overlap these regulations in the Construction Standards, final 1910.269 published today does not apply to operations involving construction work. This "exemption" is set forth in 1910.269(a)(1)(ii)(A). "Construction work" is defined in 1910.12(b) as "work for construction, alteration, and/or repair, including painting and decorating." In 1910.12(d), the term is further defined as including "the erection of new electric transmission and distribution lines and equipment, and the

FIGURE 6.3 *(Continued)*

alteration, conversion, and improvement of existing transmission and distribution lines and equipment." None of the types of work covered by these two definitions are covered by 1910.269.

OSHA believes that it is important for employees to use consistent work practices for jobs posing equivalent hazards. It may, indeed, introduce dangers if an employee has to vary the work practices used for a job depending on slightly different circumstances unrelated to safety. The Agency attempts to make its standards consistent across industries for similar situations, but it is not always possible to make them identical. The employer should ensure that the work rules are the same for similar jobs even though different regulations may apply.

(ii) Notwithstanding paragraph (a)(1)(i) of this section, 1910.269 of this Part does not apply:

(A) to construction work, as defined in 1910.12 of this Part; or

(B) to electrical installations, electrical safety-related work practices, or electrical maintenance considerations covered by Subpart S of this Part.

Note 1: Work practices conforming to 1910.332 through 1910.335 of this Part are considered as complying with the electrical safety-related work practice requirements of this section identified in Table 1 of Appendix A-2 to this section, provided the work is being performed on a generation or distribution installation meeting 1910.303 through 1910.308 of this Part. This table also identifies provisions in this section that apply to work by qualified persons directly on or associated with installations of electric power generation, transmission, and distribution lines or equipment, regardless of compliance with 1910.332 through 1910.335 of this Part.

Note 2: Work practices performed by qualified persons and conforming to 1910.269 of this Part are considered as complying with 1910.333(c) and 1910.335 of this Part.

The distinction, made under the scope of Part I of Subpart S, between installations used and those not used for the generation of electric power at utility plants is one that can be readily determined. OSHA realizes that all circuits for utilization equipment installed in generating stations must originate in the same area as the circuits for the generating installation. However, at some point,

circuits that are not an integral part of the generating installation must become independent of the generating circuits, except to the extent that they may share common cable trays or perhaps raceways. Otherwise, it would be impossible to control the lighting, for example, independently of the generator itself. With respect to the existing requirements of Part I of Subpart S, OSHA considers the "covered" installation to begin where it becomes electrically independent of conductors and equipment used for the generation of electric power. In most cases, it is a simple matter of tracing the wiring back from the utilization equipment itself until a point is reached where generation circuits are also supplied. Generally, branch circuits supplying utilization equipment (other than that used for the generation process) are covered; feeders supplying only "utilization" branch circuits are covered; feeders supplying "generation" circuits, alone or in combination with "utilization" circuits are not covered by Subpart S.

Electric utilization circuits in generating plants do pose unique hazards if the circuits are commingled with installations of power generation equipment or circuits and if the commingled generation equipment or circuits present greater electrical hazards than those posed by the utilization equipment or circuits alone (such as exposure to higher voltages or lack of overcurrent protection). Under this condition, the work practices to be used would have to conform to 1910.269 rather than 1910.332 through 1910.335, and the Subpart S work practices standard does not apply.

(iii) This section applies in addition to all other applicable standards contained in this Part 1910. Specific references in this section to other sections of Part 1910 are provided for emphasis only.

All other General Industry Standards continue to apply to installations covered by this new standard unless an exception is given in 1910.269.

(2) *Training.* (i) Employees shall be trained in and familiar with the safety-related work practices, safety procedures, and other safety requirements in this section that pertain to their respective job assignments. Employees shall also be trained in and familiar with any other safety practices, including applicable emergency procedures (such as pole top and

FIGURE 6.3   *(Continued)*

manhole rescue), that are not specifically addressed by this section but that are related to their work and are necessary for their safety.

The standard cannot specify requirements for every hazard the employee faces in performing electric power generation, transmission, or distribution work. Employers must fill in this gap by training their employees in hazards that are anticipated during the course of jobs they are expected to perform. This language clearly imparts OSHA's intent that safety training be provided in areas that are not covered by the standard but that are related to the employee's job.

(ii) Qualified employees shall also be trained and competent in:

Because qualified employees are allowed to work very close to electric power lines and equipment and because they face a high risk of electrocution, it is important that they be specially trained.

OSHA believes that qualified employees need to be extensively trained in order to perform their work safely.

(A) The skills and techniques necessary to distinguish exposed live parts from other parts of electric equipment,

(B) The skills and techniques necessary to determine the nominal voltage of exposed live parts,

(C) The minimum approach distances specified in this section corresponding to the voltages to which the qualified employee will be exposed, and

(D) The proper use of the special precautionary techniques, personal protective equipment, insulating and shielding materials, and insulated tools for working on or near exposed energized parts of electric equipment.

Note: For the purposes of this section, a person must have this training in order to be considered a qualified person.

Paragraphs (a)(2)(i) and (a)(2)(ii) require employees to be trained. They do not specifically require employers to provide this training themselves or to repeat training already provided. Clearly, the plain language of the standard allows employees to be trained by other parties or to have been trained previously by their own employers.

(iii) The employer shall determine, through regular supervision and through inspections

conducted on at least an annual basis that each employee is complying with the safety-related work practices required by this section.

(iv) An employee shall receive additional training (or retraining) under any of the following conditions:

(A) If the supervision and annual inspections required by paragraph (a)(2)(iii) of this section indicate that the employee is not complying with the safety-related work practices required by this section, or

(B) If new technology, new types of equipment, or changes in procedures necessitate the use of safety-related work practices that are different from those which the employee would normally use, or

(C) If he or she must employ safety-related work practices that are not normally used during his or her regular job duties.

Note: OSHA would consider tasks that are performed less often than once per year to necessitate retraining before the performance of the work practices involved.

The note indicates that the Agency considers tasks performed less often than once per year to require retraining before the task is actually performed. OSHA will accept instruction provided in pre-job briefings if it is detailed enough to fully inform the employee of the procedures involved in the job and to ensure that he or she can accomplish them in a safe manner. OSHA believes that this requirement will significantly improve safety for electric power generation, transmission, and distribution workers.

(v) The training required by paragraph (a)(2) of this section shall be of the classroom or on-the-job type.

This paragraph permits classroom or on-the-job training or a combination of both. This allows employers to continue the types of training programs that are currently in existence. Additionally, if an employee has already been trained (through previous job assignments, for example), the employer does not have to duplicate previous instruction.

(vi) The training shall establish employee proficiency in the work practices required by this section and shall introduce the procedures necessary for compliance with this section.

(vii) The employer shall certify that each employee has received the training required by paragraph (a)(2) of this section. This certification

FIGURE 6.3   (*Continued*)

shall be made when the employee demonstrates proficiency in the work practices involved and shall be maintained for the duration of the employee's employment.

Note: Employment records that indicate that an employee has received the required training are an acceptable means of meeting this requirement.

The employer is required, by paragraph (a)(2)(vii), to certify that each employee has been trained. This certification should not necessitate the employer's completing forms or creating new records; existing personnel records would normally suffice or the employer could simply make out a certification for each employee upon completion of training. Employers relying on training provided by previous employers are expected to take steps to verify that the employee has indeed received it.

OSHA has determined that there is a need for employees to be trained on a continuing basis. Initial instruction in safe techniques for performing specific job tasks is not sufficient to ensure that employees will use safe work practices all the time.

Continual reinforcement of this initial guidance must be provided to ensure that the employee actually uses the procedures he or she has been taught. This reinforcement can take the form of supervision, safety meetings, pre-job briefings or conferences, and retraining. Typically, adequate supervision can detect unsafe work practices with respect to tasks that are routine and are performed on a daily or regular basis. However, if an employee has to use a technique that is applied infrequently or that is based on new technology, some follow-up is needed to ensure that the employee is actually aware of the correct procedure for accomplishing the task. A detailed job briefing, as required under 1910.269(c)(2), may be adequate if the employee has previously received some instruction, but training would be necessary if the employee has never been schooled in the techniques to be used.

While there is substantial evidence in the record that electric utility employees are highly skilled and well trained, OSHA is concerned that a few employers may inaccurately "certify" the training of some employees who have not demonstrated proficiency in the work practices required by the standard. An example will help to illustrate the

need for the standard to address the overall goals of the training program. A complex tagging program and extensive training for that program, is characterized as typical for the electric utility industry as a whole. With respect to training in tagging procedures:

These detailed procedures, together with the safety manual, serve a dual purpose. They establish the specific requirements and provide the explicit direction for protection of employees from hazardous energy and they comprise the text material which is the basis for employee training in protection from hazardous energy. The training process is rigorous, including classroom presentation by qualified instructors, as well as self-study and it does include testing. Employees must demonstrate knowledge and skill in the application of hazardous energy control, consistent with established acceptance criteria, before they are qualified to either request that equipment be removed from service and tagged out, or to execute switching, valving, and tagging.

An employee who has attended a single training class on a procedure that is as complex as the lockout and tagging procedure used in an electric generating plant has generally not been fully trained in that procedure. Unless a training program establishes an employee's proficiency in safe work practices and that employee then demonstrates his or her ability to perform those work practices, there will be no assurance that safe work practices will result, and overall employee safety will not benefit nearly as much as it could.

Compliance with paragraph (a)(2)(vii) will ensure that employers do not try to comply with 1910.269 by simply handing training manuals to their employees. These provisions will require employers to take steps to assure that employees comprehend what they have been taught and that they are capable of performing the work practices mandated by the standard. OSHA believes that these two paragraphs will maximize the benefits of the training required under the standard.

OSHA believes that the training requirements contained in the final standard are sufficient to protect employees performing electric power generation, transmission, and distribution work. However, in every industry, there will be some employers who are not as faithful in following safety and health standards as others. The Agency intends to vigorously enforce the training requirements

FIGURE 6.3 (*Continued*)

of the final rule, because much of the worker's safety depends on knowledge of and skills in proper working procedures. The combination of rigorous training provisions with strict enforcement of these rules will result in increased safety to employees.

(3) *Existing conditions.* Existing conditions related to the safety of the work to be performed shall be determined before work on or near electric lines or equipment is started. Such conditions include, but are not limited to, the nominal voltages of lines and equipment, the maximum switching transient voltages, the presence of hazardous induced voltages, the presence and condition of protective grounds and equipment grounding conductors, the condition of poles, environmental conditions relative to safety, and the locations of circuits and equipment, including power and communication lines and fire protective signaling circuits.

This paragraph relates to hazards common to all types of electrical work, not just overhead line work. For example, the condition of the equipment grounding conductor that may be provided on a motor that is part of a generating installation affects the safety of anyone working on that motor.

(b) *Medical services and first aid.* The employer shall provide medical services and first aid as required in 1910.151 of this Part. In addition to the requirements of 1910.151 of this Part, the following requirements also apply:

The introductory text of paragraph (b) emphasizes that the requirements of 1910.151 apply. That existing section includes provisions for available medical personnel, first aid training and supplies, and facilities for drenching or flushing of the eyes and body in the event of exposure to corrosive materials.

Because of the hazard of electric shock when employees are performing work on or with energized lines and equipment, electric power generation, transmission, and distribution workers suffer electrocution on the job. Cardiopulmonary resuscitation (CPR) is necessary in the event of electric shock so that injured employees can be revived. CPR must be started within 4 minutes to be effective in reviving an employee whose heart has gone into fibrillation.

(1) *Cardiopulmonary resuscitation and first aid training.* When employees are performing work on or associated with exposed lines or equipment energized at 50 volts or

more, persons trained in first aid including cardiopulmonary resuscitation (CPR) shall be available as follows:

(i) For field work involving two or more employees at a work location, at least two trained persons shall be available. However, only one trained person need be available if all new employees are trained in first aid, including CPR, within 3 months of their hiring dates.

(ii) For fixed work locations such as generating stations, the number of trained persons available shall be sufficient to ensure that each employee exposed to electric shock can be reached within 4 minutes by a trained person. However, where the existing number of employees is insufficient to meet this requirement (at a remote substation, for example), all employees at the work location shall be trained.

Mr. Robert Felix of the National Arborist Association (NAA) argued that CPR was of dubious value with respect to injuries caused by electric shock. In NAA's post-hearing brief, he stated: "a study of the precise issue by medical experts (Cardiologist F. Gravino, M.D., F.A.C.C., et al.) commissioned by NAA and submitted to the Record as part of NAA's post-hearing evidence submission to the Docket, along with other related Record evidence, demonstrates the following medical assessment:

1. There is no demonstrated value of CPR in the electric injury context.
2. CPR is of no value to a person exposed to high voltage shock because of attendant "irreversible damage of either the autonomic nervous system or the cardiac tissue itself."
3. Lower voltage contacts from indirect contacts do "not respond to CPR"—see National Safety Council Newsletter of July/August 1990, at p. 1, submitted to the Docket by NAA as part of its post-hearing evidence submission. Moreover, such lower voltage contacts may induce respiratory block, rather than cardiac block, as to which artificial respiration, which is taught to line clearance tree trimmers as part of first-aid training, provides appropriate assistance, for which CPR would provide no additional benefit—a point conceded by NIOSH.
4. Even if otherwise appropriate in an electrical context—a fact not supported by the evidence—CPR is of value only if followed by defibrillation within 8 minutes of the onset of ventricular fibrillation.

FIGURE 6.3 *(Continued)*

The likelihood of getting an 8 minute ambulance response time to a line clearance job site is remote. . . . The dubious value of CPR further is attenuated by the remote likelihood of obtaining the required defibrillation within 8 minutes.

Others asserted that CPR was useful and necessary for the protection of workers exposed to electric shock. Dr. Richard Niemeier of NIOSH stated that current medical guidelines recommend CPR treatment, as follows:

The revised "Standards and Guidelines for Cardiopulmonary Resuscitation (CPR) and Emergency Cardiac Care (ECC)" recommend the same treatment for cardiopulmonary arrest, whether spontaneous or associated with electrical shock. The guidelines noted the complications of electric shock that might require CPR including tetany of the muscles used for breathing during contact with the electrical current, prolonged paralysis of breathing muscles for a period following the electric contact, and cardiac arrest. This discussion considers two categories: (1) respiratory arrest (with pulse) and (2) cardiac arrest.

NIOSH reviewed studies on the effectiveness of CPR in resuscitating electric shock victims. Regarding this review of the available evidence, Mr. Niemeier stated:

The question posed by OSHA, at this time, however, is whether sufficient evidence exists to support the recommendation that utility linemen work in pairs and be trained in CPR. Medical ethics and common sense prohibit a prospective study with random allocation of electrical shock or other cardiac arrest victims to "CPR" and "non-CPR" groups. This question must be answered, therefore, by clinical epidemiologic studies that are less than perfect. Cummins and Eisenberg[1985] reviewed the evidence regarding the relationship of early CPR and survival following cardiac arrest. The authors found nine studies that they considered credible (before 1985); all nine studies reported that early CPR had a beneficial effect. Cummins and Eisenberg[1985] concluded that the evidence clearly supported the concept that early CPR (begun on the scene by lay persons) leads to better survival rates than CPR delayed until emergency medical personnel arrive. These studies generally exclude trauma victims from analysis; this fact does not preclude the extrapolation of these results to patients with cardiac rhythm disturbances secondary to contact with electrical energy.

(2) First aid supplies. First aid supplies required by 1910.151(b) of this Part shall be placed in weatherproof containers if the supplies could be exposed to the weather.

(3) First aid kits. Each first aid kit shall be maintained, shall be readily available for use, and shall be inspected frequently enough to ensure that expended items are replaced at least once per year.

(c) Job briefing. The employer shall ensure that the employee in charge conducts a job briefing with the employees involved before they start each job. The briefing shall cover at least the following subjects: hazards associated with the job, work procedures involved, special precautions, energy source controls, and personal protective equipment requirements.

Most of the work performed under the standard requires planning in order to ensure employee safety (as well as to protect equipment and the general public).

If the work is not thoroughly planned ahead of time, the possibility of human error is increased greatly. To avoid problems, the task sequence is prescribed before work is started. For example, before climbing a pole, the employee must determine if the pole is capable of remaining in place and if minimum approach distances are sufficient, and he or she must determine what tools will be needed and what procedure should be used for performing the job. Without job planning, the worker may ignore the minimum approach distance requirements or may have to reclimb the pole to retrieve a forgotten tool or perform an overlooked task, resulting in increased exposure to the hazards of falling and contact with energized lines.

When more than one employee is involved, the job plan must be communicated to all the affected employees. If the job is planned but the plan is not discussed with the workers, one employee may perform his or her duties out of order or may otherwise not coordinate activities with the rest of the crew, endangering the entire crew. Therefore, OSHA is requiring a job briefing before work is started. The briefing would cover: hazards and work procedures involved, special precautions, energy source controls, and requirements for personal protective equipment.

FIGURE 6.3   (*Continued*)

(1) *Number of briefings.* If the work or operations to be performed during the work day or shift are repetitive and similar, at least one job briefing shall be conducted before the start of the first job of each day or shift. Additional job briefings shall be held if significant changes, which might affect the safety of the employees, occur during the course of the work.

At least one briefing is required to be conducted before the start of each shift. Only one briefing in a shift is needed if all the jobs are similar in nature. Additional planning discussions must take place for work involving significant changes in routine. For example, if the first two jobs of the day involve working on a de-energized line and the third job involves working on energized lines with live-line tools, separate briefings must be conducted for each type of job.

(2) *Extent of briefing.* A brief discussion is satisfactory if the work involved is routine and if the employee, by virtue of training and experience, can reasonably be expected to recognize and avoid the hazards involved in the job. A more extensive discussion shall be conducted:

(i) If the work is complicated or particularly hazardous, or

(ii) If the employee cannot be expected to recognize and avoid the hazards involved in the job.

Note: The briefing is always required to touch on all the subjects listed in the introductory text to paragraph (c) of this section.

The required briefing would normally consist of a concise discussion outlining the tasks to be performed. However, if the work is particularly hazardous or if the employees may not be able to recognize the hazards involved, then a more thorough discussion must take place. With this provision, OSHA recognizes that employees are familiar with the tasks and hazards involved with routine work. However, it is important to take the time to carefully discuss unusual work situations that may pose additional or different hazards to workers.

(3) *Working alone.* An employee working alone need not conduct a job briefing. However, the employer shall ensure that the tasks to be performed are planned as if a briefing were required.

(d) *Hazardous energy control (lockout/tagout) procedures.* (1) *Application.* The provisions of paragraph (d) of this section apply to the use of lockout/tagout procedures for the control

of energy sources in installations for the purpose of electric power generation, including related equipment for communication or metering. Locking and tagging procedures for the de-energizing of electric energy sources which are used exclusively for purposes of transmission and distribution are addressed by paragraph (m) of this section.

Note 1: Installations in electric power generation facilities that are not an integral part of, or inextricably commingled with, power generation processes or equipment are covered under 1910.147 and Subpart S of this Part.

Note 2: Lockout and tagging procedures that comply with paragraphs (c) through (f) of 1910.147 of this Part will also be deemed to comply with paragraph (d) of this section if the procedures address the hazards covered by paragraph (d) of this section.

(2) *General.* (i) The employer shall establish a program consisting of energy control procedures, employee training, and periodic inspections to ensure that, before any employee performs any servicing or maintenance on a machine or equipment where the unexpected energizing, start up, or release of stored energy could occur and cause injury, the machine or equipment is isolated from the energy source and rendered inoperative.

(ii) The employer's energy control program under paragraph (d)(2) of this section shall meet the following requirements:

(A) If an energy isolating device is not capable of being locked out, the employer's program shall use a tagout system.

(B) If an energy isolating device is capable of being locked out, the employer's program shall use lockout, unless the employer can demonstrate that the use of a tagout system will provide full employee protection as follows:

(1) When a tagout device is used on an energy isolating device which is capable of being locked out, the tagout device shall be attached at the same location that the lockout device would have been attached, and the employer shall demonstrate that the tagout program will provide a level of safety equivalent to that obtained by the use of a lockout program.

(2) In demonstrating that a level of safety is achieved in the tagout program equivalent to the level of safety obtained by the use of a lockout program, the employer shall demonstrate full compliance with all tagout-related provisions of this standard together with such additional elements as are necessary to provide the equivalent safety available from the use of a lockout device. Additional means to be considered as part of the

FIGURE 6.3 *(Continued)*

demonstration of full employee protection shall include the implementation of additional safety measures such as the removal of an isolating circuit element, blocking of a controlling switch, opening of an extra disconnecting device, or the removal of a valve handle to reduce the likelihood of inadvertent energizing.

(C) After November 1, 1994, whenever replacement or major repair, renovation, or modification of a machine or equipment is performed, and whenever new machines or equipment are installed, energy isolating devices for such machines or equipment shall be designed to accept a lockout device.

(iii) Procedures shall be developed, documented, and used for the control of potentially hazardous energy covered by paragraph (d) of this section.

(iv) The procedure shall clearly and specifically outline the scope, purpose, responsibility, authorization, rules, and techniques to be applied to the control of hazardous energy, and the measures to enforce compliance including, but not limited to, the following:

(A) A specific statement of the intended use of this procedure;

(B) Specific procedural steps for shutting down, isolating, blocking and securing machines or equipment to control hazardous energy;

(C) Specific procedural steps for the placement, removal, and transfer of lockout devices or tagout devices and the responsibility for them; and

(D) Specific requirements for testing a machine or equipment to determine and verify the effectiveness of lockout devices, tagout devices, and other energy control measures.

(v) The employer shall conduct a periodic inspection of the energy control procedure at least annually to ensure that the procedure and the provisions of paragraph (d) of this section are being followed.

(A) The periodic inspection shall be performed by an authorized employee who is not using the energy control procedure being inspected.

(B) The periodic inspection shall be designed to identify and correct any deviations or inadequacies.

(C) If lockout is used for energy control, the periodic inspection shall include a review between the inspector and each authorized employee, of that employee's responsibilities under the energy control procedure being inspected.

(D) Where tagout is used for energy control, the periodic inspection shall include a review between the inspector and each authorized and affected employee, of that employee's responsibilities under the energy control procedure being inspected, and the elements set forth in paragraph (d)(2)(vii) of this section.

(E) The employer shall certify that the inspections required by paragraph (d)(2)(v) of this section have been accomplished. The certification shall identify the machine or equipment on which the energy control procedure was being used, the date of the inspection, the employees included in the inspection, and the person performing the inspection.

Note: If normal work schedule and operation records demonstrate adequate inspection activity and contain the required information, no additional certification is required.

(vi) The employer shall provide training to ensure that the purpose and function of the energy control program are understood by employees and that the knowledge and skills required for the safe application, usage, and removal of energy controls are acquired by employees. The training shall include the following:

(A) Each authorized employee shall receive training in the recognition of applicable hazardous energy sources, the type and magnitude of energy available in the workplace, and in the methods and means necessary for energy isolation and control.

(B) Each affected employee shall be instructed in the purpose and use of the energy control procedure.

(C) All other employees whose work operations are or may be in an area where energy control procedures may be used shall be instructed about the procedures and about the prohibition relating to attempts to restart or reenergize machines or equipment that are locked out or tagged out.

(vii) When tagout systems are used, employees shall also be trained in the following limitations of tags:

(A) Tags are essentially warning devices affixed to energy isolating devices and do not provide the physical restraint on those devices that is provided by a lock.

(B) When a tag is attached to an energy isolating means, it is not to be removed without authorization of the authorized person responsible for it, and it is never to be bypassed, ignored, or otherwise defeated.

(C) Tags must be legible and understandable by all authorized employees, affected employees, and all other employees whose work

FIGURE 6.3 (*Continued*)

operations are or may be in the area, in order to be effective.

(D) Tags and their means of attachment must be made of materials which will withstand the environmental conditions encountered in the workplace.

(E) Tags may evoke a false sense of security, and their meaning needs to be understood as part of the overall energy control program.

(F) Tags must be securely attached to energy isolating devices so that they cannot be inadvertently or accidentally detached during use.

(viii) Retraining shall be provided by the employer as follows:

(A) Retraining shall be provided for all authorized and affected employees whenever there is a change in their job assignments, a change in machines, equipment, or processes that present a new hazard or whenever there is a change in the energy control procedures.

(B) Retraining shall also be conducted whenever a periodic inspection under paragraph (d)(2)(v) of this section reveals, or whenever the employer has reason to believe, that there are deviations from or inadequacies in an employee's knowledge or use of the energy control procedures.

(C) The retraining shall reestablish employee proficiency and shall introduce new or revised control methods and procedures, as necessary.

(ix) The employer shall certify that employee training has been accomplished and is being kept up to date. The certification shall contain each employee's name and dates of training.

(3) *Protective materials and hardware.* **(i)** Locks, tags, chains, wedges, key blocks, adapter pins, self-locking fasteners, or other hardware shall be provided by the employer for isolating, securing, or blocking of machines or equipment from energy sources.

(ii) Lockout devices and tagout devices shall be singularly identified; shall be the only devices used for controlling energy; may not be used for other purposes; and shall meet the following requirements:

(A) Lockout devices and tagout devices shall be capable of withstanding the environment to which they are exposed for the maximum period of time that exposure is expected.

(1) Tagout devices shall be constructed and printed so that exposure to weather conditions or wet and damp locations will not cause the tag to deteriorate or the message on the tag to become illegible.

(2) Tagout devices shall be so constructed as not to deteriorate when used in corrosive environments.

(B) Lockout devices and tagout devices shall be standardized within the facility in at least one of the following criteria: color, shape, and size. Additionally, in the case of tagout devices, print and format shall be standardized.

(C) Lockout devices shall be substantial enough to prevent removal without the use of excessive force or unusual techniques, such as with the use of bolt cutters or metal cutting tools.

(D) Tagout devices, including their means of attachment, shall be substantial enough to prevent inadvertent or accidental removal. Tagout device attachment means shall be of a nonreusable type, attachable by hand, self-locking, and non-releasable with a minimum unlocking strength of no less than 50 pounds and shall have the general design and basic characteristics of being at least equivalent to a one-piece, all-environment-tolerant nylon cable tie.

(E) Each lockout device or tagout device shall include provisions for the identification of the employee applying the device.

(F) Tagout devices shall warn against hazardous conditions if the machine or equipment is energized and shall include a legend such as the following: Do Not Start, Do Not Open, Do Not Close, Do Not Energize, Do Not Operate.

Note: For specific provisions covering accident prevention tags, see 1910.145 of this Part.

(4) *Energy isolation.* Lockout and tagout device application and removal may only be performed by the authorized employees who are performing the servicing or maintenance.

(5) *Notification.* Affected employees shall be notified by the employer or authorized employee of the application and removal of lockout or tagout devices. Notification shall be given before the controls are applied and after they are removed from the machine or equipment.

Note: See also paragraph (d)(7) of this section, which requires that the second notification take place before the machine or equipment is reenergized.

(6) *Lockout/tagout application.* The established procedures for the application of energy control (the lockout or tagout procedures) shall include the following elements and actions, and these procedures shall be performed in the following sequence:

(i) Before an authorized or affected employee turns off a machine or equipment, the authorized employee shall have knowledge of the type and magnitude of the energy, the hazards of the energy to be controlled, and the method or means to control the energy.

(ii) The machine or equipment shall be turned off or shut down using the procedures

FIGURE 6.3 *(Continued)*

established for the machine or equipment. An orderly shutdown shall be used to avoid any additional or increased hazards to employees as a result of the equipment stoppage.

(**iii**) All energy isolating devices that are needed to control the energy to the machine or equipment shall be physically located and operated in such a manner as to isolate the machine or equipment from energy sources.

(**iv**) Lockout or tagout devices shall be affixed to each energy isolating device by authorized employees.

(**A**) Lockout devices shall be attached in a manner that will hold the energy isolating devices in a "safe" or "off" position.

(**B**) Tagout devices shall be affixed in such a manner as will clearly indicate that the operation or movement of energy isolating devices from the "safe" or "off" position is prohibited.

(*1*) Where tagout devices are used with energy isolating devices designed with the capability of being locked out, the tag attachment shall be fastened at the same point at which the lock would have been attached.

(*2*) Where a tag cannot be affixed directly to the energy isolating device, the tag shall be located as close as safely possible to the device, in a position that will be immediately obvious to anyone attempting to operate the device.

(**v**) Following the application of lockout or tagout devices to energy isolating devices, all potentially hazardous stored or residual energy shall be relieved, disconnected, restrained, or otherwise rendered safe.

(**vi**) If there is a possibility of reaccumulation of stored energy to a hazardous level, verification of isolation shall be continued until the servicing or maintenance is completed or until the possibility of such accumulation no longer exists.

(**vii**) Before starting work on machines or equipment that have been locked out or tagged out, the authorized employee shall verify that isolation and deenergizing of the machine or equipment have been accomplished. If normally energized parts will be exposed to contact by an employee while the machine or equipment is deenergized, a test shall be performed to ensure that these parts are deenergized.

(**7**) *Release from lockout/tagout.* Before lockout or tagout devices are removed and energy is restored to the machine or equipment, procedures shall be followed and actions taken by the authorized employees to ensure the following:

(**i**) The work area shall be inspected to ensure that nonessential items have been removed

and that machine or equipment components are operationally intact.

(**ii**) The work area shall be checked to ensure that all employees have been safely positioned or removed.

(**iii**) After lockout or tagout devices have been removed and before a machine or equipment is started, affected employees shall be notified that the lockout or tagout devices have been removed.

(**iv**) Each lockout or tagout device shall be removed from each energy isolating device by the authorized employee who applied the lockout or tagout device. However, if that employee is not available to remove it, the device may be removed under the direction of the employer, provided that specific procedures and training for such removal have been developed, documented, and incorporated into the employer's energy control program. The employer shall demonstrate that the specific procedure provides a degree of safety equivalent to that provided by the removal of the device by the authorized employee who applied it. The specific procedure shall include at least the following elements:

(**A**) Verification by the employer that the authorized employee who applied the device is not at the facility;

(**B**) Making all reasonable efforts to contact the authorized employee to inform him or her that his or her lockout or tagout device has been removed; and

(**C**) Ensuring that the authorized employee has this knowledge before he or she resumes work at that facility.

(**8**) **Additional requirements. (i)** If the lockout or tagout devices must be temporarily removed from energy isolating devices and the machine or equipment must be energized to test or position the machine, equipment, or component thereof, the following sequence of actions shall be followed:

(**A**) Clear the machine or equipment of tools and materials in accordance with paragraph (d)(7)(i) of this section;

(**B**) Remove employees from the machine or equipment area in accordance with paragraphs (d)(7)(ii) and (d)(7)(iii) of this section;

(**C**) Remove the lockout or tagout devices as specified in paragraph (d)(7)(iv) of this section;

(**D**) Energize and proceed with the testing or positioning; and

(**E**) Deenergize all systems and reapply energy control measures in accordance with paragraph (d)(6) of this section to continue the servicing or maintenance.

FIGURE 6.3   (*Continued*)

(**ii**) When servicing or maintenance is performed by a crew, craft, department, or other group, they shall use a procedure which affords the employees a level of protection equivalent to that provided by the implementation of a personal lockout or tagout device. Group lockout or tagout devices shall be used in accordance with the procedures required by paragraphs (d)(2)(iii) and (d)(2)(iv) of this section including, but not limited to, the following specific requirements:

(**A**) Primary responsibility shall be vested in an authorized employee for a set number of employees working under the protection of a group lockout or tagout device (such as an operations lock);

(**B**) Provision shall be made for the authorized employee to ascertain the exposure status of all individual group members with regard to the lockout or tagout of the machine or equipment;

(**C**) When more than one crew, craft, department, or other group is involved, assignment of overall job-associated lockout or tagout control responsibility shall be given to an authorized employee designated to coordinate affected work forces and ensure continuity of protection; and

(**D**) Each authorized employee shall affix a personal lockout or tagout device to the group lockout device, group lockbox, or comparable mechanism when he or she begins work and shall remove those devices when he or she stops working on the machine or equipment being serviced or maintained.

(**iii**) Procedures shall be used during shift or personnel changes to ensure the continuity of lockout or tagout protection, including provision for the orderly transfer of lockout or tagout device protection between off-going and on-coming employees, to minimize their exposure to hazards from the unexpected energizing or start-up of the machine or equipment or from the release of stored energy.

(**iv**) Whenever outside servicing personnel are to be engaged in activities covered by paragraph (d) of this section, the on-site employer and the outside employer shall inform each other of their respective lockout or tagout procedures, and each employer shall ensure that his or her personnel understand and comply with restrictions and prohibitions of the energy control procedures being used.

(**v**) If energy isolating devices are installed in a central location and are under the exclusive control of a system operator, the following requirements apply:

(**A**) The employer shall use a procedure that affords employees a level of protection equivalent to that provided by the implementation of a personal lockout or tagout device.

(**B**) The system operator shall place and remove lockout and tagout devices in place of the authorized employee under paragraphs (d)(4), (d)(6)(iv), and (d)(7)(iv) of this section.

(**C**) Provisions shall be made to identify the authorized employee who is responsible for (that is, being protected by) the lockout or tagout device, to transfer responsibility for lockout and tagout devices, and to ensure that an authorized employee requesting removal or transfer of a lockout or tagout device is the one responsible for it before the device is removed or transferred.

(**e**) *Enclosed spaces.* This paragraph covers enclosed spaces that may be entered by employees. It does not apply to vented vaults if a determination is made that the ventilation system is operating to protect employees before they enter the space. This paragraph applies to routine entry into enclosed spaces in lieu of the permit-space entry requirements contained in paragraphs (d) through (k) of 1910.146 of this Part. If, after the precautions given in paragraphs (e) and (t) of this section are taken, the hazards remaining in the enclosed space endanger the life of an entrant or could interfere with escape from the space, then entry into the enclosed space shall meet the permit-space entry requirements of paragraphs (d) through (k) of 1910.146 of this Part.

An "enclosed space" is defined to be a space that has a limited means of entry or egress, that is designed for periodic entry by employees under normal operating conditions, and that is not expected to contain a hazardous atmosphere, but may contain one under unusual conditions. In this paragraph, OSHA intends to cover only the types of enclosed spaces that are routinely entered by employees engaged in electric power generation, transmission, and distribution work and are unique to underground utility work. Work in these spaces is part of the day-to-day activities performed by employees protected by this standard. Enclosed spaces include manholes and vaults that provide employees access to electric generation, transmission, and distribution equipment. This paragraph does not address other types of confined spaces, such as boilers, tanks, and coal bunkers, that are common to other industries as well. These

FIGURE 6.3   (*Continued*)

locations are addressed in OSHA's generic permit-required confined space standard, 1910.146, which applies to all of general industry, including industries engaged in electric power generation, transmission, and distribution work.

Section 1910.146 contains requirements that address hazards associated with entry into "permit-required confined spaces" (permit spaces). Section 1910.146 defines "confined space" and "permit-required confined space" as follows:

Confined space means a space that:

(1) is large enough and so configured that an employee can bodily enter and perform assigned work; and

(2) has limited or restricted means for entry or exit (for example, tanks, vessels, silos, storage bins, hoppers, vaults, and pits are spaces that may have limited means of entry.); and

(3) is not designed for continuous employee occupancy.

Permit-required confined space (permit space) means a confined space that has one or more of the following characteristics:

(1) Contains or has a potential to contain a hazardous atmosphere;

(2) Contains a material that has the potential for engulfing an entrant;

(3) Has an internal configuration such that an entrant could be trapped or asphyxiated by inwardly converging walls or by a floor which slopes downward and tapers to a smaller cross-section; or

(4) Contains any other recognized serious safety or health hazard.

The permit-required confined spaces standard requires employers to implement a comprehensive confined space entry program. This standard covers the wide range of permit-required confined spaces encountered throughout general industry. Because the hazards posed by these spaces vary so greatly, 1910.146 requires employers to implement a permit system for entry into them. The permit system must spell out the steps to be taken to make the space safe for entry and must include provisions for attendants stationed outside the spaces and for rescue of entrants, who could be disabled inside the space. However, an employer need not follow the permit-entry requirements of 1910.146 for spaces where the hazards have been completely eliminated

or for spaces where an alternative set of procedures are observed. The alternative procedures apply only where the space can be made safe for entry through the use of continuous forced air ventilation alone. The procedures, which are set forth in 1910.146(c)(5)(ii), ensure that conditions within the permit space do not endanger an entrant's life or ability to rescue himself or herself.

Paragraph (e) of 1910.269 applies to "enclosed spaces". By definition, an enclosed space would be a permit-required confined space in the absence of 1910.269. An enclosed space meets the definition of a confined space—it is large enough for an employee to enter; it has a limited means of access or egress; it is designed for periodic, rather than continuous, employee occupancy under normal operating conditions. An enclosed space also meets the definition of a permit space—although it is not expected to contain a hazardous atmosphere, it has the potential to contain one. The Agency notes that, if hazardous conditions which cannot be controlled through the precautions set out in paragraphs (e) and (t) of final 1910.269 are present, the enclosed space must be treated as a permit space under 1910.146.

Note: Entries into enclosed spaces conducted in accordance with the permit-space entry requirements of paragraphs (d) through (k) of 1910.146 of this Part are considered as complying with paragraph (e) of this section.

(1) Safe work practices. The employer shall ensure the use of safe work practices for entry into and work in enclosed spaces and for rescue of employees from such spaces.

Paragraph (e)(1) sets forth the general requirement that employers ensure the use of safe work practices by their employees. These safe work practices must include procedures for complying with the specific regulations contained in paragraphs (e)(4) through (e)(14) and must include safe rescue procedures.

(2) Training. Employees who enter enclosed spaces or who serve as attendants shall be trained in the hazards of enclosed space entry, in enclosed space entry procedures, and in enclosed space rescue procedures.

(3) Rescue equipment. Employers shall provide equipment to ensure the prompt and safe rescue of employees from the enclosed space.

FIGURE 6.3 *(Continued)*

The equipment must enable a rescuer to remove an injured employee from the enclosed space quickly and without injury to the rescuer or further harm to the fallen employee. A harness, a lifeline, and a self-supporting winch can normally be used in this manner.

(4) *Evaluation of potential hazards.* Before any entrance cover to an enclosed space is removed, the employer shall determine whether it is safe to do so by checking for the presence of any atmospheric pressure or temperature differences and by evaluating whether there might be a hazardous atmosphere in the space. Any conditions making it unsafe to remove the cover shall be eliminated before the cover is removed.

Note: The evaluation called for in this paragraph may take the form of a check of the conditions expected to be in the enclosed space. For example, the cover could be checked to see if it is hot and, if it is fastened in place, could be loosened gradually to release any residual pressure. A determination must also be made of whether conditions at the site could cause a hazardous atmosphere, such as an oxygen deficient or flammable atmosphere, to develop within the space.

Some conditions within an enclosed space, such as high temperature and high pressure, make it hazardous to remove any cover from the space. For example, if high pressure is present within the space, the cover could be blown off in the process of removing it. To protect employees from such hazards, paragraph (e)(4) requires a determination of whether or not it is safe to remove the cover. This determination may take the form of a quick check of the conditions expected to be in the enclosed space. For example, the cover could be checked to see if it is hot and, if it is fastened in place, could be loosened gradually to release any residual pressure. An evaluation must also be made of whether conditions at the site could cause a hazardous atmosphere to accumulate in the space. Any conditions making it unsafe for employees to remove the cover are required to be eliminated (that is, reduced to the extent that it is no longer unsafe).

(5) *Removal of covers.* When covers are removed from enclosed spaces, the opening shall be promptly guarded by a railing, temporary cover, or other barrier intended to prevent an accidental fall through the opening and to protect employees working in the space from objects entering the space.

(6) *Hazardous atmosphere.* Employees may not enter any enclosed space while it contains a hazardous atmosphere, unless the entry conforms to the generic permit-required confined spaces standard in 1910.146 of this Part.

Note: The term "entry" is defined in 1910.146(b) of this Part.

OSHA has noted earlier, paragraph (e) is intended to apply only to routine entry into enclosed spaces, where compliance with the procedures set out in paragraphs (e) and (t) adequately protects employees. If a hazardous atmosphere exists in an enclosed space after the testing and ventilation requirements in paragraphs (e)(9) through (e)(13) of 1910.269 have been met, additional measures must be taken to protect employees. When this is the case, the generic permit-spaces standard in 1910.146 contains the relevant requirements necessary to protect entrants. Paragraph (e)(6) of 1910.269 makes this clear. (It should be noted that Subpart Z of Part 1910 continues to apply to the exposure of employees to toxic substances.)

(7) *Attendants.* While work is being performed in the enclosed space, a person with first aid training meeting paragraph (b) of this section shall be immediately available outside the enclosed space to render emergency assistance if there is reason to believe that a hazard may exist in the space or if a hazard exists because of traffic patterns in the area of the opening used for entry. That person is not precluded from performing other duties outside the enclosed space if these duties do not distract the attendant from monitoring employees within the space.

Note: See paragraph (t)(3) of this section for additional requirements on attendants for work in manholes.

Paragraph (e)(7) addresses the use of an attendant outside the enclosed space to provide assistance in an emergency. An attendant is required if there is reason to believe that a hazard exists within the space or if a hazard exists because of traffic patterns near the opening. For example, a manhole containing energized electric equipment that is in danger of failing catastrophically requires an attendant under this paragraph. The purpose of the attendant would be to provide assistance in an emergency.

FIGURE 6.3 *(Continued)*

If this person were to enter the enclosed space, he or she might be unable to assist the employee already within the space. For example, if traffic hazards are present in the area of the opening to the enclosed space and if the attendant entered the space, then both the attendant and the workers he or she is intended to protect would be vulnerable upon leaving. No one would be present to minimize or control the traffic hazards. If flooding hazards are present, a person outside the space may be able to assist in a rescue attempt; an attendant inside the space would likely be another victim. Therefore, the final rule explicitly states that the attendant is required to remain outside the enclosed space.

On the other hand, if there is no reason to believe that a hazard exists inside the enclosed space and if no traffic hazards are present, an attendant would still be required under 1910.269(t)(3) while work is being performed in a manhole containing energized conductors. The major, though not the only, hazard in this case is that of electric shock. Assistance can be provided to a victim of electric shock by another person in the manhole. Therefore, the provisions of paragraph (t)(3) permit the attendant required under that paragraph to enter the manhole for brief periods of time. However, it should be noted that 1910.269(e)(7) requires the attendant to be "immediately available outside the space". Thus, an attendant required by paragraph (e)(7) (rather than by paragraph (t)(3)) is required to remain outside the space.

(8) *Calibration of test instruments.* Test instruments used to monitor atmospheres in enclosed spaces shall be kept in calibration, with a minimum accuracy of ±10 percent.

(9) *Testing for oxygen deficiency.* Before an employee enters an enclosed space, the internal atmosphere shall be tested for oxygen deficiency with a direct-reading meter or similar instrument, capable of collection and immediate analysis of data samples without the need for off-site evaluation. If continuous forced air ventilation is provided, testing is not required provided that the procedures used ensure that employees are not exposed to the hazards posed by oxygen deficiency.

(10) *Testing for flammable gases and vapors.* Before an employee enters an enclosed space, the internal atmosphere shall be tested for flammable gases and vapors with a direct-reading meter or similar instrument capable of collection and immediate analysis of data samples without the need for off-site evaluation. This test shall be performed after the oxygen testing and ventilation required by paragraph (e)(9) of this section demonstrate that there is sufficient oxygen to ensure the accuracy of the test for flammability.

(11) *Ventilation and monitoring.* If flammable gases or vapors are detected or if an oxygen deficiency is found, forced air ventilation shall be used to maintain oxygen at a safe level and to prevent a hazardous concentration of flammable gases and vapors from accumulating. A continuous monitoring program to ensure that no increase in flammable gas or vapor concentration occurs may be followed in lieu of ventilation, if flammable gases or vapors are detected at safe levels.

Note: See the definition of hazardous atmosphere for guidance in determining whether or not a given concentration of a substance is considered to be hazardous.

(12) *Specific ventilation requirements.* If continuous forced air ventilation is used, it shall begin before entry is made and shall be maintained long enough to ensure that a safe atmosphere exists before employees are allowed to enter the work area. The forced air ventilation shall be so directed as to ventilate the immediate area where employees are present within the enclosed space and shall continue until all employees leave the enclosed space.

(13) *Air supply.* The air supply for the continuous forced air ventilation shall be from a clean source and may not increase the hazards in the enclosed space.

(14) *Open flames.* If open flames are used in enclosed spaces, a test for flammable gases and vapors shall be made immediately before the open flame device is used and at least once per hour while the device is used in the space. Testing shall be conducted more frequently if conditions present in the enclosed space indicate that once per hour is insufficient to detect hazardous accumulations of flammable gases or vapors.

Note: See the definition of hazardous atmosphere for guidance in determining whether or not a given concentration of a substance is considered to be hazardous.

(f) *Excavations.* Excavation operations shall comply with Subpart P of Part 1926 of this chapter.

(g) *Personal protective equipment.* **(1)** *General.* Personal protective equipment shall meet the requirements of Subpart I of this Part.

FIGURE 6.3 (*Continued*)

Paragraph (g) sets forth requirements for personal protective equipment (PPE), which includes eye and face protection, respiratory protection, head protection, foot protection, protective clothing, electrical protective equipment, and personal fall protection equipment.

(2) *Fall protection.* (i) Personal fall arrest equipment shall meet the requirements of Subpart E of Part 1926 of this Chapter.

(ii) Body belts and safety straps for work positioning shall meet the requirements of 1926.959 of this Chapter.

(iii) Body belts, safety straps, lanyards, lifelines, and body harnesses shall be inspected before use each day to determine that the equipment is in safe working condition. Defective equipment may not be used.

(iv) Lifelines shall be protected against being cut or abraded.

(v) Fall arrest equipment, work positioning equipment, or travel restricting equipment shall be used by employees working at elevated locations more than 4 feet (1.2 m) above the ground on poles, towers, or similar structures if other fall protection has not been provided. Fall protection equipment is not required to be used by a qualified employee climbing or changing location on poles, towers, or similar structures, unless conditions, such as, but not limited to, ice, high winds, the design of the structure (for example, no provision for holding on with hands), or the presence of contaminants on the structure, could cause the employee to lose his or her grip or footing.

Note 1: This paragraph applies to structures that support overhead electric power generation, transmission and distribution lines, and equipment. It does not apply to portions of buildings, such as loading docks, to electric equipment, such as transformers and capacitors, nor to aerial lifts. Requirements for fall protection associated with walking and working surfaces are contained in Subpart D of this Part; requirements for fall protection associated with aerial lifts are contained in 1910.67 of this Part.

Note 2: Employees undergoing training are not considered "qualified employees" for the purposes of this provision. Unqualified employees (including trainees) are required to use fall protection any time they are more than 4 feet (1.2 m) above the ground.

NIOSH stated that risks associated with climbing poles are a major cause of injuries and fatalities in the electric utility industry and submitted a Canadian study that listed falls as accounting for 21.9 percent of all accidents. "Climbing up or down a pole, tower, basket, truck" accounted for 14.8 percent of all accidents in this study. The "IBEW Utility Department Survey of Fatal and Serious Occupational Accidents" for the years 1984, 1986, and 1988 report 13 fatalities from slips and falls during the period represented by these surveys. The total number of deaths was 121, and the total non-electrical accidents was 37. In this data base, falls represented about 12 percent of all fatalities and 35 percent of non-electrical deaths. Injuries due to falls from elevations were involved in 10 percent (61 of 637) of the fatality/catastrophe investigations recorded. These investigations included only electric utilities.

All of these exhibits demonstrate that electric power generation, transmission, and distribution workers face a significant risk of serious injury due to falls under current industry practices. To determine the extent to which they face hazards addressed by 1910.269(g)(2)(v), OSHA analyzed fall accidents included in various exhibits contained in the rulemaking record. The results of this analysis are presented in Table 1. As can be seen from the table, employees do fall while climbing poles, towers, or similar structures—26 percent of the falling accidents related to 1910.269 occurred in this manner. The evidence in the record indicates that climbing a pole, tower, or similar structure is not as safe, under current industry practices, as some of the hearing witnesses testified.

(vi) The following requirements apply to personal fall arrest systems:

TABLE 1 Falls by Type of Accident

Type of fall	Number of accidents
Fall from pole or tower	
Climbing or descending	10
Changing location	1
At work location	7
Other (not stated)	3
Fall from tree	6
Failure of structure	12

   Each accident involves the death or serious injury of one or more employees.

FIGURE 6.3 *(Continued)*

(A) When stopping or arresting a fall, personal fall arrest systems shall limit the maximum arresting force on an employee to 900 pounds (4 kN) if used with a body belt.

(B) When stopping or arresting a fall, personal fall arrest systems shall limit the maximum arresting force on an employee to 1800 pounds (8 kN) if used with a body harness.

(C) Personal fall arrest systems shall be rigged such that an employee can neither free fall more than 6 feet (1.8 m) nor contact any lower level.

(vii) If vertical lifelines or droplines are used, not more than one employee may be attached to any one lifeline.

This limitation recognizes that it is inherently unsafe to use a single vertical lifeline to tie off two or more employees performing separate tasks. Movement by one employee could cause the lifeline to be pulled to one side. This could, in turn, cause the other employee to lose balance. Therefore, if one employee did fall, movement of the lifeline during the arrest of the fall would very likely cause other employees connected to the lifeline to fall.

(viii) Snaphooks may not be connected to loops made in webbing-type lanyards.

(ix) Snaphooks may not be connected to each other.

(h) *Ladders, platforms, step bolts, and manhole steps* **(1)** *General.* Requirements for ladders contained in Subpart D of this Part apply, except as specifically noted in paragraph (h)(2) of this section.

(2) *Special ladders and platforms.* Portable ladders and platforms used on structures or conductors in conjunction with overhead line work need not meet paragraphs (d)(2)(i) and (d)(2)(iii) of 1910.25 of this Part or paragraph (c)(3)(iii) of 1910.26 of this Part. However, these ladders and platforms shall meet the following requirements:

(i) Ladders and platforms shall be secured to prevent their becoming accidentally dislodged.

(ii) Ladders and platforms may not be loaded in excess of the working loads for which they are designed.

(iii) Ladders and platforms may be used only in applications for which they were designed.

(iv) In the configurations in which they are used, ladders and platforms shall be capable of supporting without failure at least 2.5 times the maximum intended load.

(3) *Conductive ladders.* Portable metal ladders and other portable conductive ladders may not be used near exposed energized lines or equipment. However, in specialized high-voltage work, conductive ladders shall be used where the employer can demonstrate that nonconductive ladders would present a greater hazard than conductive ladders.

(i) *Hand and portable power tools.* **(1)** *General.* Paragraph (i)(2) of this section applies to electric equipment connected by cord and plug. Paragraph (i)(3) of this section applies to portable and vehicle-mounted generators used to supply cord- and plug-connected equipment. Paragraph (i)(4) of this section applies to hydraulic and pneumatic tools.

(2) *Cord- and plug-connected equipment.* **(i)** Cord-and plug-connected equipment supplied by premises wiring is covered by Subpart S of this Part.

(ii) Any cord- and plug-connected equipment supplied by other than premises wiring shall comply with one of the following in lieu of 1910.243(a)(5) of this Part:

(A) It shall be equipped with a cord containing an equipment grounding conductor connected to the tool frame and to a means for grounding the other end (however, this option may not be used where the introduction of the ground into the work environment increases the hazard to an employee); or

(B) It shall be of the double-insulated type conforming to Subpart S of this Part; or

(C) It shall be connected to the power supply through an isolating transformer with an ungrounded secondary.

If the equipment is supplied by the wiring of a building or other premises, existing Subpart S of Part 1910 continues to apply as it does now. If premises wiring is not involved (in which case Subpart S does not currently apply), paragraph (i)(2)(ii) requires that the tool frame be grounded or that the tool be double insulated or that the tool be supplied by an isolating transformer with ungrounded secondary. Any of these three methods can protect employees from electric shock, which could directly injure the employee or which could cause an involuntary reaction leading to a secondary injury.

OSHA received several comments suggesting that ground-fault circuit interrupter (GFCI) protection be allowed as an additional alternative. However, although a GFCI can prevent electrocution, the device cannot

FIGURE 6.3   *(Continued)*

by itself prevent an initial electric shock to an employee before it interrupts the circuit. This initial shock could lead to injury from involuntary reaction. This is of particular concern if the employee is in an elevated position, exposing him or her to a fall in the event of electric shock. For this reason, existing electrical standards require GFCI protection in addition to, not in place of, equipment grounding and double insulation. Therefore, OSHA is not allowing the use of a GFCI alone to protect employees using cord- and plug-connected equipment.

(**3**) *Portable and vehicle-mounted generators.* Portable and vehicle-mounted generators used to supply cord- and plug-connected equipment shall meet the following requirements:

(**i**) The generator may only supply equipment located on the generator or the vehicle and cord- and plug-connected equipment through receptacles mounted on the generator or the vehicle.

(**ii**) The non-current-carrying metal parts of equipment and the equipment grounding conductor terminals of the receptacles shall be bonded to the generator frame.

(**iii**) In the case of vehicle-mounted generators, the frame of the generator shall be bonded to the vehicle frame.

(**iv**) Any neutral conductor shall be bonded to the generator frame.

(**4**) *Hydraulic and pneumatic tools.* (**i**) Safe operating pressures for hydraulic and pneumatic tools, hoses, valves, pipes, filters, and fittings may not be exceeded.

Note: If any hazardous defects are present, no operating pressure would be safe, and the hydraulic or pneumatic equipment involved may not be used. In the absence of defects, the maximum rated operating pressure is the maximum safe pressure.

(**ii**) A hydraulic or pneumatic tool used where it may contact exposed live parts shall be designed and maintained for such use.

(**iii**) The hydraulic system supplying a hydraulic tool used where it may contact exposed live parts shall provide protection against loss of insulating value for the voltage involved due to the formation of a partial vacuum in the hydraulic line.

Note: Hydraulic lines without check valves having a separation of more than 35 feet (10.7 m) between the oil reservoir and the upper end of the hydraulic system promote the formation of a partial vacuum.

(**iv**) A pneumatic tool used on energized electric lines or equipment or used where it may

contact exposed live parts shall provide protection against the accumulation of moisture in the air supply.

(**v**) Pressure shall be released before connections are broken, unless quick acting, self-closing connectors are used. Hoses may not be kinked.

(**vi**) Employees may not use any part of their bodies to locate or attempt to stop a hydraulic leak.

(**j**) *Live-line tools.* (**1**) *Design of tools.* Live-line tool rods, tubes, and poles shall be designed and constructed to withstand the following minimum tests:

(**i**) 100,000 volts per foot (3281 volts per centimeter) of length for 5 minutes if the tool is made of fiberglass-reinforced plastic (FRP), or

(**ii**) 75,000 volts per foot (2461 volts per centimeter) of length for 3 minutes if the tool is made of wood, or

(**iii**) Other tests that the employer can demonstrate are equivalent.

Note: Live-line tools using rod and tube that meet ASTM F711-89, Standard Specification for Fiberglass-Reinforced Plastic (FRP) Rod and Tube Used in Live-Line Tools, conform to paragraph (j)(1)(i) of this section.

(**2**) *Condition of tools.* (**i**) Each live-line tool shall be wiped clean and visually inspected for defects before use each day.

(**ii**) If any defect or contamination that could adversely affect the insulating qualities or mechanical integrity of the live-line tool is present after wiping, the tool shall be removed from service and examined and tested according to paragraph (j)(2)(iii) of this section before being returned to service.

(**iii**) Live-line tools used for primary employee protection shall be removed from service every 2 years and whenever required under paragraph (j)(2)(ii) of this section for examination, cleaning, repair, and testing as follows:

(**A**) Each tool shall be thoroughly examined for defects.

(**B**) If a defect or contamination that could adversely affect the insulating qualities or mechanical integrity of the live-line tool is found, the tool shall be repaired and refinished or shall be permanently removed from service. If no such defect or contamination is found, the tool shall be cleaned and waxed.

(**C**) The tool shall be tested in accordance with paragraphs (j)(2)(iii)(D) and (j)(2)(iii)(E) of this section under the following conditions:

(*1*) After the tool has been repaired or refinished; and

FIGURE 6.3   *(Continued)*

(*2*) After the examination if repair or refinishing is not performed, unless the tool is made of FRP rod or foam-filled FRP tube and the employer can demonstrate that the tool has no defects that could cause it to fail in use.

(**D**) The test method used shall be designed to verify the tool's integrity along its entire working length and, if the tool is made of fiberglass-reinforced plastic, its integrity under wet conditions.

(**E**) The voltage applied during the tests shall be as follows:

(*1*) 75,000 volts per foot (2461 volts per centimeter) of length for 1 minute if the tool is made of fiberglass, or

(*2*) 50,000 volts per foot (1640 volts per centimeter) of length for 1 minute if the tool is made of wood, or

(*3*) Other tests that the employer can demonstrate are equivalent.

Note: Guidelines for the examination, cleaning, repairing, and in-service testing of live-line tools are contained in the Institute of Electrical and Electronics Engineers Guide for In-Service Maintenance and Electrical Testing of Live-Line Tools, IEEE Std. 978-1984.

(**k**) *Materials handling and storage.* (1) *General.* Material handling and storage shall conform to the requirements of Subpart N of this Part.

(2) *Materials storage near energized lines or equipment.* (**i**) In areas not restricted to qualified persons only, materials or equipment may not be stored closer to energized lines or exposed energized parts of equipment than the following distances plus an amount providing for the maximum sag and side swing of all conductors and providing for the height and movement of material handling equipment:

(**A**) For lines and equipment energized at 50 kV or less, the distance is 10 feet (305 cm).

(**B**) For lines and equipment energized at more than 50 kV, the distance is 10 feet (305 cm) plus 4 inches (10 cm) for every 10 kV over 50 kV.

(**ii**) In areas restricted to qualified employees, material may not be stored within the working space about energized lines or equipment.

Note: Requirements for the size of the working space are contained in paragraphs (u)(1) and (v)(3) of this section.

Maintaining these clearances protects unqualified employees, who are not trained in the recognition and avoidance of the hazards involved, from contacting the energized lines or equipment with materials being handled. However, the work practices these unqualified workers would employ in handling material stored near energized lines are addressed by Subpart S. The general approach taken in new 1910.269 is to provide safety-related work practices for qualified employees to follow when they are performing electric power generation, transmission, and distribution work. Safe work practices for unqualified employees are not addressed in final 1910.269, because these practices are already spelled out in Subpart S of OSHA's General Industry Standards (see in particular 1910.333(c)(3)(i) for work performed by unqualified employees near overhead power lines).

Paragraph (k)(2)(i) only regulates the storage of materials where exposure is not restricted to qualified employees. If the materials are stored where only qualified workers have access to them, the materials may be safely stored closer to the energized parts than 10 feet, providing these employees have sufficient room to perform their work.

The working space about electric equipment is the clear space to be provided around the equipment to enable qualified employees to work on the equipment. An employee enters this space to service or maintain the electric equipment. The minimum working space specifies the minimum distance an obstruction can be from the equipment. For example, if a switchboard is installed in a cabinet into which an employee will enter, the inside walls of the cabinet must provide a minimum working space to enable the employee to work safely within the cabinet.

The minimum approach distance to be maintained from a live part is the limit of the space about the equipment that a qualified employee is not permitted to enter. The minimum approach distance a qualified employee must maintain from an energized part are smaller than the working space that is required to be provided around the part. The employee must "enter" the working space and still maintain the minimum approach distance. Storing materials in this space would tempt employees to work on energized equipment in cramped quarters if access were necessary in an emergency. Alternatively, if materials stored in the working space had to be moved so that adequate

FIGURE 6.3   (*Continued*)

room could be provided, accidents could result from the movement of the material.

(l) *Working on or near exposed energized parts.* This paragraph applies to work on exposed live parts, or near enough to them, to expose the employee to any hazard they present.

(1) *General.* Only qualified employees may work on or with exposed energized lines or parts of equipment. Only qualified employees may work in areas containing unguarded, uninsulated energized lines or parts of equipment operating at 50 volts or more. Electric lines and equipment shall be considered and treated as energized unless the provisions of paragraph (d) or paragraph (m) of this section have been followed.

Paragraph (l)(1) requires employees working on or in areas containing exposed live parts of electric lines or equipment to be qualified. Without proper training in the construction and operation of the lines and equipment and in the electrical hazards involved, workers would likely be electrocuted attempting to perform this type of work and would also expose others to injury as well. In areas containing unguarded live parts energized at more than 50 volts, untrained employees would not be familiar with the practices that are necessary to recognize and avoid contact with these parts.

However, employees in training, under the direct supervision of a qualified employee, are permitted to perform work on live parts and in areas containing unguarded live parts. OSHA believes that the close supervision of trainees will reveal errors "in the act", before they cause accidents. Allowing these workers the experience of performing tasks under actual conditions may also better prepare the employees to work safely.

The Agency has added a note to the definition of "qualified employee" to indicate that employees who are undergoing on-the-job training are considered to be qualified if they have demonstrated an ability to perform duties safely and if they are under the immediate supervision of qualified employees.

(i) Except as provided in paragraph (l)(1)(ii) of this section, at least two employees shall be present while the following types of work are being performed:

(A) Installation, removal, or repair of lines that are energized at more than 600 volts,

(B) Installation, removal, or repair of deenergized lines if an employee is exposed to contact with other parts energized at more than 600 volts,

(C) Installation, removal, or repair of equipment, such as transformers, capacitors, and regulators, if an employee is exposed to contact with parts energized at more than 600 volts,

(D) Work involving the use of mechanical equipment, other than insulated aerial lifts, near parts energized at more than 600 volts, and

(E) Other work that exposes an employee to electrical hazards greater than or equal to those posed by operations that are specifically listed in paragraphs (l)(1)(i)(A) through (l)(1)(i)(D) of this section.

Many employer and employee groups commented on the issue of whether or not a minimum of two employees should be required for work involving energized electric equipment.

OSHA believes that the most relevant consideration in determining whether or not to require the presence of at least two employees is whether the hazards of the work would be significantly reduced by the presence of an additional worker. Therefore, OSHA believes it is important to determine what types of work frequently result in electric shock, regardless of the number of employees present. Electric shock accidents, in particular, necessitate the immediate availability of a person trained in CPR.

To ascertain this information, the Agency reviewed the accident data in Exhibit 9-2. The results of this analysis are presented in Table 2, which tabulates the number of accidents involving different categories of work. Accidents unrelated to work by qualified employees on energized parts are not included in this table. (Of 117 accidents, 31 are not relevant to the issue of whether or not qualified employees should be able to work alone near exposed live parts.) Data in the other categories demonstrate that working directly on energized lines causes most of the accidents and is presumably the most hazardous job performed by line workers. Even some of the jobs claimed by utilities to be safe for a single employee to perform were involved in a few of the accidents, namely, replacing fuses, opening disconnects with live-line tools, and "low voltage" (600 volts or less) work. In particular, lines operating at 600 volts or less accounted for 13 percent (11 of 86) of the relevant accidents, as shown in Table 3.

FIGURE 6.3 *(Continued)*

TABLE 2 Number of Accidents by Type of Work

Type of work	Number of accidents*
Moving or repairing lines	18
Line stringing	10
Replacing or repairing equipment	10
Rubber glove (or bare hand) work, other	17
Hot stick work, other	5
Subtotal	60
Mechanical equipment used to lift or	10
Setting poles	4
Subtotal	14
Station work, work on energized parts	9
Station work, misc.	3
Subtotal	12
Total	86

*Accidents involving one or more employees injured due to contact with energized parts.

OSHA has concluded that the following work can be performed with minimal hazard to qualified employees working by themselves:

(1) Substation work not involving direct contact with live parts or climbing on structures, and

TABLE 3 Number of Accidents by Voltage

Voltage range*	Number of accidents
120/240	10
440	1
2.4 kV	1
7.2-14.4 kV*	53
69 kV	1
115 kV	2
Unspecified	18
Total	86

*The voltage specified was in this range; however, it was not always clear whether the voltage was phase-to-phase or phase-to-ground.

(2) Opening disconnects with live-line tools, if the employee is well away from the live parts.

Other types of work, such as line installation and removal, use of mechanical apparatus to lift or position material or persons, and electric station work on energized parts, are much more hazardous.

OSHA believes that the loss of power can create public safety concerns that outweigh the safety concerns of individual employees. In such cases, action must be taken to restore power so that public safety is assured.

OSHA is also permitting an employee to work alone to effect emergency repairs to the extent necessary to safeguard the general public.

The first four work operations (A through D) are those that the record demonstrates expose employees to the greatest risk of electric shock. OSHA has included the fifth category (E) to cover types of work that, while not specifically identified in the record, pose equal or greater hazards.

(ii) Paragraph (l)(1)(i) of this section does not apply to the following operations:

(A) Routine switching of circuits, if the employer can demonstrate that conditions at the site allow this work to be performed safely,

(B) Work performed with live-line tools, if the employee is positioned so that he or she is neither within reach of nor otherwise exposed to contact with energized parts, and

(C) Emergency repairs to the extent necessary to safeguard the general public.

OSHA has placed restrictions on the use of these exceptions in view of the accidents that occurred even under these limited conditions. Accidents involving hot stick work have typically occurred only when the employee was using a live-line tool but was close enough to energized parts to be injured - sometimes through direct contact, other times by contact through conductors being handled. Employees have been injured during switching operations when unusual conditions, such as poor lighting, bad weather, and hazardous configuration or state of repair of the switching equipment, were present. Because such conditions make the work unsafe, paragraph (l)(1)(ii)(A) would not permit switching operations to be performed by an employee working alone.

The requirement for at least two employees to be present during certain operations

FIGURE 6.3 (*Continued*)

does not apply if the voltage of the energized parts involved is 600 volts or less. The record contains conflicting data regarding the safety of performing work at these voltages. Many have said that it was safe to perform such work, but the data in Table 3 strongly suggests that this is not true.

Unfortunately, the types of work involving voltages of 600 volts or less are not clearly defined in the rulemaking record, at least with respect to the degree of risk they present. For example, electric meter work, which typically involves these lower voltages, is one type of work commonly performed by electric utility workers. However, there are very few accidents involving this type of work. It appears that many of the lower voltage accidents in the record involved qualified employees working on service drops, but there may be conditions making even this type of work safe.

There is insufficient evidence in the record as to whether or not it is safe for qualified employees to work alone on live parts energized at these lower voltages.

(2) *Minimum approach distances.* The employer shall ensure that no employee approaches or takes any conductive object closer to exposed energized parts than set forth in Table R-6 through Table R-10, unless:

(i) The employee is insulated from the energized part (insulating gloves or insulating gloves and sleeves worn in accordance with paragraph (l)(3) of this section are considered insulation of the employee only with regard to the energized part upon which work is being performed), or

(ii) The energized part is insulated from the employee and from any other conductive object at a different potential, or

(iii) The employee is insulated from any other exposed conductive object, as during live-line bare-hand work.

Note: Paragraphs (u)(5)(i) and (v)(5)(i) of this section contain requirements for the guarding and isolation of live parts. Parts of electric circuits that meet these two provisions are not considered as "exposed" unless a guard is removed or an employee enters the space intended to provide isolation from the live parts.

(3) *Type of insulation.* If the employee is to be insulated from energized parts by the use of insulating gloves (under paragraph (l)(2)(i) of this section), insulating sleeves shall also be used. However, insulating sleeves need not be used under the following conditions:

TABLE R-6   AC Live-Line Work Minimum Approach Distance

Nominal voltage in kilovolts phase to phase	Phase-to-ground exposure (ft-in)	(m)	Phase-to-phase exposure (ft-in)	(m)
0.05 to 1.0	Avoid Contact		Avoid Contact	
1.1 to 15.0	2-1	0.64	2-2	0.66
15.1 to 36.0	2-4	0.72	2-7	0.77
36.1 to 46.0	2-7	0.77	2-10	0.85
46.1 to 72.5	3-0	0.90	3-6	1.05
72.6 to 121	3-2	0.95	4-3	1.29
138 to 145	3-7	1.09	4-11	1.50
161 to 169	4-0	1.22	5-8	1.71
230 to 242	5-3	1.59	7-6	2.27
345 to 362	8-6	2.59	12-6	3.80
500 to 550	11-3	3.42	18-1	5.50
765 to 800	14-11	4.53	26-0	7.91

These distances take into consideration the highest switching surge an employee will be exposed to on any system with air as the insulating medium and the maximum voltages shown.

The clear live-line tool distance shall equal or exceed the values for the indicated voltage ranges.

See Appendix B to this section for information on how the minimum approach distances listed in the tables were derived.

FIGURE 6.3   *(Continued)*

TABLE R-7 AC Live-Line Work Minimum Approach Distance With Overvoltage Factor Phase-to-Ground Exposure

Maximum anticipated	Distance in feet-inches						
Per-unit transient	Maximum phase-to-phase voltage in kilovolts						
Overvoltage	121	145	169	242	362	552	800
1.5	—	—	—	—	—	6-0	9-8
1.6	—	—	—	—	—	6-6	10-8
1.7	—	—	—	—	—	7-0	11-8
1.8	—	—	—	—	—	7-7	12-8
1.9	—	—	—	—	—	8-1	13-9
2.0	2-5	2-9	3-0	3-10	5-3	8-9	14-11
2.1	2-6	2-10	3-2	4-0	5-5	9-4	—
2.2	2-7	2-11	3-3	4-1	5-9	9-11	—
2.3	2-8	3-0	3-4	4-3	6-1	10-6	—
2.4	2-9	3-1	3-5	4-5	6-4	11-3	—
2.5	2-9	3-2	3-6	4-6	6-8	—	—
2.6	2-10	3-3	3-8	4-8	7-1	—	—
2.7	2-11	3-4	3-9	4-10	7-5	—	—
2.8	3-0	3-5	3-10	4-11	7-9	—	—
2.9	3-1	3-6	3-11	5-1	8-2	—	—
3.0	3-2	3-7	4-0	5-3	8-6	—	—

The distances specified in this table may be applied only where the maximum anticipated per-unit transient overvoltage has been determined by engineering analysis and has been supplied by the employer. Table R-6 applies otherwise.

The distances specified in this table are the air, bare-hand, and live-line tool distances.

See Appendix B to this section for information on how the minimum approach distances listed in the tables were derived and on how to calculate revised minimum approach distances based on the control of transient overvoltages.

(i) If exposed energized parts on which work is not being performed are insulated from the employee and

(ii) If such insulation is placed from a position not exposing the employee's upper arm to contact with other energized parts.

(4) *Working position.* The employer shall ensure that each employee, to the extent that other safety-related conditions at the worksite permit, works in a position from which a slip or shock will not bring the employee's body into contact with exposed, uninsulated parts energized at a potential different from the employee.

(5) *Making connections.* The employer shall ensure that connections are made as follows:

(i) In connecting deenergized equipment or lines to an energized circuit by means of a conducting wire or device, an employee shall first attach the wire to the deenergized part;

(ii) When disconnecting equipment or lines from an energized circuit by means of a conducting wire or device, an employee shall remove the source end first; and

(iii) When lines or equipment are connected to or disconnected from energized circuits, loose conductors shall be kept away from exposed energized parts.

(6) *Apparel.* (i) When work is performed within reaching distance of exposed energized parts of equipment, the employer shall ensure that each employee removes or renders nonconductive all exposed conductive articles, such as key or watch chains, rings, or wrist watches or bands, unless such articles do not increase the hazards associated with contact with the energized parts.

(ii) The employer shall train each employee who is exposed to the hazards of flames or electric arcs in the hazards involved.

(iii) The employer shall ensure that each employee who is exposed to the hazards of flames or electric arcs does not wear clothing

FIGURE 6.3 (*Continued*)

TABLE R-8 AC Live-Line Work Minimum Approach Distance With Overvoltage Factor Phase-to-Phase Exposure

Maximum anticipated	Distance in feet-inches						
Per-unit transient	Maximum phase-to-phase voltage in kilovolts						
Overvoltage	121	145	169	242	362	552	800
1.5	—	—	—	—	—	7-4	12-1
1.6	—	—	—	—	—	8-9	14-6
1.7	—	—	—	—	—	10-2	17-2
1.8	—	—	—	—	—	11-7	19-11
1.9	—	—	—	—	—	13-2	22-11
2.0	3-7	4-1	4-8	6-1	8-7	14-10	26-0
2.1	3-7	4-2	4-9	6-3	8-10	15-7	—
2.2	3-8	4-3	4-10	6-4	9-2	16-4	—
2.3	3-9	4-4	4-11	6-6	9-6	17-2	—
2.4	3-10	4-5	5-0	6-7	9-11	18-1	—
2.5	3-11	4-6	5-2	6-9	10-4	—	—
2.6	4-0	4-7	5-3	6-11	10-9	—	—
2.7	4-1	4-8	5-4	7-0	11-2	—	—
2.8	4-1	4-9	5-5	7-2	11-7	—	—
2.9	4-2	4-10	5-6	7-4	12-1	—	—
3.0	4-3	4-11	5-8	7-6	12-6	—	—

The distances specified in this table may be applied only where the maximum anticipated per-unit transient overvoltage has been determined by engineering analysis and has been supplied by the employer. Table R-6 applies otherwise.

The distances specified in this table are the air, bare-hand, and live-line tool distances.

See Appendix B to this section for information on how the minimum approach distances listed in the tables were derived and on how to calculate revised minimum approach distances based on the control of transient overvoltages.

TABLE R-9 DC Live-Line Work Minimum Approach Distance With Overvoltage Factor

Maximum anticipated	Distance in feet-inches				
Per-unit transient	Maximum line-to-ground voltage in kilovolts				
Transient overvoltage	250	400	500	600	750
1.5 or lower	3-8	5-3	6-9	8-7	11-10
1.6	3-10	5-7	7-4	9-5	13-1
1.7	4-1	6-0	7-11	10-3	14-4
1.8	4-3	6-5	8-7	11-2	15-9

The distances specified in this table may be applied only where the maximum anticipated per-unit transient overvoltage has been determined by engineering analysis and has been supplied by the employer. However, if the transient overvoltage factor is not known, a factor of 1.8 shall be assumed.

The distances specified in this table are the air, bare-hand, and live-line tool distances.

FIGURE 6.3 (*Continued*)

TABLE R-10   Altitude Correction Factor

Altitude		Correction factor
ft	m	
3000	900	1.00
4000	1200	1.02
5000	1500	1.05
6000	1800	1.08
7000	2100	1.11
8000	2400	1.14
9000	2700	1.17
10000	3000	1.20
12000	3600	1.25
14000	4200	1.30
16000	4800	1.35
18000	5400	1.39
20000	6000	1.44

If the work is performed at elevations greater than 3000 ft (900 m) above mean sea level, the minimum approach distance shall be determined by multiplying the distances in 1 through 1 by the correction factor corresponding to the altitude at which work is performed.

that, when exposed to flames or electric arcs, could increase the extent of injury that would be sustained by the employee.

Note: Clothing made from the following types of fabrics, either alone or in blends, is prohibited by this paragraph, unless the employer can demonstrate that the fabric has been treated to withstand the conditions that may be encountered or that the clothing is worn in such a manner as to eliminate the hazard involved: acetate, nylon, polyester, rayon.

(7) *Fuse handling.* When fuses must be installed or removed with one or both terminals energized at more than 300 volts or with exposed parts energized at more than 50 volts, the employer shall ensure that tools or gloves rated for the voltage are used. When expulsion-type fuses are installed with one or both terminals energized at more than 300 volts, the employer shall ensure that each employee wears eye protection meeting the requirements of Subpart I of this Part, uses a tool rated for the voltage, and is clear of the exhaust path of the fuse barrel.

(8) *Covered (noninsulated) conductors.* The requirements of this section which pertain to the hazards of exposed live parts also apply

when work is performed in the proximity of covered (noninsulated) wires.

(9) *Noncurrent-carrying metal parts.* Noncurrent-carrying metal parts of equipment or devices, such as transformer cases and circuit breaker housings, shall be treated as energized at the highest voltage to which they are exposed, unless the employer inspects the installation and determines that these parts are grounded before work is performed.

(10) *Opening circuits under load.* Devices used to open circuits under load conditions shall be designed to interrupt the current involved.

(m) *Deenergizing lines and equipment for employee protection.* **(1)** *Application.* Paragraph (m) of this section applies to the deenergizing of transmission and distribution lines and equipment for the purpose of protecting employees. Control of hazardous energy sources used in the generation of electric energy is covered in paragraph (d) of this section. Conductors and parts of electric equipment that have been deenergized under procedures other than those required by paragraph (d) or (m) of this section, as applicable, shall be treated as energized.

Transmission and distribution systems are different from other energy systems found in general industry or even in the electric utility industry itself. The hazardous energy control methods for these systems are necessarily different from those covered under 1910.269(d). Transmission and distribution lines and equipment are installed outdoors and are subject to being reenergized by means other than the normal energy sources. For example, lightning can strike a line and energize an otherwise deenergized conductor, or a line could be energized by unknown cogeneration sources not under the control of the employer. Additionally, some deenergized transmission and distribution lines are subject to being reenergized by induced voltage from nearby energized conductors or by contact with other energized sources of electrical energy. Another difference is that energy control devices are often very remote from the worksite and are frequently under the centralized control of a system operator.

For these reasons, OSHA proposed to cover the control of hazardous energy sources related to transmission and distribution systems separately. Because paragraph (m) covers this area, the general

FIGURE 6.3   (*Continued*)

requirements for hazardous energy control in paragraph (d) of final 1910.269 do not apply to the disconnection of transmission and distribution lines and equipment from sources of electrical energy.

OSHA firmly believes that certain procedures must be followed for deenergizing live parts at any voltage over 50 volts if employees will be in contact with the parts during the course of work. Contact with electric circuit parts energized at 600 volts or less can be as fatal as contact with higher voltages. The basic steps necessary for deenergizing electric circuits are the same regardless of voltage - first, the disconnecting means for the circuit must be opened; second, a method of securing the disconnecting means from accidental closure must be used; third, the circuit must be tested to ensure that it is in fact deenergized; and, fourth, measures (such as grounding) must be used to ensure that no hazardous voltage can be impressed on the circuit while employees are working. These are the same steps that are set forth, without a voltage limitation, in the NESC Section 423.

(2) *General.* (i) If a system operator is in charge of the lines or equipment and their means of disconnection, all of the requirements of paragraph (m)(3) of this section shall be observed, in the order given.

(ii) If no system operator is in charge of the lines or equipment and their means of disconnection, one employee in the crew shall be designated as being in charge of the clearance. All of the requirements of paragraph (m)(3) of this section apply, in the order given, except as provided in paragraph (m)(2)(iii) of this section. The employee in charge of the clearance shall take the place of the system operator, as necessary.

(iii) If only one crew will be working on the lines or equipment and if the means of disconnection is accessible and visible to and under the sole control of the employee in charge of the clearance, paragraphs (m)(3)(i), (m)(3)(iii), (m)(3)(iv), (m)(3)(viii), and (m)(3)(xii) of this section do not apply. Additionally, tags required by the remaining provisions of paragraph (m)(3) of this section need not be used.

(iv) Any disconnecting means that are accessible to persons outside the employer's control (for example, the general public) shall be rendered inoperable while they are open for the purpose of protecting employees.

(3) *De-energizing lines and equipment.* (i) A designated employee shall make a request of the system operator to have the particular section

of line or equipment deenergized. The designated employee becomes the employee in charge (as this term is used in paragraph (m)(3) of this section) and is responsible for the clearance.

(ii) All switches, disconnectors, jumpers, taps, and other means through which known sources of electric energy may be supplied to the particular lines and equipment to be deenergized shall be opened. Such means shall be rendered inoperable, unless its design does not so permit, and tagged to indicate that employees are at work.

(iii) Automatically and remotely controlled switches that could cause the opened disconnecting means to close shall also be tagged at the point of control. The automatic or remote control feature shall be rendered inoperable, unless its design does not so permit.

(iv) Tags shall prohibit operation of the disconnecting means and shall indicate that employees are at work.

(v) After the applicable requirements in paragraphs (m)(3)(i) through (m)(3)(iv) of this section have been followed and the employee in charge of the work has been given a clearance by the system operator, the lines and equipment to be worked shall be tested to ensure that they are deenergized.

(vi) Protective grounds shall be installed as required by paragraph (n) of this section.

(vii) After the applicable requirements of paragraphs (m)(3)(i) through (m)(3)(vi) of this section have been followed, the lines and equipment involved may be worked as deenergized.

(viii) If two or more independent crews will be working on the same lines or equipment, each crew shall independently comply with the requirements in paragraph (m)(3) of this section.

(ix) To transfer the clearance, the employee in charge (or, if the employee in charge is forced to leave the worksite due to illness or other emergency, the employee's supervisor) shall inform the system operator; employees in the crew shall be informed of the transfer; and the new employee in charge shall be responsible for the clearance.

(x) To release a clearance, the employee in charge shall:

(A) Notify employees under his or her direction that the clearance is to be released;

(B) Determine that all employees in the crew are clear of the lines and equipment;

(C) Determine that all protective grounds installed by the crew have been removed; and

(D) Report this information to the system operator and release the clearance.

FIGURE 6.3 (*Continued*)

(xi) The person releasing a clearance shall be the same person that requested the clearance, unless responsibility has been transferred under paragraph (m)(3)(ix) of this section.

(xii) Tags may not be removed unless the associated clearance has been released under paragraph (m)(3)(x) of this section.

(xiii) Only after all protective grounds have been removed, after all crews working on the lines or equipment have released their clearances, after all employees are clear of the lines and equipment, and after all protective tags have been removed from a given point of disconnection, may action be initiated to reenergize the lines or equipment at that point of disconnection.

Action may be taken to reenergize the lines or equipment only after grounds and tags have been removed, after all clearances have been released, and after all employees are in the clear. This protects employees from the possibility that the line or equipment could be reenergized while employees are still at work.

The Agency did not intend for this provision to require the removal of all tags from all disconnecting means before any of them could be reclosed. It was intended to require that all tags for any particular switch be removed before that switch was closed. It is very important in a tagging system that no energy isolating device be returned to a position allowing energy flow if there are any tags on it that are protecting employees.

(n) *Grounding for the protection of employees.* **(1)** *Application.* Paragraph (n) of this section applies to the grounding of transmission and distribution lines and equipment for the purpose of protecting employees. Paragraph (n)(4) of this section also applies to the protective grounding of other equipment as required elsewhere in this section.

Sometimes, normally energized lines and equipment which have been deenergized to permit employees to work become accidentally energized. This can happen in several ways, for example, by contact with another energized circuit, by voltage backfeed from a customer's cogeneration installation, by lightning contact, or by failure of the clearance system outlined in 1910.269(m).

Transmission and distribution lines and equipment are normally installed outdoors where they are exposed to damage from the weather and from actions taken by members of the general public. Many utility poles are installed alongside roadways where they may be struck by motor vehicles. Distribution lines have been damaged by falling trees, and transmission line insulators have been used for target practice. Additionally, customers fed by a utility company's distribution line may have cogeneration or backup generation capability, sometimes without the utility company's knowledge. All these factors can reenergize a deenergized transmission or distribution line or equipment. Energized lines can be knocked down onto deenergized lines. A backup generator or a cogenerator can cause voltage backfeed on the deenergized power line. Lastly, lightning, even miles from the worksite, can reenergize a line. All of these problems pose hazards to employees working on deenergized transmission and distribution lines and equipment. In fact, these problems have been a factor in 14 of the accidents in the record.

Grounding the lines and equipment is used to protect employees from injury, should such reenergizing occur. Grounding also provides protection against induced voltages and static charges on a line. (These induced and static voltages can be high enough to endanger employees, either directly from electric shock or indirectly from involuntary reaction.)

Grounding, as a temporary protective measure, involves connecting the deenergized lines and equipment to earth through conductors. As long as the conductors remain deenergized, this maintains the lines and equipment at the same potential as the earth. However, if voltage is impressed on a line, the voltage on the grounded line rises to a value dependent upon the impressed voltage, the impedance between its source and the grounding point, and the impedance of the grounding conductor.

Various techniques are used to limit the voltage to which an employee working on a grounded line would be exposed. Bonding is one of these techniques. Conductive objects within the reach of the employee are bonded together to create an equipotential work area for the employee. Within this area of equal potentials, voltage differences are limited to a safe value.

As noted in paragraph (n)(1), entire paragraph (n) applies to the grounding of deenergized transmission and distribution lines and

FIGURE 6.3 (*Continued*)

equipment for the purpose of protecting employees. Additionally, paragraph (n)(1) indicates that paragraph (n)(4) applies to the protective grounding of non-electrical equipment, such as aerial lift trucks, as well. Under normal conditions, such equipment would not be connected to a source of electric energy. However, to protect employees in case of accidental contact of the equipment with live parts, protective grounding is required elsewhere in the; and, to ensure the adequacy of this grounding, the provisions of paragraph (n)(4) must be followed.

Additional comments did not provide any information indicating that work on ungrounded deenergized equipment normally operating at 600 volts or less is safe. The Agency is particularly concerned that undetected voltage from a customer's generating system may backfeed the low voltage circuit and energize the line while the employee is working. Several of the accidents in the record occurred in this manner. Although the employee usually happened to be working on the high voltage side of a transformer in these cases, a similar result would have occurred had the worker been contacting the low voltage side. For these reasons, no voltage limitation has been included in paragraph (n)(1) of final 1910.269.

(2) *General.* For the employee to work lines or equipment as deenergized, the lines or equipment shall be deenergized under the provisions of paragraph (m) of this section and shall be grounded as specified in paragraphs (n)(3) through (n)(9) of this section. However, if the employer can demonstrate that installation of a ground is impracticable or that the conditions resulting from the installation of a ground would present greater hazards than working without grounds, the lines and equipment may be treated as deenergized provided all of the following conditions are met:

(i) The lines and equipment have been deenergized under the provisions of paragraph (m) of this section.

(ii) There is no possibility of contact with another energized source.

(iii) The hazard of induced voltage is not present.

Grounding may be omitted only if the installation of a ground is impracticable (such as during the initial stages of work on underground cables, when the conductor is not exposed for grounding) or if the conditions resulting from the installation of a ground would introduce more serious hazards than work without grounds. It is expected that conditions warranting the absence of protective grounds will be relatively rare.

If grounds are not installed and the lines and equipment are to be treated as deenergized, however, precautions have to be observed, and certain conditions must be met. Obviously, the lines and equipment still must be deenergized by the procedures of 1910.269(m). Also, there may be no possibility of contact with another source of voltage, and the hazard of induced voltage may not be present. Since these precautions and conditions do not protect against the possible reenergizing of the lines or equipment under all conditions, the omission of grounding is permitted only in very limited circumstances.

(3) *Equipotential zone.* Temporary protective grounds shall be placed at such locations and arranged in such a manner as to prevent each employee from being exposed to hazardous differences in electrical potential.

Paragraph (n)(3) allows employers and employees to use whatever grounding method they prefer as long as employees are protected. For employees working at elevated positions on poles and towers, single point grounding may be necessary, together with grounding straps to provide an equipotential zone for the worker. Employees in insulated aerial lifts working at midspan between two conductor supporting structures may be protected by grounding at convenient points on both sides of the work area. Bonding the aerial lift to the grounded conductor will ensure that the employee remains at the potential of the conductor in case of a fault. Other methods may be necessary to protect workers on the ground, including grounding mats and insulating platforms. The Agency believes that this performance-oriented approach will provide the flexibility needed by employers, but will afford the best protection to employees.

(4) *Protective grounding equipment.* (i) Protective grounding equipment shall be capable of conducting the maximum fault current that could flow at the point of grounding for the time necessary to clear the fault. This equipment shall have an ampacity greater than or equal to that of No. 2 AWG copper.

So that the protective grounding equipment does not fail, it is required to have an

FIGURE 6.3   *(Continued)*

ampacity high enough so that the fault current would be carried for the amount of time necessary to allow protective devices to interrupt the circuit.

Note: Guidelines for protective grounding equipment are contained in American Society for Testing and Materials Standard Specifications for Temporary Grounding Systems to be Used on De-Energized Electric Power Lines and Equipment, ASTM F855-1990.

(ii) Protective grounds shall have an impedance low enough to cause immediate operation of protective devices in case of accidental energizing of the lines or equipment.

Paragraphs (n)(4)(i) and (n)(4)(ii) help ensure the prompt clearing of the circuit supplying voltage to the point where the employee is working. Thus, the grounding equipment limits the duration and reduces the severity of any electric shock, though it does not itself prevent shock from occurring.

(5) *Testing.* Before any ground is installed, lines and equipment shall be tested and found absent of nominal voltage, unless a previously installed ground is present.

This requirement prevents energized equipment from being grounded, which could result in injury to the employee installing the ground.

(6) *Order of connection.* When a ground is to be attached to a line or to equipment, the ground-end connection shall be attached first, and then the other end shall be attached by means of a live-line tool.

(7) *Order of removal.* When a ground is to be removed, the grounding device shall be removed from the line or equipment using a live-line tool before the ground-end connection is removed.

Paragraphs (n)(6) and (n)(7) set forth the procedure for installing and removing grounds. To protect employees in the event that the "deenergized" equipment to be grounded is or becomes energized, the standard requires the "equipment end" of the grounding device to be applied last and removed first and that a live-line tool be used for both procedures in order to protect workers.

OSHA will consider any device that is insulated for the voltage and that allows an employee to apply or remove the ground from a safe position to be a live-line tool for the purposes of 1910.269(n)(6) and (n)(7). It should be noted that, during the periods

before the ground is installed and after it is removed, the line or equipment involved must be considered as energized (under paragraph (l)(1)). As a result, the minimum approach distances specified in paragraph (l)(2) apply when grounds are installed or removed.

(8) *Additional precautions.* When work is performed on a cable at a location remote from the cable terminal, the cable may not be grounded at the cable terminal if there is a possibility of hazardous transfer of potential should a fault occur.

With certain underground cable installations, a fault at one location along the cable can create a substantial potential difference between the earth at that location and the earth at other locations. Under normal conditions, this is not a hazard. However, if an employee is in contact with a remote ground (by being in contact with a conductor that is grounded at a remote station), he or she can be exposed to the difference in potential (because he or she is also in contact with the local ground). To protect employees in such situations, paragraph (n)(8) prohibits grounding cables at remote locations if a hazardous potential transfer could occur under fault conditions.

(9) *Removal of grounds for test.* Grounds may be removed temporarily during tests. During the test procedure, the employer shall ensure that each employee uses insulating equipment and is isolated from any hazards involved, and the employer shall institute any additional measures as may be necessary to protect each exposed employee in case the previously grounded lines and equipment become energized.

(o) *Testing and test facilities.* **(1)** *Application.* Paragraph (o) of this section provides for safe work practices for high-voltage and high-power testing performed in laboratories, shops, and substations, and in the field and on electric transmission and distribution lines and equipment. It applies only to testing involving interim measurements utilizing high voltage, high power, or combinations of both, and not to testing involving continuous measurements as in routine metering, relaying, and normal line work.

Note: Routine inspection and maintenance measurements made by qualified employees are considered to be routine line work and are not included in the scope of paragraph (o) of this

FIGURE 6.3   (*Continued*)

section, as long as the hazards related to the use of intrinsic high-voltage or high-power sources require only the normal precautions associated with routine operation and maintenance work required in the other paragraphs of this section. Two typical examples of such excluded test work procedures are "phasing-out" testing and testing for a "no-voltage" condition.

For the purposes of these requirements, high-voltage testing is assumed to involve voltage sources having sufficient energy to cause injury and having magnitudes generally in excess of 1000 volts, nominal. High-power testing involves sources where fault currents, load currents, magnetizing currents, or line dropping currents are used for testing, either at the rated voltage of the equipment under test or at lower voltages. Paragraph (o) covers such testing in laboratories, in shops and substations, and in the field and on transmission and distribution lines.

Examples of typical special tests in which either high-voltage sources or high-power sources are used as part of operation and maintenance of electric power generation, transmission, and distribution systems include cable-fault locating, large capacitive load tests, high current fault-closure tests, insulation resistance and leakage tests, direct-current proof tests, and other tests requiring direct connection to power lines.

(2) General requirements. (i) The employer shall establish and enforce work practices for the protection of each worker from the hazards of high-voltage or high-power testing at all test areas to temporary and permanent. Such work practices shall include, as a minimum, test area guarding, grounding, and the safe use of measuring and control circuits. A means providing for periodic safety checks of field test areas shall also be included. (See paragraph (o)(6) of this section.)

Paragraph (o)(2)(i) requires employers to establish work practices governing employees engaged in certain testing activities. These work practices are intended to delineate precautions that employees must observe for protection from the hazards of high-voltage or high-power testing. For example, if high-voltage sources are used in the testing, employees are required to follow the safety practices established under paragraph (o)(2)(i) to protect against such typical hazards as inadvertent arcing or voltage over-

stress destruction, as well as accidental contact with objects which have become residually charged by induced voltage from electric field exposure. If high-power sources are used in the testing, employees are required to follow established safety practices to protect against such typical hazards as ground voltage rise as well as exposure to excessive electromagnetically-caused physical forces associated with the passage of heavy current.

These practices apply to work performed at both permanent and temporary test areas (that is, areas permanently located in the controlled environment of a laboratory or shop and in areas temporarily located in a noncontrolled field environment). At a minimum, the safety work practices are required to cover the following types of test-associated activities:

(1) Guarding the test area to prevent inadvertent contact with energized parts,
(2) Safe grounding practices to be observed,
(3) Precautions to be taken in the use of control and measuring circuits, and
(4) Periodic checks of field test areas.

(ii) Employees shall be trained in safe work practices upon their initial assignment to the test area, with periodic reviews and updates provided as required by paragraph (a)(2) of this section.

Paragraph (o)(2)(ii) complements the general rule on the use of safe work practices in test areas with a requirement that all employees involved in this type of work be trained in these safety test practices. This paragraph further requires a periodic review of these practices to be conducted from time to time as a means of providing reemphasis and updating.

Although specific work practices used in test areas are generally unique to the particular test being conducted, three basic elements affecting safety are commonly found to some degree at all test sites: guarding, grounding, and the safe utilization of control and measuring circuits. By considering safe work practices in these three categories, OSHA has attempted to achieve a performance-oriented standard applicable to high-voltage and high-power testing and test facilities.

(3) Guarding of test areas. (i) Permanent test areas shall be guarded by walls, fences, or barriers designed to keep employees out of the test areas.

FIGURE 6.3   *(Continued)*

OSHA believes that guarding can best be achieved when it is provided both around and within test areas. By controlling access to all parts that are likely to become energized by either direct or inductive coupling, the standard will prevent accidental contact by employees. Paragraph (o)(3)(i) requires permanent test areas to be guarded by having them completely enclosed by walls or some other type of physical barrier.

(ii) In field testing, or at a temporary test site where permanent fences and gates are not provided, one of the following means shall be used to prevent unauthorized employees from entering:

(A) The test area shall be guarded by the use of distinctively colored safety tape that is supported approximately waist high and to which safety signs are attached,

(B) The test area shall be guarded by a barrier or barricade that limits access to the test area to a degree equivalent, physically and visually, to the barricade specified in paragraph (o)(3)(ii)(A) of this section, or

(C) The test area shall be guarded by one or more test observers stationed so that the entire area can be monitored.

In the case of field testing, paragraph (o)(3)(ii) attempts to achieve a level of safety for temporary test sites comparable to that achieved in laboratory test areas. For these areas, a barricade of tapes and cones or observation by an attendant are acceptable methods of guarding.

(iii) The barriers required by paragraph (o)(3)(ii) of this section shall be removed when the protection they provide is no longer needed.

Since the effectiveness of the temporary guarding means can be severely compromised by failing to remove it when it is not required, frequent safety checks must be made to monitor its use. For example, leaving barriers in place for a week at a time when testing is performed only an hour or two per day is likely to result in disregard for the barriers. For this reason, paragraph (o)(3)(iii) requires the temporary barriers to be removed when they are no longer needed.

(iv) Guarding shall be provided within test areas to control access to test equipment or to apparatus under test that may become energized as part of the testing by either direct or inductive coupling, in order to prevent accidental employee contact with energized parts.

Within test areas, whether temporary or permanent, additional safety can be achieved by observing the guarding practices that control access to test areas. Paragraph (o)(3)(iv) therefore requires that such guarding be provided if the test equipment or apparatus under test may become energized as part of the testing by either direct or inductive coupling. A combination of guards and barriers, preferably interlocked, is intended to provide protection to all employees in the vicinity.

(4) *Grounding practices.* **(i)** The employer shall establish and implement safe grounding practices for the test facility.

(A) All conductive parts accessible to the test operator during the time the equipment is operating at high voltage shall be maintained at ground potential except for portions of the equipment that are isolated from the test operator by guarding.

(B) Wherever ungrounded terminals of test equipment or apparatus under test may be present, they shall be treated as energized until determined by tests to be deenergized.

Suitable grounding is another important work practice that can be employed for the protection of personnel from the hazards of high-voltage or high-power testing. If high currents are intentionally employed in the testing, an isolated ground-return conductor, adequate for the service, is required so that no intentional passage of heavy current, with its attendant voltage rise, will occur in the ground grid or in the earth. Another safety consideration involving grounding is that all conductive parts accessible to the test operator during the time that the equipment is operating at high voltage be maintained at ground potential, except portions of the equipment that are isolated from the test operator by suitable guarding. Paragraph (o)(4) of final 1910.269 contains requirements for proper grounding at test sites.

(ii) Visible grounds shall be applied, either automatically or manually with properly insulated tools, to the high-voltage circuits after they are deenergized and before work is performed on the circuit or item or apparatus under test. Common ground connections shall be solidly connected to the test equipment and the apparatus under test.

(iii) In high-power testing, an isolated ground-return conductor system shall be provided so that no intentional passage of current,

FIGURE 6.3   *(Continued)*

with its attendant voltage rise, can occur in the ground grid or in the earth. However, an isolated ground-return conductor need not be provided if the employer can demonstrate that both the following conditions are met:

(A) An isolated ground-return conductor cannot be provided due to the distance of the test site from the electric energy source, and

(B) Employees are protected from any hazardous step and touch potentials that may develop during the test.

Note: See Appendix C to this section for information on measures that can be taken to protect employees from hazardous step and touch potentials.

Consideration must always be given to the possibility of voltage gradients developing in the earth during impulse, short-circuit, inrush, or oscillatory conditions. Such voltages may appear between the feet of an observer, or between his or her body and a grounded object, and are usually referred to as "step" and "touch" potentials. Examples of acceptable protection from step and touch potentials include suitable electrical protective equipment and the removal of employees from areas that may expose them to hazardous potentials.

(iv) In tests where grounding of test equipment by means of the equipment grounding conductor located in the equipment power cord cannot be used due to increased hazards to test personnel or the prevention of satisfactory measurements, a ground that the employer can demonstrate affords equivalent safety shall be provided, and the safety ground shall be clearly indicated in the test set-up.

Another grounding situation is recognized by paragraph (o)(4)(iv) in which grounding through the power cord of test equipment may be inadequate and actually increase the hazard to test operators. Normally, an equipment grounding conductor is required in the power cord of test equipment to connect it to a grounding connection in the power receptacle. However, in some circumstances, this practice can prevent satisfactory measurements, or current induced in the grounding conductor can cause a hazard to personnel. If these conditions exist, the use of the equipment grounding conductor within the cord is not mandatory, and paragraph (o)(4)(iv) requires that an equivalent safety ground be provided.

(v) When the test area is entered after equipment is deenergized, a ground shall be placed on the high-voltage terminal and any other exposed terminals.

(A) High capacitance equipment or apparatus shall be discharged through a resistor rated for the available energy.

(B) A direct ground shall be applied to the exposed terminals when the stored energy drops to a level at which it is safe to do so.

(vi) If a test trailer or test vehicle is used in field testing, its chassis shall be grounded. Protection against hazardous touch potentials with respect to the vehicle, instrument panels, and other conductive parts accessible to employees shall be provided by bonding, insulation, or isolation.

Paragraph (o)(4)(vi) recognizes the hazards associated with field testing in which test trailers or test vehicles are used. In addition to requiring the chassis of such vehicles to be grounded, paragraph (o)(4)(vi) provides for a performance-oriented approach by requiring that protection be provided against hazardous touch potentials by bonding, by insulation, or by isolation. The protection provided by each of these methods is described in the following examples:

(1) Protection by bonding can be effected by providing, around the vehicle, an area covered by a metallic mat or mesh of substantial cross-section and low impedance which is bonded to the vehicle at several points and is also bonded to an adequate number of driven ground rods or, where available, to an adequate number of accessible points on the station ground grid. All bonding conductors must be of sufficient electrical size to keep the voltage developed during maximum anticipated current tests at a safe value. The mat must be of a size which precludes simultaneous contact with the vehicle and with the earth or with metallic structures not adequately bonded to the mat.

(2) Protection by insulation can be accomplished, for example, by providing around the vehicle an area of dry wooden planks covered with rubber insulating blankets. The physical extent of the insulated area must be sufficient to prevent simultaneous contact with the vehicle, or the ground lead of the vehicle, and with the earth or with metallic structures in the vicinity.

(3) Protection by isolation can be implemented by providing an effective means to exclude personnel from any area where

FIGURE 6.3 *(Continued)*

simultaneous contact could be made with the vehicle (or conductive parts electrically connected to the vehicle) and with other conductive materials. A combination of barriers together with effective, interlocked restraints may be employed to prevent the inadvertent exit from the vehicle during the testing.

(5) *Control and measuring circuits.* **(i)** Control wiring, meter connections, test leads, and cables may not be run from a test area unless they are contained in a grounded metallic sheath and terminated in a grounded metallic enclosure or unless other precautions are taken that the employer can demonstrate as ensuring equivalent safety.

(ii) Meters and other instruments with accessible terminals or parts shall be isolated from test personnel to protect against hazards arising from such terminals and parts becoming energized during testing. If this isolation is provided by locating test equipment in metal compartments with viewing windows, interlocks shall be provided to interrupt the power supply if the compartment cover is opened.

(iii) The routing and connections of temporary wiring shall be made secure against damage, accidental interruptions, and other hazards. To the maximum extent possible, signal, control, ground, and power cables shall be kept separate.

(iv) If employees will be present in the test area during testing, a test observer shall be present. The test observer shall be capable of implementing the immediate deenergizing of test circuits for safety purposes.

A third category of safe work practices applicable to employees performing testing work, which complements the first two safety work practices of guarding and grounding, involves work practices associated with the installation of control and measurement circuits utilized at test facilities. Practices necessary for the protection of personnel and equipment from the hazards of high-voltage or high-power testing must be observed for every test where special signal-gathering equipment is used (that is, meters, oscilloscopes, and other special instruments). In addition, special settings of protective relays and the reexamination of backup schemes may be necessary to ensure an adequate level of safety during the tests or to minimize the effects of the testing on other parts of the system under test. As a consequence, paragraphs (o)(5)(i) through

(o)(5)(iii) address the principal safe work practices involving control and measuring circuit utilization within the test area.

Generally control and measuring circuit wiring should remain within the test area. If this is not possible, however, paragraph (o)(5)(i) covers requirements to minimize hazards should it become necessary to have the test wiring routed outside the test area. Cables and other wiring must be contained within a grounded metallic sheath and terminated in a grounded metal enclosure, or other precautions must be taken to provide equivalent safety.

Since the environment in which field tests are conducted differs in important respects from that of laboratory tests, extra care must be taken to ensure appropriate levels of safety. Permanent fences and gates for isolating the field test area are not usually provided, nor is there a permanent conduit for the instrumentation and control wiring. As a further hazard, there may be other sources of high-voltage electric energy in the vicinity in addition to the source of test voltage.

It is not always possible in the field to prevent ingress of persons into a test area physically, as is accomplished by the fences and interlocked gates of the laboratory environment. Consequently, readily recognizable means are required to discourage such ingress; and, before test potential or current is applied to a test area, the test operator in charge must ensure that all necessary barriers are in place.

As a consequence of these safety considerations, paragraph (o)(6)(i) calls for a safety check to be made at temporary or field test areas at the beginning of each group of continuous tests (that is, a series of tests conducted one immediately after another).

(6) *Safety check.* **(i)** Safety practices governing employee work at temporary or field test areas shall provide for a routine check of such test areas for safety at the beginning of each series of tests.

(ii) The test operator in charge shall conduct these routine safety checks before each series of tests and shall verify at least the following conditions:

(A) That barriers and guards are in workable condition and are properly placed to isolate hazardous areas;

(B) That system test status signals, if used, are in operable condition;

FIGURE 6.3   (*Continued*)

(C) That test power disconnects are clearly marked and readily available in an emergency;

(D) That ground connections are clearly identifiable;

(E) That personal protective equipment is provided and used as required by Subpart I of this Part and by this section; and

(F) That signal, ground, and power cables are properly separated.

(p) *Mechanical equipment.* (1) *General requirements.* (i) The critical safety components of mechanical elevating and rotating equipment shall receive a thorough visual inspection before use on each shift.

Note: Critical safety components of mechanical elevating and rotating equipment are components whose failure would result in a free fall or free rotation of the boom.

OSHA has worded the provision in the final rule to make it clear that a thorough visual inspection is required. It is not necessary to disassemble equipment to perform a visual inspection.

(ii) No vehicular equipment having an obstructed view to the rear may be operated on off-highway jobsites where any employee is exposed to the hazards created by the moving vehicle, unless:

(A) The vehicle has a reverse signal alarm audible above the surrounding noise level, or

(B) The vehicle is backed up only when a designated employee signals that it is safe to do so.

It is not intended for this provision to require the presence of a second employee. If the driver of the equipment is the only employee present and if no employees would be exposed to the hazards of vehicle backup, the standard would not apply.

(iii) The operator of an electric line truck may not leave his or her position at the controls while a load is suspended, unless the employer can demonstrate that no employee (including the operator) might be endangered.

This ensures that the operator will be at the controls if an emergency arises that necessitates moving the suspended load. For example, due to wind or unstable soil, the equipment might start to tip over. Having the operator at the controls ensures that corrective action can be taken quickly enough to prevent an accident.

(iv) Rubber-tired, self-propelled scrapers, rubber-tired front-end loaders, rubber-tired dozers, wheel-type agricultural and industrial tractors, crawler-type tractors, crawler-type loaders, and motor graders, with or without attachments, shall have roll-over protective structures that meet the requirements of Subpart W of Part 1926 of this chapter.

The equipment listed in this paragraph is frequently used for electric power generation, transmission, and distribution work during construction, and Subpart W of Part 1926, which contains the same list, already requires this equipment to have such protection. The final rule extends the protection afforded by the construction standards to operations that do not involve construction work. The roll-over protective structures must conform to Subpart W of Part 1926.

(2) *Outriggers.* (i) Vehicular equipment, if provided with outriggers, shall be operated with the outriggers extended and firmly set as necessary for the stability of the specific configuration of the equipment. Outriggers may not be extended or retracted outside of clear view of the operator unless all employees are outside the range of possible equipment motion.

The stability of the equipment in various configurations is normally provided by the manufacturer, but it can also be derived through engineering analysis.

(ii) If the work area or the terrain precludes the use of outriggers, the equipment may be operated only within its maximum load ratings for the particular configuration of the equipment without outriggers.

These two paragraphs help ensure the stability of the equipment while loads are being handled and prevent injuries caused by extending outriggers into employees.

A few of the accident descriptions submitted into the record by OSHA indicated that fatalities are occurring because of the use of aerial lift buckets to move overhead power lines. The employees in the aerial lift were killed when the unrestrained line slid up the bucket and contacted the employee (in two cases) or when current passed through a leakage hole in the bottom of the bucket (in the other case).

(3) *Applied loads.* Mechanical equipment used to lift or move lines or other material shall be used within its maximum load rating and

FIGURE 6.3 *(Continued)*

other design limitations for the conditions under which the work is being performed.

It is important for mechanical equipment to be used within its design limitations so that the lifting equipment does not fail during use and so that employees are not otherwise endangered.

Even in electric-utility operations, contact with live parts through mechanical equipment causes many fatalities each year. A sample of typical accidents involving the operation of mechanical equipment near overhead lines is given in Table 5. Industry practice and existing rules in Subpart V of the Construction Standards require aerial lifts and truck-mounted booms to be kept away from exposed energized lines and equipment at distances greater than or approximately equal to those in set forth in Table R-6. However, some contact with the energized parts does occur during the hundreds of thousands of operations carried out near overhead power lines each year. If the equipment operator is distracted briefly or if the distances involved or the speed of the equipment towards the line is misjudged, contact with the lines is the expected result, rather than simple coincidence, especially when the minimum approach distances are relatively small.

(4) *Operations near energized lines or equipment.* **(i)** Mechanical equipment shall be operated so that the minimum approach distances of Table R-6 through Table R-10 are maintained from exposed energized lines and equipment. However, the insulated portion of an aerial lift operated by a qualified employee in the lift is exempt from this requirement.

(ii) A designated employee other than the equipment operator shall observe the approach distance to exposed lines and equipment and give timely warnings before the minimum approach distance required by paragraph (p)(4)(i) is reached, unless the employer can demonstrate that the operator can accurately determine that the minimum approach distance is being maintained.

(iii) If, during operation of the mechanical equipment, the equipment could become energized, the operation shall also comply with at least one of the paragraphs (p)(4)(iii)(A) through (p)(4)(iii)(C) of this section.

(A) The energized lines exposed to contact shall be covered with insulating protective material that will withstand the type of contact that might be made during the operation.

(B) The equipment shall be insulated for the voltage involved. The equipment shall be positioned so that its uninsulated portions cannot approach the lines or equipment any closer

TABLE 5 Accidents Involving the Operation of Mechanical Equipment Near Overhead Lines

Type of equipment	Number of fatalities				Type of accident
	Total	Grounded			
		Yes	No	?	
Boom truck/ Derrick truck	7	1		6	Boom contact with energized line
	2	1		1	Pole contact with energized line
Aerial lift	1			1	Boom contact with energized line
	3		1	2	Lower boom contact with energized line
	3			3	Employee working on deenergized line when upper boom contacted energized line
	1			1	Winch on lift used on energized line arced to nearby ground
Vehicle	1		1		Line fell on vehicle
	1			1	Unknown type of vehicle and type of accident
Total	19	2	2	15	

FIGURE 6.3 *(Continued)*

than the minimum approach distances specified in Table R-6 through Table R-10.

(**C**) Each employee shall be protected from hazards that might arise from equipment contact with the energized lines. The measures used shall ensure that employees will not be exposed to hazardous differences in potential. Unless the employer can demonstrate that the methods in use protect each employee from the hazards that might arise if the equipment contacts the energized line, the measures used shall include all of the following techniques:

(*1*) Using the best available ground to minimize the time the lines remain energized,

(*2*) Bonding equipment together to minimize potential differences,

(*3*) Providing ground mats to extend areas of equipotential, and

(*4*) Employing insulating protective equipment or barricades to guard against any remaining hazardous potential differences.

Note: Appendix C to this section contains information on hazardous step and touch potentials and on methods of protecting employees from hazards resulting from such potentials.

(**q**) *Overhead lines.* This paragraph provides additional requirements for work performed on or near overhead lines and equipment.

(**1**) *General.* (**i**) Before elevated structures, such as poles or towers, are subjected to such stresses as climbing or the installation or removal of equipment may impose, the employer shall ascertain that the structures are capable of sustaining the additional or unbalanced stresses. If the pole or other structure cannot withstand the loads which will be imposed, it shall be braced or otherwise supported so as to prevent failure.

Note: Appendix D to this section contains test methods that can be used in ascertaining whether a wood pole is capable of sustaining the forces that would be imposed by an employee climbing the pole. This paragraph also requires the employer to ascertain that the pole can sustain all other forces that will be imposed by the work to be performed.

Paragraph (q)(1)(i) requires the employer to determine that elevated structures such as poles and towers are of adequate strength to withstand the stresses which will be imposed by the work to be performed. For example, if the work involves removing and reinstalling an existing line on a utility pole, the pole will be subjected to the weight of the employee (a vertical force) and to the release and replacement of the force imposed by the overhead line (a vertical and possibly a horizontal force). The additional stress involved may cause the pole to break, particularly if the pole has rotted at its base. If the pole or structure cannot withstand the loads to be imposed, it must be reinforced so that failure does not occur. This rule protects employees from falling to the ground upon failure of the pole or other elevated structure.

As the last step in ascertaining whether a wood pole is safe to climb, as required under paragraph (q)(1)(i), checking the actual condition of the pole is important because of the possibility of decay and other conditions adversely affecting the strength of the pole.

OSHA realizes that the employee at the worksite will be the one to inspect the structure for deterioration and will also determine whether it is safe to climb. However, under the OSH Act, it is the employer's responsibility to ensure that this is accomplished, regardless of who performs the work. (See the discussion of this issue under the summary and explanation of the introductory text of paragraph (c), earlier in the preamble.) Additionally, some work involves changing the loading on the structure. For example, replacement transformers may be heavier, and the equipment needed to perform the work will impose extra stress on the pole. The employee in the field is not necessarily skilled in structural engineering, and a determination as to whether or not the pole could withstand the stresses involved would need to be performed by the employer's engineering staff.

(**ii**) When poles are set, moved, or removed near exposed energized overhead conductors, the pole may not contact the conductors.

(**iii**) When a pole is set, moved, or removed near an exposed energized overhead conductor, the employer shall ensure that each employee wears electrical protective equipment or uses insulated devices when handling the pole and that no employee contacts the pole with uninsulated parts of his or her body.

Poles are often conductive. They can be made of metal or concrete, which OSHA considers to be conductive, as well as wood. Even wood poles pose an electric shock hazard when being moved near electric power lines. Wet poles and poles with ground wires running along their length are both highly conductive. Some of the accidents described

FIGURE 6.3   *(Continued)*

in the record involve wood poles with installed ground wires being placed between energized conductors. Even though the voltage was greater than 600 volts or was unspecified, these accidents show the dangers, regardless of the voltage involved. (Any voltage greater than 50 volts is normally considered lethal.)

(iv) To protect employees from falling into holes into which poles are to be placed, the holes shall be attended by employees or physically guarded whenever anyone is working nearby.

(2) *Installing and removing overhead lines.* The following provisions apply to the installation and removal of overhead conductors or cable.

(i) The employer shall use the tension stringing method, barriers, or other equivalent measures to minimize the possibility that conductors and cables being installed or removed will contact energized power lines or equipment.

Common methods of accomplishing this include the use of the following techniques: stringing conductors by means of the tension stringing method (which keeps the conductors off the ground and clear of energized circuits) and the use of rope nets and guards (which physically prevent one line from contacting another). These precautions, or equivalent measures, are necessary to protect employees against electric shock and against the effects of equipment damage resulting from accidental contact of the line being installed with energized parts.

(ii) The protective measures required by paragraph (p)(4)(iii) of this section for mechanical equipment shall also be provided for conductors, cables, and pulling and tensioning equipment when the conductor or cable is being installed or removed close enough to energized conductors that any of the following failures could energize the pulling or tensioning equipment or the wire or cable being installed or removed:

(A) Failure of the pulling or tensioning equipment,

(B) Failure of the wire or cable being pulled, or

(C) Failure of the previously installed lines or equipment.

(iii) If the conductors being installed or removed cross over energized conductors in excess of 600 volts and if the design of the circuit-interrupting devices protecting the lines so permits, the automatic-reclosing feature of these devices shall be made inoperative.

(iv) Before lines are installed parallel to existing energized lines, the employer shall make a determination of the approximate voltage to be induced in the new lines, or work shall proceed on the assumption that the induced voltage is hazardous. Unless the employer can demonstrate that the lines being installed are not subject to the induction of a hazardous voltage or unless the lines are treated as energized, the following requirements also apply:

(A) Each bare conductor shall be grounded in increments so that no point along the conductor is more than 2 miles (3.22 km) from a ground.

(B) The grounds required in paragraph (q)(2)(iv)(A) of this section shall be left in place until the conductor installation is completed between dead ends.

(C) The grounds required in paragraph (q)(2)(iv)(A) of this section shall be removed as the last phase of aerial cleanup.

(D) If employees are working on bare conductors, grounds shall also be installed at each location where these employees are working, and grounds shall be installed at all open dead-end or catch-off points or the next adjacent structure.

(E) If two bare conductors are to be spliced, the conductors shall be bonded and grounded before being spliced.

The standard does not provide guidelines for determining whether or not a hazard exists due to induced voltage. The hazard depends not only on the voltage of the existing line, but also on the length of the line being installed and the distance between the existing line and the new one. Electric shock from induced voltage poses two different hazards. First, the electric shock could cause an involuntary reaction, which could cause a fall or other injury. Second, the electric shock itself could cause respiratory or cardiac arrest. If no precautions are taken to protect employees from hazards associated with involuntary reactions from electric shock, a hazard is presumed to exist if the induced voltage is sufficient to pass a current of 1 milliampere through a 500 ohm resistor. (The 500 ohm resistor represents the resistance of an employee. The 1 milliampere current is the threshold of perception.) If employees are protected from injury due to involuntary reactions from electric shock, a hazard is presumed to exist if the resultant current would be more than 6 milliamperes (the let-go threshold for women). It is up to the employer to ensure that employees are protected against

FIGURE 6.3 *(Continued)*

serious injury from any voltages induced on lines being installed and to determine whether the voltages are high enough to warrant the adoption of the additional provisions on grounding spelled out in paragraphs (q)(2) (iv)(A) through (q)(2)(iv)(E).

(v) Reel handling equipment, including pulling and tensioning devices, shall be in safe operating condition and shall be leveled and aligned.

Proper alignment of the stringing machines will help prevent failure of the equipment, conductors, and supporting structures, which could result in injury to workers.

(vi) Load ratings of stringing lines, pulling lines, conductor grips, load-bearing hardware and accessories, rigging, and hoists may not be exceeded.

(vii) Pulling lines and accessories shall be repaired or replaced when defective.

(viii) Conductor grips may not be used on wire rope, unless the grip is specifically designed for this application.

Equipment that has been damaged beyond manufacturing specifications or that has been damaged to the extent that its load ratings would be reduced are considered to be defective. Load limits and design specifications are normally provided by the manufacturer, but they can also be found in engineering and materials handbooks.

(ix) Reliable communications, through two-way radios or other equivalent means, shall be maintained between the reel tender and the pulling rig operator.

When the tension stringing method is used, the pulling rig (which takes up the pulling rope and thereby pulls the conductors into place) is separated from the reel stands and tensioner (which pay out the conductors and apply tension to them) by one or more spans (the distance between the structures supporting the conductors). In an emergency, the pulling equipment operator may have to shut down the operation. Paragraph (q)(2)(ix) requires communication to be maintained between the reel tender and the pulling rig operator, so that in case of emergency at the conductor supply end, the pulling rig operator can shut the equipment down before injury-causing damage occurs.

(x) The pulling rig may only be operated when it is safe to do so.

Note: Examples of unsafe conditions include employees in locations prohibited by paragraph (q)(2)(xi) of this section, conductor and pulling line hang-ups, and slipping of the conductor grip.

(xi) While the conductor or pulling line is being pulled (in motion) with a power-driven device, employees are not permitted directly under overhead operations or on the cross arm, except as necessary to guide the stringing sock or board over or through the stringing sheave.

(3) *Live-line bare-hand work.* In addition to other applicable provisions contained in this section, the following requirements apply to live-line bare-hand work:

Under certain conditions, work must be performed on transmission and distribution lines while they remain energized. Sometimes, this work is accomplished using rubber insulating equipment or live-line tools. However, this equipment has voltage and other limitations which make it impossible to insulate the employee performing work on live lines under all conditions. In such cases, usually on medium- and high-voltage transmission lines, the work is performed using the live-line bare-hand technique. If work is to be performed "bare-handed", the employee works from an insulated aerial platform and is electrically bonded to the energized line. Since there is essentially no potential difference across the worker's body, he or she is protected from electric shock.

(i) Before using or supervising the use of the live-line bare-hand technique on energized circuits, employees shall be trained in the technique and in the safety requirements of paragraph (q)(3) of this section. Employees shall receive refresher training as required by paragraph (a)(2) of this section.

(ii) Before any employee uses the live-line bare-hand technique on energized high-voltage conductors or parts, the following information shall be ascertained:

(A) The nominal voltage rating of the circuit on which the work is to be performed,

(B) The minimum approach distances to ground of lines and other energized parts on which work is to be performed, and

(C) The voltage limitations of equipment to be used.

(iii) The insulated equipment, insulated tools, and aerial devices and platforms used shall be designed, tested, and intended for live-line

FIGURE 6.3   *(Continued)*

bare-hand work. Tools and equipment shall be kept clean and dry while they are in use.

(iv) The automatic-reclosing feature of circuit-interrupting devices protecting the lines shall be made inoperative, if the design of the devices permits.

(v) Work may not be performed when adverse weather conditions would make the work hazardous even after the work practices required by this section are employed. Additionally, work may not be performed when winds reduce the phase-to-phase or phase-to-ground minimum approach distances at the work location below that specified in paragraph (q)(3)(xiii) of this section, unless the grounded objects and other lines and equipment are covered by insulating guards.

Note: Thunderstorms in the immediate vicinity, high winds, snow storms, and ice storms are examples of adverse weather conditions that are presumed to make live-line bare-hand work too hazardous to perform safely.

(vi) A conductive bucket liner or other conductive device shall be provided for bonding the insulated aerial device to the energized line or equipment.

(A) The employee shall be connected to the bucket liner or other conductive device by the use of conductive shoes, leg clips, or other means.

(B) Where differences in potentials at the worksite pose a hazard to employees, electrostatic shielding designed for the voltage being worked shall be provided.

(vii) Before the employee contacts the energized part, the conductive bucket liner or other conductive device shall be bonded to the energized conductor by means of a positive connection. This connection shall remain attached to the energized conductor until the work on the energized circuit is completed.

(viii) Aerial lifts to be used for live-line bare-hand work shall have dual controls (lower and upper) as follows:

(A) The upper controls shall be within easy reach of the employee in the bucket. On a two-bucket-type lift, access to the controls shall be within easy reach from either bucket.

(B) The lower set of controls shall be located near the base of the boom, and they shall be so designed that they can override operation of the equipment at any time.

(ix) Lower (ground-level) lift controls may not be operated with an employee in the lift, except in case of emergency.

(x) Before employees are elevated into the work position, all controls (ground level and bucket) shall be checked to determine that they are in proper working condition.

(xi) Before the boom of an aerial lift is elevated, the body of the truck shall be grounded, or the body of the truck shall be barricaded and treated as energized.

(xii) A boom-current test shall be made before work is started each day, each time during the day when higher voltage is encountered, and when changed conditions indicate a need for an additional test. This test shall consist of placing the bucket in contact with an energized source equal to the voltage to be encountered for a minimum of 3 minutes. The leakage current may not exceed 1 microampere per kilovolt of nominal phase-to-ground voltage. Work from the aerial lift shall be immediately suspended upon indication of a malfunction in the equipment.

(xiii) The minimum approach distances specified in Table-R-6 through Table R-10 shall be maintained from all grounded objects and from lines and equipment at a potential different from that to which the live-line bare-hand equipment is bonded, unless such grounded objects and other lines and equipment are covered by insulating guards.

(xiv) While an employee is approaching, leaving, or bonding to an energized circuit, the minimum approach distances in Table R-6 through Table R-10 shall be maintained between the employee and any grounded parts, including the lower boom and portions of the truck.

(xv) While the bucket is positioned alongside an energized bushing or insulator string, the phase-to-ground minimum approach distances of Table R-6 through Table R-10 shall be maintained between all parts of the bucket and the grounded end of the bushing or insulator string or any other grounded surface.

(xvi) Hand lines may not be used between the bucket and the boom or between the bucket and the ground. However, non-conductive-type hand lines may be used from conductor to ground if not supported from the bucket. Ropes used for live-line bare-hand work may not be used for other purposes.

(xvii) Uninsulated equipment or material may not be passed between a pole or structure and an aerial lift while an employee working from the bucket is bonded to an energized part.

(xviii) A minimum approach distance table reflecting the minimum approach distances listed in Table R-6 through Table R-10 shall be printed on a plate of durable non-conductive material. This table shall be mounted so as to be visible to the operator of the boom.

FIGURE 6.3 (*Continued*)

(xix) A non-conductive measuring device shall be readily accessible to assist employees in maintaining the required minimum approach distance.

(4) *Towers and structures.* The following requirements apply to work performed on towers or other structures which support overhead lines.

(i) The employer shall ensure that no employee is under a tower or structure while work is in progress, except where the employer can demonstrate that such a working position is necessary to assist employees working above.

(ii) Tag lines or other similar devices shall be used to maintain control of tower sections being raised or positioned, unless the employer can demonstrate that the use of such devices would create a greater hazard.

(iii) The loadline may not be detached from a member or section until the load is safely secured.

(iv) Except during emergency restoration procedures, work shall be discontinued when adverse weather conditions would make the work hazardous in spite of the work practices required by this section.

Note: Thunderstorms in the immediate vicinity, high winds, snow storms, and ice storms are examples of adverse weather conditions that are presumed to make this work too hazardous to perform, except under emergency conditions.

Some weather conditions can make work from towers and other overhead structures more hazardous than usual. For example, icy conditions may make slips and falls much more likely, in fact even unavoidable. Under such conditions, work from towers and other structures would generally be prohibited.

(r) *Line-clearance tree trimming operations.* This paragraph provides additional requirements for line-clearance tree trimming operations and for equipment used in these operations.

(1) *Electrical hazards.* This paragraph does not apply to qualified employees.

(i) Before an employee climbs, enters, or works around any tree, a determination shall be made of the nominal voltage of electric power lines posing a hazard to employees. However, a determination of the maximum nominal voltage to which an employee will be exposed may be made instead, if all lines are considered as energized at this maximum voltage.

(ii) There shall be a second line-clearance tree trimmer within normal (that is, unassisted)

voice communication under any of the following conditions:

(A) If a line-clearance tree trimmer is to approach more closely than 10 feet (305 cm) any conductor or electric apparatus energized at more than 750 volts or

(B) If branches or limbs being removed are closer to lines energized at more than 750 volts than the distances listed in Table R-6, Table R-9, and Table R-10 or

(C) If roping is necessary to remove branches or limbs from such conductors or apparatus.

(iii) Line-clearance tree trimmers shall maintain the minimum approach distances from energized conductors given in Table R-6, Table R-9, and Table R-10.

(iv) Branches that are contacting exposed energized conductors or equipment or that are within the distances specified in Table R-6, Table R-9, and Table R-10 may be removed only through the use of insulating equipment.

Note: A tool constructed of a material that the employer can demonstrate has insulating qualities meeting paragraph (j)(1) of this section is considered as insulated under this paragraph if the tool is clean and dry.

(v) Ladders, platforms, and aerial devices may not be brought closer to an energized part than the distances listed in. Table R-6, Table R-9, and Table R-10.

(vi) Line-clearance tree-trimming work may not be performed when adverse weather conditions make the work hazardous in spite of the work practices required by this section. Each employee performing line-clearance tree trimming work in the aftermath of a storm or under similar emergency conditions shall be trained in the special hazards related to this type of work.

Note: Thunderstorms in the immediate vicinity, high winds, snow storms, and ice storms are examples of adverse weather conditions that are presumed to make line-clearance tree trimming work too hazardous to perform safely.

(2) *Brush chippers.* **(i)** Brush chippers shall be equipped with a locking device in the ignition system.

(ii) Access panels for maintenance and adjustment of the chipper blades and associated drive train shall be in place and secure during operation of the equipment.

(iii) Brush chippers not equipped with a mechanical infeed system shall be equipped with an infeed hopper of length sufficient to

FIGURE 6.3 *(Continued)*

prevent employees from contacting the blades or knives of the machine during operation.

(iv) Trailer chippers detached from trucks shall be chocked or otherwise secured.

(v) Each employee in the immediate area of an operating chipper feed table shall wear personal protective equipment as required by Subpart I of this Part.

(3) *Sprayers and related equipment.* (i) Walking and working surfaces of sprayers and related equipment shall be covered with slip-resistant material. If slipping hazards cannot be eliminated, slip-resistant footwear or handrails and stair rails meeting the requirements of Subpart D may be used instead of slip-resistant material.

(ii) Equipment on which employees stand to spray while the vehicle is in motion shall be equipped with guardrails around the working area. The guardrail shall be constructed in accordance with Subpart D of this Part.

(4) *Stump cutters.* (i) Stump cutters shall be equipped with enclosures or guards to protect employees.

(ii) Each employee in the immediate area of stump grinding operations (including the stump cutter operator) shall wear personal protective equipment as required by Subpart I of this Part.

(5) *Gasoline-engine power saws.* Gasoline-engine power saw operations shall meet the requirements of 1910.266(c)(5) of this Part and the following:

(i) Each power saw weighing more than 15 pounds (6.8 kilograms, service weight) that is used in trees shall be supported by a separate line, except when work is performed from an aerial lift and except during topping or removing operations where no supporting limb will be available.

(ii) Each power saw shall be equipped with a control that will return the saw to idling speed when released.

(iii) Each power saw shall be equipped with a clutch and shall be so adjusted that the clutch will not engage the chain drive at idling speed.

(iv) A power saw shall be started on the ground or where it is otherwise firmly supported. Drop starting of saws over 15 pounds (6.8 kg) is permitted outside of the bucket of an aerial lift only if the area below the lift is clear of personnel.

(v) A power saw engine may be started and operated only when all employees other than the operator are clear of the saw.

(vi) A power saw may not be running when the saw is being carried up into a tree by an employee.

(vii) Power saw engines shall be stopped for all cleaning, refueling, adjustments, and repairs to the saw or motor, except as the manufacturer's servicing procedures require otherwise.

(6) *Backpack power units for use in pruning and clearing.* (i) While a backpack power unit is running, no one other than the operator may be within 10 feet (305 cm) of the cutting head of a brush saw.

(ii) A backpack power unit shall be equipped with a quick shutoff switch readily accessible to the operator.

(iii) Backpack power unit engines shall be stopped for all cleaning, refueling, adjustments, and repairs to the saw or motor, except as the manufacturer's servicing procedures require otherwise.

(7) *Rope.* (i) Climbing ropes shall be used by employees working aloft in trees. These ropes shall have a minimum diameter of 0.5 inch (1.2 cm) with a minimum breaking strength of 2300 pounds (10.2 kN). Synthetic rope shall have elasticity of not more than 7 percent.

(ii) Rope shall be inspected before each use and, if unsafe (for example, because of damage or defect), may not be used.

(iii) Rope shall be stored away from cutting edges and sharp tools. Rope contact with corrosive chemicals, gas, and oil shall be avoided.

(iv) When stored, rope shall be coiled and piled, or shall be suspended, so that air can circulate through the coils.

(v) Rope ends shall be secured to prevent their unraveling.

(vi) Climbing rope may not be spliced to effect repair.

(vii) A ropes that is wet, that is contaminated to the extent that its insulating capacity is impaired, or that is otherwise not considered to be insulated for the voltage involved may not be used near exposed energized lines.

(8) *Fall protection.* Each employee shall be tied in with a climbing rope and safety saddle when the employee is working above the ground in a tree, unless he or she is ascending into the tree.

(s) *Communication facilities.* (1) *Microwave transmission.* (i) The employer shall ensure that no employee looks into an open waveguide or antenna that is connected to an energized microwave source.

(ii) If the electromagnetic radiation level within an accessible area associated with microwave communications systems exceeds the

FIGURE 6.3 (*Continued*)

radiation protection guide given in 1910.97(a)(2) of this Part, the area shall be posted with the warning symbol described in 1910.97(a)(3) of this Part. The lower half of the warning symbol shall include the following statements or ones that the employer can demonstrate are equivalent:

Radiation in this area may exceed hazard limitations and special precautions are required. Obtain specific instruction before entering.

(iii) When an employee works in an area where the electromagnetic radiation could exceed the radiation protection guide, the employer shall institute measures that ensure that the employee's exposure is not greater than that permitted by that guide. Such measures may include administrative and engineering controls and personal protective equipment.

(2) *Power line carrier.* Power line carrier work, including work on equipment used for coupling carrier current to power line conductors, shall be performed in accordance with the requirements of this section pertaining to work on energized lines.

(t) *Underground electrical installations.* This paragraph provides additional requirements for work on underground electrical installations.

(1) *Access.* A ladder or other climbing device shall be used to enter and exit a manhole or subsurface vault exceeding 4 feet (122 cm) in depth. No employee may climb into or out of a manhole or vault by stepping on cables or hangers.

Employees can easily be injured in the course of jumping into subsurface enclosures or in climbing on the cables and hangers which have been installed in these enclosures, the standard requires the use of appropriate devices for employees entering and exiting manholes and vaults. The practice of climbing on equipment such as cables and cable hangers is specifically prohibited.

(2) *Lowering equipment into manholes.* Equipment used to lower materials and tools into manholes or vaults shall be capable of supporting the weight to be lowered and shall be checked for defects before use. Before tools or material are lowered into the opening for a manhole or vault, each employee working in the manhole or vault shall be clear of the area directly under the opening.

(3) *Attendants for manholes.* (i) While work is being performed in a manhole containing energized electric equipment, an employee

with first aid and CPR training meeting paragraph (b)(1) of this section shall be available on the surface in the immediate vicinity to render emergency assistance.

(ii) Occasionally, the employee on the surface may briefly enter a manhole to provide assistance, other than emergency.

Note 1: An attendant may also be required under paragraph (e)(7) of this section. One person may serve to fulfill both requirements. However, attendants required under paragraph (e)(7) of this section are not permitted to enter the manhole.

Note 2: Employees entering manholes containing unguarded, uninsulated energized lines or parts of electric equipment operating at 50 volts or more are required to be qualified under paragraph (l)(1) of this section.

(iii) For the purpose of inspection, housekeeping, taking readings, or similar work, an employee working alone may enter, for brief periods of time, a manhole where energized cables or equipment are in service, if the employer can demonstrate that the employee will be protected from all electrical hazards.

(iv) Reliable communications, through two-way radios or other equivalent means, shall be maintained among all employees involved in the job.

Because the hazards addressed by paragraph (t)(3) are primarily related to electric shock, allowing the attendant to enter the manhole briefly has no significant effect on the safety of the employee he or she is protecting. In case of electric shock, the attendant would still be able to provide assistance.

The attendant is permitted to remain within the manhole only for the short period of time necessary to assist the employee inside the manhole with a task that one employee cannot perform alone. For example, if a second employee is needed to help lift a piece of equipment into place, the attendant could enter only for the amount of time that is needed to accomplish this task. However, if significant portions of the job require the assistance of a second worker in the manhole, the attendant would not be permitted to remain in the manhole for the length of time that would be necessary, and a third employee would be required.

(4) *Duct rods.* If duct rods are used, they shall be installed in the direction presenting the least hazard to employees. An employee shall be stationed at the far end of the duct line being

FIGURE 6.3 *(Continued)*

rodded to ensure that the required minimum approach distances are maintained.

To install cables into the underground ducts, or conduits, that will contain them, employees use a series of short jointed rods or a long flexible rod inserted into the ducts. The insertion of these rods into the ducts is known as "rodding." The rods are used to thread the cable-pulling rope through the conduit. After the rods have been withdrawn and the cable-pulling ropes have been inserted, the cables can then be pulled through by mechanical means.

(5) Multiple cables. When multiple cables are present in a work area, the cable to be worked shall be identified by electrical means, unless its identity is obvious by reason of distinctive appearance or location or by other readily apparent means of identification. Cables other than the one being worked shall be protected from damage.

(6) Moving cables. Energized cables that are to be moved shall be inspected for defects.

(7) Defective cables. Where a cable in a manhole has one or more abnormalities that could lead to or be an indication of an impending fault, the defective cable shall be deenergized before any employee may work in the manhole, except when service load conditions and a lack of feasible alternatives require that the cable remain energized. In that case, employees may enter the manhole provided they are protected from the possible effects of a failure by shields or other devices that are capable of containing the adverse effects of a fault in the joint.

Note: Abnormalities such as oil or compound leaking from cable or joints, broken cable sheaths or joint sleeves, hot localized surface temperatures of cables or joints, or joints that are swollen beyond normal tolerance are presumed to lead to or be an indication of an impending fault.

Paragraph (t)(7) requires employees to be protected by shields capable of containing the adverse effects of a failure. The energy that could be released in case of a fault is known, and the energy absorbing capability of a shield can be obtained from the manufacturer or can be calculated. As long as the energy absorbing capability of the shield exceeds the available fault energy, the shield will protect employees. Employees are required to be protected, regardless of the type of shielding device used and of how it is applied. Additionally, the standard permits this

option to be used only "if the defective cable or splice cannot be de-energized due to service load conditions".

(8) Sheath continuity. When work is performed on buried cable or on cable in man holes, metallic sheath continuity shall be maintained or the cable sheath shall be treated as energized.

Paragraph (t)(8) requires metallic sheath continuity to be maintained while work is performed on underground cables. Bonding across an opening in a cable's sheath protects employees against shock from a difference in potential between the two sides of the opening.

(u) Substations. This paragraph provides additional requirements for substations and for work performed in them.

(1) Access and working space. Sufficient access and working space shall be provided and maintained about electric equipment to permit ready and safe operation and maintenance of such equipment.

Note: Guidelines for the dimensions of access and working space about electric equipment in substations are contained in American National Standard, National Electrical Safety Code, ANSI C2-1987. Installations meeting the ANSI provisions comply with paragraph (u)(1) of this section. An installation that does not conform to this ANSI standard will, nonetheless, be considered as complying with paragraph (u)(1) of this section if the employer can demonstrate that the installation provides ready and safe access based on the following evidence:

(1) That the installation conforms to the edition of ANSI C2 that was in effect at the time the installation was made,

(2) That the configuration of the installation enables employees to maintain the minimum approach distances required by paragraph (l)(2) of this section while they are working on exposed, energized parts, and

(3) That the precautions taken when work is performed on the installation provide protection equivalent to the protection that would be provide by access and working space meeting ANSI C2-1987.

(2) Draw-out-type circuit breakers. When draw-out-type circuit breakers are removed or inserted, the breaker shall be in the open position. The control circuit shall also be rendered inoperative, if the design of the equipment permits.

FIGURE 6.3   *(Continued)*

A draw-out-type circuit breaker is one in which the removable portion may be withdrawn from the stationary portion without the necessity of unbolting connections or mounting supports.

Some circuit breaker and control device designs do not incorporate a feature allowing the control circuit for the breaker to be rendered inoperable. These provisions are intended to prevent arcing, which could injure employees.

(3) Substation fences. Conductive fences around substations shall be grounded. When a substation fence is expanded or a section is removed, fence grounding continuity shall be maintained, and bonding shall be used to prevent electrical discontinuity.

(4) Guarding of rooms containing electric supply equipment. (i) Rooms and spaces in which electric supply lines or equipment are installed shall meet the requirements of paragraphs (u)(4)(ii) through (u)(4)(v) of this section under the following conditions:

(A) If exposed live parts operating at 50 to 150 volts to ground are located within 8 feet of the ground or other working surface inside the room or space,

(B) If live parts operating at 151 to 600 volts and located within 8 feet of the ground or other working surface inside the room or space are guarded only by location, as permitted under paragraph (u)(5)(i) of this section, or

(C) If live parts operating at more than 600 volts are located within the room or space, unless:

(1) The live parts are enclosed within grounded, metal-enclosed equipment whose only openings are designed so that foreign objects inserted in these openings will be deflected from energized parts, or

(2) The live parts are installed at a height above ground and any other working surface that provides protection at the voltage to which they are energized corresponding to the protection provided by an 8-foot height at 50 volts.

(ii) The rooms and spaces shall be so enclosed within fences, screens, partitions, or walls as to minimize the possibility that unqualified persons will enter.

(iii) Signs warning unqualified persons to keep out shall be displayed at entrances to the rooms and spaces.

(iv) Entrances to rooms and spaces that are not under the observation of an attendant shall be kept locked.

(v) Unqualified persons may not enter the rooms or spaces while the electric supply lines or equipment are energized.

(5) Guarding of energized parts. (i) Guards shall be provided around all live parts operating at more than 150 volts to ground without an insulating covering, unless the location of the live parts gives sufficient horizontal or vertical or a combination of these clearances to minimize the possibility of accidental employee contact.

Note: Guidelines for the dimensions of clearance distances about electric equipment in substations are contained in American National Standard, National Electrical Safety Code, ANSI C2-1987. Installations meeting the ANSI provisions comply with paragraph (u)(5)(i) of this section. An installation that does not conform to this ANSI standard will, nonetheless, be considered as complying with paragraph (u)(5)(i) of this section if the employer can demonstrate that the installation provides sufficient clearance based on the following evidence:

(1) That the installation conforms to the edition of ANSI C2 that was in effect at the time the installation was made,

(2) That each employee is isolated from energized parts at the point of closest approach, and

(3) That the precautions taken when work is performed on the installation provide protection equivalent to the protection that would be provided by horizontal and vertical clearances meeting ANSI C2-1987.

(ii) Except for fuse replacement and other necessary access by qualified persons, the guarding of energized parts within a compartment shall be maintained during operation and maintenance functions to prevent accidental contact with energized parts and to prevent tools or other equipment from being dropped on energized parts.

(iii) When guards are removed from energized equipment, barriers shall be installed around the work area to prevent employees who are not working on the equipment, but who are in the area, from contacting the exposed live parts.

(6) Substation entry. (i) Upon entering an attended substation, each employee other than those regularly working in the station shall report his or her presence to the employee in charge in order to receive information on special system conditions affecting employee safety.

(ii) The job briefing required by paragraph (c) of this section shall cover such additional subjects as the location of energized equipment in or adjacent to the work area and the limits of any de-energized work area.

(v) Power generation. This paragraph provides additional requirements and related work practices for power generating plants.

FIGURE 6.3 *(Continued)*

(1) *Interlocks and other safety devices.*
(i) Interlocks and other safety devices shall be maintained in a safe, operable condition.

(ii) No interlock or other safety device may be modified to defeat its function, except for test, repair, or adjustment of the device.

Paragraph (v)(1)(i) requires the employer to maintain interlocks and other safety devices (such as relief valves) in a safe and operable condition. This requirement ensures that these devices perform their intended function of protecting workers when called upon to do so.

(2) *Changing brushes.* Before exciter or generator brushes are changed while the generator is in service, the exciter or generator field shall be checked to determine whether a ground condition exists. The brushes may not be changed while the generator is energized if a ground condition exists.

Sometimes the brushes on a generator or exciter must be replaced while the machine is in operation. This work is unusually hazardous, and extreme caution must be observed by employees performing the job. To protect these workers, paragraph (v)(2) contains requirements for replacing brushes while the generator is in service. Since field windings and exciters are operated in an ungrounded condition, there is no voltage with respect to ground on the brushes as long as there is no ground fault in the circuit. So that no voltage to ground is present while employees are changing the brushes, paragraph (v)(2) requires the exciter-field circuit to be checked to ensure that a ground condition does not exist.

(3) *Access and working space.* Sufficient access and working space shall be provided and maintained about electric equipment to permit ready and safe operation and maintenance of such equipment.

Note: Guidelines for the dimensions of access and working space about electric equipment in generating stations are contained in American National Standard, National Electrical Safety Code, ANSI C2-1987. Installations meeting the ANSI provisions comply with paragraph (v)(3) of this section. An installation that does not conform to this ANSI standard will, nonetheless, be considered as complying with paragraph (v)(3) of this section if the employer can demonstrate that the installation provides ready and safe access based on the following evidence:

(1) That the installation conforms to the edition of ANSI C2 that was in effect at the time the installation was made,

(2) That the configuration of the installation enables employees to maintain the minimum approach distances required by paragraph (l)(2) of this section while they are working on exposed, energized parts, and

(3) That the precautions taken when work is performed on the installation provide protection equivalent to the protection that would be provide by access and working space meeting ANSI C2-1987.

(4) *Guarding of rooms containing electric supply equipment.* **(i)** Rooms and spaces in which electric supply lines or equipment are installed shall meet the requirements of paragraphs (v)(4)(ii) through (v)(4)(v) of this section under the following conditions:

(A) If exposed live parts operating at 50 to 150 volts to ground are located within 8 feet of the ground or other working surface inside the room or space,

(B) If live parts operating at 151 to 600 volts and located within 8 feet of the ground or other working surface inside the room or space are guarded only by location, as permitted under paragraph (v)(5)(i) of this section, or

(C) If live parts operating at more than 600 volts are located within the room or space, unless:

(1) The live parts are enclosed within grounded, metal-enclosed equipment whose only openings are designed so that foreign objects inserted in these openings will be deflected from energized parts, or

(2) The live parts are installed at a height above ground and any other working surface that provides protection at the voltage to which they are energized corresponding to the protection provided by an 8-foot height at 50 volts.

(ii) The rooms and spaces shall be so enclosed within fences, screens, partitions, or walls as to minimize the possibility that unqualified persons will enter.

(iii) Signs warning unqualified persons to keep out shall be displayed at entrances to the rooms and spaces.

(iv) Entrances to rooms and spaces that are not under the observation of an attendant shall be kept locked.

(v) Unqualified persons may not enter the rooms or spaces while the electric supply lines or equipment are energized.

(5) Guarding of energized parts. (i) Guards shall be provided around all live parts operating at more than 150 volts to ground without an

FIGURE 6.3   *(Continued)*

insulating covering, unless the location of the live parts gives sufficient horizontal or vertical or a combination of these clearances to minimize the possibility of accidental employee contact.

Note: Guidelines for the dimensions of clearance distances about electric equipment in generating stations are contained in American National Standard National Electrical Safety Code, ANSI C2-1987. Installations meeting the ANSI provisions comply with paragraph (v)(5)(i) of this section. An installation that does not conform to this ANSI standard will, nonetheless, be considered as complying with paragraph (v)(5)(i) of this section if the employer can demonstrate that the installation provides sufficient clearance based on the following evidence:

(1) That the installation conforms to the edition of ANSI C2 that was in effect at the time the installation was made,

(2) That each employee is isolated from energized parts at the point of closest approach, and

(3) That the precautions taken when work is performed on the installation provide protection equivalent to the protection that would be provided by horizontal and vertical clearances meeting ANSI C2-1987.

(ii) Except for fuse replacement or other necessary access by qualified persons, the guarding of energized parts within a compartment shall be maintained during operation and maintenance functions to prevent accidental contact with energized parts and to prevent tools or other equipment from being dropped on energized parts.

(iii) When guards are removed from energized equipment, barriers shall be installed around the work area to prevent employees who are not working on the equipment, but who are in the area, from contacting the exposed live parts.

(6) Water or steam spaces. The following requirements apply to work in water and steam spaces associated with boilers:

(i) A designated employee shall inspect conditions before work is permitted and after its completion. Eye protection, or full face protection if necessary, shall be worn at all times when condenser, heater, or boiler tubes are being cleaned.

(ii) Where it is necessary for employees to work near tube ends during cleaning, shielding shall be installed at the tube ends.

(7) Chemical cleaning of boilers and pressure vessels. The following requirements apply to chemical cleaning of boilers and pressure vessels:

(i) Areas where chemical cleaning is in progress shall be cordoned off to restrict access during cleaning. If flammable liquids, gases, or vapors or combustible materials will be used or might be produced during the cleaning process, the following requirements also apply:

(A) The area shall be posted with signs restricting entry and warning of the hazards of fire and explosion; and

(B) Smoking, welding, and other possible ignition sources are prohibited in these restricted areas.

(ii) The number of personnel in the restricted area shall be limited to those necessary to accomplish the task safely.

(iii) There shall be ready access to water or showers for emergency use.

Note: See 1910.141 of this Part for requirements that apply to the water supply and to washing facilities.

(iv) Employees in restricted areas shall wear protective equipment meeting the requirements of Subpart I of this Part and including, but not limited to, protective clothing, boots, goggles, and gloves.

(8) Chlorine systems. (i) Chlorine system enclosures shall be posted with signs restricting entry and warning of the hazard to health and the hazards of fire and explosion.

Note: See Subpart Z of this Part for requirements necessary to protect the health of employees from the effects of chlorine.

(ii) Only designated employees may enter the restricted area. Additionally, the number of personnel shall be limited to those necessary to accomplish the task safely.

(iii) Emergency repair kits shall be available near the shelter or enclosure to allow for the prompt repair of leaks in chlorine lines, equipment, or containers.

(iv) Before repair procedures are started, chlorine tanks, pipes, and equipment shall be purged with dry air and isolated from other sources of chlorine.

(v) The employer shall ensure that chlorine is not mixed with materials that would react with the chlorine in a dangerously exothermic or other hazardous manner.

(9) Boilers. (i) Before internal furnace or ash hopper repair work is started, overhead areas shall be inspected for possible falling objects. If the hazard of falling objects exists, overhead protection such as planking or nets shall be provided.

(ii) When opening an operating boiler door, employees shall stand clear of the opening of

FIGURE 6.3 *(Continued)*

the door to avoid the heat blast and gases which may escape from the boiler.

(10) Turbine generators. (i) Smoking and other ignition sources are prohibited near hydrogen or hydrogen sealing systems, and signs warning of the danger of explosion and fire shall be posted.

(ii) Excessive hydrogen makeup or abnormal loss of pressure shall be considered as an emergency and shall be corrected immediately.

(iii) A sufficient quantity of inert gas shall be available to purge the hydrogen from the largest generator.

(11) Coal and ash handling. (i) Only designated persons may operate railroad equipment.

(ii) Before a locomotive or locomotive crane is moved, a warning shall be given to employees in the area.

(iii) Employees engaged in switching or dumping cars may not use their feet to line up drawheads.

(iv) Drawheads and knuckles may not be shifted while locomotives or cars are in motion.

(v) When a railroad car is stopped for unloading, the car shall be secured from displacement that could endanger employees.

(vi) An emergency means of stopping dump operations shall be provided at railcar dumps.

(vii) The employer shall ensure that employees who work in coal- or ash-handling conveyor areas are trained and knowledgeable in conveyor operation and in the requirements of paragraphs (v)(11)(viii) through (v)(11)(xii) of this section.

(viii) Employees may not ride a coal- or ash-handling conveyor belt at any time. Employees may not cross over the conveyor belt, except at walkways, unless the conveyor's energy source has been deenergized and has been locked out or tagged in accordance with paragraph (d) of this section.

(ix) A conveyor that could cause injury when started may not be started until personnel in the area are alerted by a signal or by a designated person that the conveyor is about to start.

(x) If a conveyor that could cause injury when started is automatically controlled or is controlled from a remote location, an audible device shall be provided that sounds an alarm that will be recognized by each employee as a warning that the conveyor will start and that can be clearly heard at all points along the conveyor where personnel may be present. The warning device shall be actuated by the device starting the conveyor and shall continue for a period of time before the conveyor starts that is long enough to allow employees to move clear of the conveyor system. A visual warning may be used in place of the audible device if the employer can demonstrate that it will provide an equally effective warning in the particular circumstances involved.

Exception: If the employer can demonstrate that the system's function would be seriously hindered by the required time delay, warning signs may be provided in place of the audible warning device. If the system was installed before January 31, 1995, warning signs may be provided in place of the audible warning device until such time as the conveyor or its control system is rebuilt or rewired. These warning signs shall be clear, concise, and legible and shall indicate that conveyors and allied equipment may be started at any time, that danger exists, and that personnel must keep clear. These warning signs shall be provided along the conveyor at areas not guarded by position or location.

(xi) Remotely and automatically controlled conveyors, and conveyors that have operating stations which are not manned or which are beyond voice and visual contact from drive areas, loading areas, transfer points, and other locations on the conveyor path not guarded by location, position, or guards shall be furnished with emergency stop buttons, pull cords, limit switches, or similar emergency stop devices. However, if the employer can demonstrate that the design, function, and operation of the conveyor do not expose an employee to hazards, an emergency stop device is not required.

(A) Emergency stop devices shall be easily identifiable in the immediate vicinity of such locations.

(B) An emergency stop device shall act directly on the control of the conveyor involved and may not depend on the stopping of any other equipment.

(C) Emergency stop devices shall be installed so that they cannot be overridden from other locations.

(xii) Where coal-handling operations may produce a combustible atmosphere from fuel sources or from flammable gases or dust, sources of ignition shall be eliminated or safely controlled to prevent ignition of the combustible atmosphere.

Note: Locations that are hazardous because of the presence of combustible dust are classified as Class II hazardous locations. See 1910.307 of this Part.

(xiii) An employee may not work on or beneath overhanging coal in coal bunkers, coal

FIGURE 6.3 (*Continued*)

silos, or coal storage areas, unless the employee is protected from all hazards posed by shifting coal.

(xiv) An employee entering a bunker or silo to dislodge the contents shall wear a body harness with lifeline attached. The lifeline shall be secured to a fixed support outside the bunker and shall be attended at all times by an employee located outside the bunker or facility.

(12) Hydroplants and equipment. Employees working on or close to water gates, valves, intakes, forebays, flumes, or other locations where increased or decreased water flow or levels may pose a significant hazard shall be warned and shall vacate such dangerous areas before water flow changes are made.

(w) Special conditions. (1) Capacitors. The following additional requirements apply to work on capacitors and on lines connected to capacitors.

Note: See paragraphs (m) and (n) of this section for requirements pertaining to the deenergizing and grounding of capacitor installations.

(i) Before employees work on capacitors, the capacitors shall be disconnected from energized sources and, after a wait of at least 5 minutes from the time of disconnection, short-circuited.

(ii) Before the units are handled, each unit in series-parallel capacitor banks shall be short-circuited between all terminals and the capacitor case or its rack. If the cases of capacitors are on ungrounded substation racks, the racks shall be bonded to ground.

(iii) Any line to which capacitors are connected shall be short-circuited before it is considered deenergized.

Since capacitors store electric charge and can release electrical energy even when disconnected from their sources of supply, some precautions may be necessary, in addition to those contained in §1910.269(m) (deenergizing lines and equipment) and §1910.269(n) (grounding), when work is performed on capacitors or on lines which are connected to capacitors. Paragraph (w)(1) sets forth precautions which will enable this equipment to be considered as deenergized. Under paragraph (w)(1)(i), capacitors on which work is to be performed must be disconnected from their sources of supply and short-circuited. This not only removes the sources of electric current but relieves the capacitors of their charge as well.

For work on individual capacitors in a series-parallel capacitor bank, each unit must be short-circuited between its terminals and the capacitor tank or rack; otherwise, individual capacitors could retain a charge.

(2) Current transformer secondaries. The secondary of a current transformer may not be opened while the transformer is energized. If the primary of the current transformer cannot be deenergized before work is performed on an instrument, a relay, or other section of a current transformer secondary circuit, the circuit shall be bridged so that the current transformer secondary will not be opened.

Although the magnetic flux density in the core of a current transformer is usually very low, resulting in a low secondary voltage, it will rise to saturation if the secondary circuit is opened while the transformer primary is energized. If this occurs, the magnetic flux will induce a voltage in the secondary winding high enough to be hazardous to the insulation in the secondary circuit and to personnel. Because of this hazard to workers, paragraph (w)(2) prohibits the opening of the secondary circuit of a current transformer while the primary is energized.

(3) Series streetlighting. (i) If the open-circuit voltage exceeds 600 volts, the series streetlighting circuit shall be worked in accordance with paragraph (q) or (t) of this section, as appropriate.

(ii) A series loop may only be opened after the streetlighting transformer has been deenergized and isolated from the source of supply or after the loop is bridged to avoid an open-circuit condition.

In a series streetlighting circuit, the lamps are connected in series, and the same current flows in each lamp. This current is supplied by a constant-current transformer, which provides a constant current at a variable voltage from a source of constant voltage and variable current. Like the current transformer, the constant current source attempts to supply current even when the secondary circuit is open. The resultant open-circuit voltage can be very high and hazardous to employees.

(4) Illumination. Sufficient illumination shall be provided to enable the employee to perform the work safely.

Frequently, electric power generation, transmission, and distribution employees must work at night or in enclosed places, such as manholes, that are not illuminated by the

FIGURE 6.3 *(Continued)*

sun. Since inadvertent contact with live parts can be fatal, good lighting is important to the safety of these workers.

(5) Protection against drowning. (i) Whenever an employee may be pulled or pushed or may fall into water where the danger of drowning exists, the employee shall be provided with and shall use U.S. Coast Guard approved personal flotation devices.

(ii) Each personal flotation device shall be maintained in safe condition and shall be inspected frequently enough to ensure that it does not have rot, mildew, water saturation, or any other condition that could render the device unsuitable for use.

(iii) An employee may cross streams or other bodies of water only if a safe means of passage, such as a bridge, is provided.

(6) Employee protection in public work areas. (i) Traffic control signs and traffic control devices used for the protection of employees shall meet the requirements of 1926.200(g)(2) of this Chapter.

(ii) Before work is begun in the vicinity of vehicular or pedestrian traffic that may endanger employees, warning signs or flags, and other traffic control devices shall be placed in conspicuous locations to alert and channel approaching traffic.

(iii) Where additional employee protection is necessary, barricades shall be used.

(iv) Excavated areas shall be protected with barricades.

(v) At night, warning lights shall be prominently displayed.

(7) Backfeed. If there is a possibility of voltage backfeed from sources of cogeneration or from the secondary system (for example, backfeed from more than one energized phase feeding a common load), the requirements of paragraph (l) of this section apply if the lines or equipment are to be worked as energized, and the requirements of paragraphs (m) and (n) of this section apply if the lines or equipment are to be worked as deenergized.

(8) Lasers. Laser equipment shall be installed, adjusted, and operated in accordance with 1926.54 of this Chapter.

(9) Hydraulic fluids. Hydraulic fluids used for the insulated sections of equipment shall provide insulation for the voltage involved.

(x) Definitions.

Affected employee. An employee whose job requires him or her to operate or use a machine or equipment on which servicing or maintenance is being performed under lockout/tagout, or whose job requires him or her to work in an area in which such servicing or maintenance is being performed.

Attendant. An employee assigned to remain immediately outside the entrance to an enclosed or other space to render assistance as needed to employees inside the space.

Authorized employee. An employee who locks out or tags out machines or equipment in order to perform servicing or maintenance on that machine or equipment. An affected employee becomes an authorized employee when that employee's duties include performing servicing or maintenance covered under this section.

Automatic circuit recloser. A self-controlled device for interrupting and reclosing an alternating current circuit with a predetermined sequence of opening and reclosing followed by resetting, hold-closed, or lockout operation.

Barricade. A physical obstruction such as tapes, cones, or A-frame type wood or metal structures intended to provide a warning about and to limit access to a hazardous area.

Barrier. A physical obstruction which is intended to prevent contact with energized lines or equipment or to prevent unauthorized access to a work area.

Bond. The electrical interconnection of conductive parts designed to maintain a common electrical potential.

Bus. A conductor or a group of conductors that serve as a common connection for two or more circuits.

Bushing. An insulating structure, including a through conductor or providing a passageway for such a conductor, with provision for mounting on a barrier, conducting or otherwise, for the purposes of insulating the conductor from the barrier and conducting current from one side of the barrier to the other.

Cable. A conductor with insulation, or a stranded conductor with or without insulation and other coverings (single-conductor cable), or a combination of conductors insulated from one another (multiple-conductor cable).

Cable sheath. A conductive protective covering applied to cables.

Note: A cable sheath may consist of multiple layers of which one or more is conductive.

Circuit. A conductor or system of conductors through which an electric current is intended to flow.

Clearance (for work). Authorization to perform specified work or permission to enter a restricted area.

FIGURE 6.3 (*Continued*)

Clearance (between objects). The clear distance between two objects measured surface to surface.

Communication lines. (See Lines, communication.)

Conductor. A material, usually in the form of a wire, cable, or bus bar, used for carrying an electric current.

Covered conductor. A conductor covered with a dielectric having no rated insulating strength or having a rated insulating strength less than the voltage of the circuit in which the conductor is used.

Current-carrying part. A conducting part intended to be connected in an electric circuit to a source of voltage. Non-current-carrying parts are those not intended to be so connected.

De-energized. Free from any electrical connection to a source of potential difference and from electric charge; not having a potential different from that of the earth.

Note: The term is used only with reference to current-carrying parts, which are sometimes energized (alive).

Designated employee (designated person). An employee (or person) who is designated by the employer to perform specific duties under the terms of this section and who is knowledgeable in the construction and operation of the equipment and the hazards involved.

Electric line truck. A truck used to transport personnel, tools, and material for electric supply line work.

Electric supply equipment. Equipment that produces, modifies, regulates, controls, or safeguards a supply of electric energy.

Electric supply lines. (See Lines, electric supply.)

Electric utility. An organization responsible for the installation, operation, or maintenance of an electric supply system.

Enclosed space. A working space, such as a manhole, vault, tunnel, or shaft that has a limited means of egress or entry that is designed for periodic employee entry under normal operating conditions and that under normal conditions does not contain a hazardous atmosphere, but that may contain a hazardous atmosphere under abnormal conditions.

Note: Spaces that are enclosed but not designed for employee entry under normal operating conditions are not considered to be enclosed spaces for the purposes of this section. Similarly, spaces that are enclosed and that are expected to contain a hazardous atmosphere are not considered to be enclosed spaces for the

purposes of this section. Such spaces meet the definition of permit spaces in 1910.146 of this Part, and entry into them must be performed in accordance with that standard.

Energized (alive, live). Electrically connected to a source of potential difference, or electrically charged so as to have a potential significantly different from that of earth in the vicinity.

Energy isolating device. A physical device that prevents the transmission or release of energy, including, but not limited to, the following: a manually operated electric circuit breaker, a disconnect switch, a manually operated switch, a slide gate, a slip blind, a line valve, blocks, and any similar device with a visible indication of the position of the device. (Push buttons, selector switches, and other control-circuit-type devices are not energy isolating devices.)

Energy source. Any electrical, mechanical, hydraulic, pneumatic, chemical, nuclear, thermal, or other energy source that could cause injury to personnel.

Equipment (electric). A general term including material, fittings, devices, appliances, fixtures, apparatus, and the like used as part of or in connection with an electrical installation.

Exposed. Not isolated or guarded.

Ground. A conducting connection, whether intentional or accidental, between an electric circuit or equipment and the earth, or to some conducting body that serves in place of the earth.

Grounded. Connected to earth or to some conducting body that serves in place of the earth.

Guarded. Covered, fenced, enclosed, or otherwise protected by means of suitable covers or casings, barrier rails or screens, mats, or platforms, designed to minimize the possibility under normal conditions of dangerous approach or accidental contact by persons or objects.

Note: Wires which are insulated, but not otherwise protected are not considered as guarded.

Hazardous atmosphere means an atmosphere that may expose employees to the risk of death, incapacitation, impairment of ability to self-rescue (that is, escape unaided from an enclosed space), injury, or acute illness from one or more of the following causes:

(1) Flammable gas, vapor, or mist in excess of 10 percent of its lower flammable limit (LFL);

(2) Airborne combustible dust at a concentration that meets or exceeds its LFL;

Note: This concentration may be approximated as a condition in which the dust obscures vision at a distance of 5 feet (1.52 m) or less.

FIGURE 6.3 *(Continued)*

(3) Atmospheric oxygen concentration below 19.5 percent or above 23.5 percent;

(4) Atmospheric concentration of any substance for which a dose or a permissible exposure limit is published in Subpart G, Occupational Health and Environmental Control, or in Subpart Z, Toxic and Hazardous Substances, of this Part and which could result in employee exposure in excess of its dose or permissible exposure limit;

Note: An atmospheric concentration of any substance that is not capable of causing death, incapacitation, impairment of ability to self-rescue, injury, or acute illness due to its health effects is not covered by this provision.

(5) Any other atmospheric condition that is immediately dangerous to life or health.

Note: For air contaminants for which OSHA has not determined a dose or permissible exposure limit, other sources of information, such as Material Safety Data Sheets that comply with the Hazard Communication Standard, 1910.1200 of this Part, published information, and internal documents can provide guidance in establishing acceptable atmospheric conditions.

High-power tests. Tests in which fault currents, load currents, magnetizing currents, and line-dropping currents are used to test equipment, either at the equipment's rated voltage or at lower voltages.

High-voltage tests. Tests in which voltages of approximately 1000 volts are used as a practical minimum and in which the voltage source has sufficient energy to cause injury.

High wind. A wind of such velocity that the following hazards would be present:

(1) An employee would be exposed to being blown from elevated locations, or

(2) An employee or material handling equipment could lose control of material being handled, or

(3) An employee would be exposed to other hazards not controlled by the standard involved.

Note: Winds exceeding 40 miles per hour (64.4 kilometers per hour) or 30 miles per hour (48.3 kilometers per hour), if material handling is involved, are normally considered as meeting this criteria unless precautions are taken to protect employees from the hazardous effects of the wind.

Immediately dangerous to life or health (IDLH) means any condition that poses an immediate or delayed threat to life or that would cause irreversible adverse health effects or that would interfere with an individual's ability to escape unaided from a permit space.

Note: Some materials - hydrogen fluoride gas and cadmium vapor, for example - may produce immediate transient effects that, even if severe, may pass without medical attention, but are followed by sudden, possibly fatal collapse 12-72 hours after exposure. The victim "feels normal" from recovery from transient effects until collapse. Such materials in hazardous quantities are considered to be "immediately" dangerous to life or health.

Insulated. Separated from other conducting surfaces by a dielectric (including air space) offering a high resistance to the passage of current.

Note: When any object is said to be insulated, it is understood to be insulated for the conditions to which it is normally subjected. Otherwise, it is, within the purpose of this section, uninsulated.

Insulation (cable). That which is relied upon to insulate the conductor from other conductors or conducting parts or from ground.

Line-clearance tree trimmer. An employee who, through related training or on-the-job experience or both, is familiar with the special techniques and hazards involved in line-clearance tree trimming.

Note 1: An employee who is regularly assigned to a line-clearance tree-trimming crew and who is undergoing on-the-job training and who, in the course of such training, has demonstrated an ability to perform duties safely at his or her level of training and who is under the direct supervision of a line-clearance tree trimmer is considered to be a line-clearance tree trimmer for the performance of those duties.

Note 2: A line-clearance tree trimmer is not considered to be a "qualified employee" under this section unless he or she has the training required for a qualified employee under paragraph (a)(2)(ii) of this section. However, under the electrical safety-related work practices standard in Subpart S of this part, a line-clearance tree trimmer is considered to be a "qualified employee". Tree trimming performed by such "qualified employees" is not subject to the electrical safety-related work practice requirements contained in 1910.331 through 1910.335 of this Part. (See also the note following 1910.332(b)(3) of this Part for information regarding the training an employee must have to be considered a qualified employee under 1910.331 through 1910.335.)

Line-clearance tree trimming. The pruning, trimming, repairing, maintaining, removing, or clearing of trees or the cutting of brush

FIGURE 6.3 *(Continued)*

that is within 10 feet (305 cm) of electric supply lines and equipment.

Lines. (1) Communication lines. The conductors and their supporting or containing structures which are used for public or private signal or communication service, and which operate at potentials not exceeding 400 volts to ground or 750 volts between any two points of the circuit, and the transmitted power of which does not exceed 150 watts. If the lines are operating at less than 150 volts, no limit is placed on the transmitted power of the system. Under certain conditions, communication cables may include communication circuits exceeding these limitations where such circuits are also used to supply power solely to communication equipment.

Note: Telephone, telegraph, railroad signal, data, clock, fire, police alarm, cable television, and other systems conforming to this definition are included. Lines used for signaling purposes, but not included under this definition, are considered as electric supply lines of the same voltage.

(2) Electric supply lines. Conductors used to transmit electric energy and their necessary supporting or containing structures. Signal lines of more than 400 volts are always supply lines within this section, and those of less than 400 volts are considered as supply lines, if so run and operated throughout.

Manhole. A subsurface enclosure which personnel may enter and which is used for the purpose of installing, operating, and maintaining submersible equipment or cable.

Manhole steps. A series of steps individually attached to or set into the walls of a manhole structure.

Minimum approach distance. The closest distance an employee is permitted to approach an energized or a grounded object.

Qualified employee (qualified person). One knowledgeable in the construction and operation of the electric power generation, transmission, and distribution equipment involved, along with the associated hazards.

Note 1: An employee must have the training required by paragraph (a)(2)(ii) of this section in order to be considered a qualified employee.

Note 2: Except under paragraph (g)(2)(v) of this section, an employee who is undergoing

on-the-job training and who, in the course of such training, has demonstrated an ability to perform duties safely at his or her level of training and who is under the direct supervision of a qualified person is considered to be a qualified person for the performance of those duties.

Step bolt. A bolt or rung attached at intervals along a structural member and used for foot placement during climbing or standing.

Switch. A device for opening and closing or for changing the connection of a circuit. In this section, a switch is understood to be manually operable, unless otherwise stated.

System operator. A qualified person designated to operate the system or its parts.

Vault. An enclosure, above or below ground, which personnel may enter and which is used for the purpose of installing, operating, or maintaining equipment or cable.

Vented vault. A vault that has provision for air changes using exhaust flue stacks and low level air intakes operating on differentials of pressure and temperature providing for airflow which precludes a hazardous atmosphere from developing.

Voltage. The effective (rms) potential difference between any two conductors or between a conductor and ground. Voltages are expressed in nominal values unless otherwise indicated. The nominal voltage of a system or circuit is the value assigned to a system or circuit of a given voltage class for the purpose of convenient designation. The operating voltage of the system may vary above or below this value.

APPENDIX A TO §1910.269—FLOW CHARTS

This appendix presents information, in the form of flow charts that illustrates the scope and application of 1910.269. This appendix addresses the interface between 1910.269 and Subpart S of this part (electrical), between 1910.269 and 1910.146 of this part (permit-required confined spaces), and between 1910.269 and 1910.147 of this part [The control of hazardous energy (lockout/tagout)]. These flow charts provide guidance for employers trying to implement the requirements of 1910.269 in combination with other general industry Standards contained in Part 1910.

FIGURE 6.3 *(Continued)*

Appendix A-1 to §1910.269—Application of §1910.269 and Subpart S of This Part to Electrical Installations

[1]Electrical installation design requirements only. See Appendix 1B for electrical safety-related work practices. Supplementary electric generating equipment that is used to supply a workplace for emergency, standby, or similar purposes only is not considered to be an electric power generation installation.

[2]See Table 1 of Appendix A-2 for requirements that can be met through compliance with subpart S.

FIGURE 6.3   (*Continued*)

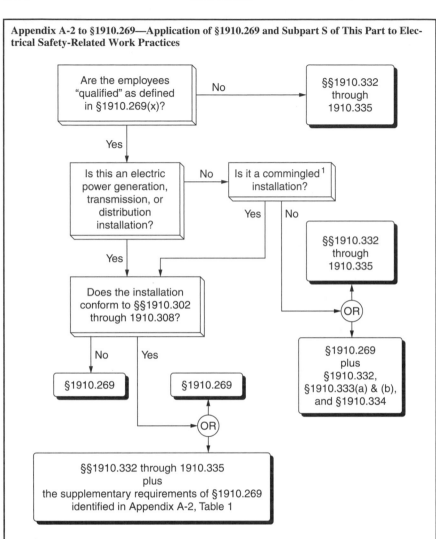

Appendix A-2 to §1910.269—Application of §1910.269 and Subpart S of This Part to Electrical Safety-Related Work Practices

[1]Commingled to the extent that the electric power generation, transmission, or distribution installation poses the greater hazard.

FIGURE 6.3 *(Continued)*

TABLE 1   Electrical Safety-Related Work Practices in § 1910.269

Compliance with subpart S is considered as compliance with §1910.269[1]	Paragraphs that apply regardless of compliance with subpart S
{d}, electric shock hazards only	{a}{2}[2] and (a)(3)[2].
{h}(3)	(b)[2]
(i)(2)	(c)[2]
(k)	(d). other than electric shock hazards.
(i)(1) through (1)(4), (1)(6)(i), and (1)(8) through (1)(10)	(e).
(m)	(f).
(p)(4)	(g).
(s)(2)	(h)(1) and (h)(2).
(u)(1) and (u)(3) through (u)(5)	(1)(3)[2] and (1)(4)[2].
(v)(3) through (v)(5)	(j)[2]
(w)(1) and (w)(7)	(1)(5)[2], (1)(6)(ii)[2], (1)(6)(iii)[2], and (1)(7)(2).
	(n)[2].
	(o)[2].
	(p)(1) through (p)(3).
	(q)[2]
	(r)[2].
	(s)(1).
	(1)[2].
	(u)(2)[2] and (u)(6)[2].
	(v)(1), (v)(2)[2], and (v)(5) through (v)(12).
	(w)(2) through (w)(6)[2], (w)(8), and (w)(9)[2].

[1] If the electrical installation meets the requirements of §§1910.303 through 1910.308 of this part, then the electrical installation and any associated electrical safety-related work practices conforming to §§1910.332 through 1910.335 of this part are considered to comply with these provisions of §§1910.269 of this part.

[2] These provisions include electrical safety requirements that must be met regardless of compliance with subpart S of this part.

FIGURE 6.3   (*Continued*)

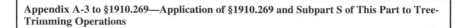

Appendix A-3 to §1910.269—Application of §1910.269 and Subpart S of This Part to Tree-Trimming Operations

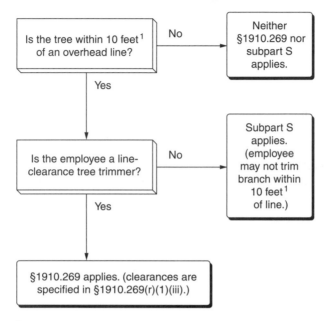

[1] 10 feet plus 4 inches for every 10 kilovolts over 50 kilovolts.

FIGURE 6.3 *(Continued)*

Appendix A-4 to §1910.269—Application of §§1910.147, 1910.269, and 1910.333 to Hazardous Energy Control Procedures (Lockout/Tagout)

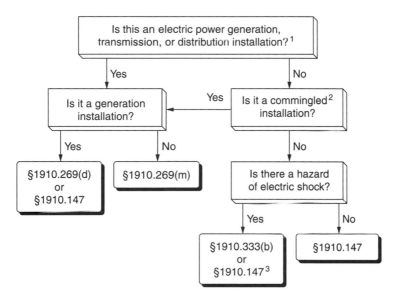

[1]If the installation conforms to §§1910.303 through 1910.308, the lockout and tagging procedures of 1910.333(b) may be followed for electric shock hazards.

[2]Commingled to the extent that the electric power generation, transmission, or distribution installation poses the greater hazard.

[3]§1910.333(b)(2)(iii)(D) and (b)(2)(iv)(B) still apply.

FIGURE 6.3 *(Continued)*

Appendix A-5 to §1910.269—Application of §§1910.146 and 1910.269 to Permit-Required Confined Spaces

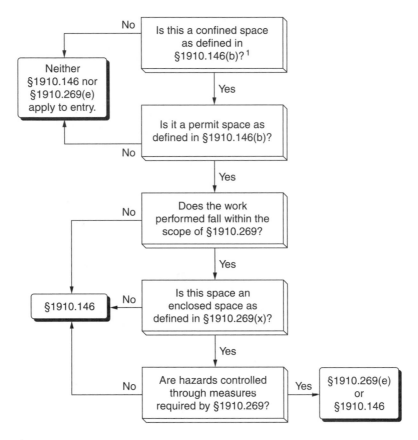

^{1}See §1910.146(c) for general non-entry requirements that apply to all confined spaces.

APPENDIX B TO §1910.269—WORKING ON EXPOSED ENERGIZED PARTS

I. Introduction

Electric transmission and distribution line installations have been designed to meet National Electrical Safety Code (NESC), ANSI C2, requirements and to provide the level of line outage performance required by system reliability criteria. Transmission and distribution lines are also designed to withstand the maximum overvolt-

ages expected to be impressed on the system. Such overvoltages can be caused by such conditions as switching surges, faults, or lightning. Insulator design and lengths and the clearances to structural parts (which, for low voltage through extra-high voltage, or EHV, facilities, are generally based on the performance of the line as a result of contamination of the insulation or during storms) have, over the years, come closer to the minimum approach distances used by workers (which are generally based on non-storm conditions). Thus, as minimum approach (working)

FIGURE 6.3   (*Continued*)

distances and structural distances (clearances) converge, it is increasingly important that basic considerations for establishing safe approach distances for performing work be understood by the designers and the operating and maintenance personnel involved.

The information in this appendix will assist employers in complying with the minimum approach distance requirements contained in paragraphs (l),(2), and (q)(3) of this section. The technical criteria and methodology presented herein is mandatory for employers using reduced minimum approach distances as permitted in Table R-7 and Table R-8. This Appendix is intended to provide essential background information and technical criteria for the development or modification, if possible, of the safe minimum approach distances for electric transmission and distribution of live-line work. The development of these safe distances must be undertaken by persons knowledgeable in the techniques discussed in this appendix, and competent in the field of electric transmission and distribution system design.

II. General

A. Definitions. The following definitions from 1910.269(x) relate to work on or near transmission and distribution lines and equipment, and the electrical hazards they present.

Exposed. Not isolated or guarded.

Guarded. Covered, fenced, enclosed, or otherwise protected, by means of suitable covers or casings, barrier rails or screens, mats, or platforms designed to minimize the possibility, under normal conditions, of dangerous approach or accidental contact by persons or objects.

Note: Wires which are insulated, but not otherwise protected, are not considered as guarded.

Insulated. Separated from other conducting surfaces by a dielectric (including air space) offering a high resistance to the passage of current.

Note: When any object is said to be insulated, it is understood to be insulated for the conditions to which it is normally subjected. Otherwise, it is, within the purpose of this section, uninsulated.

B. Installations Energized at 50 to 300 Volts. The hazards posed by installations energized at 50 to 300 volts are the same as those found in many other workplaces. That is not to say that there is no hazard, but the complexity of

electrical protection required does not compare to that required for high voltage systems. The employee must avoid contact with the exposed parts, and the protective equipment used (such as, rubber insulating gloves) must provide insulation for the voltages involved.

C. Exposed Energized Parts Over 300 Volts AC. Table R-6, Table R-7, and Table R-8 of 1910.269 provide safe approach and working distances in the vicinity of energized electric apparatus so that work can be done safely without risk of electrical flashover.

The working distances must withstand the maximum transient overvoltage that can reach the work site under the working conditions and practices in use. Normal system design may provide or include a means to control transient overvoltages, or temporary devices may be employed to achieve the same result. The use of technically correct practices or procedures to control overvoltages (for example, portable gaps, or preventing the automatic control from initiating breaker reclosing) enables line design and operation to be based on reduced transient overvoltage values. Technical information for U.S. electrical systems indicates that current design provides for the following maximum transient overvoltage values (usually produced by switching surges): 362 kV and less than 3.0 per unit; 552 kV-2.4 per unit; 800 kV-2.0 per unit.

Additional discussion of maximum transient overvoltages can be found in paragraph IV.A.2, later in this appendix.

III. Determination of the Electrical Component of Minimum Approach Distances

A. Voltages of 1.1 kV to 72.5 kV. For voltages of 1.1 kV to 72.5 kV, the electrical component of minimum approach distances is based on American National Standards Institute (ANSI)/American Institute of Electrical Engineers (AIEE) Standard No. 4, March 1943, Tables III and IV. [AIEE is the predecessor technical society to the Institute of Electrical and Electronic Engineers (IEEE).] These distances are calculated by the following formula:

Equation (1)—For voltages of 1.1 kV to 72.5 kV

$$D = \left(\frac{V_{max} \times pu}{124} \right)^{1.63}$$

where D = electrical component of the minimum approach distance in air in feet

FIGURE 6.3 (*Continued*)

TABLE 1   AC Energized Line-Work Phase-to-Ground Electrical Component of the minimum approach Distance—1.1 to 72.5 kV

Maximum anticipated per-unit transient overvoltage	Phase to phase voltage			
	15,000	36,000	46,000	72,500
3.0.....................	0.08	0.33	0.49	1.03

Note: The distances given (in feet) are for air as the insulating medium and provide no additional clearance for inadvertent movement.

V_{max} = maximum rated line-to-ground rms voltage in kV

pu = maximum transient overvoltage factor in per unit

Source: AIEE Standard No. 4, 1943.

This formula has been used to generate Table 1.

B. Voltages of 72.6 kV to 800 kV.   For voltages of 72.6 kV to 800 kV, the electrical component of minimum approach distances is based on ANSI/IEEE Standard 516-1987, "IEEE Guide for Maintenance Methods on Energized Power Lines." This standard gives the electrical component of the minimum approach distance based on power frequency rod-gap data, supplemented with transient overvoltage information and a saturation factor for high voltages. The distances listed in ANSI/IEEE Standard 516 have been calculated according to the following formula:

Equation (2)—For voltages of 72.6 kV to 800 kV

$$D = (C + a)\text{pu } V(\text{MAX})$$

where D = electrical component of the minimum approach distance in air (in Feet)

C = 0.01 to take care of correction factors associated with the variation of gap spark over with voltage

a = a factor relating to the saturation of air at voltages of 345 kV or higher

pu = maximum anticipated transient overvoltage, in per unit (p.u.)

$V(\text{MAX})$ = maximum rms system line-to-ground voltage in kilovolts-M it should be the "actual" maximum, or the normal highest voltage for the range (for example, 10 percent above the nominal voltage)

Source: Formula developed from ANSI/IEEE Standard No. 516, 1987.

This formula is used to calculate the electrical component of the minimum approach distances in air and is used in the development of Table 2 and Table 3.

C. Provisions for Inadvertent Movement. The minimum approach distances (working distances) must include an "adder" to compensate for the inadvertent movement of the worker relative to an energized part or the movement of the part relative to the worker. A certain allowance must be made to account for this possible inadvertent movement and to provide the worker with a comfortable and safe zone in which to work. A distance for inadvertent movement (called the "ergonomic component of the minimum approach distance") must be added to the electrical component to determine the total safe minimum approach distances used in live-line work.

One approach that can be used to estimate the ergonomic component of the minimum approach distance is response time-distance analysis. When this technique is used, the total response time to a hazardous incident is estimated and converted to distance traveled. For example, the driver of a car takes a given amount of time to respond to a "stimulus" and stop the vehicle. The elapsed time involved results in a distance being traveled before the car comes to a complete stop. This distance is dependent on the speed of the car at the time the stimulus appears.

In the case of live-line work, the employee must first perceive that he or she is approaching the danger zone. Then, the worker responds to the danger and must decelerate and stop all motion toward the energized part. During the time it takes to stop, a distance will have been traversed. It is this distance that must be added to the electrical component of the minimum approach distance to obtain the total safe minimum approach distance.

FIGURE 6.3   *(Continued)*

TABLE 2 AC Energized Line-Work Phase-to-Ground Electrical
Component of the Minimum Approach Distance—121 to 242 kV

Maximum anticipated per-unit transient overvoltage	Phase to phase voltage			
	121,000	145,000	169,000	242,000
2.0..................	1.40	1.70	2.00	2.80
2.1..................	1.47	1.79	2.10	2.94
2.2..................	1.54	1.87	2.20	3.08
2.3..................	1.61	1.96	2.30	3.22
2.4..................	1.68	2.04	2.40	3.35
2.5..................	1.75	2.13	2.50	3.50
2.6..................	1.82	2.21	2.60	3.64
2.7..................	1.89	2.30	2.70	3.76
2.8..................	1.96	2.38	2.38	3.92
2.9..................	2.03	2.47	2.90	4.05
3.0..................	2.10	2.55	3.00	4.29

Note: The distances given (in feet) are for air as the insulating medium and provide no additional clearance for inadvertent movement.

At voltages below 72.5 kV, the electrical component of the minimum approach distance is smaller than the ergonomic component. At 72.5 kV the electrical component is only a little more than 1 foot. An ergonomic component of the minimum approach distance is needed that will provide for all the worker's unexpected movements. The usual live-line work method for these voltages

TABLE 3 AC Energized Line-Work Phase-to-Ground Electrical
Component of the Minimum Approach Distance—362 to 800 kv

Maximum anticipated per-unit transient overvoltage	Phase to phase voltage		
	362,000	552,000	800,000
1.5..................	....	4.97	8.66
1.6..................	....	5.46	9.60
1.7..................	....	5.98	10.60
1.8..................	....	6.51	11.64
1.9..................	....	7.08	12.73
2.0..................	4.20	7.68	13.86
2.1..................	4.41	8.27	....
2.2..................	4.70	8.87	....
2.3..................	5.01	9.49	....
2.4..................	5.34	10.21	....
2.5..................	5.67	....	....
2.6..................	6.01	....	....
2.7..................	6.36	....	....
2.8..................	6.73	....	....
2.9..................	7.10	....	....
3.0..................	7.48	....	....

Note: The distances given (in feet) are for air as the insulating medium and provide no additional clearance for inadvertent movement.

FIGURE 6.3 (*Continued*)

is the use of rubber insulating equipment, frequently rubber gloves. The energized object needs to be far enough away to provide the worker's face with a safe approach distance, as his or her hands and arms are insulated. In this case, 2 feet has been accepted as a sufficient and practical value.

For voltages between 72.6 and 800 kV, there is a change in the work practices employed during energized line work. Generally, live-line tools (hot sticks) are employed to perform work while equipment is energized. These tools, by design, keep the energized part at a constant distance from the employee and thus maintain the appropriate minimum approach distance automatically.

The length of the ergonomic component of the minimum approach distance is also influenced by the location of the worker and by the nature of the work. In these higher voltage ranges, the employees use work methods that more tightly control their movements than when the workers perform rubber glove work. The worker is farther from energized line or equipment and needs to be more precise in his or her movements just to perform the work.

For these reasons, a smaller ergonomic component of the minimum approach distance is needed, and a distance of 1 foot has been selected for voltages between 72.6 and 800 kV.

Table 4 summarizes the ergonomic component of the minimum approach distance for the two voltage ranges.

D. Bare-Hand Live-Line Minimum Approach Distances.

Calculating the strength of phase-to-phase transient overvoltages is complicated by the varying time displacement between overvoltages on parallel conductors (electrodes) and by the varying ratio between the positive and negative voltages on the two electrodes. The time displacement causes the maximum voltage between phases to be less than the sum of the phase-to-ground voltages. The International

TABLE 4 Ergonomic Component of Minimum Approach Distance

Voltage range (kV)	Distance (feet)
1.1 to 72.5.....................	2.0
72.6 to 800.....................	1.0

Note: This distance must be added to the electrical component of the minimum approach distance to obtain the full minimum approach distance.

Electrotechnical Commission (IEC) Technical Committee 28, Working Group 2, has developed the following formula for determining the phase-to-phase maximum transient overvoltage, based on per unit (p.u.) of the system nominal voltage phase-to-ground crest:

$$pu(p) = pu(g) + 1.6.$$

where pu(g) = p.u. phase-to-ground maximum transient overvoltage

pu(p) = p.u. phase-to-phase maximum transient overvoltage

This value of maximum anticipated transient overvoltage must be used in Equation (2) to calculate the phase-to-phase minimum approach distances for live-line bare-hand work.

E. Compiling the Minimum Approach Distance Tables.

For each voltage involved, the distance in Table 4 in this appendix has been added to the distance in Table 1, Table 2, or Table 3 in this appendix to determine the resulting minimum approach distances in Table R-6, Table R-7, and Table R-8 in 1910.269.

F. Miscellaneous Correction Factors.

The strength of an air gap is influenced by the changes in the air medium that forms the insulation. A brief discussion of each factor follows, with a summary at the end.

1. *Dielectric strength of air.* The dielectric strength of air in a uniform electric field at standard atmospheric conditions is approximately 31 kV (crest) per cm at 60 Hz. The disruptive gradient is affected by the air pressure, temperature, and humidity, by the shape, dimensions, and separation of the electrodes, and by the characteristics of the applied voltage (wave shape).

2. *Atmospheric effect.* Flashover for a given air gap is inhibited by an increase in the density (humidity) of the air. The empirically determined electrical strength of a given gap is normally applicable at standard atmospheric conditions (20°C, 101.3 kPa, 11 g/cm^3 humidity).

The combination of temperature and air pressure that gives the lowest gap flashover voltage is high temperature and low pressure. These are conditions not likely to occur simultaneously. Low air pressure is generally associated with high humidity, and this causes increased electrical strength. An average air pressure is more likely to be associated with low humidity. Hot

FIGURE 6.3 *(Continued)*

and dry working conditions are thus normally associated with reduced electrical strength.

The electrical component of the minimum approach distances in Table 1, Table 2, and Table 3 has been calculated using the maximum transient overvoltages to determine withstand voltages at standard atmospheric conditions.

3. *Altitude.* The electrical strength of an air gap is reduced at high altitude, principally due to the reduced air pressure. An increase of 3 percent per 300 meters in the minimum approach distance for altitudes above 900 meters is required. Table R-10 of 1910.269 presents this information in tabular form.

Summary. After taking all these correction factors into account and after considering their interrelationships relative to the air gap insulation strength and the conditions under which live work is performed, one finds that only a correction for altitude needs to be made. An elevation of 900 meters is established as the base elevation, and the values of the electrical component of the minimum approach distances have been derived with this correction factor in mind. Thus, the values used for elevations below 900 meters are conservative without any change; corrections have to be made only above this base elevation.

IV. Determination of Reduced Minimum Approach Distances

A. *Factors Affecting Voltage Stress at the Work Site*

1. *System voltage (nominal).* The nominal system voltage range sets the absolute lower limit for the minimum approach distance. The highest value within the range, as given in the relevant table, is selected and used as a reference for per unit calculations.

2. *Transient overvoltages.* Transient overvoltages may be generated on an electrical system by the operation of switches or breakers, by the occurrence of a fault on the line or circuit being worked or on an adjacent circuit, and by similar activities. Most of the overvoltages are caused by switching, and the term "switching surge" is often used to refer generically to all types of overvoltages. However, each overvoltage has an associated transient voltage wave shape. The wave shape arriving at the site and its magnitude vary considerably.

The information used in the development of the minimum approach distances takes into consideration the most common wave shapes; thus, the required minimum approach distances are appropriate for any transient overvoltage level usually found on electric power generation, transmission, and distribution systems. The values of the per unit (p.u.) voltage relative to the nominal maximum voltage are used in the calculation of these distances.

3. *Typical magnitude of overvoltages.* The magnitude of typical transient overvoltages is given in Table 5.

4. *Standard deviation—air-gap withstand.* For each air-gap length, and under the same atmospheric conditions, there is a statistical variation in the breakdown voltage. The probability of the breakdown voltage is assumed to have a normal (Gaussian) distribution. The standard deviation of this distribution varies with the wave shape, gap geometry, and atmospheric

TABLE 5 Magnitude of Typical Transient Overvoltages

Cause	Magnitude (per unit)
Energized 200 mile line without closing resistors......................	3.5
Energized 200 mile line with one step closing resistor.......................	2.1
Energized 200 mile line with multi-step resistor............................	2.5
Reclosed with trapped charge one step resistor............................	2.2
Opening surge with single restrike.......	3.0
Fault initiation unfaulted phase.........	2.1
Fault initiation adjacent circuit........	2.5
Fault clearing..........................	1.7–1.9

Source: ANSI/IEEE Standard No. 516, 1987.

FIGURE 6.3 (*Continued*)

conditions. The withstand voltage of the air gap used in calculating the electrical component of the minimum approach distance has been set at three standard deviations (3 sigma*) below the critical flashover voltage. (The critical flashover voltage is the crest value of the impulse wave that, under specified conditions, causes flashover on 50 percent of the applications. An impulse wave of three standard deviations below this value, that is, the withstand voltage, has a probability of flashover of approximately 1 in 1000.)

> *Note:* Sigma is the symbol for standard deviation. http://www.osha.gov/pls/oshaweb/1#1

5. *Broken Insulators.* Tests have shown that the insulation strength of an insulator string with broken skirts is reduced. Broken units may have lost up to 70 percent of their withstand capacity. Because the insulating capability of a broken unit cannot be determined without testing it, damaged units in an insulator are usually considered to have no insulating value. Additionally, the overall insulating strength of a string with broken units may be further reduced in the presence of a live-line tool alongside it. The number of good units that must be present in a string is based on the maximum overvoltage possible at the worksite.

B. Minimum Approach Distances Based on Known Maximum Anticipated Per-Unit Transient Overvoltages

1. *Reduction of the minimum approach distance for AC systems.* When the transient overvoltage values are known and supplied by the employer, Table R-7 and Table R-8 of 1910.269 allow the minimum approach distances from energized parts to be reduced. In order to determine what this maximum overvoltage is, the employer must undertake an engineering analysis of the system. As a result of this engineering study, the employer must provide new live work procedures, reflecting the new minimum approach distances, the conditions and limitations of application of the new minimum approach distances, and the specific practices to be used when these procedures are implemented.

2. *Calculation of reduced approach distance values.* The following method of calculating reduced minimum approach distances is based on ANSI/IEEE Standard 516:

Step 1. Determine the maximum voltage (with respect to a given nominal voltage range) for the energized part.

Step 2. Determine the maximum transient overvoltage (normally a switching surge) that can be present at the work site during work operation.

Step 3. Determine the technique to be used to control the maximum transient overvoltage. (See paragraphs IV.C and IV.D of this appendix.) Determine the maximum voltage that can exist at the work site with that form of control in place and with a confidence level of 3 sigma. This voltage is considered to be the withstand voltage for the purpose of calculating the appropriate minimum approach distance.

Step 4. Specify in detail the control technique to be used, and direct its implementation during the course of the work.

Step 5. Using the new value of transient overvoltage in per unit (p.u.), determine the required phase-to-ground minimum approach distance from Table R-7 or Table R-8 of 1910.269.

C. Methods of Controlling Possible Transient Overvoltage Stress Found on a System

1. *Introduction.* There are several means of controlling overvoltages that occur on transmission systems. First, the operation of circuit breakers or other switching devices may be modified to reduce switching transient overvoltages. Second, the overvoltage itself may be forcibly held to an acceptable level by means of installation of surge arresters at the specific location to be protected. Third, the transmission system may be changed to minimize the effect of switching operations.

2. *Operation of circuit breakers.*[†] The maximum transient overvoltage that can reach the work site is often due to switching on the line on which work is being performed. If the automatic-reclosing is removed during energized line work so that the line will not be reenergized after being opened for any reason, the maximum switching surge overvoltage is then limited to the larger of the opening surge or the greatest possible fault-generated surge, provided that the devices (for example, insertion resistors) are operable and will function to limit the transient overvoltage. It is essential that the operating ability of such devices be assured when they are employed to limit the overvoltage level. If it is prudent not to remove the reclosing feature because of system operating conditions, other methods of controlling the switching surge level may be necessary.

> [†]*Note:* The detailed design of a circuit interrupter, such as the design of the contacts, of

FIGURE 6.3   *(Continued)*

resistor insertion, and of breaker timing control, are beyond the scope of this appendix. These features are routinely provided as part of the design for the system. Only features that can limit the maximum switching transient overvoltage on a system are discussed in this appendix. http://www.osha.gov/pls/oshaweb/2#2

Transient surges on an adjacent line, particularly for double circuit construction, may cause a significant overvoltage on the line on which work is being performed. The coupling to adjacent lines must be accounted for when minimum approach distances are calculated based on the maximum transient overvoltage.

3. *Surge arresters.* The use of modern surge arresters has permitted a reduction in the basic impulse-insulation levels of much transmission system equipment. The primary function of early arresters was to protect the system insulation from the effects of lightning. Modern arresters not only dissipate lightning-caused transients, but may also control many other system transients that may be caused by switching or faults.

It is possible to use properly designed arresters to control transient overvoltages along a transmission line and thereby reduce the requisite length of the insulator string. On the other hand, if the installation of arresters has not been used to reduce the length of the insulator string, it may be used to reduce the minimum approach distance instead.‡

‡**Note:** Surge arrestor application is beyond the scope of this appendix. However, if the arrestor is installed near the work site, the application would be similar to protective gaps as discussed in paragraph IV.D. of this appendix. http://www.osha.gov/pls/oshaweb/3#3

4. *Switching Restrictions.* Another form of overvoltage control is the establishment of switching restrictions, under which breakers are not permitted to be operated until certain system conditions are satisfied. Restriction of switching is achieved by the use of a tagging system, similar to that used for a "permit," except that the common term used for this activity is a "hold-off" or "restriction." These terms are used to indicate that operation is not prevented, but only modified during the live-work activity.

D. *Minimum Approach Distance Based on Control of Voltage Stress (Overvoltages) at the Work Site.*
Reduced minimum approach distances can be calculated as follows:

1. *First Method—Determining the reduced minimum approach distance from a given withstand voltage.*§

§**Note:** Since a given rod gap of a given configuration corresponds to a certain withstand voltage, this method can also be used to determine the minimum approach distance for a known gap. http://www.osha.gov/pls/oshaweb/4#4

Step 1. Select the appropriate withstand voltage for the protective gap based on system requirements and an acceptable probability of actual gap flashover.

Step 2. Determine a gap distance that provides a withstand voltage** greater than or equal to the one selected in the first step. ††

Note: The withstand voltage for the gap is equal to 85 percent of its critical flashover voltage. http://www.osha.gov/pls/oshaweb/5#5

††**Note:** Switch steps 1 and 2 if the length of the protective gap is known. The withstand voltage must then be checked to ensure that it provides an acceptable probability of gap flashover. In general, it should be at least 1.25 times the maximum crest operating voltage. http://www. osha.gov/pls/oshaweb/6#6

Step 3. Using 110 percent of the gap's critical flashover voltage, determine the electrical component of the minimum approach distance from Equation (2) or Table 6, which is a tabulation of distance vs. withstand voltage based on Equation (2).

Step 4. Add the 1-foot ergonomic component to obtain the total minimum approach distance to be maintained by the employee.

2. *Second Method—Determining the necessary protective gap length from a desired (reduced) minimum approach distance.*

Step 1. Determine the desired minimum approach distance for the employee. Subtract the 1-foot ergonomic component of the minimum approach distance.

Step 2. Using this distance, calculate the air gap withstand voltage from Equation (2). Alternatively, find the voltage corresponding to the distance in Table 6.‡‡

‡‡**Note:** Since the value of the saturation factor, a, in Equation (2) is dependent on the maximum voltage, several iterative computations may be necessary to determine the correct withstand voltage using the equation.

FIGURE 6.3 (*Continued*)

TABLE 6   Withstand Distances for Transient Overvoltages

Crest voltage (kV)	Withstand distance (in feet) air gap
100......................	0.71
150......................	1.06
200......................	1.41
250......................	1.77
300......................	2.12
350......................	2.47
400......................	2.83
450......................	3.18
500......................	3.54
550......................	3.89
600......................	4.24
650......................	4.60
700......................	5.17
750......................	5.73
800......................	6.31
850......................	6.91
900......................	7.57
950......................	8.23
1000......................	8.94
1050......................	9.65
1100......................	10.42
1150......................	11.18
1200......................	12.05
1250......................	12.90
1300......................	13.79
1350......................	14.70
1400......................	15.64
1450......................	16.61
1500......................	17.61
1550......................	18.63

Source: Calculations are based on Equation (2).
Note: The air gap is based on the 60-Hz rod-gap withstand distance.

A graph of withstand voltage versus distance is given in ANSI/IEEE Std. 516, 1987. This graph could also be used to determine the appropriate withstand voltage for the minimum approach distance involved. http://www.osha.gov/pls/oshaweb/7#7

Step 3. Select a protective gap distance corresponding to a critical flashover voltage that, when multiplied by 110 percent, is less than or equal to the withstand voltage from Step 2.

Step 4. Calculate the withstand voltage of the protective gap (85 percent of the critical flashover voltage) to ensure that it provides an acceptable risk of flashover during the time the gap is installed.

    3. *Sample protective gap calculations.*

    Problem 1: Work is to be performed on a 500-kV transmission line that is subject to transient overvoltages of 2.4 p.u. The maximum operating voltage of the line is 552 kV. Determine the length of the protective gap that will provide the minimum practical safe approach distance. Also, determine what that minimum approach distance is.

    Step 1. Calculate the smallest practical maximum transient overvoltage (1.25 times the crest line-to-ground voltage):[§§]

    [§§]**Note:** To eliminate unwanted flashovers due to minor system disturbances, it is desirable to have the crest withstand voltage not lower than 1.25 p.u.http://www.osha.gov/pls/oshaweb/ 8#8

Equation in Step 1:

$$552 \text{ kV} \times \frac{\sqrt{2}}{\sqrt{3}} \times 1.25 = 563 \text{ kV}$$

This will be the withstand voltage of the protective gap.

    Step 2. Using test data for a particular protective gap, select a gap that has a critical flashover voltage greater than or equal to:

$$563 \text{ kV}/0.85 = 662 \text{ kV}$$

For example, if a protective gap with a 4.0 foot spacing tested to a critical flashover voltage of 665 kV, crest, select this gap spacing.

    Step 3. This protective gap corresponds to a 110 percent of critical flashover voltage value of:

$$665 \text{ kV} \times 1.10 = 732 \text{ kV}$$

This corresponds to the withstand voltage of the electrical component of the minimum approach distance.

    Step 4. Using this voltage in Equation (2) results in an electrical component of the minimum approach distance of:

$$D = (0.01 + 0.0006) \times \frac{552 \text{ kV}}{\sqrt{3}} = 5.5 \text{ ft}$$

The preceding is the Equation in Step 4.

    Step 5. Add 1 foot to the distance calculated in step 4, resulting in a total minimum approach distance of 6.5 feet.

FIGURE 6.3   *(Continued)*

Problem 2: For a line operating at a maximum voltage of 552 kV subject to a maximum transient overvoltage of 2.4 p.u., find a protective gap distance that will permit the use of a 9.0 foot minimum approach distance. (A minimum approach distance of 11 feet, 3 inches is normally required.)

Step 1. The electrical component of the minimum approach distance is 8.0 feet (9.0-1.0).

Step 2. From Table 6, select the withstand voltage corresponding to a distance of 8.0 feet. By interpolation:

$$900 \ \text{kV} + \left[50 \times \frac{(8.00 - 7.57)}{(8.23 - 7.57)}\right] = 933 \ \text{kV}$$

Step 3. The voltage calculated in Step 2 corresponds to 110 percent of the critical flashover voltage of the gap that should be employed. Using test data for a particular protective gap, select a gap that has a critical flashover voltage less than or equal to:

$$D = \frac{(0.01 + 0.0006)}{\sqrt{2}} \times 732 \ \text{kV}$$

For example, if a protective gap with a 5.8 foot spacing tested to a critical flashover voltage of 820 kV, crest, select this gap spacing.

Step 4. The withstand voltage of this protective gap would be:

$$820 \ \text{kV} \times 0.85 = 697 \ \text{kV}.$$

The maximum operating crest voltage would be:

$$552 \ \text{kV} \times \frac{\sqrt{2}}{\sqrt{3}} = 449 \ \text{kV}$$

The crest withstand voltage of the protective gap in per unit is thus:

$$697 \ \text{kV} + 449 \ \text{kV} = 1.55 \ \text{p.u}$$

If this is acceptable, the protective gap could be installed with a 5.8 foot spacing, and the minimum approach distance could then be reduced to 9.0 feet.

4. *Comments and variations.* The 1 foot ergonomic component of the minimum approach distance must be added to the electrical component of the minimum approach distance calculated under paragraph IV.D of this appendix.

The calculations may be varied by starting with the protective gap distance or by starting with the minimum approach distance.

E. *Location of Protective Gaps*

1. Installation of the protective gap on a structure adjacent to the work site is an acceptable practice, as this does not significantly reduce the protection afforded by the gap.

2. Gaps installed at terminal stations of lines or circuits provide a given level of protection. The level may not, however, extend throughout the length of the line to the worksite. The use of gaps at terminal stations must be studied in depth. The use of substation terminal gaps raises the possibility that separate surges could enter the line at opposite ends, each with low enough magnitude to pass the terminal gaps without flashover. When voltage surges are initiated simultaneously at each end of a line and travel toward each other, the total voltage on the line at the point where they meet is the arithmetic sum of the two surges. A gap that is installed within 0.5 mile of the work site will protect against such intersecting waves. Engineering studies of a particular line or system may indicate that adequate protection can be provided by even more distant gaps.

3. If protective gaps are used at the work site, the work site impulse insulation strength is established by the gap setting. Lightning strikes as much as 6 miles away from the worksite may cause a voltage surge greater than the insulation withstand voltage, and a gap flashover may occur. The flashover will not occur between the employee and the line, but across the protective gap instead.

4. There are two reasons to disable the automatic-reclosing feature of circuit-interrupting devices while employees are performing live-line maintenance:

   • To prevent the reenergizing of a circuit faulted by actions of a worker, which could possibly create a hazard or compound injuries or damage produced by the original fault;
   • To prevent any transient overvoltage caused by the switching surge that would occur if the circuit were reenergized.

However, due to system stability considerations, it may not always be feasible to disable the automatic-reclosing feature.

FIGURE 6.3 (*Continued*)

APPENDIX C TO §1910.269—PROTECTION FROM STEP AND TOUCH POTENTIALS

I. Introduction

When a ground fault occurs on a power line, voltage is impressed on the "grounded" object faulting the line. The voltage to which this object rises depends largely on the voltage on the line, on the impedance of the faulted conductor, and on the impedance to "true," or "absolute," ground represented by the object. If the object causing the fault represents a relatively large impedance, the voltage impressed on it is essentially the phase-to-ground system voltage. However, even faults to well-grounded transmission towers or substation structures can result in hazardous voltages.*** The degree of the hazard depends upon the magnitude of the fault current and the time of exposure.

***Note: This appendix provides information primarily with respect to employee protection from contact between equipment being used and an energized power line. The information presented is also relevant to ground faults to transmission towers and substation structures; however, grounding systems for these structures should be designed to minimize the step and touch potentials involved. http://www.osha.gov/pls/oshaweb/1#1

II. Voltage-Gradient Distribution

A. *Voltage-Gradient Distribution Curve.* The dissipation of voltage from a grounding electrode (or from the grounded end of an energized grounded object) is called the ground potential gradient. Voltage drops associated with this dissipation of voltage are called ground potentials. Figure 1 is a typical voltage-gradient distribution curve (assuming a uniform soil texture). This graph shows that voltage decreases rapidly with increasing distance from the grounding electrode.

B. *Step and Touch Potentials.* "Step potential" is the voltage between the feet of a person

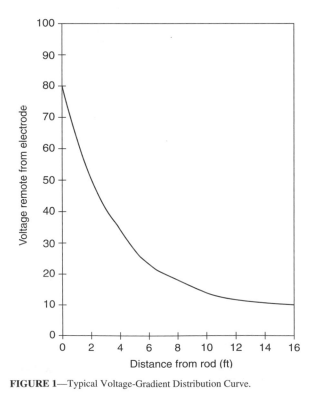

FIGURE 1—Typical Voltage-Gradient Distribution Curve.

FIGURE 6.3 (*Continued*)

standing near an energized grounded object. It is equal to the difference in voltage, given by the voltage distribution curve, between two points at different distances from the "electrode." A person could be at risk of injury during a fault simply by standing near the grounding point.

"Touch potential" is the voltage between the energized object and the feet of a person in contact with the object. It is equal to the difference in voltage between the object (which is at a distance of 0 feet) and a point some distance away. It should be noted that the touch potential could be nearly the full voltage across the grounded object if that object is grounded at a point remote from the place where the person is in contact with it. For example, a crane that was grounded to the system neutral and that contacted an energized line would expose any person in contact with the crane or its uninsulated load line to a touch potential nearly equal to the full fault voltage.

Step and touch potentials are illustrated in Fig. 2.

C. Protection From the Hazards of Ground-Potential Gradients.

An engineering analysis of the power system under fault conditions can be used to determine whether or not hazardous step and touch voltages will develop. The result of this analysis can ascertain the need for protective measures and can guide the selection of appropriate precautions.

Several methods may be used to protect employees from hazardous ground-potential gradients, including equipotential zones, insulating equipment, and restricted work areas.

1. The creation of an equipotential zone will protect a worker standing within it from hazardous step and touch potentials (See Fig. 3). Such a zone can be produced through the use of a metal mat connected to the grounded object. In some cases, a grounding grid can be used to equalize the voltage within the grid. Equipotential zones will not, however, protect employees who are either wholly or partially outside the protected area. Bonding conductive objects in the immediate work area can also be used to minimize the potential between the objects and between each object and ground. (Bonding an object outside the work area can increase the touch potential to that object in some cases, however.)

2. The use of insulating equipment, such as rubber gloves, can protect employees handling grounded equipment and conductors from hazardous touch potentials. The insulating equipment must be rated for the highest voltage that can be impressed on the grounded objects under fault conditions (rather than for the full system voltage).

3. Restricting employees from areas where hazardous step or touch potentials could arise can protect employees not directly involved in

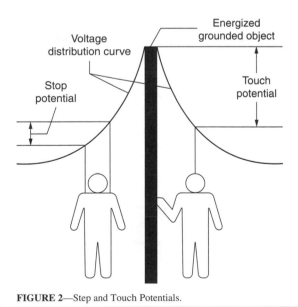

FIGURE 2—Step and Touch Potentials.

FIGURE 6.3   (*Continued*)

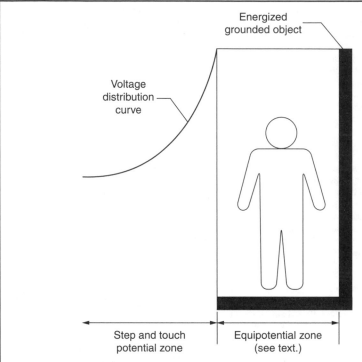

FIGURE 3—Protection from Ground-Potential Gradients.

the operation being performed. Employees on the ground in the vicinity of transmission structures should be kept at a distance where step voltages would be insufficient to cause injury. Employees should not handle grounded conductors or equipment likely to become energized to hazardous voltages unless the employees are within an equipotential zone or are protected by insulating equipment.

APPENDIX D TO §1910.269—METHODS OF INSPECTING AND TESTING WOOD POLES

I. "Introduction"

When work is to be performed on a wood pole, it is important to determine the condition of the pole before it is climbed. The weight of the employee, the weight of equipment being installed, and other working stresses (such as the removal or retensioning of conductors) can lead to the failure of a defective pole or one that is not designed to handle the additional

stresses.[†††] For these reasons, it is essential that an inspection and test of the condition of a wood pole be performed before it is climbed.

[†††]**Note:** A properly guyed pole in good condition should, at a minimum, be able to handle the weight of an employee climbing it. If the pole is found to be unsafe to climb or to work from, it must be secured so that it does not fall while an employee is on it. The pole can be secured by a line truck boom, by ropes or guys, or by lashing a new pole alongside it. If a new one is lashed alongside the defective pole, work should be performed from the new one.

II. "Inspection of Wood Poles"

Wood poles should be inspected by a qualified employee for the following conditions:[‡‡‡]

[‡‡‡]**Note:** The presence of any of these conditions is an indication that the pole may not be safe to climb or to work from. The employee performing the inspection must

FIGURE 6.3   (*Continued*)

be qualified to make a determination as to whether or not it is safe to perform the work without taking additional precautions.

A. General Condition. The pole should be inspected for buckling at the ground line and for an unusual angle with respect to the ground. Buckling and odd angles may indicate that the pole has rotted or is broken.

B. Cracks. The pole should be inspected for cracks. Horizontal cracks perpendicular to the grain of the wood may weaken the pole. Vertical ones, although not considered to be a sign of a defective pole, can pose a hazard to the climber, and the employee should keep his or her gaffs away from them while climbing.

C. Holes. Hollow spots and woodpecker holes can reduce the strength of a wood pole.

D. Shell Rot and Decay. Rotting and decay are cutout hazards and are possible indications of the age and internal condition of the pole.

E. Knots. One large knot or several smaller ones at the same height on the pole may be evidence of a weak point on the pole.

F. Depth of Setting. Evidence of the existence of a former ground line substantially above the existing ground level may be an indication that the pole is no longer buried to a sufficient extent.

G. Soil Conditions. Soft, wet, or loose soil may not support any changes of stress on the pole.

H. Burn Marks. Burning from transformer failures or conductor faults could damage the pole so that it cannot withstand mechanical stress changes.

III. "Testing of Wood Poles"

The following tests, which have been taken from 1910.268(n)(3), are recognized as acceptable methods of testing wood poles:

A. Hammer Test. Rap the pole sharply with a hammer weighing about 3 pounds, starting near the ground line and continuing upwards circumferentially around the pole to a height of approximately 6 feet. The hammer will produce a clear sound and rebound sharply when

striking sound wood. Decay pockets will be indicated by a dull sound or a less pronounced hammer rebound. Also, prod the pole as near the ground line as possible using a pole prod or a screwdriver with a blade at least 5 inches long. If substantial decay is encountered, the pole is considered unsafe.

B. Rocking Test. Apply a horizontal force to the pole and attempt to rock it back and forth in a direction perpendicular to the line. Caution must be exercised to avoid causing power lines to swing together. The force may be applied either by pushing with a pike pole or pulling with a rope. If the pole cracks during the test, it shall be considered unsafe.

**APPENDIX E TO §1910.269—
REFERENCE DOCUMENTS**

The references contained in this appendix provide information that can be helpful in understanding and complying with the requirements contained in 1910.269. The national consensus standards referenced in this appendix contain detailed specifications that employers may follow in complying with the more performance-oriented requirements of OSHA's final rule. Except as specifically noted in 1910.269, however, compliance with the national consensus standards is not a substitute for compliance with the provisions of the OSHA standard.

ANSI/SIA A92.2-1990, American National Standard for Vehicle-Mounted Elevating and Rotating Aerial Devices.

ANSI C2-1993, National Electrical Safety Code.

ANSI Z133.1-1988, American National Standard Safety Requirements for Pruning, Trimming, Repairing, Maintaining, and Removing Trees, and for Cutting Brush.

ANSI/ASME B20.1-1990, Safety Standard for Conveyors and Related Equipment.

ANSI/IEEE Std. 4 1978 (Fifth Printing), IEEE Standard Techniques for High-Voltage Testing.

ANSI/IEEE Std. 100-1988, IEEE Standard Dictionary of Electrical and Electronic Terms.

ANSI/IEEE Std. 516-1987, IEEE Guide for Maintenance Methods on Energized Power-Lines.

ANSI/IEEE Std. 935-1989, IEEE Guide on Terminology for Tools and Equipment to Be Used in Live Line Working.

FIGURE 6.3 *(Continued)*

ANSI/IEEE Std. 957-1987, IEEE Guide for Cleaning Insulators.

ANSI/IEEE Std. 978-1984 (R1991), IEEE Guide for In-Service Maintenance and Electrical Testing of Live-Line Tools.

ASTM D 120-87, Specification for Rubber Insulating Gloves.

ASTM D 149-92, Test Method for Dielectric Breakdown Voltage and Dielectric Strength of Solid Electrical Insulating Materials at Commercial Power Frequencies.

ASTM D 178-93, Specification for Rubber Insulating Matting.

ASTM D 1048-93, Specification for Rubber Insulating Blankets.

ASTM D 1049-93, Specification for Rubber Insulating Covers.

ASTM D 1050-90, Specification for Rubber Insulating Line Hose.

ASTM D 1051-87, Specification for Rubber Insulating Sleeves.

ASTM F 478-92, Specification for In-Service Care of Insulating Line Hose and Covers.

ASTM F 479-93, Specification for In-Service Care of Insulating Blankets.

ASTM F 496-93B, Specification for In-Service Care of Insulating Gloves and Sleeves.

ASTM F 711-89, Specification for Fiberglass-Reinforced Plastic (FRP) Rod and Tube Used in Live Line Tools.

ASTM F 712-88, Test Methods for Electrically Insulating Plastic Guard Equipment for Protection of Workers.

ASTM F 819-83a (1988), Definitions of Terms Relating to Electrical Protective Equipment for Workers.

ASTM F 855-90, Specifications for Temporary Grounding Systems to Be Used on De-Energized Electric Power Lines and Equipment.

ASTM F 887-91a, Specifications for Personal Climbing Equipment.

ASTM F 914-91, Test Method for Acoustic Emission for Insulated Aerial Personnel Devices.

ASTM F 968-93, Specification for Electrically Insulating Plastic Guard Equipment for Protection of Workers.

ASTM F 1116-88, Test Method for Determining Dielectric Strength of Overshoe Footwear.

ASTM F 1117-87, Specification for Dielectric Overshoe Footwear.

ASTM F 1236-89, Guide for Visual Inspection of Electrical Protective Rubber Products.

ASTM F 1505-94, Standard Specification for Insulated and Insulating Hand Tools.

ASTM F 1506-94, Standard Performance Specification for Textile Materials for Wearing Apparel for Use by Electrical Workers Exposed to Momentary Electric Arc and Related Thermal Hazards.

IEEE Std. 62-1978, IEEE Guide for Field Testing Power Apparatus Insulation.

IEEE Std. 524-1992, IEEE Guide to the Installation of Overhead Transmission Line Conductors.

IEEE Std. 1048-1990, IEEE Guide for Protective Grounding of Power Lines.

IEEE Std. 1067-1990, IEEE Guide for the In-Service Use, Care, Maintenance, and Testing of Conductive Clothing for Use on Voltages up to 765 kV AC.

FIGURE 6.3   (*Continued*)

(Indented statements are quoted from the preamble to this regulation and from the Directorate of Compliance.)

(a) Scope, application and purpose
(1) Scope
(i) This standard covers the servicing and maintenance of machines and equipment in which the unexpected energization or start up of the machines or equipment, or release of stored energy could cause injury to employees. This standard establishes minimum performance requirements for the control of such hazardous energy.
(ii) This standard does not cover the following:
(A) Construction, agriculture, and maritime employment;
(B) Installations under the exclusive control of electric utilities for the purpose of power generation, transmission, and distribution, including related equipment for communication or metering; and

The lockout/tagout requirements for utilities are contained in 29 CFR 1910.269, "Electric Power Generation, Transmission, and Distribution" Final Rule dated January 31, 1994.

(C) Exposure to electrical hazards from work on, near, or with conductors or equipment in electric utilization installations, which is covered by Subpart S of this part; and

The lockout/tagout requirements electrical conductors or equipment are contained in 29 CFR 1910.331-.335, "Electrical Safety-Related Work Practices" Final Rule dated August 6, 1990.

OSHA states that the lockout/tagout provisions of this standard are for the protection of general industry workers while performing servicing and maintenance functions and augment the safeguards specified in Subparts O, S, and other applicable portions of 29 CFR 1910.

(D) Oil and gas well drilling and servicing.

The lockout/tagout requirements oil and gas well drilling and servicing are contained in 29 CFR 1910.290, "Oil and Gas Well Drilling and Servicing."

(2) Application. (i) This standard applies to the control of energy during servicing and/or maintenance of machines and equipment.

(ii) Normal production operations are not covered by this standard (See Subpart O of this Part). Servicing and/or maintenance which takes place during normal production operations is covered by this standard only if:
The standards, in Subpart O, that apply here are 1910.212 (general machine guarding) and 1910.219 (guarding power transmission apparatus).
(A) An employee is required to remove or bypass a guard or other safety device; or
(B) An employee is required to place any part of his or her body into an area on a machine or piece of equipment where work is actually performed upon the material being processed (point of operation) or where an associated danger zone exists during a machine operating cycle.
Note: Exception to paragraph (a)(2)(ii): Minor tool changes and adjustments, and other minor servicing activities, which take place during normal production operations, are not covered by this standard if they are routine, repetitive, and integral to the use of the equipment for production, provided that the work is performed using alternative measures which provide effective protection (See Subpart O of this Part).
(iii) This standard does not apply to the following:
(A) Work on cord and plug connected electric equipment for which exposure to the hazards of unexpected energization or start up of the equipment is controlled by the unplugging of the equipment from the energy source and by the plug being under the exclusive control of the employee performing the servicing or maintenance.

This paragraph encompasses the many varieties of portable hand tools that are found in the workplace, as well as cord and plug connected equipment which is intended for use at a fixed location. If the plug is not near the employee doing the maintenance or servicing of the equipment then the plug must be locked out or tagged out to prevent another employee from inadvertently plugging it in. There are plug locking devices available.
OSHA states that "The plug is under the exclusive control of the employee if it is physically in the position of the employee, or in arm's reach and in the line of sight of the employee, or if the employee has affixed a lockout/tagout device on the plug."
OSHA further states that "The company lockout/tagout procedures required by the

FIGURE 6.4 Control of Hazardous Energy Source (Lockout/Tagout) (OSHA-CFR, Title 29, Part 1910, Paragraph 147).

standard at 29 CFR 1910 .147(c)(4)(i) shall specify the acceptable procedure for handling cord and plug connected equipment."

(B) Hot tap operations involving transmission and distribution systems for substances such as gas, steam, water, or petroleum products when they are performed on pressurized pipelines, provided that the employer demonstrates that *{1}* continuity of service is essential; *{2}* shutdown of the system is impractical; and *{3}* documented procedures are followed, and special equipment is used which will provide proven effective protection for employees.

(3) *Purpose.* **(i)** This section requires employers to establish a program and utilize procedures for affixing appropriate lockout devices or tagout devices to energy isolating devices, and to otherwise disable machines or equipment to prevent unexpected energization, start up or release of stored energy in order to prevent injury to employees.

(ii) When other standards in this part require the use of lockout/tagout, they shall be used and supplemented by the procedural and training requirements of this section.

"This part" means 1910. The requirements of this standard also apply to 1910.331-.335 and 1910.269.

(b) *Definitions applicable to this section.*

Affected employee. An employee whose job requires him/her to operate or use a machine or equipment on which servicing or maintenance is being performed under lockout/tagout, or whose job requires him/her to work in an area in which such servicing or maintenance is being performed.

An affected employee is one who does not perform the servicing or implement the energy control procedure, but whose responsibilities are performed in an area in which the energy control procedure and servicing operations are performed under that procedure. The affected employee does not need to know how to perform lockout/tagout, nor does that employee need to be trained in the detailed implementation of the energy control procedure. Rather, the affected employee need only be able to recognize when the energy control procedure is being implemented, to identify the locks or tags being used, and to understand the purpose of the procedures and the importance of not attempting to start up or use the equipment which has been locked out or tagged out.

The definition of affected employee also recognizes that an affected person and an authorized person may be one and the same person when a machine operator or user must also perform servicing or maintenance on the machine or equipment. In this case, the employee must have the requisite knowledge of an authorized employee.

Authorized employee. A person who locks out or tags out machines or equipment in order to perform servicing or maintenance on that machine or equipment. An affected employee becomes an authorized employee when that employee's duties include performing servicing or maintenance covered under this section.

OSHA agrees that as long as an employee is involved in performing an element of servicing and maintenance which is covered by the energy control procedure, that employee should be considered and authorized employee for the purpose of this standard. This is particularly important in the context of the requirement in paragraph (d)(3) "Machine or equipment isolation" as well as (f)(3)(ii)(D) of this standard, which requires the authorized person to affix a personal lockout or tagout device on the energy isolating device as part of the energy control procedure. This definition assures that when a servicing task is performed by a team or group of employees, each employee who is directly exposed to the hazards of the servicing operation will have the responsibility to affix his/her personal lockout/tagout device before beginning the work and to remove it when he/she completes the work.

Paragraph (c)(5)(ii)(D) of this standard provides additional accountability by requiring such lockout and tagout devices to identify the authorized person responsible for applying them.

"Capable of being locked out." An energy isolating device is capable of being locked out if it has a hasp or other means of attachment to which, or through which, a lock can be affixed, or it has a locking mechanism built into it. Other energy isolating devices are capable of being locked out, if lockout can be achieved without the need to dismantle, rebuild, or replace the energy isolating device or permanently alter its energy control capability.

The capability for lockout does not necessarily mean that the equipment has an actual hasp or other physical attachment point for a

FIGURE 6.4 *(Continued)*

lock. For example, the use of chains can be an effective means of facilitating lockout of many types of valves, even if the valve does not have a specific locking point. Many examples of equipment which was lockable with minor modifications have been provided to the record. For equipment of this type, OSHA believes that the lockout capability should be used in order to maximize the protection afforded by this standard.

Energized. Connected to an energy source or containing residual or stored energy.

Energy isolating device. A mechanical device that physically prevents the transmission or release of energy, including but not limited to the following: A manually operated electrical circuit breaker, a disconnect switch, a manually operated switch by which the conductors of a circuit can be disconnected from all ungrounded supply conductors and, in addition, no pole can be operated independently; a line valve; a block; and any similar device used to block or isolate energy. Push buttons, selector switches, and other control circuit type devices are not energy isolating devices.

Energy source. Any source of electrical, mechanical, hydraulic, pneumatic, chemical, thermal, or other energy.

The identification of energy sources, as defined here, is complicated by three very important considerations: (1) Energy is always present in machinery, equipment, or processes; (2) energy is not necessarily dangerous; and (3) danger is only present when energy is released in quantities or at rates that would harm an employee. Generally speaking, however, potentially hazardous energy sources are defined as those that can cause injury to employees working in, on, or around machines or equipment.

Energy as used in this standard means mechanical motion; potential energy due to pressure, gravity, or springs; electrical energy; or thermal energy resulting from high or low temperatures. Some energy sources can be turned on and off, some can be dissipated, some can be eliminated, and some can only be controlled.

The following brief analysis of energy sources may provide the reader with a better understanding of the provisions of this standard.

1. Mechanical motion can be linear translation or rotation, or it can produce work which, in turn, produces changes in temperature. This type of energy can be turned off or left on.

2. Potential energy can be due to pressure (above or below atmospheric) as in hydraulic, pneumatic, or vacuum systems. Potential energy manifested as pressure or in springs can be dissipated or controlled; it cannot be turned off or on.

3. Electrical energy refers to generated electrical power or static electricity. In the case of generated electricity, the electrical power can be turned on or off. Static electricity cannot be turned off; it can only be dissipated or controlled.

4. Thermal energy is manifested by high or low temperatures. This type of energy is the result of mechanical work, radiation, chemical reaction, or electrical resistance. It cannot be turned off or eliminated; however, it can be dissipated or controlled.

Hot tap. A procedure used in the repair maintenance and services activities which involves welding on a piece of equipment (pipelines, vessels, or tanks) under pressure, in order to install connections or appurtenances. It is commonly used to replace or add sections of pipeline without the interruption of service for air, gas, water, steam, and petrochemical distribution systems.

Lockout. The placement of a lockout device on an energy isolating device, in accordance with an established procedure, ensuring that the energy isolating device and the equipment being controlled cannot be operated until the lockout device is removed.

Lockout device. A device that utilizes a positive means such as a lock, either key or combination type, to hold an energy isolating device in the safe position and prevent the energizing of a machine or equipment.

Normal production operations. The utilization of a machine or equipment to perform its intended production function.

Servicing and/or maintenance. Workplace activities such as constructing, installing, setting up, adjusting, inspecting, modifying, and maintaining and/or servicing machines or equipment. These activities include lubrication, cleaning, or unjamming of machines or equipment, and making adjustments or tool changes, where the employee may be exposed to the unexpected energization or startup of the equipment or release of hazardous energy.

Setting up. Any work performed to prepare a machine or equipment to perform its normal production operation.

FIGURE 6.4 *(Continued)*

Tagout. The placement of a tagout device on an energy isolating device, in accordance with an established procedure, to indicate that the energy isolating device and the equipment being controlled may not be operated until the tagout device is removed.

Tagout device. A prominent warning device, such as a tag and a means of attachment, which can be securely fastened to an energy isolating device in accordance with an established procedure, to indicate that the energy isolating device and the equipment being controlled may not be operated until the tagout device is removed.

(c) *General* **(1)** *Energy control program.* The employer shall establish a program consisting of energy control procedures, employee training, and to periodic inspections to ensure that before any employee performs any servicing or maintenance on a machine or equipment where the unexpected energizing, startup or release of stored energy could occur and cause injury, the machine or equipment shall be isolated from the energy source and rendered inoperative.

The OSHA compliance officer is directed to ask the employer for the documentation including:

1. procedures for the control of hazardous energy including:
   a. shutdown,
   b. equipment isolation,
   c. lockout/tagout application,
   d. release of stored energy, and
   e. verification of isolation;
2. certification of periodic inspections; and
3. certification of training.

The documented procedure must identify the specific types of energy to be controlled and, in instances where a common procedure is to be used, the specific equipment covered by the common procedure must be identified at least by type and location. The identification of the energy to be controlled may be by magnitude and type of energy. See the exception to 147(c)(4)(i), "Note". The employer need not document the required procedure for a particular machine or equipment when all eight (8) elements listed in the "Note" exist.

(2) *Lockout/tagout.* **(i)** If an energy isolating device is not capable of being locked out, the employer's energy control program under paragraph (c)(1) of this section shall utilize a tagout system.

(ii) If an energy isolating device is capable of being locked out, the employer's energy control program under paragraph (c)(1) of this section shall utilize lockout, unless the employer can demonstrate that the utilization of a tagout system will provide full employee protection as set forth in paragraph (c)(3) of this section.

(iii) After January 2, 1990, whenever replacement or major repair, renovation or modification of a machine or equipment is performed, and whenever new machines or equipment are installed, energy isolating devices for such machine or equipment shall be designed to accept a lockout device.

(3) *Full employee protection.* **(i)** When a tagout device is used on an energy isolating device which is capable of being locked out, the tagout device shall be attached at the same location that the lockout device would have been attached, and the employer shall demonstrate that the tagout program will provide a level of safety equivalent to that obtained by using a lockout program.

(ii) In demonstrating that a level of safety is achieved in the tagout program which is equivalent to the level of safety obtained by using a lockout program, the employer shall demonstrate full compliance with all tagout-related provisions of this standard together with such additional elements as are necessary to provide the equivalent safety available from the use of a lockout device.

Additional means to be considered as part of the demonstration of full employee protection shall include the implementation of additional safety measures such as the removal of an isolating circuit element, blocking of a controlling switch, opening of an extra disconnecting device, or the removal of a valve handle to reduce the likelihood of inadvertent energization.

(4) *Energy control procedure.* **(i)** Procedures shall be developed, documented, and utilized for the control of potentially hazardous energy when employees are engaged in the activities covered by this section.

Note: *Exception:* The employer need not document the required procedure for a particular machine or equipment, when all of the following elements exist: **(1)** The machine or equipment has no potential for stored or residual energy or reaccumulation of stored energy after shut down which could endanger employees: **(2)** the machine or equipment has a single energy source which can be readily identified and isolated: **(3)** the isolation and locking out of that energy source will completely de-energize and deactivate

FIGURE 6.4 *(Continued)*

the machine or equipment: **(4)** the machine or equipment is isolated from that energy source and locked out during servicing or maintenance: **(5)** a single lockout device will achieve a locked-out condition: **(6)** the lockout device is under the exclusive control of the authorized employee performing the servicing or maintenance: **(7)** the servicing or maintenance does not create hazards for other employees; and **(8)** the employer, in utilizing this exception, has had no accidents involving the unexpected activation or reenergization of the machine or equipment during servicing or maintenance.

The exception is intended to apply to situations in which the procedure for deenergization, servicing, and reenergization can be carried out without detailed interactions of energy sources, machines, and employees. For example, a motor in a small machine shop is wired into a single electrical disconnect, with no other source, and the motor does not present the hazards of stored or residual energy. When the motor needs repair, the authorized employee can isolate the motor from the single energy source and lock it out, using his/her personal lockout device on the disconnect, in accordance with the procedures set forth in the standard.

Under these conditions, and provided that no other employees are exposed to hazards from the servicing operation, the servicing may be performed without the need to document the energy control procedure.

(ii) The procedures shall clearly and specifically outline the scope, purpose, authorization, rules, and techniques to be utilized for the control of hazardous energy, and the means to enforce compliance including, but not limited to, the following:

(A) A specific statement of the intended use of the procedure;

(B) Specific procedural steps for shutting down, isolating, blocking, and securing machines or equipment to control hazardous energy;

(C) Specific procedural steps for the placement, removal, and transfer of lockout/tagout devices and the responsibility for them; and

(D) Specific requirements for testing a machine or equipment to determine and verify the effectiveness of lockout devices, tagout devices, and other energy control measures.

OSHA uses the word "specific" to emphasize the need to have a detailed procedure, one which clearly and specifically outlines the steps to be followed. Overgeneralization can result in a document which has little or no meaning to the employee who must follow the procedure. However, whereas the procedure is required to be written in detail, this does not mean that a separate procedure must be written for each and every machine or piece of equipment. Similar machines and/or equipment (those using the same type and magnitude energy) which have the same or similar types of controls can be covered with a single procedure.

The written energy control procedure required by this standard need not be overly complicated or detailed, depending on the complexity of the equipment and the control measures to be utilized.

For example, if there is a single machine with a single energy source that must be isolated, and the control means is simple, such as opening an electrical disconnect and locking out that energy source during servicing, the written procedure could be very simple. The steps set forth in the standard can be incorporated into the procedure with very little detail, reflecting the lack of complexity of the control measure. In addition, the employer's procedures may not need to be unique for a single machine or task, but can apply to a group of similar machines, types of energy and tasks if a single procedure can address the hazards and the steps to be taken satisfactorily.

OSHA believes that because of the need to follow the steps in the energy control procedure carefully and specifically, and the number of variables involved in controlling hazardous energy, a documented procedure is necessary for most energy control situations. However, OSHA has determined that in certain limited situations, documentation of the procedure will not add markedly to the projections otherwise provided by the standard.

These situations incorporate several common elements:

1. There is a single source of hazardous energy which can be easily identified and isolated, and there is no potential for stored or residual energy in the equipment. This greatly simplifies the procedure for controlling the energy, since the single energy source is all that needs to be isolated.

2. The isolation and locking out of that single energy source will totally deenergize

FIGURE 6.4 (*Continued*)

and deactivate the machine or equipment. There are no collateral sources of energy which need to be addressed.

3. A full lockout of the energy source is achieved by a single lockout device which is under the exclusive control of the authorized employee performing the servicing or maintenance. As used in this provision, exclusive control means that the authorized employee is the only person who can affix or remove the device. The authorized employee follows all steps necessary for deenergizing the equipment, verifying the deenergization, performing the work, and reenergizing the equipment upon completion of servicing. Because the energy control elements are simple, with a single energy source being locked out and no other potential sources of unexpected activation or energization, the authorized employee can perform them without referring to a written document

4. While the equipment is locked out, the servicing or maintenance cannot expose other employees to hazards. For example, shutdown and lockout of a conveyer cannot cause jams or other hazards at other conveyers which feed into the conveyer being serviced.

(5) *Protective materials and hardware.* **(i)** Locks, tags, chains, wedges, key blocks, adapter pins, self-locking fasteners, or other hardware shall be provided by the employer for isolating, securing, or blocking of machines or equipment from energy sources.

To meet this requirement, the employer can either issue devices to each employee responsible for implementing energy control measures, or can exercise the option of simply having a sufficient quantity of the devices on hand at any given time and assign or distribute them to the employees as the need arises. All authorized employees will need to have these devices available to attach to energy isolating devices whenever they perform servicing or maintenance using the energy control procedure.

(ii) Lockout/tagout devices shall be singularly identified; shall be the only device(s) used for controlling energy; shall not be used for other purposes; and shall meet the following requirements:

This restriction was imposed to ensure that the sight of a distinctive lock or tag will provide a constant message of the use that the device is being put to and the restrictions which this device is intended to covey. If lockout or tagout devices are used for other purposes, they can lose their significance in the workplace.

For the energy control procedure to be effective, these devices must have a single meaning to employees: "Do not energize the equipment when such a device is affixed to it."

(A) *Durable.* *{1}* Lockout and tagout devices shall be capable of withstanding the environment to which they are exposed for the maximum period of time that exposure is expected.

{2} Tagout devices shall be constructed and printed so that exposure to weather conditions or wet and damp locations will not cause the tag to deteriorate or the message on the tag to become illegible.

{3} Tags shall not deteriorate when used in corrosive environments such as areas where acid and alkali chemicals are handled and stored.

OSHA requires that tagout devices must be constructed and printed so that exposure to weather or other environmental conditions which exist in the workplace will not cause the tag to become unserviceable and/or the message on the tag to become illegible. For any sign, tag, or other message bearing item, the message must remain legible for the employees to be able to ascertain the meaning and intent of the message. Devices which are not exposed to harsh environments need not be capable of withstanding such exposure.

(B) *Standardized.* Lockout and tagout devices shall be standardized within the facility in at least one of the following criteria: Color; shape; or size; and additionally, in the case of tagout devices, print and format shall be standardized.

OSHA requires that lockout or tagout devices be standardized in order to be readily identifiable and distinguished from other similar devices found in the workplace. In addition, there is a requirement for the use of a standardized print and format for tagout devices. This is done to ensure that the tagout devices, which rely exclusively on employee recognition for their effectiveness, will be so unique as to minimize the chances of their being misidentified or their message misinterpreted.

FIGURE 6.4 *(Continued)*

(C) *Substantial. {1} Lockout devices.* Lockout devices shall be substantial enough to prevent removal without the use of excessive force or unusual techniques, such as with the use of bolt cutters or other metal cutting tools.

{2} Tagout devices. Tagout devices, including their means of attachment, shall be substantial enough to prevent inadvertent or accidental removal. Tagout device attachment means shall be of a non-reusable type, attachable by hand, self-locking, and non-releasable with a minimum unlocking strength of no less than 50 pounds and having the general design and basic characteristics of being at least equivalent to a one-piece all environment-tolerant nylon cable tie.

OSHA requires that "Tagout devices having reusable, non-locking, easily detachable means of attachment (such as string, cord, or adhesive) are not permitted."

The "tagout device" addressed here is when a tagout device is used instead of a lockout device. If a tag is used in conjunction with a lock then these restrictions do not apply.

OSHA requires that lockout and tagout devices be substantial enough to minimize the possibility of premature removal.

The additional requirements for a tagout device are being imposed to ensure that tags do not become disconnected or lost during use, thereby negating their effectiveness.

(D) *Identifiable.* Lockout and tagout devices shall indicate the identity of the employee applying the device(s).

Identification of the user provides an additional degree of accountability to the overall program. It enables the employer to inspect the application of the energy control procedure and determine which employees are properly implementing its requirements. If locks or tags are not being properly attached by an employee, Identification on the locks and tags will enable the employer to locate that employee and correct the problem promptly, including additional training, as necessary. For other employees, this requirement will enable them to determine at a glance which authorized employees are performing a given servicing operation.

It puts them on notice that if questions arise about the servicing or the energy control procedure, the persons listed on the lockout and tagout devices are the appropriate persons to ask. The authorized employee has the additional assurance that other employees know of his/her involvement in the servicing, and that only he/she is allowed to remove the device.

OSHA believes that knowing who applied a lockout device to a machine or equipment can save time and lives. If an employee, upon completing a job, forgets to remove a lockout device, the identity of the employee can be immediately determined and the employee made available to complete the procedure. If that employee cannot be located, it is possible that he/she is still working on the equipment. It would then be possible to check out the area and assure that the employee and others are out of the danger area before the device is removed. Marking a lockout or tagout device is a simple way of identifying the person who applies it, and can prevent the inadvertent reenergization or reactivation of equipment before that employee has been located and has moved clear of the equipment. Thus, marking the identity of the employee who uses a lockout or tagout device is an appropriate safeguard.

Marking of the lockout or tagout devices can also promote a sense of security in employees, in that each device is the individual employee's device, used only for his or her protection. This sense of identity also can be used to encourage willing utilization of the energy control procedure. When an employee can identify with a part of the program he/she controls for his/her own protection, that employee will likely be an active participant in making the program work.

(iii) Tagout devices shall warn against hazardous conditions if the machine or equipment is energized and shall include a legend such as the following: Do Not Start. Do Not Open. Do Not Close. Do Not Energize. Do Not Operate.

OSHA states that the legend (major message) on the tagout devices must warn against hazardous conditions if the equipment is reenergized.

OSHA recognizes, however, that these messages may not be sufficient to cover all conditions involving hazardous energy control. For that reason, the above stated legends are only examples of what must be stated. The use of graphics, pictographs, or other symbols to convey the message which the tag represents serves the same purpose as the written message and therefore would be acceptable to OSHA.

FIGURE 6.4 *(Continued)*

The significance of this message is imparted through the training of employees and enforcement of the program. The backbone of a tagout system is that when a tagout device is placed on an energy isolating device, it informs employees that the energy isolating device is not to be turned on or otherwise moved to a position which will allow the flow of energy.

The printed message on the tag provides information about what the tag stands for and what it prohibits, and indicates the name of the employee who affixed it to the energy isolating device.

(6) *Periodic inspection.* **(i)** The employer shall conduct a periodic inspection of the energy control procedure at least annually to ensure that the procedure and the requirements of this standard are being followed.

(A) The periodic inspection shall be performed by an authorized employee other than the one(s) utilizing the energy control procedure being inspected.

(B) The periodic inspection shall be conducted to correct any deviations or inadequacies identified.

(C) Where lockout is used for energy control, the periodic inspection shall include a review, between the inspector and each authorized employee, of that employee's responsibilities under the energy control procedure being inspected.

(D) Where tagout is used for energy control, the periodic inspection shall include a review, between the inspector and each authorized and affected employee, of that employee's responsibilities under the energy control procedure being inspected, and the elements set forth in paragraph (c)(7)(ii) of this section.

(ii) The employer shall certify that the periodic inspections have been performed. The certification shall identify the machine or equipment on which the energy control procedure was being utilized, the date of the inspection, the employees included in the inspection, and the person performing the inspection.

The periodic inspection is intended to assure that the energy control procedures continue to be implemented properly, and that the employees involved are familiar with their responsibilities under those procedures.

The inspector, who is required to be an authorized person not involved in the energy control procedure being inspected, must be able to determine three things:

1. Whether the steps in the energy control procedure are being followed;
2. Whether the employees involved know their responsibilities under the procedure; and
3. Whether the procedure is adequate to provide the necessary protection, and what changes, if any, are needed.

The inspector will need to observe and talk with the employees in order to make these determinations. This regulation provides some additional guidance as to the inspector's duties in performing periodic inspections, to assure that he or she obtains the necessary information about the energy control procedure and its effectiveness.

Whenever lockout is used. The inspector must review each authorized employee's responsibilities under the procedure with that employee. This does not necessarily require separate one-on-one meetings, but can involve the inspector meeting with the whole servicing crew at one time. Indeed, group meetings can be the most effective way of dealing with this situation, because it reinforces the employee's and that they need to follow the procedure carefully.

Where tagout is used. The inspector's review of responsibilities extends to affected employees as well, because of the increased importance of their role in avoiding accidental or inadvertent activation of the equipment or machinery being serviced.

OSHA believes that periodic inspections by the employer are necessary to ensure continued compliance with the procedure.

The next section addresses the training requirements for the control of hazardous energy.

OSHA states "The standard recognizes three types of employees: (1) "authorized", (2) "affected' and (3) "other". Different levels of training are required based upon the respective roles of employees in the control of energy and the knowledge which they must posses to accomplish their tasks safely and to ensure the safety of fellow workers as related to the lockout/tagout procedures."

(7) *Training and communication.*

OSHA specifies that the employer provide effective initial training, periodic retraining, and certification of such training of employees. OSHA considers these requirements to be of critical importance in helping

FIGURE 6.4   *(Continued)*

to ensure that the applicable provisions of the hazardous energy control procedure(s) are known, understood, and strictly adhered to by employees.

In order to provide adequate information, any training program under this standard will need to cover at least three areas; (1) The employer's energy control program; (2) the elements of the energy control procedure which is relevant to the employee's duties; and (3) the requirements of this regulation.

(i) The employer shall provide training to ensure that the purpose and function of the energy control program are understood by employees and that the knowledge and skills required for the safe application, usage, and removal of the energy controls are required by employees. The training shall include the following:

OSHA has determined that the marking or labeling of energy isolating devices is not reasonably necessary for the effectiveness of the energy control program. When employees need to know details on energy sources for protection under the standard, the energy control procedure is required to spell out this information, and the training must incorporate it, as well. For example, authorized employees, in order to perform their servicing or maintenance duties under the energy control procedure, are required to know the type and magnitude of the energy sources which must be controlled. The marking or labeling of the sources themselves will not provide the authorized employees with any additional information. Second, as far as affected or other employees are concerned, their role in the energy control program is essentially to understand what the program is designed to accomplish, and to recognize that when they see an energy isolating device with a tag and/or lock on it, they are not to touch the equipment, regardless of what the type and magnitude of the energy might be. OSHA believes that marking the equipment with this information would not enhance the protection of these employees, because their compliance with the energy control procedure does not depend upon knowledge of these details.

(A) Each authorized employee shall receive training in the recognition of applicable hazardous energy sources, the type and magnitude of the energy available in the workplace, and the methods and means necessary for energy isolation and control.

Because authorized employees are charged with the responsibility for implementing energy control procedures, it is important that they receive training in recognizing and understanding all potentially hazardous energy sources that they might be exposed to during their work assignments, but that they also be trained in the use of adequate methods and means for the control of such energy sources. These employees are the ones authorized to implement the energy control procedure and to perform servicing of the machine or equipment. Therefore, they need extensive training in aspects of the procedure and its proper utilization, together with all relevant information about the equipment being serviced.

OSHA believes that employee knowledge of this information is essential to ensure that the correct energy control devices are used on the proper energy isolating devices and in the proper manner. This provision requires the employee to have that specific information prior to deenergizing the equipment, in order to control the energy and render the machine or equipment safe to work on.

(B) Each affected employee shall be instructed in the purpose and use of the energy control procedure.

Affected employees are important to the overall protection provided in the energy control program, however, because such employees work in areas where the program is being utilized by authorized employees, it is vital to the safety of the authorized employees that the affected employees recognize lockout/tagout devices immediately that they know not to disturb the lockout/tagout devices or the equipment to which the devices are affixed.

The instruction needs to be sufficient to enable the employees to determine if a control measure is in use. The instructions also need to make affected employees aware that disregarding or violating the prohibitions imposed by the energy control program could endanger their own lives or the lives of others.

(C) All other employees whose work operations are or may be in an area where energy control procedures may be utilized, shall be instructed about the procedure, and about the

FIGURE 6.4   *(Continued)*

prohibition relating to attempts to restart or reenergize machines or equipment which are locked out or tagged out.

OSHA requires that all other employees shall be instructed about the restrictions imposed upon all employees by the energy control program. This instruction as the employer's lockout/tagout procedure can be conveyed during new employee orientation sessions, by the use of employee handbooks, or through regularly scheduled safety meetings. The training of employees other than authorized and affected employees is considered by OSHA to be essential since other employees working in the plant or facility have been known to have turned on the power to a machine or equipment on which another employee is performing a servicing or maintenance activity. Inadvertent and intentional activation of machines or equipment by employees other than those working on the machine or equipment is not limited to affected employees. The training requirements for those other employees are minimal, essentially required only that these employees know what the energy control program does and that they are not to touch any locks, tags, or equipment covered by this program.

(ii) When tagout systems are used, employees shall also be trained in the following limitations of tags:

(A) Tags are essentially warning devices affixed to energy isolating devices, and do not provide the physical restraint on those devices that is provided by a lock.

(B) When a tag is attached to an energy isolating means, it is not to be removed without authorization of the authorized person responsible for it, and it is never to be bypassed, ignored, or otherwise defeated.

(C) Tags must be legible and understandable by all authorized employees, affected employees, and all other employees whose work operations are or may be in the area, in order to be effective.

(D) Tags and their means of attachment must be made of materials which will withstand the environmental conditions encountered in the workplace.

(E) Tags may evoke a false sense of security, and their meaning needs to be understood as a part of the overall energy control program.

(F) Tags must be securely attached to energy isolating devices so that they cannot be inadvertently or accidentally detached during use.

(iii) Employee retraining.

OSHA recommends that periodic retraining be provided for authorized employees be conducted at least annually. This retraining may need to be conducted more frequently, that is, whenever an inspection under paragraph (c)(6) reveals or whenever the employer has reason to believe that there are deviations from or inadequacies in the energy control procedure.

OSHA believes that the effectiveness of training diminishes as the time from the last training session increases. OSHA has determined that the provisions of this standard simply rely upon the finding of a problem with the program to trigger the retraining program. Retraining is intended to provide for continued proficiency, and not merely to remedy situations in which such proficiency has been found wanting.

In addition to the periodic retraining, additional retraining is to be conducted whenever a problem is identified during periodic inspections, or whenever the employer has reason to believe that there are problems with the energy control procedure itself or with its implementation.

This retraining should be more concentrated or more encompassing than the routine retraining, based upon the severity of the problem encountered with the use of the energy control program in the workplace.

OSHA is of the opinion that full and uniform utilization of an energy control procedure is necessary in order for that procedure to maintain effectiveness.

Every effort should be made during the periodic inspection to determine whether or not the procedure is being used properly. If deviations are observed, retraining would be required. However, retraining could be triggered by events separate from the findings of the periodic inspection. For example, an employee working with an energy control procedure might be injured in the course of his duties, or there might be a "near miss" where no one is actually injured, but where the energy control program has failed nonetheless. If a subsequent investigation indicates that an employee failed to operate within the guidelines of the control procedure, retraining would be required.

In addition, the investigation might also reveal that the procedure itself was not adequate. Such inadequacies in the procedure could be the result of using a general

FIGURE 6.4 (*Continued*)

procedure that does not handle effectively a specific application, or they may arise because changes have been made to the equipment or process that did not take the existing energy control procedure into consideration. In such cases when changes to the energy control procedure must be made, the employer is required to retrain employees in the new or revised procedure.

(A) Retraining shall be provided for all authorized and affected employees whenever there is a change in their job assignments, a change in machines, equipment or processes that present a new hazard, or when there is a change in the energy control procedures.

(B) Additional retraining shall also be conducted whenever a periodic inspection under paragraph (c)(6) of this section reveals, or whenever the employer has reason to believe that there are deviations from or inadequacies in the employee's knowledge or use of the energy control procedures.

(C) The retraining shall reestablish employee proficiency and introduce new or revised control methods and procedures, as necessary.

In this Final Rule, when lockout is implemented, OSHA is limiting the annual retraining recommendations to "Authorized" employees. These are the employees who must implement the energy control procedure, and their protection is the primary consideration under this standard. Because their safety requires them to follow the steps of the procedure precisely, these employees must be properly trained, and that training must be reinforced to assure their continued proficiency. By contrast, affected employees are not provided with annual retraining under this standard when lockout is used. In these situations, affected employees are initially trained about the energy control procedure and its implementation, and the relevance of that procedure to his/her work. Under lockout conditions, the essential element of the affected employee's training is a simple one: Locks are not to be defeated or bypassed, and locked out equipment must remain deenergized.

This message is reinforced whenever the affected employees work in an area where energy control procedures are being implemented, because this standard requires that such employees be notified before the energy control devices are applied. Further, when a lockout device is attached to a piece of equipment by an authorized employee, an affected employee should not be able to remove the lock, and thus will not have the potential of placing the authorized employee in danger.

OSHA notes that annual retraining should be provided for authorized and affected employees in the use of the tagout system. This additional training is necessary because of the inherent difficulties of tagout systems as opposed to lockout.

The use of tags relies uniquely upon the knowledge and training of the employees involved, and the continued reinforcement of the meaning of the tags. In a lockout system, even if an affected employee has not been adequately trained, the lock will prevent that employee from reenergizing the equipment. Tags, on the other hand, can be inadvertently or intentionally bypassed or ignored by an affected employee, because the tags do not actually prevent the activation of the tagged out equipment. Employees operating under a tagging system must be constantly vigilant, and their awareness of the importance of the tagout device must be frequently reinforced. OSHA believes that when tagout is used, retraining must be provided on at least an annual basis, in order to maximize its effectiveness.

(iv) The employer shall certify that employee training has been accomplished and is being kept up-to-date. The certification shall contain each employee's name and dates of training.

Certifications are intended to cover both the initial training and the periodic retraining. In addition to certifications, the employer must be able to demonstrate that the training includes all elements of the energy control procedure which is directly relevant to the duties of the employee. The adequacy of the training can be evaluated by the employer, employee, and OSHA alike by comparing the elements of the training to the elements of the procedure, which is required to be in written form.

It should be noted that the certification is not intended as a means of evaluating the completeness or efficacy of the training; it only provides an indication that training has been performed. The quality and content of the training are not evaluated through the certification of performance. In evaluating whether an employee has been adequately trained, OSHA will examine the employee's responsibilities

FIGURE 6.4   (*Continued*)

under the energy control program in relation to the elements of the standard.

(8) *Energy isolation.* Lockout/tagout shall be performed only by the authorized employees who are performing the servicing or maintenance.

These are the only employees who are required to be trained to know in detail about the types of energy available in the workplace and how to control the hazards of that energy. Only properly trained and qualified employees can be relied on to deenergize and to properly lockout/tagout machines or equipment which are being serviced or maintained, in order to ensure that the work will be accomplished safely.

(9) *Notification of employees.* Affected employees shall be notified by the employer or authorized employee of the application and removal of lockout devices or tagout devices. Notification shall be given before the controls are applied, and after they are removed from the machine or equipment.

OSHA requires that whenever lockout/tagout control might directly affect another employee's work activities, the employer or authorized employee must notify the affected employee before taking any action to apply or to remove lockout/tagout devices.

OSHA believes that this requirement is an essential component of the total energy control program. Notification of affected employees when lockout/tagout is going to be applied provides the perfect opportunity for the employer or authorized employee who notifies them of the impending interruption of the normal production operation to remind them and reinforce the importance of the restrictions imposed upon them by the energy control program.

OSHA believes that these measures are important to ensure that employees who operate or use machines or equipment do not unknowingly attempt to reenergize those machines or equipment that have been taken out of service and deenergized for the performance of activities covered by this standard. The lack of information regarding the status of the equipment could endanger both the servicing employees and the employees attempting to reenergize or operate the equipment. Such notification is also needed after servicing is completed to assure that employees know when the control measures have

been removed. Without such information, employees might mistakenly believe that a system is still deenergized and that it is safe to continue working on or around it.

The main thrust of the standard is to mandate the development, documentation and implementation of control procedures, and this is to be accomplished as outlined in paragraph (d) of this standard. The employer is given considerable flexibility in developing a control program, and such a program will be evaluated by OSHA compliance officers to determine whether it meets all the criteria in this standard.

(d) *Application of control.* The established procedures for the application of energy control (the lockout/tagout procedures) shall cover the following elements and actions, and shall be done in the following sequence:

This section provides that six separate and distinct steps be followed in meeting the procedural requirements of paragraph (c)(1) and the application of energy control (lockout or tagout) measures, and that the actions be taken in the sequence presented.

(1) *Preparation for shutdown.* Before an authorized or affected employee turns off a machine or equipment, the authorized employee shall have knowledge of the type and magnitude of the energy, the hazards of the energy to be controlled, and the method or means to control the energy.

This paragraph requires that in preparation for shutdown of machinery or equipment, the authorized employee must know about the type and magnitude of the energy, the hazards involved and the means of controlling them.

(This would require the authorized person to be qualified on the equipment or system.)

(2) *Machine or equipment shutdown.* The machine or equipment shall be turned off or shut down using the procedures established for the machine or equipment. An orderly shutdown must be utilized to avoid any additional or increased hazard(s) to employees as a result of the equipment stoppage.

This paragraph requires that the machine be turned off or shut down by an authorized employee according to established procedures. This is the starting point for all subsequent actions necessary to put the machine

FIGURE 6.4   *(Continued)*

or equipment in a state that will permit employees to work on it safely.

In many operations, activation of an electrical push-button control or the movement of a simple throw switch (electrical, hydraulic, or pneumatic) to the "stop" or "off" mode is sufficient to meet this provision. In other cases, however, such as those found typically in a refining or chemical process, there are control devices that do not necessarily address an "off-on" or "start-stop" condition (i.e., level controls, pressure controllers, etc.). In these instances, a series of predetermined steps may be necessary to achieve a shutdown of the machine or equipment.

OSHA is aware that although an authorized employee would usually have the necessary knowledge and capability to shut down machines or equipment, a machine or equipment operator or user should also be in a position and know how to shut down the machine or equipment he or she is utilizing.

In many cases, allowing a machine or equipment operator or user to shut it down when something goes wrong may save time and money, and may possibly avoid an accident. In many cases, the affected employee may be infinitely more familiar with the shutdown procedure for the machine or equipment, and would be able to accomplish the shutdown more rapidly and safely than an authorized person who does not work with that particular machine or equipment every day.

In the event that a machine or equipment malfunctions, the wise and prudent thing to do in most cases is to require the machine or equipment be immediately shut down. Shutting down a machine or equipment is analogous to stopping the production operation. OSHA believes that stopping the machine's production function is the necessary and appropriate first step in the procedure. OSHA also believes that the necessary first step is to interrupt the production process to allow non-servicing (affected) personnel to move clear of the machine or equipment. Once this is done and employees are not exposed to a hazard, the machine or equipment can be restarted by the authorized employee under the guidelines of paragraph (f)(1) when necessary to allow positioning of the machine or equipment, or components thereof.

(3) *Machine or equipment isolation.* All energy isolating devices that are needed to control the energy to the machine or equipment shall be physically located and operated in such a manner as to isolate the machine or equipment from the energy source(s).

Following the shutdown of the machine or equipment, paragraph (d)(3), as the next step in the procedure, provides that energy isolation devices be physically located and operated in such a manner as to isolate the machine or equipment from the energy source(s). For example, once an electrical push-button control has been utilized to stop the movement of machine or equipment parts as the first step of the shutdown procedure, isolation can then be accomplished by assuring that the push-button circuitry cannot be supplied with additional electrical energy. For such equipment, the isolation requirement can be accomplished by the employee's actions in tracing the path from the control toward the energy source until he/she locates the energy isolating device, and moving the energy isolating device control lever to the "safe," "off," or "open" position. Performing these actions will prevent the reintroduction of energy to the push-button circuitry and will isolate the operating control and the machine or equipment from the energy source.

(4) *Lockout or tagout device application.* **(i)** Lockout or tagout devices shall be affixed to each energy isolating device by authorized employees.

(ii) Lockout devices, where used, shall be affixed in a manner that will hold the energy isolating devices in a "safe" or "off" position.

(iii) Tagout devices, where used, shall be affixed in such a manner as will clearly indicate that the operation or movement of energy isolating devices from the "safe" or "off" position is prohibited.

(A) Where tagout devices are used with energy isolating devices designed with the capability of being locked, the tag attachment shall be fastened at the same point at which the lock would have been attached.

(B) Where a tag cannot be affixed directly to the energy isolating device, the tag shall be located as close as safely possible to the device, in a position that will be immediately obvious to anyone attempting to operate the device.

As the fourth step in the procedure, paragraph (d)(4) provides that action be taken to secure the energy isolating devices in the "safe" or "off" position. This paragraph requires that appropriate and effective lockout/tagout devices be affixed to each energy isolating device by an authorized employee, and that they

FIGURE 6.4 (*Continued*)

be attached so as to prevent reactivation of the machine or equipment.

OSHA is of the opinion that, as a general rule, when it is feasible, the physical protection offered by the use of a lock, when supported by the information provided on a tag used in conjunction with the lock, provides the greatest assurance of employee protection from the release of hazardous energy.

(5) *Stored energy.* (i) Following the application of lockout or tagout devices to energy isolating devices, all potentially hazardous stored or residual energy shall be relieved, disconnected, restrained, and otherwise rendered safe.

(ii) If there is a possibility of reaccumulation of stored energy to a hazardous level, verification of isolation shall be continued until the servicing or maintenance is completed, or until the possibility of such accumulation no longer exists.

Paragraph (d)(5) provides that the next step taken in the energy control procedure is to determine the presence of, and relieve, disconnect and/or restrain all potentially hazardous, stored or residual energy in the machine or equipment. Up to this point, the purpose of following all the steps of the procedure has been to enable the employee to isolate and block the source of energy feeding the machine or equipment to be worked on, at a point beyond which it can not be bypassed. However, energy can very easily be trapped in a system down-stream from an energy isolating device, or can be present in the form of potential energy from gravity or from spring action.

Stored or residual energy of this sort cannot be turned on or off; it must be dissipated or controlled.

When energy may still be present in a system that has been isolated from the energy source, this paragraph requires that energy to be controlled before an employee attempts to perform any work covered by the scope of this standard. Compliance with this provision might require, for example, the use of blocks or other physical restraints to immobilize the machine, machine components, or equipment where necessary for control of the hazard. In the case of electrical circuits, grounding might be necessary to discharge hazardous energy. Hydraulic or pneumatic systems

might necessitate the use of bleed valves to relieve the pressure.

OSHA believes that these requirements must be utilized since there is no manner to ensure that some leakage or drainage of energy or energy containing substances, such as super cooled or cryogenic fluids, can occur. In the case of one of those substances being present in a piping, containment, or transport system, a certain amount of leakage may occur without endangering employees. However, if servicing or maintenance must be performed on such a system, the standard requires the employer to continue to verify the isolation of energy sources which may be hazardous, in order to assure that such leakage does not approach a dangerous level. This may involve means, such as continuous monitoring for the displacement of oxygen or the buildup of the concentration of the substance toward the lower explosive limit of the substance, such as could occur with a hydrogen system.

(6) *Verification of isolation.* Prior to starting work on machines or equipment that have been locked/tagged out, the authorized employee shall verify that isolation and deenergization of the machine or equipment have been accomplished.

In this paragraph, as the sixth step in the energy control procedure, the authorized employee must ensure that the previous steps of the procedure have been taken to isolate the machine or equipment effectively. This must be done prior to starting the servicing or maintenance work. The authorized employee needs to verify that the machine or equipment has been turned off or shut down properly; that all energy isolating devices were identified, located and operated; that the lockout or tagout devices have been attached to energy isolating devices; and that stored energy has been rendered safe.

(e) *Release from lockout or tagout.* Before lockout or tagout devices are removed and energy is restored to the machine or equipment, procedures shall be followed and actions taken by the authorized employee(s) to ensure the following:

(1) *The machine or equipment.* The work area shall be inspected to ensure that nonessential items have been removed and to ensure that machine or equipment components are operationally intact.

FIGURE 6.4 *(Continued)*

(2) *Employees.* **(i)** The work area shall be checked to ensure that all employees have been safely positioned or removed.

(ii) Before lockout or tagout devices are removed and before machines or equipment are energized, affected employees shall be notified that the lockout or tagout devices have been removed.

(iii) After lockout or tagout devices have been removed and before a machine or equipment is started, affected employees shall be notified that the lockout or tagout device(s) have been removed.

(3) *Lockout or tagout devices removal.* Each lockout or tagout device shall be removed from each energy isolating device by the employee who applied the device. *Exception to paragraph (e)(3):* When the authorized employee who applied the lockout or tagout device is not available to remove it, that device may be removed under the direction of the employer, provided that specific procedures and training for such removal have been developed, documented and incorporated into the employer's energy control program. The employer shall demonstrate that the specific procedure shall include at least the following elements:

(i) Verification by the employer that the authorized employee who applied the device is not at the facility:

(ii) Making all reasonable efforts to contact the authorized employee to inform him/her that his/her lockout or tagout device has been removed; and

(iii) Ensuring that the authorized employee has this knowledge before he/she resumes work at that facility.

(f) *Additional requirements.* **(1)** *Testing or positioning of machines, equipment or components thereof.* In situations in which lockout or tagout devices must be temporarily removed from the energy isolating device and the machine or equipment energized to test or position the machine, equipment or component thereof, the following sequence of actions shall be followed:

(i) Clear the machine or equipment of tools and materials in accordance with paragraph (e)(1) of this section;

(ii) Remove employees from the machine or equipment area in accordance with paragraph (e)(2) of this section;

(iii) Remove the lockout or tagout devices as specified in paragraph (e)(3) of this section;

(iv) Energize and proceed with testing or positioning;

(v) De-energize all systems and reapply energy control measures in accordance with

paragraph (d) of this section to continue the servicing and/or maintenance.

(2) *Outside personnel (contractors, etc.).* **(i)** Whenever outside servicing personnel are to be engaged in activities covered by the scope and application of this standard, the on-site employer and the outside employer shall inform each other of their respective lockout or tagout procedures.

(ii) The on-site employer shall ensure that his/her employees understand and comply with the restrictions and prohibitions of the outside employer's energy control program.

(3) *Group lockout or tagout.* **(i)** When servicing and/or maintenance is performed by a crew, craft, department, or other group, they shall utilize a procedure which affords the employees a level of protection equivalent to that provided by the implementation of a personal lockout or tagout device.

(ii) Group lockout or tagout devices shall be used in accordance with the procedures required by paragraph (c)(4) of this section including, but not necessarily limited to, the following specific requirements:

(A) Primary responsibility is vested in an authorized employee for a set number of employees working under the protection of a group lockout or tagout device (such as an operations lock);

(B) Provision for the authorized employee to ascertain the exposure status of individual group members with regard to the lockout or tagout of the machine or equipment and

(C) When more than one crew, craft, department, etc. is involved, assignment of overall job-associated lockout or tagout control responsibility to an authorized employee designated to coordinate affected work forces and ensure continuity of protection; and

(D) Each authorized employee shall affix a personal lockout or tagout device to the group lockout device, group lockbox, or comparable mechanism when he or she begins work, and shall remove those devices when he or she stops working on the machine or equipment being serviced or maintained.

(4) *Shift or personnel changes.* Specific procedures shall be utilized during shift or personnel changes to ensure the continuity of lockout or tagout protection, including provision for the orderly transfer of lockout or tagout device protection between off-going and oncoming employees, to minimize exposure to hazards from the unexpected energization or start-up of the machine or equipment, or the release of stored energy.

FIGURE 6.4 *(Continued)*

APPENDIX A—TYPICAL MINIMAL LOCKOUT OR TAGOUT SYSTEM PROCEDURES

General

The following simple lockout procedure is provided to assist employers in developing their procedures so they meet the requirements of this standard. When the energy isolating devices are not lockable, tagout may be used, provided the employer complies with the provisions of the standard which require additional training and more rigorous periodic inspections. When tagout is used and the energy isolating devices are lockable, the employer must provide full employee protection (see paragraph (c)(3)) and additional training and more rigorous periodic inspections are required. For more complex systems, more comprehensive procedures may need to be developed, documented, and utilized.

Lockout/Tagout Procedure

Lockout Procedure for

_____

(Name of Company for single procedure or identification of equipment if multiple procedures are used).

Purpose

This procedure establishes the minimum requirements for the lockout of energy isolating devices whenever maintenance or servicing is done on machines or equipment. It shall be used to ensure that the machine or equipment is stopped, isolated from all potentially hazardous energy sources, and locked out before employees perform any servicing or maintenance where the unexpected energization, or start-up of the machine or equipment, or release of stored energy could cause injury.

Compliance With This Program

All employees are required to comply with the restrictions and limitations imposed upon them during the use of lockout. The authorized employees are required to perform the lockout in accordance with this procedure. All employees, upon observing a machine or piece of equipment which is locked out to perform servicing or maintenance shall not attempt to start, energize, or use that machine or equipment.

_____

Type of compliance enforcement to be taken for violation of the above.

Sequence of Lockout

(1) Notify all affected employees that servicing or maintenance is required on a machine or equipment and that the machine or equipment must be shut down and locked out to perform the servicing or maintenance.

_____

Name(s)/Job Title(s) of affected employees and how to notify.

(2) The authorized employee shall refer to the company procedure to identify the type and magnitude of the energy that the machine or equipment utilizes, shall understand the hazards of the energy, and shall know the methods to control the energy.

_____

Type(s) and magnitude(s) of energy, its hazards and the methods to control the energy.

(3) If the machine or equipment is operating, shut it down by the normal stopping procedure (depress the stop button, open switch, close valve, etc.).

_____

Type(s) and location(s) of machine or equipment operating controls.

(4) De-activate the energy isolating device(s) so that the machine or equipment is isolated from the energy source(s).

_____

Type(s) and location(s) of energy isolating devices.

(5) Lock out the energy isolating device(s) with assigned individual lock(s).

(6) Stored or residual energy (such as that in capacitors, springs, elevated machine members, rotating flywheels, hydraulic systems, and air, gas, steam, or water pressure, etc.) must be dissipated or restrained by methods such as grounding, repositioning, blocking, bleeding down, etc.

_____

Type(s) of stored energy—methods to dissipate or restrain.

(7) Ensure that the equipment is disconnected from the energy source(s) by first checking that no personnel are exposed, then verify the isolation of the equipment by operating the push button or other normal operating control(s) or by testing to make certain the equipment will not operate.

Caution: Return operating control(s) to neutral or "off" position after verifying the isolation of the equipment.

FIGURE 6.4   *(Continued)*

Method of verifying the isolation of the equipment.

(8) The machine or equipment is now locked out.

Restoring Equipment to Service

When the servicing or maintenance is completed and the machine or equipment is ready to return to normal operating condition, the following steps shall be taken.

(1) Check the machine or equipment and the immediate area around the machine to ensure that non-essential items have been removed and that the machine or equipment components are operationally intact.

(2) Check the work area to ensure that all employees have been safely positioned or removed from the area.

(3) Verify that the controls are in neutral.

(4) Remove the lockout devices and reenergize the machine or equipment.

Note: The removal of some forms of blocking may require reenergization of the machine before safe removal.

(5) Notify affected employees that the servicing or maintenance is completed and the machine or equipment is ready for use.

FIGURE 6.4   (*Continued*)

§ 1910.399 Definitions applicable to this subpart.

(a) Definitions applicable to 1910.302 through 1910.330—

Acceptable. An installation or equipment is acceptable to the Assistant Secretary of Labor, and approved within the meaning of this Subpart S:

(i) If it is accepted, or certified, or listed, or labeled, or otherwise determined to be safe by a nationally recognized testing laboratory; or

(ii) With respect to an installation or equipment of a kind which no nationally recognized testing laboratory accepts, certifies, lists, labels, or determines to be safe, if it is inspected or tested by another Federal agency, or by a State, municipal, or other local authority responsible for enforcing occupational safety provisions of the National Electrical Code, and found in compliance with the provisions of the National Electrical Code as applied in this subpart; or

(iii) With respect to custom-made equipment or related installations which are designed, fabricated for, and intended for use by a particular customer, if it is determined to be safe for its intended use by its manufacturer on the basis of test data which the employer keeps and makes available for inspection to the Assistant Secretary and his authorized representatives. Refer to 1910.7 for definition of nationally recognized testing laboratory.

Accepted. An installation is "accepted" if it has been inspected and found by a nationally recognized testing laboratory to conform to specified plans or to procedures of applicable codes.

Accessible. (As applied to wiring methods.) Capable of being removed or exposed without damaging the building structure or finish, or not permanently closed in by the structure or finish of the building. (See "*concealed*" and "*exposed.*")

Accessible. (As applied to equipment.) Admitting close approach; not guarded by locked doors, elevation, or other effective means. (See "*Readily accessible.*")

Ampacity. Current-carrying capacity of electric conductors expressed in amperes.

Appliances. Utilization equipment, generally other than industrial, normally built in standardized sizes or types, which is installed or connected as a unit to perform one or more functions such as clothes washing, air conditioning, food mixing, deep frying, etc.

Approved. Acceptable to the authority enforcing this subpart. The authority enforcing this subpart is the Assistant Secretary of Labor for Occupational Safety and Health. The definition of "acceptable" indicates what is acceptable to the Assistant Secretary of Labor, and therefore approved within the meaning of this Subpart.

Approved for the purpose. Approved for a specific purpose, environment, or application described in a particular standard requirement.

Suitability of equipment or materials for a specific purpose, environment or application may be determined by a nationally recognized testing laboratory, inspection agency or other organization concerned with product evaluation as part of its listing and labeling program. (See *Labeled* or *Listed.*)

Armored cable. Type AC armored cable is a fabricated assembly of insulated conductors in a flexible metallic enclosure.

Askarel. A generic term for a group of nonflammable synthetic chlorinated hydrocarbons used as electrical insulating media. Askarels of various compositional types are used. Under arcing conditions the gases produced, while consisting predominantly of noncombustible hydrogen chloride, can include varying amounts of combustible gases depending upon the askarel type.

Attachment plug (Plug cap)(Cap). A device which, by insertion in a receptacle, establishes connection between the conductors of the attached flexible cord and the conductors connected permanently to the receptacle.

Automatic. Self-acting, operating by its own mechanism when actuated by some impersonal influence, as, for example, a change in current strength, pressure, temperature, or mechanical configuration.

Bare conductor. See *Conductor.*

Bonding. The permanent joining of metallic parts to form an electrically conductive path which will assure electrical continuity and the capacity to conduct safely any current likely to be imposed.

Bonding jumper. A reliable conductor to assure the required electrical conductivity between metal parts required to be electrically connected.

Branch circuit. The circuit conductors between the final overcurrent device protecting the circuit and the outlet(s).

Building. A structure which stands alone or which is cut off from adjoining structures by

FIGURE 6.5 Definitions Applying to Part 1910, Subpart S (OSHA-CFR, Title 29, Part 1910, Paragraph 399).

fire walls with all openings therein protected by approved fire doors.

Cabinet. An enclosure designed either for surface or flush mounting, and provided with a frame, mat, or trim in which a swinging door or doors are or may be hung.

Cable tray system. A cable tray system is a unit or assembly of units or sections, and associated fittings, made of metal or other noncombustible materials forming a rigid structural system used to support cables. Cable tray systems include ladders, troughs, channels, solid bottom trays, and other similar structures.

Cablebus. Cablebus is an approved assembly of insulated conductors with fittings and conductor terminations in a completely enclosed, ventilated, protective metal housing.

Center pivot irrigation machine. A center pivot irrigation machine is a multi-motored irrigation machine which revolves around a central pivot and employs alignment switches or similar devices to control individual motors.

Certified. Equipment is "certified" if it (a) has been tested and found by a nationally recognized testing laboratory to meet nationally recognized standards or to be safe for use in a specified manner, or (b) is of a kind whose production is periodically inspected by a nationally recognized testing laboratory, and (c) it bears a label, tag, or other record of certification.

Circuit breaker. (i) (600 volts nominal, or less). A device designed to open and close a circuit by nonautomatic means and to open the circuit automatically on a predetermined overcurrent without injury to itself when properly applied within its rating.

(ii) (Over 600 volts, nominal). A switching device capable of making, carrying, and breaking currents under normal circuit conditions, and also making, carrying for a specified time, and breaking currents under specified abnormal circuit conditions, such as those of short circuit.

Class I locations. Class I locations are those in which flammable gases or vapors are or may be present in the air in quantities sufficient to produce explosive or ignitable mixtures. Class I locations include the following:

(i) *Class I, Division 1.* A Class I, Division 1 location is a location: (a) in which hazardous concentrations of flammable gases or vapors may exist under normal operating conditions; or (b) in which hazardous concentrations of such gases or vapors may exist frequently because of repair or maintenance operations or because of

leakage; or (c) in which breakdown or faulty operation of equipment or processes might release hazardous concentrations of flammable gases or vapors, and might also cause simultaneous failure of electric equipment.

NOTE: This classification usually includes locations where volatile flammable liquids or liquefied flammable gases are transferred from one container to another; interiors of spray booths and areas in the vicinity of spraying and painting operations where volatile flammable solvents are used; locations containing open tanks or vats of volatile flammable liquids; drying rooms or compartments for the evaporation of flammable solvents; locations containing fat and oil extraction equipment using volatile flammable solvents; portions of cleaning and dyeing plants where flammable liquids are used; gas generator rooms and other portions of gas manufacturing plants where flammable gas may escape; inadequately ventilated pump rooms for flammable gas or for volatile flammable liquids; the interiors of refrigerators and freezers in which volatile flammable materials are stored in open, lightly stoppered, or easily ruptured containers; and all other locations where ignitable concentrations of flammable vapors or gases are likely to occur in the course of normal operations.

(ii) *Class I, Division 2.* A Class I, Division 2 location is a location: (a) in which volatile flammable liquids or flammable gases are handled, processed, or used, but in which the hazardous liquids, vapors, or gases will normally be confined within closed containers or closed systems from which they can escape only in case of accidental rupture or breakdown of such containers or systems, or in case of abnormal operation of equipment; or (b) in which hazardous concentrations of gases or vapors are normally prevented by positive mechanical ventilation, and which might become hazardous through failure or abnormal operations of the ventilating equipment; or (c) that is adjacent to a Class I, Division 1 location, and to which hazardous concentrations of gases or vapors might occasionally be communicated unless such communication is prevented by adequate positive-pressure ventilation from a source of clean air, and effective safeguards against ventilation failure are provided.

NOTE: This classification usually includes locations where volatile flammable liquids or flammable gases or vapors are used, but which would become

FIGURE 6.5 (*Continued*)

hazardous only in case of an accident or of some unusual operating condition. The quantity of flammable material that might escape in case of accident, the adequacy of ventilating equipment, the total area involved, and the record of the industry or business with respect to explosions or fires are all factors that merit consideration in determining the classification and extent of each location.

Piping without valves, checks, meters, and similar devices would not ordinarily introduce a hazardous condition even though used for flammable liquids or gases. Locations used for the storage of flammable liquids or a liquefied or compressed gases in sealed containers would not normally be considered hazardous unless also subject to other hazardous conditions.

Electrical conduits and their associated enclosures separated from process fluids by a single seal or barrier are classed as a Division 2 location if the outside of the conduit and enclosures is a nonhazardous location.

Class II locations. Class II locations are those that are hazardous because of the presence of combustible dust. Class II locations include the following:

(i) *Class II, Division 1.* A Class II, Division 1 location is a location: (a) In which combustible dust is or may be in suspension in the air under normal operating conditions, in quantities sufficient to produce explosive or ignitable mixtures; or (b) where mechanical failure or abnormal operation of machinery or equipment might cause such explosive or ignitable mixtures to be produced, and might also provide a source of ignition through simultaneous failure of electric equipment, operation of protection devices, or from other causes, or (c) in which combustible dusts of an electrically conductive nature may be present.

NOTE: This classification may include areas of grain handling and processing plants, starch plants, sugar-pulverizing plants, malting plants, hay-grinding plants, coal pulverizing plants, areas where metal dusts and powders are produced or processed, and other similar locations which contain dust producing machinery and equipment (except where the equipment is dust-tight or vented to the outside). These areas would have combustible dust in the air, under normal operating conditions, in quantities sufficient to produce explosive or ignitable mixtures. Combustible dusts which are electrically nonconductive include dusts produced in the handling and processing of grain and grain products, pulverized sugar

and cocoa, dried egg and milk powders, pulverized spices, starch and pastes, potato and woodflour, oil meal from beans and seed, dried hay, and other organic materials which may produce combustible dusts when processed or handled. Dusts containing magnesium or aluminum are particularly hazardous and the use of extreme caution is necessary to avoid ignition and explosion.

(ii) *Class II, Division 2.* A Class II, Division 2 location is a location in which: (a) combustible dust will not normally be in suspension in the air in quantities sufficient to produce explosive or ignitable mixtures, and dust accumulations are normally insufficient to interfere with the normal operation of electrical equipment or other apparatus; or (b) dust may be in suspension in the air as a result of infrequent malfunctioning of handling or processing equipment, and dust accumulations resulting therefrom may be ignitable by abnormal operation or failure of electrical equipment or other apparatus.

NOTE: This classification includes locations where dangerous concentrations of suspended dust would not be likely but where dust accumulations might form on or in the vicinity of electric equipment. These areas may contain equipment from which appreciable quantities of dust would escape under abnormal operating conditions or be adjacent to a Class II Division 1 location, as described above, into which an explosive or ignitable concentration of dust may be put into suspension under abnormal operating conditions.

Class III locations. Class III locations are those that are hazardous because of the presence of easily ignitable fibers or flyings but in which such fibers or flyings are not likely to be in suspension in the air in quantities sufficient to produce ignitable mixtures. Class III locations include the following:

(i) *Class III, Division 1.* A Class III, Division 1 location is a location in which easily ignitable fibers or materials producing combustible flyings are handled, manufactured, or used.

NOTE: Such locations usually include some parts of rayon, cotton, and other textile mills; combustible fiber manufacturing and processing plants; cotton gins and cotton-seed mills; flax-processing plants; clothing manufacturing plants; woodworking plants, and establishments; and industries involving similar hazardous processes or conditions.

Easily ignitable fibers and flyings include rayon, cotton (including cotton linters and cotton waste),

FIGURE 6.5 *(Continued)*

sisal or henequen, istle, jute, hemp, tow, cocoa fiber, oakum, baled waste kapok, Spanish moss, excelsior, and other materials of similar nature.

(ii) *Class III, Division 2.* A Class III, Division 2 location is a location in which easily ignitable fibers are stored or handled, except in process of manufacture.

Collector ring. A collector ring is an assembly of slip rings for transferring electrical energy from a stationary to a rotating member.

Concealed. Rendered inaccessible by the structure or finish of the building. Wires in concealed raceways are considered concealed, even though they may become accessible by withdrawing them. [See *Accessible. (As applied to wiring methods.)*]

Conductor. (i) Bare. A conductor having no covering or electrical insulation whatsoever.

(ii) *Covered.* A conductor encased within material of composition or thickness that is not recognized as electrical insulation.

(iii) *Insulated.* A conductor encased within material of composition and thickness that is recognized as electrical insulation.

Conduit body. A separate portion of a conduit or tubing system that provides access through a removable cover(s) to the interior of the system at a junction of two or more sections of the system or at a terminal point of the system. Boxes such as FS and FD or larger cast or sheet metal boxes are not classified as conduit bodies.

Controller. A device or group of devices that serves to govern, in some predetermined manner, the electric power delivered to the apparatus to which it is connected.

Cooking unit, counter-mounted. A cooking appliance designed for mounting in or on a counter and consisting of one or more heating elements, internal wiring, and built-in or separately mountable controls. (See *Oven, wall-mounted.*)

Covered conductor. See *Conductor.*

Cutout. (Over 600 volts, nominal.) An assembly of a fuse support with either a fuseholder, fuse carrier, or disconnecting blade. The fuseholder or fuse carrier may include a conducting element (fuse link), or may act as the disconnecting blade by the inclusion of a nonfusible member.

Cutout box. An enclosure designed for surface mounting and having swinging doors or covers secured directly to and telescoping with the walls of the box proper. (See *Cabinet.*)

Damp location. See *Location.*

Dead front. Without live parts exposed to a person on the operating side of the equipment.

Device. A unit of an electrical system which is intended to carry but not utilize electric energy.

Dielectric heating. Dielectric heating is the heating of a nominally insulating material due to its own dielectric losses when the material is placed in a varying electric field.

Disconnecting means. A device, or group of devices, or other means by which the conductors of a circuit can be disconnected from their source of supply.

Disconnecting (or Isolating) switch. (Over 600 volts, nominal.) A mechanical switching device used for isolating a circuit or equipment from a source of power.

Dry location. See *Location.*

Electric sign. A fixed, stationary, or portable self-contained, electrically illuminated utilization equipment with words or symbols designed to convey information or attract attention.

Enclosed. Surrounded by a case, housing, fence or walls which will prevent persons from accidentally contacting energized parts.

Enclosure. The case or housing of apparatus, or the fence or walls surrounding an installation to prevent personnel from accidentally contacting energized parts, or to protect the equipment from physical damage.

Equipment. A general term including material, fittings, devices, appliances, fixtures, apparatus, and the like, used as a part of, or in connection with, an electrical installation.

Equipment grounding conductor. See *Grounding conductor, equipment.*

Explosion-proof apparatus. Apparatus enclosed in a case that is capable of withstanding an explosion of a specified gas or vapor which may occur within it and of preventing the ignition of a specified gas or vapor surrounding the enclosure by sparks, flashes, or explosion of the gas or vapor within, and which operates at such an external temperature that it will not ignite a surrounding flammable atmosphere.

Exposed. (As applied to live parts.) Capable of being inadvertently touched or approached nearer than a safe distance by a person. It is applied to parts not suitably guarded, isolated, or insulated. (See *Accessible.* and *Concealed.*)

Exposed. (As applied to wiring methods.) On or attached to the surface or behind panels designed to allow access. [See *Accessible. (As applied to wiring methods.)*]

Exposed. (For the purposes of 1910.308 (e), Communications systems.) Where the circuit is in such a position that in case of failure

FIGURE 6.5 (*Continued*)

of supports or insulation, contact with another circuit may result.

Externally operable. Capable of being operated without exposing the operator to contact with live parts.

Feeder. All circuit conductors between the service equipment, or the generator switchboard of an isolated plant, and the final branch-circuit overcurrent device.

Fitting. An accessory such as a locknut, bushing, or other part of a wiring system that is intended primarily to perform a mechanical rather than an electrical function.

Fuse. (Over 600 volts, nominal.) An overcurrent protective device with a circuit opening fusible part that is heated and severed by the passage of overcurrent through it. A fuse comprises all the parts that form a unit capable of performing the prescribed functions. It may or may not be the complete device necessary to connect it into an electrical circuit.

Ground. A conducting connection, whether intentional or accidental, between an electrical circuit or equipment and the earth, or to some conducting body that serves in place of the earth.

Grounded. Connected to earth or to some conducting body that serves in place of the earth.

Grounded, effectively. (Over 600 volts, nominal.) Permanently connected to earth through a ground connection of sufficiently low impedance and having sufficient ampacity that ground fault current which may occur cannot build up to voltages dangerous to personnel.

Grounded conductor. A system or circuit conductor that is intentionally grounded.

Grounding conductor. A conductor used to connect equipment or the grounded circuit of a wiring system to a grounding electrode or electrodes.

Grounding conductor, equipment. The conductor used to connect the non-current-carrying metal parts of equipment, raceways, and other enclosures to the system grounded conductor and/or the grounding electrode conductor at the service equipment or at the source of a separately derived system.

Grounding electrode conductor. The conductor used to connect the grounding electrode to the equipment grounding conductor and/or to the grounded conductor of the circuit at the service equipment or at the source of a separately derived system.

Ground-fault circuit-interrupter. A device whose function is to interrupt the electric circuit to the load when a fault current to ground exceeds some predetermined value that is less than that required to operate the overcurrent protective device of the supply circuit.

Guarded. Covered, shielded, fenced, enclosed, or otherwise protected by means of suitable covers, casings, barriers, rails, screens, mats, or platforms to remove the likelihood of approach to a point of danger or contact by persons or objects.

Health care facilities. Buildings or portions of buildings and mobile homes that contain, but are not limited to, hospitals, nursing homes, extended care facilities, clinics, and medical and dental offices, whether fixed or mobile.

Heating equipment. For the purposes of 1910.306(g), the term "heating equipment" includes any equipment used for heating purposes if heat is generated by induction or dielectric methods.

Hoistway. Any shaftway, hatchway, well hole, or other vertical opening or space in which an elevator or dumbwaiter is designed to operate.

Identified. Identified, as used in reference to a conductor or its terminal, means that such conductor or terminal can be readily recognized as grounded.

Induction heating. Induction heating is the heating of a nominally conductive material due to its own I^2R losses when the material is placed in a varying electromagnetic field.

Insulated conductor. See *Conductor.*

Interrupter switch. (Over 600 volts, nominal.) A switch capable of making, carrying, and interrupting specified currents.

Irrigation machine. An irrigation machine is an electrically driven or controlled machine, with one or more motors, not hand portable, and used primarily to transport and distribute water for agricultural purposes.

Isolated. Not readily accessible to persons unless special means for access are used.

Isolated power system. A system comprising an isolating transformer or it equivalent, a line isolation monitor, and its ungrounded circuit conductors.

Labeled. Equipment is "labeled" if there is attached to it a label, symbol, or other identifying mark of a nationally recognized testing laboratory which, (a) makes periodic inspections of the production of such equipment, and (b) whose labeling indicates compliance with nationally recognized standards or tests to determine safe use in a specified manner.

Lighting outlet. An outlet intended for the direct connection of a lampholder, a lighting

FIGURE 6.4 *(Continued)*

fixture, or a pendant cord terminating in a lampholder.

Line-clearance tree trimming. The pruning, trimming, repairing, maintaining, removing, or clearing of trees or cutting of brush that is within 10 feet (305 cm) of electric supply lines and equipment.

Listed. Equipment is "listed" if it is of a kind mentioned in a list which, (a) is published by a nationally recognized laboratory which makes periodic inspection of the production of such equipment, and (b) states such equipment meets nationally recognized standards or has been tested and found safe for use in a specified manner.

Location—(i) *Damp location.* Partially protected locations under canopies, marquees, roofed open porches, and like locations, and interior locations subject to moderate degrees of moisture, such as some basements, some barns, and some cold-storage warehouses.

(ii) *Dry location.* A location not normally subject to dampness or wetness. A location classified as dry may be temporarily subject to dampness or wetness, as in the case of a building under construction.

(iii) *Wet location.* Installations underground or in concrete slabs or masonry in direct contact with the earth, and locations subject to saturation with water or other liquids, such as vehicle-washing areas, and locations exposed to weather and unprotected.

May. If a discretionary right, privilege, or power is abridged or if an obligation to abstain from acting is imposed, the word "may" is used with a restrictive "no," "not," or "only." (E.g., no employer may . . . ; an employer may not . . . ; only qualified persons may . . .)

Medium voltage cable. Type MV medium voltage cable is a single or multiconductor solid dielectric insulated cable rated 2000 volts or higher.

Metal-clad cable. Type MC cable is a factory assembly of one or more conductors, each individually insulated and enclosed in a metallic sheath of interlocking tape, or a smooth or corrugated tube.

Mineral-insulated metal-sheathed cable. Type MI mineral-insulated metal-sheathed cable is a factory assembly of one or more conductors insulated with a highly compressed refractory mineral insulation and enclosed in a liquidtight and gastight continuous copper sheath.

Mobile X-ray. X-ray equipment mounted on a permanent base with wheels and/or casters for moving while completely assembled.

Nonmetallic-sheathed cable. Nonmetallic-sheathed cable is a factory assembly of two or more insulated conductors having an outer sheath of moisture resistant, flame-retardant, nonmetallic material. Nonmetallic sheathed cable is manufactured in the following types:

(i) *Type NM.* The overall covering has a flame-retardant and moisture-resistant finish.

(ii) *Type NMC.* The overall covering is flame-retardant, moisture-resistant, fungus-resistant, and corrosion-resistant.

Oil (filled) cutout. (Over 600 volts, nominal.) A cutout in which all or part of the fuse support and its fuse link or disconnecting blade are mounted in oil with complete immersion of the contacts and the fusible portion of the conducting element (fuse link), so that arc interruption by severing of the fuse link or by opening of the contacts will occur under oil.

Open wiring on insulators. Open wiring on insulators is an exposed wiring method using cleats, knobs, tubes, and flexible tubing for the protection and support of single insulated conductors run in or on buildings, and not concealed by the building structure.

Outlet. A point on the wiring system at which current is taken to supply utilization equipment.

Outline lighting. An arrangement of incandescent lamps or electric discharge tubing to outline or call attention to certain features such as the shape of a building or the decoration of a window.

Oven, wall-mounted. An oven for cooking purposes designed for mounting in or on a wall or other surface and consisting of one of more heating elements, internal wiring, and built-in or separately mountable controls. (See *Cooking unit, counter-mounted.*)

Overcurrent. Any current in excess of the rated current of equipment or the ampacity of a conductor. It may result from overload (see definition), short circuit, or ground fault. A current in excess of rating may be accommodated by certain equipment and conductors for a given set of conditions. Hence the rules for overcurrent protection are specific for particular situations.

Overload. Operation of equipment in excess of normal, full load rating, or of a conductor in excess of rated ampacity which, when it persists for a sufficient length of time, would cause damage or dangerous overheating. A fault, such as a short circuit or ground fault, is not an overload. (See *Overcurrent.*)

Panelboard. A single panel or group of panel units designed for assembly in the form

FIGURE 6.4   (*Continued*)

of a single panel; including buses, automatic overcurrent devices, and with or without switches for the control of light, heat, or power circuits; designed to be placed in a cabinet or cutout box placed in or against a wall or partition and accessible only from the front. (See *Switchboard.*)

Permanently installed decorative fountains and reflection pools. Those that are constructed in the ground, on the ground or in a building in such a manner that the pool cannot be readily disassembled for storage and are served by electrical circuits of any nature. These units are primarily constructed for their aesthetic value and not intended for swimming or wading.

Permanently installed swimming pools, wading and therapeutic pools. Those that are constructed in the ground, on the ground, or in a building in such a manner that the pool cannot be readily disassembled for storage whether or not served by electrical circuits of any nature.

Portable X-ray. X-ray equipment designed to be hand-carried.

Power and control tray cable. Type TC power and control tray cable is a factory assembly of two or more insulated conductors, with or without associated bare or covered grounding conductors under a nonmetallic sheath, approved for installation in cable trays, in raceways, or where supported by a messenger wire.

Power fuse. (Over 600 volts, nominal.) See *Fuse.*

Power-limited tray cable. Type PLTC nonmetallic-sheathed power limited tray cable is a factory assembly of two or more insulated conductors under a nonmetallic jacket.

Power outlet. An enclosed assembly which may include receptacles, circuit breakers, fuseholders, fused switches, buses and watt-hour meter mounting means; intended to supply and control power to mobile homes, recreational vehicles or boats, or to serve as a means for distributing power required to operate mobile or temporarily installed equipment.

Premises wiring system. That interior and exterior wiring, including power, lighting, control, and signal circuit wiring together with all of its associated hardware, fittings, and wiring devices, both permanently and temporarily installed, which extends from the load end of the service drop, or load end of the service lateral conductors to the outlet(s). Such wiring does not include wiring internal to appliances, fixtures, motors, controllers, motor control centers, and similar equipment.

Qualified person. One familiar with the construction and operation of the equipment and the hazards involved.

NOTE 1: Whether an employee is considered to be a "qualified person" will depend upon various circumstances in the workplace. It is possible and, in fact, likely for an individual to be considered qualified" with regard to certain equipment in the workplace, but "unqualified" as to other equipment. (See 1910.332(b)(3) for training requirements that specifically apply to qualified persons.)

NOTE 2: An employee who is undergoing on-the-job training and who, in the course of such training, has demonstrated an ability to perform duties safely at his or her level of training and who is under the direct supervision of a qualified person is considered to be a qualified person for the performance of those duties.

Raceway. A channel designed expressly for holding wires, cables, or busbars, with additional functions as permitted in this subpart. Raceways may be of metal or insulating material, and the term includes rigid metal conduit, rigid nonmetallic conduit, intermediate metal conduit, liquidtight flexible metal conduit, flexible metallic tubing, flexible metal conduit, electrical metallic tubing, underfloor raceways, cellular concrete floor raceways, cellular metal floor raceways, surface raceways, wireways, and busways.

Readily accessible. Capable of being reached quickly for operation, renewal, or inspections, without requiring those to whom ready access is requisite to climb over or remove obstacles or to resort to portable ladders, chairs, etc. (See *Accessible.*)

Receptacle. A receptacle is a contact device installed at the outlet for the connection of a single attachment plug. A single receptacle is a single contact device with no other contact device on the same yoke. A multiple receptacle is a single device containing two or more receptacles.

Receptacle outlet. An outlet where one or more receptacles are installed.

Remote-control circuit. Any electric circuit that controls any other circuit through a relay or an equivalent device.

Sealable equipment. Equipment enclosed in a case or cabinet that is provided with a means of sealing or locking so that live parts cannot be made accessible without opening the enclosure.

FIGURE 6.5   (*Continued*)

The equipment may or may not be operable without opening the enclosure.

Separately derived system. A premises wiring system whose power is derived from generator, transformer, or converter winding and has no direct electrical connection, including a solidly connected grounded circuit conductor, to supply conductors originating in another system.

Service. The conductors and equipment for delivering energy from the electricity supply system to the wiring system of the premises served.

Service cable. Service conductors made up in the form of a cable.

Service conductors. The supply conductors that extend from the street main or from transformers to the service equipment of the premises supplied.

Service drop. The overhead service conductors from the last pole or other aerial support to and including the splices, if any, connecting to the service-entrance conductors at the building or other structure.

Service-entrance cable. Service-entrance cable is a single conductor or multiconductor assembly provided with or without an overall covering, primarily used for services and of the following types:

(i) Type SE, having a flame-retardant, moisture-resistant covering, but not required to have inherent protection against mechanical abuse.

(ii) Type USE, recognized for underground use, having a moisture-resistant covering, but not required to have a flame-retardant covering or inherent protection against mechanical abuse. Single-conductor cables having an insulation specifically approved for the purpose do not require an outer covering.

Service-entrance conductors, overhead system. The service conductors between the terminals of the service equipment and a point usually outside the building, clear of building walls, where joined by tap or splice to the service drop.

Service entrance conductors, underground system. The service conductors between the terminals of the service equipment and the point of connection to the service lateral. Where service equipment is located outside the building walls, there may be no service-entrance conductors, or they may be entirely outside the building.

Service equipment. The necessary equipment, usually consisting of a circuit breaker or switch and fuses, and their accessories, located near the point of entrance of supply conductors to a building or other structure, or an otherwise defined area, and intended to constitute the main control and means of cutoff of the supply.

Service raceway. The raceway that encloses the service-entrance conductors.

Shielded nonmetallic-sheathed cable. Type SNM, shielded nonmetallic-sheathed cable is a factory assembly of two or more insulated conductors in an extruded core of moisture-resistant, flame-resistant nonmetallic material, covered with an overlapping spiral metal tape and wire shield and jacketed with an extruded moisture-, flame-, oil-, corrosion-, fungus-, and sunlight-resistant nonmetallic material.

Show window. Any window used or designed to be used for the display of goods or advertising material, whether it is fully or partly enclosed or entirely open at the rear and whether or not it has a platform raised higher than the street floor level.

Sign. See *Electric Sign.*

Signaling circuit. Any electric circuit that energizes signaling equipment.

Special permission. The written consent of the authority having jurisdiction.

Storable swimming or wading pool. A pool with a maximum dimension of 15 feet and a maximum wall height of 3 feet and is so constructed that it may be readily disassembled for storage and reassembled to its original integrity.

Switchboard. A large single panel, frame, or assembly of panels which have switches, buses, instruments, overcurrent and other protective devices mounted on the face or back or both. Switchboards are generally accessible from the rear as well as from the front and are not intended to be installed in cabinets. (See *Panelboard.*)

Switches.

General-use switch. A switch intended for use in general distribution and branch circuits. It is rated in amperes, and it is capable of interrupting its rated current at its rated voltage.

(ii) *General-use snap switch.* A form of general-use switch so constructed that it can be installed in flush device boxes or on outlet box covers, or otherwise used in conjunction with wiring systems recognized by this subpart.

(iii) *Isolating switch.* A switch intended for isolating an electric circuit from the source of power. It has no interrupting rating, and it is intended to be operated only after the circuit has been opened by some other means.

(iv) *Motor-circuit switch.* A switch, rated in horsepower, capable of interrupting the

FIGURE 6.5   (*Continued*)

maximum operating overload current of a motor of the same horsepower rating as the switch at the rated voltage.

Switching devices. (Over 600 volts, nominal.) Devices designed to close and/or open one or more electric circuits. Included in this category are circuit breakers, cutouts, disconnecting (or isolating) switches, disconnecting means, interrupter switches, and oil (filled) cutouts.

Transportable X-ray. X-ray equipment installed in a vehicle or that may readily be disassembled for transport in a vehicle.

Utilization equipment. Utilization equipment means equipment which utilizes electric energy for mechanical, chemical, heating, lighting, or similar useful purpose.

Utilization system. A utilization system is a system which provides electric power and light for employee workplaces, and includes the premises wiring system and utilization equipment.

Ventilated. Provided with a means to permit circulation of air sufficient to remove an excess of heat, fumes, or vapors.

Volatile flammable liquid. A flammable liquid having a flash point below 38 degrees C (100 degrees F) or whose temperature is above its flash point.

Voltage (of a circuit). The greatest root-mean-square (effective) difference of potential between any two conductors of the circuit concerned.

Voltage, nominal. A nominal value assigned to a circuit or system for the purpose of conveniently designating its voltage class (as 120/240, 480Y/277, 600, etc.). The actual voltage at which a circuit operates can vary from the nominal within a range that permits satisfactory operation of equipment.

Voltage to ground. For grounded circuits, the voltage between the given conductor and that point or conductor of the circuit that is grounded; for ungrounded circuits, the greatest voltage between the given conductor and any other conductor of the circuit.

Watertight. So constructed that moisture will not enter the enclosure.

Weatherproof. So constructed or protected that exposure to the weather will not interfere with successful operation. Rainproof, raintight, or watertight equipment can fulfill the requirements for weatherproof where varying weather conditions other than wetness, such as snow, ice, dust, or temperature extremes, are not a factor.

Wet location. See *Location.*

Wireways. Wireways are sheet-metal troughs with hinged or removable covers for housing and protecting electric wires and cable and in which conductors are laid in place after the wireway has been installed as a complete system.

(b) Definitions applicable to 1910.331 through 1910.360 [Reserved]

(c) Definitions applicable to 1910.360 through 1910.380 [Reserved]

(d) Definitions applicable to 1910.381 through 1910.398 [Reserved]

[46 FR 4056, Jan. 16, 1981; 46 FR 40185, Aug. 7, 1981; as amended at 53 FR 12123, Apr. 12, 1988; 55 FR 32020, Aug. 6, 1990; 55 FR 46054, Nov. 1, 1990]

FIGURE 6.5 *(Continued)*

§ 1926.402 Applicability.

(a) Covered. Sections 1926.402 through 1926.408 contain installation safety requirements for electrical equipment and installations used to provide electric power and light at the jobsite. These sections apply to installations, both temporary and permanent, used on the jobsite; but these sections do not apply to existing permanent installations that were in place before the construction activity commenced.

NOTE: If the electrical installation is made in accordance with the National Electrical Code ANSI/NFPA 70-1984, exclusive of Formal Interpretations and Tentative Interim Amendments, it will be deemed to be in compliance with 1926.403 through 1926.408, except for 1926.404(b)(1) and 1926.405(a)(2)(ii)(E), (F), (G), and (J).

(b) Not covered. Sections 1926.402 through 1926.408 do not cover installations used for the generation, transmission, and distribution of electric energy, including related communication, metering, control, and transformation installations. (However, these regulations do cover portable and vehicle-mounted generators used to provide power for equipment used at the jobsite.) See Subpart V of this Part for the construction of power distribution and transmission lines.

§ 1926.403 General requirements.

(a) Approval. All electrical conductors and equipment shall be approved.

(b) Examination, installation, and use of equipment—

(b)(1) Examination. The employer shall ensure that electrical equipment is free from recognized hazards that are likely to cause death or serious physical harm to employees. Safety of equipment shall be determined on the basis of the following considerations:

(b)(1)(i) Suitability for installation and use in conformity with the provisions of this subpart. Suitability of equipment for an identified purpose may be evidenced by listing, labeling, or certification for that identified purpose.

(b)(1)(ii) Mechanical strength and durability, including, for parts designed to enclose and protect other equipment, the adequacy of the protection thus provided.

(b)(1)(iii) Electrical insulation.

(b)(1)(iv) Heating effects under conditions of use.

§ 1926.403(b)(1)(v)

(b)(1)(v) Arcing effects.

(b)(1)(vi) Classification by type, size, voltage, current capacity, specific use.

(b)(1)(vii) Other factors which contribute to the practical safeguarding of employees using or likely to come in contact with the equipment.

(b)(2) Installation and use. Listed, labeled, or certified equipment shall be installed and used in accordance with instructions included in the listing, labeling, or certification.

(c) Interrupting rating. Equipment intended to break current shall have an interrupting rating at system voltage sufficient for the current that must be interrupted.

(d) Mounting and cooling of equipment—

(d)(1) Mounting. Electric equipment shall be firmly secured to the surface on which it is mounted. Wooden plugs driven into holes in masonry, concrete, plaster, or similar materials shall not be used.

§ 1926.403(d)(2)

(d)(2) Cooling. Electrical equipment which depends upon the natural circulation of air and convection principles for cooling of exposed surfaces shall be installed so that room air flow over such surfaces is not prevented by walls or by adjacent installed equipment. For equipment designed for floor mounting, clearance between top surfaces and adjacent surfaces shall be provided to dissipate rising warm air. Electrical equipment provided with ventilating openings shall be installed so that walls or other obstructions do not prevent the free circulation of air through the equipment.

(e) Splices. Conductors shall be spliced or joined with splicing devices designed for the use or by brazing, welding, or soldering with a fusible metal or alloy. Soldered splices shall first be so spliced or joined as to be mechanically and electrically secure without solder and then soldered. All splices and joints and the free ends of conductors shall be covered with an insulation equivalent to that of the conductors or with an insulating device designed for the purpose.

(f) Arcing parts. Parts of electric equipment which in ordinary operation produce arcs, sparks, flames, or molten metal shall be enclosed or separated and isolated from all combustible material.

(g) Marking. Electrical equipment shall not be used unless the manufacturer's name, trademark, or other descriptive marking by which the organization responsible for the product may be

FIGURE 6.6 Installation Safety Requirements (OSHA-CFR, Title 29, Part 1926, Paragraphs 402–408).

identified is placed on the equipment and unless other markings are provided giving voltage, current, wattage, or other ratings as necessary. The marking shall be of sufficient durability to withstand the environment involved.

§ 1926.403(h)

(h) Identification of disconnecting means and circuits. Each disconnecting means required by this subpart for motors and appliances shall be legibly marked to indicate its purpose, unless located and arranged so the purpose is evident. Each service, feeder, and branch circuit, at its disconnecting means or overcurrent device, shall be legibly marked to indicate its purpose, unless located and arranged so the purpose is evident. These markings shall be of sufficient durability to withstand the environment involved.

(i) 600 Volts, nominal, or less. This paragraph applies to equipment operating at 600 volts, nominal, or less.

(i)(1) Working space about electric equipment. Sufficient access and working space shall be provided and maintained about all electric equipment to permit ready and safe operation and maintenance of such equipment.

(i)(1)(i) Working clearances. Except as required or permitted elsewhere in this subpart, the dimension of the working space in the direction of access to live parts operating at 600 volts or less and likely to require examination, adjustment, servicing, or maintenance while alive shall not be less than indicated in Table K-1. In addition to the dimensions shown in Table K-1, workspace shall not be less than 30 inches (762 mm) wide in front of the electric equip-

ment. Distances shall be measured from the live parts if they are exposed, or from the enclosure front or opening if the live parts are enclosed. Walls constructed of concrete, brick, or tile are considered to be grounded. Working space is not required in back of assemblies such as dead-front switchboards or motor control centers where there are no renewable or adjustable parts such as fuses or switches on the back and where all connections are accessible from locations other than the back.

§ 1926.403(i)(1)(ii)

(i)(1)(ii) Clear spaces. Working space required by this subpart shall not be used for storage. When normally enclosed live parts are exposed for inspection or servicing, the working space, if in a passageway or general open space, shall be guarded.

(i)(1)(iii) Access and entrance to working space. At least one entrance shall be provided to give access to the working space about electric equipment.

(i)(1)(iv) Front working space. Where there are live parts normally exposed on the front of switchboards or motor control centers, the working space in front of such equipment shall not be less than 3 feet (914 mm).

(i)(1)(v) Headroom. The minimum headroom of working spaces about service equipment, switchboards, panelboards, or motor control centers shall be 6 feet 3 inches (1.91 m).

(i)(2) Guarding of live parts.

(i)(2)(i) Except as required or permitted elsewhere in this subpart, live parts of electric equipment operating at 50 volts or more shall be guarded against accidental contact by cabinets

TABLE K-1 Working Clearances

	Minimum clear distance for conditions (1)		
	(a) Feet (2)	(b) Feet (2)	(c) Feet (2)
Nominal voltage to ground			
0–150	3	3	3
151–600	3	$3\frac{1}{2}$	4

Footnote(1) Conditions (a), (b), and (c) are as follows: [a] Exposed live parts on one side and no live or grounded parts on the other side of the working space, or exposed live parts on both sides effectively guarded by insulating material. Insulated wire or insulated busbars operating at not over 300 volts are not considered live parts. [b] Exposed live parts on one side and grounded parts on the other side. [c] Exposed live parts on both sides of the workplace [not guarded as provided in Condition (a)] with the operator between.

Footnote(2) Note: For International System of Units (SI): one foot = 0.3048 m.

FIGURE 6.6   (*Continued*)

or other forms of enclosures, or by any of the following means:

(i)(2)(i)(A) By location in a room, vault, or similar enclosure that is accessible only to qualified persons.

§ **1926.403(i)(2)(i)(B)**

(i)(2)(i)(B) By partitions or screens so arranged that only qualified persons will have access to the space within reach of the live parts. Any openings in such partitions or screens shall be so sized and located that persons are not likely to come into accidental contact with the live parts or to bring conducting objects into contact with them.

(i)(2)(i)(C) By location on a balcony, gallery, or platform so elevated and arranged as to exclude unqualified persons.

(i)(2)(i)(D) By elevation of 8 feet (2.44 m) or more above the floor or other working surface and so installed as to exclude unqualified persons.

(i)(2)(ii) In locations where electric equipment would be exposed to physical damage, enclosures or guards shall be so arranged and of such strength as to prevent such damage.

(i)(2)(iii) Entrances to rooms and other guarded locations containing exposed live parts shall be marked with conspicuous warning signs forbidding unqualified persons to enter.

(j) Over 600 volts, nominal.

(j)(1) General. Conductors and equipment used on circuits exceeding 600 volts, nominal, shall comply with all applicable provisions of paragraphs (a) through (g) of this section and with the following provisions which supplement or modify those requirements. The provisions of paragraphs (j)(2), (j)(3), and (j)(4) of this section do not apply to equipment on the supply side of the service conductors.

§ **1926.403(j)(2)**

(j)(2) Enclosure for electrical installations. Electrical installations in a vault, room, closet or in an area surrounded by a wall, screen, or fence, access to which is controlled by lock and key or other equivalent means, are considered to be accessible to qualified persons only. A wall, screen, or fence less than 8 feet (2.44 m) in height is not considered adequate to prevent access unless it has other features that provide a degree of isolation equivalent to an 8-foot (2.44-m) fence. The entrances to all buildings, rooms or enclosures containing exposed live parts or exposed conductors operating at over 600 volts, nominal, shall be kept locked or shall be under the observation of a qualified person at all times.

(j)(2)(i) Installations accessible to qualified persons only. Electrical installations having exposed live parts shall be accessible to qualified persons only and shall comply with the applicable provisions of paragraph (j)(3) of this section.

(j)(2)(ii) Installations accessible to unqualified persons. Electrical installations that are open to unqualified persons shall be made with metal-enclosed equipment or shall be enclosed in a vault or in an area, access to which is controlled by a lock. Metal-enclosed switchgear, unit substations, transformers, pull boxes, connection boxes, and other similar associated equipment shall be marked with appropriate caution signs. If equipment is exposed to physical damage from vehicular traffic, guards shall be provided to prevent such damage. Ventilating or similar openings in metal-enclosed equipment shall be designed so that foreign objects inserted through these openings will be deflected from energized parts.

§ **1926.403(j)(3)**

(j)(3) Workspace about equipment. Sufficient space shall be provided and maintained about electric equipment to permit ready and safe operation and maintenance of such equipment. Where energized parts are exposed, the minimum clear workspace shall not be less than 6 feet 6 inches (1.98 m) high (measured vertically from the floor or platform), or less than 3 feet (914 mm) wide (measured parallel to the equipment). The depth shall be as required in Table K-2. The workspace shall be adequate to permit at least a 90-degree opening of doors or hinged panels.

(j)(3)(i) Working space. The minimum clear working space in front of electric equipment such as switchboards, control panels, switches, circuit breakers, motor controllers, relays, and similar equipment shall not be less than specified in Table K-2 unless otherwise specified in this subpart. Distances shall be measured from the live parts if they are exposed, or from the enclosure front or opening if the live parts are enclosed. However, working space is not required in back of equipment such as deadfront switchboards or control assemblies where there are no renewable or adjustable parts (such as fuses or switches) on the back and where all connections are accessible from locations other than the back. Where rear access is required to work on de-energized parts on the back of enclosed equipment, a minimum working space of 30 inches (762 mm) horizontally shall be provided.

FIGURE 6.6 *(Continued)*

TABLE K-2 Minimum Depth of Clear Working Space in Front of Electric Equipment

Nominal voltage to ground	Conditions (1)		
	(a) Feet (2)	(b) Feet (2)	(c) Feet (2)
601 to 2,500	3	4	5
2,501 to 9,000	4	5	6
9,001 to 25,000	5	6	9
25,001 to 75 kV	6	8	10
Above 75kV	8	10	12

Footnote(1) Conditions (a), (b), and (c) are as follows:

(j)(3)(i)(a) Exposed live parts on one side and no live or grounded parts on the other side of the working space, or exposed live parts on both sides effectively guarded by insulating materials. Insulated wire or insulated busbars operating at not over 300 volts are not considered live parts.

(j)(3)(i)(b) Exposed live parts on one side and grounded parts on the other side. Walls constructed of concrete, brick, or tile are considered to be grounded surfaces.

(j)(3)(i)(c) Exposed live parts on both sides of the workspace [not guarded as provided in Condition (a)] with the operator between.

Footnote(2) NOTE: For SI units: one foot = 0.3048 m.

§ 1926.403(j)(3)(ii)

(j)(3)(ii) Lighting outlets and points of control. The lighting outlets shall be so arranged that persons changing lamps or making repairs on the lighting system will not be endangered by live parts or other equipment. The points of control shall be so located that persons are not likely to come in contact with any live part or moving part of the equipment while turning on the lights.

(j)(3)(iii) Elevation of unguarded live parts. Unguarded live parts above working space shall be maintained at elevations not less than specified in Table K-3.

(j)(4) Entrance and access to workspace. At least one entrance not less than 24 inches (610 mm) wide and 6 feet 6 inches (1.98 m) high shall be provided to give access to the working space about electric equipment. On switchboard and control panels exceeding 48 inches (1.22 m) in width, there shall be one entrance at each end of such board where practicable. Where bare energized parts at any voltage or insulated energized parts above 600 volts are located adjacent to such entrance, they shall be guarded.

[61 FR 5507, Feb. 13, 1996]

§ 1926.404 Wiring design and protection.

(a) Use and identification of grounded and grounding conductors—

(a)(1) Identification of conductors. A conductor used as a grounded conductor shall be identifiable and distinguishable from all other conductors. A conductor used as an equipment grounding conductor shall be identifiable and distinguishable from all other conductors.

TABLE K-3 Elevation of Unguarded Energized Parts Above Working Space

Nominal voltage between phases	Minimum Elevation
601–7,500	8 feet 6 inches.[1]
7,501–35,000	9 feet.
Over 35kV	9 feet + 0.37 inches per kV above 35kV.

Footnote(1) NOTE: For SI units: one inch = 25.4 mm; one foot = 0.3048 m

FIGURE 6.6 (*Continued*)

(a)(2) Polarity of connections. No grounded conductor shall be attached to any terminal or lead so as to reverse designated polarity.

(a)(3) Use of grounding terminals and devices. A grounding terminal or grounding-type device on a receptacle, cord connector, or attachment plug shall not be used for purposes other than grounding.

(b) Branch circuits—

(b)(1) Ground-fault protection—

(b)(1)(i) General. The employer shall use either ground fault circuit interrupters as specified in paragraph (b)(1)(ii) of this section or an assured equipment grounding conductor program as specified in paragraph (b)(1)(iii) of this section to protect employees on construction sites. These requirements are in addition to any other requirements for equipment grounding conductors.

§ 1926.404(b)(1)(ii)

(b)(1)(ii) Ground-fault circuit interrupters. All 120-volt, single-phase 15- and 20-ampere receptacle outlets on construction sites, which are not a part of the permanent wiring of the building or structure and which are in use by employees, shall have approved ground-fault circuit interrupters for personnel protection. Receptacles on a two-wire, single-phase portable or vehicle-mounted generator rated not more than 5kW, where the circuit conductors of the generator are insulated from the generator frame and all other grounded surfaces, need not be protected with ground-fault circuit interrupters.

(b)(1)(iii) Assured equipment grounding conductor program. The employer shall establish and implement an assured equipment grounding conductor program on construction sites covering all cord sets, receptacles which are not a part of the building or structure, and equipment connected by cord and plug which are available for use or used by employees. This program shall comply with the following minimum requirements:

(b)(1)(iii)(A) A written description of the program, including the specific procedures adopted by the employer, shall be available at the jobsite for inspection and copying by the Assistant Secretary and any affected employee.

(b)(1)(iii)(B) The employer shall designate one or more competent persons (as defined in 1926.32(f)) to implement the program.

(b)(1)(iii)(C) Each cord set, attachment cap, plug and receptacle of cord sets, and any equipment connected by cord and plug, except cord sets and receptacles which are fixed and not exposed to damage, shall be visually inspected before each day's use for external defects, such as

deformed or missing pins or insulation damage, and for indications of possible internal damage. Equipment found damaged or defective shall not be used until repaired.

§ 1926.404(b)(1)(iii)(D)

(b)(1)(iii)(D) The following tests shall be performed on all cord sets, receptacles which are not a part of the permanent wiring of the building or structure, and cord- and plug-connected equipment required to be grounded:

(b)(1)(iii)(D)(*1*) All equipment grounding conductors shall be tested for continuity and shall be electrically continuous.

(b)(1)(iii)(D)(*2*) Each receptacle and attachment cap or plug shall be tested for correct attachment of the equipment grounding conductor. The equipment grounding conductor shall be connected to its proper terminal.

(b)(1)(iii)(E) All required tests shall be performed:

(b)(1)(iii)(E)(*1*) Before first use;

(b)(1)(iii)(E)(*2*) Before equipment is returned to service following any repairs;

(b)(1)(iii)(E)(*3*) Before equipment is used after any incident which can be reasonably suspected to have caused damage (for example, when a cord set is run over); and

(b)(1)(iii)(E)(*4*) At intervals not to exceed 3 months, except that cord sets and receptacles which are fixed and not exposed to damage shall be tested at intervals not exceeding 6 months.

§ 1926.404(b)(1)(iii)(F)

(b)(1)(iii)(F) The employer shall not make available or permit the use by employees of any equipment which has not met the requirements of this paragraph (b)(1)(iii) of this section.

(b)(1)(iii)(G) Tests performed as required in this paragraph shall be recorded. This test record shall identify each receptacle, cord set, and cord- and plug-connected equipment that passed the test and shall indicate the last date it was tested or the interval for which it was tested. This record shall be kept by means of logs, color coding, or other effective means and shall be maintained until replaced by a more current record. The record shall be made available on the jobsite for inspection by the Assistant Secretary and any affected employee.

(b)(2) Outlet devices. Outlet devices shall have an ampere rating not less than the load to be served and shall comply with the following:

(b)(2)(i) Single receptacles. A single receptacle installed on an individual branch circuit

FIGURE 6.6 (*Continued*)

TABLE K-4 Receptacle Ratings for Various Size Circuits

Circuit rating amperes	Receptacle rating amperes
15	Not over 15.
20	15 or 20.
30	30.
40	40 or 50.
50	50.

shall have an ampere rating of not less than that of the branch circuit.

(b)(2)(ii) Two or more receptacles. Where connected to a branch circuit supplying two or more receptacles or outlets, receptacle ratings shall conform to the values listed in Table K-4.

(b)(2)(iii) Receptacles used for the connection of motors. The rating of an attachment plug or receptacle used for cord- and plug-connection of a motor to a branch circuit shall not exceed 15 amperes at 125 volts or 10 amperes at 250 volts if individual overload protection is omitted.

§ 1926.404(c)

(c) Outside conductors and lamps—

(c)(1) 600 volts, nominal, or less. Paragraphs (c)(1)(i) through (c)(1)(iv) of this section apply to branch circuit, feeder, and service conductors rated 600 volts, nominal, or less and run outdoors as open conductors.

(c)(1)(i) Conductors on poles. Conductors supported on poles shall provide a horizontal climbing space not less than the following:

(c)(1)(i)(A) Power conductors below communication conductors—30 inches (762 mm).

(c)(1)(i)(B) Power conductors alone or above communication conductors: 300 volts or less—24 inches (610 mm); more than 300 volts—30 inches (762 mm).

(c)(1)(i)(C) Communication conductors below power conductors: with power conductors 300 volts or less—24 inches (610 mm); more than 300 volts—30 inches (762 mm).

(c)(1)(ii) Clearance from ground. Open conductors shall conform to the following minimum clearances:

(c)(1)(ii)(A) 10 feet (3.05 m)—above finished grade, sidewalks, or from any platform or projection from which they might be reached.

(c)(1)(ii)(B) 12 feet (3.66 m)—over areas subject to vehicular traffic other than truck traffic.

(c)(1)(ii)(C) 15 feet (4.57 m)—over areas other than those specified in paragraph

(c)(1)(ii)(D) of this section that are subject to truck traffic.

§ 1926.404(c)(1)(ii)(D)

(c)(1)(ii)(D) 18 feet (5.49 m)—over public streets, alleys, roads, and driveways.

(c)(1)(iii) Clearance from building openings. Conductors shall have a clearance of at least 3 feet (914 mm) from windows, doors, fire escapes, or similar locations. Conductors run above the top level of a window are considered to be out of reach from that window and, therefore, do not have to be 3 feet (914 mm) away.

(c)(1)(iv) Clearance over roofs. Conductors above roof space accessible to employees on foot shall have a clearance from the highest point of the roof surface of not less than 8 feet (2.44 m) vertical clearance for insulated conductors, not less than 10 feet (3.05 m) vertical or diagonal clearance for covered conductors, and not less than 15 feet (4.57 m) for bare conductors, except that:

(c)(1)(iv)(A) Where the roof space is also accessible to vehicular traffic, the vertical clearance shall not be less than 18 feet (5.49 m), or

(c)(1)(iv)(B) Where the roof space is not normally accessible to employees on foot, fully insulated conductors shall have a vertical or diagonal clearance of not less than 3 feet (914 mm), or

(c)(1)(iv)(C) Where the voltage between conductors is 300 volts or less and the roof has a slope of not less than 4 inches (102 mm) in 12 inches (305 mm), the clearance from roofs shall be at least 3 feet (914 mm), or

§ 1926.404(c)(1)(iv)(D)

(c)(1)(iv)(D) Where the voltage between conductors is 300 volts or less and the conductors do not pass over more than 4 feet (1.22 m) of the overhang portion of the roof and they are terminated at a through-the-roof raceway or support, the clearance from roofs shall be at least 18 inches (457 mm).

(c)(2) Location of outdoor lamps. Lamps for outdoor lighting shall be located below all live conductors, transformers, or other electric equipment, unless such equipment is controlled by a disconnecting means that can be locked in the open position or unless adequate clearances or other safeguards are provided for relamping operations.

(d) Services—

(d)(1) Disconnecting means—

(d)(1)(i) General. Means shall be provided to disconnect all conductors in a building or other structure from the service-entrance conductors. The disconnecting means shall plainly

FIGURE 6.6 (*Continued*)

indicate whether it is in the open or closed position and shall be installed at a readily accessible location nearest the point of entrance of the service-entrance conductors.

(d)(1)(ii) Simultaneous opening of poles. Each service disconnecting means shall simultaneously disconnect all ungrounded conductors.

(d)(2) Services over 600 volts, nominal. The following additional requirements apply to services over 600 volts, nominal.

(d)(2)(i) Guarding. Service-entrance conductors installed as open wires shall be guarded to make them accessible only to qualified persons.

§ 1926.404(d)(2)(ii)

(d)(2)(ii) Warning signs. Signs warning of high voltage shall be posted where unauthorized employees might come in contact with live parts.

(e) Overcurrent protection—

(e)(1) 600 volts, nominal, or less. The following requirements apply to overcurrent protection of circuits rated 600 volts, nominal, or less.

(e)(1)(i) Protection of conductors and equipment. Conductors and equipment shall be protected from overcurrent in accordance with their ability to safely conduct current. Conductors shall have sufficient ampacity to carry the load.

(e)(1)(ii) Grounded conductors. Except for motor-running overload protection, overcurrent devices shall not interrupt the continuity of the grounded conductor unless all conductors of the circuit are opened simultaneously.

(e)(1)(iii) Disconnection of fuses and thermal cutouts. Except for devices provided for current-limiting on the supply side of the service disconnecting means, all cartridge fuses which are accessible to other than qualified persons and all fuses and thermal cutouts on circuits over 150 volts to ground shall be provided with disconnecting means. This disconnecting means shall be installed so that the fuse or thermal cutout can be disconnected from its supply without disrupting service to equipment and circuits unrelated to those protected by the overcurrent device.

§ 1926.404(e)(1)(iv)

(e)(1)(iv) Location in or on premises. Overcurrent devices shall be readily accessible. Overcurrent devices shall not be located where they could create an employee safety hazard by being exposed to physical damage or located in the vicinity of easily ignitable material.

(e)(1)(v) Arcing or suddenly moving parts. Fuses and circuit breakers shall be so located or shielded that employees will not be burned or otherwise injured by their operation.

(e)(1)(vi) Circuit breakers—

(e)(1)(vi)(A) Circuit breakers shall clearly indicate whether they are in the open (off) or closed (on) position.

(e)(1)(vi)(B) Where circuit breaker handles on switchboards are operated vertically rather than horizontally or rotationally, the up position of the handle shall be the closed (on) position.

(e)(1)(vi)(C) If used as switches in 120-volt, fluorescent lighting circuits, circuit breakers shall be marked "SWD."

(e)(2) Over 600 volts, nominal. Feeders and branch circuits over 600 volts, nominal, shall have short-circuit protection.

(f) Grounding. Paragraphs (f)(1) through (f)(11) of this section contain grounding requirements for systems, circuits, and equipment.

(f)(1) Systems to be grounded. The following systems which supply premises wiring shall be grounded:

§ 1926.404(f)(1)(i)

(f)(1)(i) Three-wire DC systems. All 3-wire DC systems shall have their neutral conductor grounded.

(f)(1)(ii) Two-wire DC systems. Two-wire DC systems operating at over 50 volts through 300 volts between conductors shall be grounded unless they are rectifier-derived from an AC system complying with paragraphs (f)(1)(iii), (f)(1)(iv), and (f)(1)(v) of this section.

(f)(1)(iii) AC circuits, less than 50 volts. AC circuits of less than 50 volts shall be grounded if they are installed as overhead conductors outside of buildings or if they are supplied by transformers and the transformer primary supply system is ungrounded or exceeds 150 volts to ground.

(f)(1)(iv) AC systems, 50 volts to 1000 volts. AC systems of 50 volts to 1000 volts shall be grounded under any of the following conditions, unless exempted by paragraph (f)(1)(v) of this section:

(f)(1)(iv)(A) If the system can be so grounded that the maximum voltage to ground on the ungrounded conductors does not exceed 150 volts;

(f)(1)(iv)(B) If the system is nominally rated 480Y/277 volt, 3-phase, 4-wire in which the neutral is used as a circuit conductor;

(f)(1)(iv)(C) If the system is nominally rated 240/120 volt, 3-phase, 4-wire in which

FIGURE 6.6   (*Continued*)

the midpoint of one phase is used as a circuit conductor; or

§ 1926.404(f)(1)(iv)(D)

(f)(1)(iv)(D) If a service conductor is uninsulated.

(f)(1)(v) Exceptions. AC systems of 50 volts to 1000 volts are not required to be grounded if the system is separately derived and is supplied by a transformer that has a primary voltage rating less than 1000 volts, provided all of the following conditions are met:

(f)(1)(v)(A) The system is used exclusively for control circuits,

(f)(1)(v)(B) The conditions of maintenance and supervision assure that only qualified persons will service the installation,

(f)(1)(v)(C) Continuity of control power is required, and

(f)(1)(v)(D) Ground detectors are installed on the control system.

(f)(2) Separately derived systems. Where paragraph (f)(1) of this section requires grounding of wiring systems whose power is derived from generator, transformer, or converter windings and has no direct electrical connection, including a solidly connected grounded circuit conductor, to supply conductors originating in another system, paragraph (f)(5) of this section shall also apply.

§ 1926.404(f)(3)

(f)(3) Portable and vehicle-mounted generators—

(f)(3)(i) Portable generators. Under the following conditions, the frame of a portable generator need not be grounded and may serve as the grounding electrode for a system supplied by the generator:

(f)(3)(i)(A) The generator supplies only equipment mounted on the generator and/or cord- and plug-connected equipment through receptacles mounted on the generator, and

(f)(3)(i)(B) The noncurrent-carrying metal parts of equipment and the equipment grounding conductor terminals of the receptacles are bonded to the generator frame.

(f)(3)(ii) Vehicle-mounted generators. Under the following conditions the frame of a vehicle may serve as the grounding electrode for a system supplied by a generator located on the vehicle:

(f)(3)(ii)(A) The frame of the generator is bonded to the vehicle frame, and

(f)(3)(ii)(B) The generator supplies only equipment located on the vehicle and/or cord- and plug-connected equipment through receptacles mounted on the vehicle or on the generator, and

(f)(3)(ii)(C) The noncurrent-carrying metal parts of equipment and the equipment grounding conductor terminals of the receptacles are bonded to the generator frame, and

(f)(3)(ii)(D) The system complies with all other provisions of this section.

§ 1926.404(f)(3)(iii)

(f)(3)(iii) Neutral conductor bonding. A neutral conductor shall be bonded to the generator frame if the generator is a component of a separately derived system. No other conductor need be bonded to the generator frame.

(f)(4) Conductors to be grounded. For AC premises wiring systems the identified conductor shall be grounded.

(f)(5) Grounding connections—

(f)(5)(i) Grounded system. For a grounded system, a grounding electrode conductor shall be used to connect both the equipment grounding conductor and the grounded circuit conductor to the grounding electrode. Both the equipment grounding conductor and the grounding electrode conductor shall be connected to the grounded circuit conductor on the supply side of the service disconnecting means, or on the supply side of the system disconnecting means or overcurrent devices if the system is separately derived.

(f)(5)(ii) Ungrounded systems. For an ungrounded service-supplied system, the equipment grounding conductor shall be connected to the grounding electrode conductor at the service equipment. For an ungrounded separately derived system, the equipment grounding conductor shall be connected to the grounding electrode conductor at, or ahead of, the system disconnecting means or overcurrent devices.

(f)(6) Grounding path. The path to ground from circuits, equipment, and enclosures shall be permanent and continuous.

§ 1926.404(f)(7)

(f)(7) Supports, enclosures, and equipment to be grounded—

(f)(7)(i) Supports and enclosures for conductors. Metal cable trays, metal raceways, and metal enclosures for conductors shall be grounded, except that:

(f)(7)(i)(A) Metal enclosures such as sleeves that are used to protect cable assemblies from physical damage need not be grounded; and

(f)(7)(i)(B) Metal enclosures for conductors added to existing installations of open wire,

FIGURE 6.6 (*Continued*)

knob-and-tube wiring, and nonmetallic-sheathed cable need not be grounded if all of the following conditions are met:

(f)(7)(i)(B)(*1*) Runs are less than 25 feet (7.62 m);

(f)(7)(i)(B)(*2*) Enclosures are free from probable contact with ground, grounded metal, metal laths, or other conductive materials; and

(f)(7)(i)(B)(*3*) Enclosures are guarded against employee contact.

(f)(7)(ii) Service equipment enclosures. Metal enclosures for service equipment shall be grounded.

(f)(7)(iii) Fixed equipment. Exposed noncurrent-carrying metal parts of fixed equipment which may become energized shall be grounded under any of the following conditions:

(f)(7)(iii)(A) If within 8 feet (2.44 m) vertically or 5 feet (1.52 m) horizontally of ground or grounded metal objects and subject to employee contact.

§ 1926.404(f)(7)(iii)(B)

(f)(7)(iii)(B) If located in a wet or damp location and subject to employee contact.

(f)(7)(iii)(C) If in electrical contact with metal.

(f)(7)(iii)(D) If in a hazardous (classified) location.

(f)(7)(iii)(E) If supplied by a metal-clad, metal-sheathed, or grounded metal raceway wiring method.

(f)(7)(iii)(F) If equipment operates with any terminal at over 150 volts to ground; however, the following need not be grounded:

(f)(7)(iii)(F)(*1*) Enclosures for switches or circuit breakers used for other than service equipment and accessible to qualified persons only;

(f)(7)(iii)(F)(*2*) Metal frames of electrically heated appliances which are permanently and effectively insulated from ground; and

(f)(7)(iii)(F)(*3*) The cases of distribution apparatus such as transformers and capacitors mounted on wooden poles at a height exceeding 8 feet (2.44 m) above ground or grade level.

§ 1926.404(f)(7)(iv)

(f)(7)(iv) Equipment connected by cord and plug. Under any of the conditions described in paragraphs (f)(7)(iv)(A) through (f)(7)(iv)(C) of this section, exposed noncurrent-carrying metal parts of cord- and plug-connected equipment which may become energized shall be grounded:

(f)(7)(iv)(A) If in a hazardous (classified) location (see 1926.407).

(f)(7)(iv)(B) If operated at over 150 volts to ground, except for guarded motors and metal frames of electrically heated appliances if the appliance frames are permanently and effectively insulated from ground.

(f)(7)(iv)(C) If the equipment is one of the types listed in paragraphs (f)(7)(iv)(C)(*1*) through (f)(7)(iv)(C)(*5*) of this section. However, even though the equipment may be one of these types, it need not be grounded if it is exempted by paragraph (f)(7)(iv)(C)(6).

(f)(7)(iv)(C)(*1*) Hand held motor-operated tools;

(f)(7)(iv)(C)(*2*) Cord- and plug-connected equipment used in damp or wet locations or by employees standing on the ground or on metal floors or working inside of metal tanks or boilers;

(f)(7)(iv)(C)(*3*) Portable and mobile X-ray and associated equipment;

(f)(7)(iv)(C)(*4*) Tools likely to be used in wet and/or conductive locations;

(f)(7)(iv)(C)(*5*) Portable hand lamps.

§ 1926.404(f)(7)(iv)(C)(6)

(f)(7)(iv)(C)(*6*) Tools likely to be used in wet and/or conductive locations need not be grounded if supplied through an isolating transformer with an ungrounded secondary of not over 50 volts. Listed or labeled portable tools and appliances protected by a system of double insulation, or its equivalent, need not be grounded. If such a system is employed, the equipment shall be distinctively marked to indicate that the tool or appliance utilizes a system of double insulation.

(f)(7)(v) Nonelectrical equipment. The metal parts of the following nonelectrical equipment shall be grounded: Frames and tracks of electrically operated cranes; frames of nonelectrically driven elevator cars to which electric conductors are attached; hand-operated metal shifting ropes or cables of electric elevators, and metal partitions, grill work, and similar metal enclosures around equipment of over IkV between conductors.

(f)(8) Methods of grounding equipment—

(f)(8)(i) With circuit conductors. Noncurrent-carrying metal parts of fixed equipment, if required to be grounded by this subpart, shall be grounded by an equipment grounding conductor which is contained within the same raceway, cable, or cord, or runs with or encloses the circuit conductors. For DC circuits only, the equipment grounding conductor may be run separately from the circuit conductors.

FIGURE 6.6 (*Continued*)

(f)(8)(ii) Grounding conductor. A conductor used for grounding fixed or movable equipment shall have capacity to conduct safely any fault current which may be imposed on it.

§ 1926.404(f)(8)(iii)

(f)(8)(iii) Equipment considered effectively grounded. Electric equipment is considered to be effectively grounded if it is secured to, and in electrical contact with, a metal rack or structure that is provided for its support and the metal rack or structure is grounded by the method specified for the noncurrent-carrying metal parts of fixed equipment in paragraph (f)(8)(i) of this section. Metal car frames supported by metal hoisting cables attached to or running over metal sheaves or drums of grounded elevator machines are also considered to be effectively grounded.

(f)(9) Bonding. If bonding conductors are used to assure electrical continuity, they shall have the capacity to conduct any fault current which may be imposed.

(f)(10) Made electrodes. If made electrodes are used, they shall be free from nonconductive coatings, such as paint or enamel; and, if practicable, they shall be embedded below permanent moisture level. A single electrode consisting of a rod, pipe or plate which has a resistance to ground greater than 25 ohms shall be augmented by one additional electrode installed no closer than 6 feet (1.83 m) to the first electrode.

(f)(11) Grounding of systems and circuits of 1000 volts and over (high voltage)—

(f)(11)(i) General. If high voltage systems are grounded, they shall comply with all applicable provisions of paragraphs (f)(1) through (f)(10) of this section as supplemented and modified by this paragraph (f)(11).

(f)(11)(ii) Grounding of systems supplying portable or mobile equipment. Systems supplying portable or mobile high voltage equipment, other than substations installed on a temporary basis, shall comply with the following:

§ 1926.404(f)(11)(ii)(A)

(f)(11)(ii)(A) Portable and mobile high voltage equipment shall be supplied from a system having its neutral grounded through an impedance. If a delta-connected high voltage system is used to supply the equipment, a system neutral shall be derived.

(f)(11)(ii)(B) Exposed noncurrent-carrying metal parts of portable and mobile equipment shall be connected by an equipment grounding conductor to the point at which the system neutral impedance is grounded.

(f)(11)(ii)(C) Ground-fault detection and relaying shall be provided to automatically de-energize any high voltage system component which has developed a ground fault. The continuity of the equipment grounding conductor shall be continuously monitored so as to de-energize automatically the high voltage feeder to the portable equipment upon loss of continuity of the equipment grounding conductor.

(f)(11)(ii)(D) The grounding electrode to which the portable or mobile equipment system neutral impedance is connected shall be isolated from and separated in the ground by at least 20 feet (6.1 m) from any other system or equipment grounding electrode, and there shall be no direct connection between the grounding electrodes, such as buried pipe, fence or like objects.

§ 1926.404(f)(11)(iii)

(f)(11)(iii) Grounding of equipment. All noncurrent-carrying metal parts of portable equipment and fixed equipment including their associated fences, housings, enclosures, and supporting structures shall be grounded. However, equipment which is guarded by location and isolated from ground need not be grounded. Additionally, pole-mounted distribution apparatus at a height exceeding 8 feet (2.44 m) above ground or grade level need not be grounded.

[54 FR 24334, June 7, 1989; 61 FR 5507, Feb. 13, 1996]

§ 1926.405 Wiring methods, components, and equipment for general use.

(a) Wiring methods. The provisions of this paragraph do not apply to conductors which form an integral part of equipment such as motors, controllers, motor control centers and like equipment.

(a)(1) General requirements—

(a)(1)(i) Electrical continuity of metal raceways and enclosures. Metal raceways, cable armor, and other metal enclosures for conductors shall be metallically joined together into a continuous electric conductor and shall be so connected to all boxes, fittings, and cabinets as to provide effective electrical continuity.

(a)(1)(ii) Wiring in ducts. No wiring systems of any type shall be installed in ducts used to transport dust, loose stock or flammable vapors. No wiring system of any type shall be installed in any duct used for vapor removal or in any shaft containing only such ducts.

FIGURE 6.6 *(Continued)*

§ 1926.405(a)(2)

(a)(2) Temporary wiring—

(a)(2)(i) Scope. The provisions of paragraph (a)(2) of this section apply to temporary electrical power and lighting wiring methods which may be of a class less than would be required for a permanent installation. Except as specifically modified in paragraph (a)(2) of this section, all other requirements of this subpart for permanent wiring shall apply to temporary wiring installations. Temporary wiring shall be removed immediately upon completion of construction or the purpose for which the wiring was installed.

(a)(2)(ii) General requirements for temporary wiring—

(a)(2)(ii)(A) Feeders shall originate in a distribution center. The conductors shall be run as multiconductor cord or cable assemblies or within raceways; or, where not subject to physical damage, they may be run as open conductors on insulators not more than 10 feet (3.05 m) apart.

(a)(2)(ii)(B) Branch circuits shall originate in a power outlet or panelboard. Conductors shall be run as multiconductor cord or cable assemblies or open conductors, or shall be run in raceways. All conductors shall be protected by overcurrent devices at their ampacity. Runs of open conductors shall be located where the conductors will not be subject to physical damage, and the conductors shall be fastened at intervals not exceeding 10 feet (3.05 m). No branch-circuit conductors shall be laid on the floor. Each branch circuit that supplies receptacles or fixed equipment shall contain a separate equipment grounding conductor if the branch circuit is run as open conductors.

§ 1926.405(a)(2)(ii)(C)

(a)(2)(ii)(C) Receptacles shall be of the grounding type. Unless installed in a complete metallic raceway, each branch circuit shall contain a separate equipment grounding conductor, and all receptacles shall be electrically connected to the grounding conductor. Receptacles for uses other than temporary lighting shall not be installed on branch circuits which supply temporary lighting. Receptacles shall not be connected to the same ungrounded conductor of multiwire circuits which supply temporary lighting.

(a)(2)(ii)(D) Disconnecting switches or plug connectors shall be installed to permit the disconnection of all ungrounded conductors of each temporary circuit.

(a)(2)(ii)(E) All lamps for general illumination shall be protected from accidental contact or breakage. Metal-case sockets shall be grounded.

(a)(2)(ii)(F) Temporary lights shall not be suspended by their electric cords unless cords and lights are designed for this means of suspension.

(a)(2)(ii)(G) Portable electric lighting used in wet and/or other conductive locations, as for example, drums, tanks, and vessels, shall be operated at 12 volts or less. However, 120-volt lights may be used if protected by a ground-fault circuit interrupter.

(a)(2)(ii)(H) A box shall be used wherever a change is made to a raceway system or a cable system which is metal clad or metal sheathed.

(a)(2)(ii)(I) Flexible cords and cables shall be protected from damage. Sharp corners and projections shall be avoided. Flexible cords and cables may pass through doorways or other pinch points, if protection is provided to avoid damage.

§ 1926.405(a)(2)(ii)(J)

(a)(2)(ii)(J) Extension cord sets used with portable electric tools and appliances shall be of three-wire type and shall be designed for hard or extra-hard usage. Flexible cords used with temporary and portable lights shall be designed for hard or extra-hard usage.

NOTE: The National Electrical Code, ANSI/NFPA 70, in Article 400, Table 400-4, lists various types of flexible cords, some of which are noted as being designed for hard or extra-hard usage. Examples of these types of flexible cords include hard service cord (types S, ST, SO, STO) and junior hard service cord (types SJ, SJO, SJT, SJTO).

(a)(2)(iii) Guarding. For temporary wiring over 600 volts, nominal, fencing, barriers, or other effective means shall be provided to prevent access of other than authorized and qualified personnel.

(b) Cabinets, boxes, and fittings.

(b)(1) Conductors entering boxes, cabinets, or fittings. Conductors entering boxes, cabinets, or fittings shall be protected from abrasion, and openings through which conductors enter shall be effectively closed. Unused openings in cabinets, boxes, and fittings shall also be effectively closed.

(b)(2) Covers and canopies. All pull boxes, junction boxes, and fittings shall be provided with covers. If metal covers are used, they shall be grounded. In energized installations each outlet box shall have a cover, faceplate, or fixture canopy. Covers of outlet boxes having holes through which flexible cord pendants

FIGURE 6.6 *(Continued)*

pass shall be provided with bushings designed for the purpose or shall have smooth, well-rounded surfaces on which the cords may bear.

§ 1926.405(b)(3)

(b)(3) Pull and junction boxes for systems over 600 volts, nominal. In addition to other requirements in this section for pull and junction boxes, the following shall apply to these boxes for systems over 600 volts, nominal:

(b)(3)(i) Complete enclosure. Boxes shall provide a complete enclosure for the contained conductors or cables.

(b)(3)(ii) Covers. Boxes shall be closed by covers securely fastened in place. Underground box covers that weigh over 100 pounds (43.6 kg) meet this requirement. Covers for boxes shall be permanently marked "HIGH VOLTAGE." The marking shall be on the outside of the box cover and shall be readily visible and legible.

(c) Knife switches. Single-throw knife switches shall be so connected that the blades are dead when the switch is in the open position. Single-throw knife switches shall be so placed that gravity will not tend to close them. Single-throw knife switches approved for use in the inverted position shall be provided with a locking device that will ensure that the blades remain in the open position when so set. Double-throw knife switches may be mounted so that the throw will be either vertical or horizontal. However, if the throw is vertical, a locking device shall be provided to ensure that the blades remain in the open position when so set.

§ 1926.405(d)

(d) Switchboards and panelboards. Switchboards that have any exposed live parts shall be located in permanently dry locations and accessible only to qualified persons. Panelboards shall be mounted in cabinets, cutout boxes, or enclosures designed for the purpose and shall be dead front. However, panelboards other than the dead front externally-operable type are permitted where accessible only to qualified persons. Exposed blades of knife switches shall be dead when open.

(e) Enclosures for damp or wet locations.

(e)(1) Cabinets, fittings, and boxes. Cabinets, cutout boxes, fittings, boxes, and panelboard enclosures in damp or wet locations shall be installed so as to prevent moisture or water from entering and accumulating within the enclosures. In wet locations the enclosures shall be weatherproof.

(e)(2) Switches and circuit breakers. Switches, circuit breakers, and switchboards installed in wet locations shall be enclosed in weatherproof enclosures.

(f) Conductors for general wiring. All conductors used for general wiring shall be insulated unless otherwise permitted in this Subpart. The conductor insulation shall be of a type that is suitable for the voltage, operating temperature, and location of use. Insulated conductors shall be distinguishable by appropriate color or other means as being grounded conductors, ungrounded conductors, or equipment grounding conductors.

(g) Flexible cords and cables—

(g)(1) Use of flexible cords and cables—

(g)(1)(i) Permitted uses. Flexible cords and cables shall be suitable for conditions of use and location. Flexible cords and cables shall be used only for:

§ 1926.405(g)(1)(i)(A)

(g)(1)(i)(A) Pendants;

(g)(1)(i)(B) Wiring of fixtures;

(g)(1)(i)(C) Connection of portable lamps or appliances;

(g)(1)(i)(D) Elevator cables;

(g)(1)(i)(E) Wiring of cranes and hoists;

(g)(1)(i)(F) Connection of stationary equipment to facilitate their frequent interchange;

(g)(1)(i)(G) Prevention of the transmission of noise or vibration; or

(g)(1)(i)(H) Appliances where the fastening means and mechanical connections are designed to permit removal for maintenance and repair.

(g)(1)(ii) Attachment plugs for cords. If used as permitted in paragraphs (g)(1)(i)(C), (g)(1)(i)(F), or (g)(1)(i)(H) of this section, the flexible cord shall be equipped with an attachment plug and shall be energized from a receptacle outlet.

(g)(1)(iii) Prohibited uses. Unless necessary for a use permitted in paragraph (g)(1)(i) of this section, flexible cords and cables shall not be used:

§ 1926.405(g)(1)(iii)(A)

(g)(1)(iii)(A) As a substitute for the fixed wiring of a structure;

(g)(1)(iii)(B) Where run through holes in walls, ceilings, or floors;

(g)(1)(iii)(C) Where run through doorways, windows, or similar openings, except as permitted in paragraph (a)(2)(ii)(1) of this section;

(g)(1)(iii)(D) Where attached to building surfaces; or

(g)(1)(iii)(E) Where concealed behind building walls, ceilings, or floors.

FIGURE 6.6 (*Continued*)

(g)(2) Identification, splices, and terminations—

(g)(2)(i) Identification. A conductor of a flexible cord or cable that is used as a grounded conductor or an equipment grounding conductor shall be distinguishable from other conductors.

(g)(2)(ii) Marking. Type SJ, SJO, SJT, SJTO, S, SO, ST, and STO cords shall not be used unless durably marked on the surface with the type designation, size, and number of conductors.

(g)(2)(iii) Splices. Flexible cords shall be used only in continuous lengths without splice or tap. Hard service flexible cords No. 12 or larger may be repaired if spliced so that the splice retains the insulation, outer sheath properties, and usage characteristics of the cord being spliced.

§ 1926.405(g)(2)(iv)

(g)(2)(iv) Strain relief. Flexible cords shall be connected to devices and fittings so that strain relief is provided which will prevent pull from being directly transmitted to joints or terminal screws.

(g)(2)(v) Cords passing through holes. Flexible cords and cables shall be protected by bushings or fittings where passing through holes in covers, outlet boxes, or similar enclosures.

(h) Portable cables over 600 volts, nominal. Multiconductor portable cable for use in supplying power to portable or mobile equipment at over 600 volts, nominal, shall consist of No. 8 or larger conductors employing flexible stranding. Cables operated at over 2000 volts shall be shielded for the purpose of confining the voltage stresses to the insulation. Grounding conductors shall be provided. Connectors for these cables shall be of a locking type with provisions to prevent their opening or closing while energized. Strain relief shall be provided at connections and terminations. Portable cables shall not be operated with splices unless the splices are of the permanent molded, vulcanized, or other equivalent type. Termination enclosures shall be marked with a high voltage hazard warning, and terminations shall be accessible only to authorized and qualified personnel.

(i) Fixture wires—

(i)(1) General. Fixture wires shall be suitable for the voltage, temperature, and location of use. A fixture wire which is used as a grounded conductor shall be identified.

§ 1926.405(i)(2)

(i)(2) Uses permitted. Fixture wires may be used:

(i)(2)(i) For installation in lighting, fixtures and in similar equipment where enclosed or protected and not subject to bending or twisting in use; or

(i)(2)(ii) For connecting lighting fixtures to the branch-circuit conductors supplying the fixtures.

(i)(3) Uses not permitted. Fixture wires shall not be used as branch-circuit conductors except as permitted for Class 1 power-limited circuits.

(j) Equipment for general use—

(j)(1) Lighting fixtures, lampholders, lamps, and receptacles—

(j)(1)(i) Live parts. Fixtures, lampholders, lamps, rosettes, and receptacles shall have no live parts normally exposed to employee contact. However, rosettes and cleat-type lampholders and receptacles located at least 8 feet (2.44 m) above the floor may have exposed parts.

§ 1926.405(j)(1)(ii)

(j)(1)(ii) Support. Fixtures, lampholders, rosettes, and receptacles shall be securely supported. A fixture that weighs more than 6 pounds (2.72 kg) or exceeds 16 inches (406 mm) in any dimension shall not be supported by the screw shell of a lampholder.

(j)(1)(iii) Portable lamps. Portable lamps shall be wired with flexible cord and an attachment plug of the polarized or grounding type. If the portable lamp uses an Edison-based lampholder, the grounded conductor shall be identified and attached to the screw shell and the identified blade of the attachment plug. In addition, portable handlamps shall comply with the following:

(j)(1)(iii)(A) Metal shell, paperlined lampholders shall not be used;

(j)(1)(iii)(B) Handlamps shall be equipped with a handle of molded composition or other insulating material;

(j)(1)(iii)(C) Handlamps shall be equipped with a substantial guard attached to the lampholder or handle;

(j)(1)(iii)(D) Metallic guards shall be grounded by the means of an equipment grounding conductor run within the power supply cord.

(j)(1)(iv) Lampholders. Lampholders of the screw-shell type shall be installed for use as lampholders only. Lampholders installed in wet or damp locations shall be of the weatherproof type.

§ 1926.405(j)(1)(v)

(j)(1)(v) Fixtures. Fixtures installed in wet or damp locations shall be identified for the

FIGURE 6.6 (*Continued*)

purpose and shall be installed so that water cannot enter or accumulate in wireways, lampholders, or other electrical parts.

(j)(2) Receptacles, cord connectors, and attachment plugs (caps)—

(j)(2)(i) Configuration. Receptacles, cord connectors, and attachment plugs shall be constructed so that no receptacle or cord connector will accept an attachment plug with a different voltage or current rating than that for which the device is intended. However, a 20-ampere T-slot receptacle or cord connector may accept a 15-ampere attachment plug of the same voltage rating. Receptacles connected to circuits having different voltages, frequencies, or types of current (ac or dc) on the same premises shall be of such design that the attachment plugs used on these circuits are not interchangeable.

(j)(2)(ii) Damp and wet locations. A receptacle installed in a wet or damp location shall be designed for the location.

(j)(3) Appliances—

(j)(3)(i) Live parts. Appliances, other than those in which the current-carrying parts at high temperatures are necessarily exposed, shall have no live parts normally exposed to employee contact.

(j)(3)(ii) Disconnecting means. A means shall be provided to disconnect each appliance.

§ **1926.405(j)(3)(iii)**

(j)(3)(iii) Rating. Each appliance shall be marked with its rating in volts and amperes or volts and watts.

(j)(4) Motors. This paragraph applies to motors, motor circuits, and controllers.

(j)(4)(i) In sight from. If specified that one piece of equipment shall be "in sight from" another piece of equipment, one shall be visible and not more than 50 feet (15.2 m) from the other.

(j)(4)(ii) Disconnecting means—

(j)(4)(ii)(A) A disconnecting means shall be located in sight from the controller location. The controller disconnecting means for motor branch circuits over 600 volts, nominal, may be out of sight of the controller, if the controller is marked with a warning label giving the location and identification of the disconnecting means which is to be locked in the open position.

(j)(4)(ii)(B) The disconnecting means shall disconnect the motor and the controller from all ungrounded supply conductors and shall be so designed that no pole can be operated independently.

(j)(4)(ii)(C) If a motor and the driven machinery are not in sight from the controller location, the installation shall comply with one of the following conditions:

§ **1926.405(j)(4)(ii)(C)(***1***)**

(j)(4)(ii)(C)(*1*) The controller disconnecting means shall be capable of being locked in the open position.

(j)(4)(ii)(C)(*2*) A manually operable switch that will disconnect the motor from its source of supply shall be placed in sight from the motor location.

(j)(4)(ii)(D) The disconnecting means shall plainly indicate whether it is in the open (off) or closed (on) position.

(j)(4)(ii)(E) The disconnecting means shall be readily accessible. If more than one disconnect is provided for the same equipment, only one need be readily accessible.

(j)(4)(ii)(F) An individual disconnecting means shall be provided for each motor, but a single disconnecting means may be used for a group of motors under any one of the following conditions:

(j)(4)(ii)(F)(*1*) If a number of motors drive special parts of a single machine or piece of apparatus, such as a metal or woodworking machine, crane, or hoist;

(j)(4)(ii)(F)(*2*) If a group of motors is under the protection of one set of branch-circuit protective devices; or

(j)(4)(ii)(F)(*3*) If a group of motors is in a single room in sight from the location of the disconnecting means.

§ **1926.405(j)(4)(iii)**

(j)(4)(iii) Motor overload, short-circuit, and ground-fault protection. Motors, motor-control apparatus, and motor branch-circuit conductors shall be protected against overheating due to motor overloads or failure to start, and against short-circuits or ground faults. These provisions do not require overload protection that will stop a motor where a shutdown is likely to introduce additional or increased hazards, as in the case of fire pumps, or where continued operation of a motor is necessary for a safe shutdown of equipment or process and motor overload sensing devices are connected to a supervised alarm.

(j)(4)(iv) Protection of live parts—all voltages—

(j)(4)(iv)(A) Stationary motors having commutators, collectors, and brush rigging located inside of motor end brackets and not conductively connected to supply circuits operating at more than 150 volts to ground need not have such parts guarded. Exposed live parts of motors

FIGURE 6.6　(*Continued*)

and controllers operating at 50 volts or more between terminals shall be guarded against accidental contact by any of the following:

(j)(4)(iv)(A)(*1*) By installation in a room or enclosure that is accessible only to qualified persons;

(j)(4)(iv)(A)(*2*) By installation on a balcony, gallery, or platform, so elevated and arranged as to exclude unqualified persons; or

(j)(4)(iv)(A)(*3*) By elevation 8 feet (2.44 m) or more above the floor.

(j)(4)(iv)(B) Where live parts of motors or controllers operating at over 150 volts to ground are guarded against accidental contact only by location, and where adjustment or other attendance may be necessary during the operation of the apparatus, insulating mats or platforms shall be provided so that the attendant cannot readily touch live parts unless standing on the mats or platforms.

§ 1926.405(j)(5)

(j)(5) Transformers—

(j)(5)(i) Application. The following paragraphs cover the installation of all transformers, except:

(j)(5)(i)(A) Current transformers;

(j)(5)(i)(B) Dry-type transformers installed as a component part of other apparatus;

(j)(5)(i)(C) Transformers which are an integral part of an X-ray, high frequency, or electrostatic-coating apparatus;

(j)(5)(i)(D) Transformers used with Class 2 and Class 3 circuits, sign and outline lighting, electric discharge lighting, and power-limited fire-protective signaling circuits.

(j)(5)(ii) Operating voltage. The operating voltage of exposed live parts of transformer installations shall be indicated by warning signs or visible markings on the equipment or structure.

(j)(5)(iii) Transformers over 35 kV. Dry-type, high fire point liquid-insulated, and askarel-insulated transformers installed indoors and rated over 35 kV shall be in a vault.

(j)(5)(iv) Oil-insulated transformers. If they present a fire hazard to employees, oil-insulated transformers installed indoors shall be in a vault.

§ 1926.405(j)(5)(v)

(j)(5)(v) Fire protection. Combustible material, combustible buildings and parts of buildings, fire escapes, and door and window openings shall be safeguarded from fires which may originate in oil-insulated transformers

attached to or adjacent to a building or combustible material.

(j)(5)(vi) Transformer vaults. Transformer vaults shall be constructed so as to contain fire and combustible liquids within the vault and to prevent unauthorized access. Locks and latches shall be so arranged that a vault door can be readily opened from the inside.

(j)(5)(vii) Pipes and ducts. Any pipe or duct system foreign to the vault installation shall not enter or pass through a transformer vault.

(j)(5)(viii) Material storage. Materials shall not be stored in transformer vaults.

(j)(6) Capacitors—

(j)(6)(i) Drainage of stored charge. All capacitors, except surge capacitors or capacitors included as a component part of other apparatus, shall be provided with an automatic means of draining the stored charge and maintaining the discharged state after the capacitor is disconnected from its source of supply.

(j)(6)(ii) Over 600 volts. Capacitors rated over 600 volts, nominal, shall comply with the following additional requirements:

§ 1926.405(j)(6)(ii)(A)

(j)(6)(ii)(A) Isolating or disconnecting switches (with no interrupting rating) shall be interlocked with the load interrupting device or shall be provided with prominently displayed caution signs to prevent switching load current.

(j)(6)(ii)(B) For series capacitors the proper switching shall be assured by use of at least one of the following:

(j)(6)(ii)(B)(*1*) Mechanically sequenced isolating and bypass switches,

(j)(6)(ii)(B)(*2*) Interlocks, or

(j)(6)(ii)(B)(*3*) Switching procedure prominently displayed at the switching location.

[61 FR 5507, Feb. 13, 1996]

§ 1926.406 Specific purpose equipment and installations.

(a) Cranes and hoists. This paragraph applies to the installation of electric equipment and wiring used in connection with cranes, monorail hoists, hoists, and all runways.

(a)(1) Disconnecting means—

(a)(1)(i) Runway conductor disconnecting means. A readily accessible disconnecting means shall be provided between the runway contact conductors and the power supply.

(a)(1)(ii) Disconnecting means for cranes and monorail hoists. A disconnecting means, capable of being locked in the open position, shall

FIGURE 6.6 (*Continued*)

be provided in the leads from the runway contact conductors or other power supply on any crane or monorail hoist.

(a)(1)(ii)(A) If this additional disconnecting means is not readily accessible from the crane or monorail hoist operating station, means shall be provided at the operating station to open the power circuit to all motors of the crane or monorail hoist.

(a)(1)(ii)(B) The additional disconnect may be omitted if a monorail hoist or hand-propelled crane bridge installation meets all of the following:

§ 1926.406(a)(1)(ii)(B)(*1*)

(a)(1)(ii)(B)(*1*) The unit is floor controlled;
(a)(1)(ii)(B)(*2*) The unit is within view of the power supply disconnecting means; and
(a)(1)(ii)(B)(*3*) No fixed work platform has been provided for servicing the unit.

(a)(2) Control. A limit switch or other device shall be provided to prevent the load block from passing the safe upper limit of travel of any hoisting mechanism.

(a)(3) Clearance. The dimension of the working space in the direction of access to live parts which may require examination, adjustment, servicing, or maintenance while alive shall be a minimum of 2 feet 6 inches (762 mm). Where controls are enclosed in cabinets, the door(s) shall open at least 90 degrees or be removable, or the installation shall provide equivalent access.

§ 1926.406(a)(4)

(a)(4) Grounding. All exposed metal parts of cranes, monorail hoists, hoists and accessories including pendant controls shall be metallically joined together into a continuous electrical conductor so that the entire crane or hoist will be grounded in accordance with 1926.404(f). Moving parts, other than removable accessories or attachments, having metal-to-metal bearing surfaces shall be considered to be electrically connected to each other through the bearing surfaces for grounding purposes. The trolley frame and bridge frame shall be considered as electrically grounded through the bridge and trolley wheels and its respective tracks unless conditions such as paint or other insulating materials prevent reliable metal-to-metal contact. In this case a separate bonding conductor shall be provided.

(b) Elevators, escalators, and moving walks—

(b)(1) Disconnecting means. Elevators, escalators, and moving walks shall have a single means for disconnecting all ungrounded main power supply conductors for each unit.

(b)(2) Control panels. If control panels are not located in the same space as the drive machine, they shall be located in cabinets with doors or panels capable of being locked closed.

(c) Electric welders—disconnecting means—

(c)(1) Motor-generator, AC transformer, and DC rectifier arc welders. A disconnecting means shall be provided in the supply circuit for each motor-generator arc welder, and for each AC transformer and DC rectifier arc welder which is not equipped with a disconnect mounted as an integral part of the welder.

(c)(2) Resistance welders. A switch or circuit breaker shall be provided by which each resistance welder and its control equipment can be isolated from the supply circuit. The ampere rating of this disconnecting means shall not be less than the supply conductor ampacity.

§ 1926.406(d)

(d) X-Ray equipment—
(d)(1) Disconnecting means—

(d)(1)(i) General. A disconnecting means shall be provided in the supply circuit. The disconnecting means shall be operable from a location readily accessible from the X-ray control. For equipment connected to a 120-volt branch circuit of 30 amperes or less, a grounding-type attachment plug cap and receptacle of proper rating may serve as a disconnecting means.

(d)(1)(ii) More than one piece of equipment. If more than one piece of equipment is operated from the same high-voltage circuit, each piece or each group of equipment as a unit shall be provided with a high-voltage switch or equivalent disconnecting means. This disconnecting means shall be constructed, enclosed, or located so as to avoid contact by employees with its live parts.

(d)(2) Control—Radiographic and fluoroscopic types. Radiographic and fluoroscopic-type equipment shall be effectively enclosed or shall have interlocks that deenergize the equipment automatically to prevent ready access to live current-carrying parts.

§ 1926.407 Hazardous (classified) locations.

(a) Scope. This section sets forth requirements for electric equipment and wiring in locations which are classified depending on the properties of the flammable vapors, liquids or gases, or combustible dusts or fibers which may be present therein and the likelihood that a flammable or combustible concentration or quantity

FIGURE 6.6 (*Continued*)

is present. Each room, section or area shall be considered individually in determining its classification. These hazardous (classified) locations are assigned six designations as follows:

Class I, Division 1 Class I, Division 2 Class II, Division 1 Class II, Division 2 Class III, Division 1 Class III, Division 2

For definitions of these locations see 1926.449. All applicable requirements in this subpart apply to all hazardous (classified) locations, unless modified by provisions of this section.

(b) Electrical installations. Equipment, wiring methods, and installations of equipment in hazardous (classified) locations shall be approved as intrinsically safe or approved for the hazardous (classified) location or safe for the hazardous (classified) location. Requirements for each of these options are as follows:

(b)(1) Intrinsically safe. Equipment and associated wiring approved as intrinsically safe is permitted in any hazardous (classified) location included in its listing or labeling.

§ 1926.407(b)(2)

(b)(2) Approved for the hazardous (classified) location—

(b)(2)(i) General. Equipment shall be approved not only for the class of location but also for the ignitable or combustible properties of the specific gas, vapor, dust, or fiber that will be present.

NOTE: NFPA 70, the National Electrical Code, lists or defines hazardous gases, vapors, and dusts by "Groups" characterized by their ignitable or combustible properties.

(b)(2)(ii) Marking. Equipment shall not be used unless it is marked to show the class, group, and operating temperature or temperature range, based on operation in a 40-degree C ambient, for which it is approved. The temperature marking shall not exceed the ignition temperature of the specific gas, vapor, or dust to be encountered. However, the following provisions modify this marking requirement for specific equipment:

(b)(2)(ii)(A) Equipment of the non-heat-producing type (such as junction boxes, conduit, and fitting) and equipment of the heat-producing type having a maximum temperature of not more than 100 degrees C (212 degrees F) need not have a marked operating temperature or temperature range.

(b)(2)(ii)(B) Fixed lighting fixtures marked for use only in Class I, Division 2 locations need not be marked to indicate the group.

(b)(2)(ii)(C) Fixed general-purpose equipment in Class I locations, other than lighting fixtures, which is acceptable for use in Class I, Division 2 locations need not be marked with the class, group, division, or operating temperature.

§ 1926.407(b)(2)(ii)(D)

(b)(2)(ii)(D) Fixed dust-tight equipment, other than lighting fixtures, which is acceptable for use in Class II, Division 2 and Class III locations need not be marked with the class, group, division, or operating temperature.

(b)(3) Safe for the hazardous (classified) location. Equipment which is safe for the location shall be of a type and design which the employer demonstrates will provide protection from the hazards arising from the combustibility and flammability of vapors, liquids, gases, dusts, or fibers.

NOTE: The National Electrical Code, NFPA 70, contains guidelines for determining the type and design of equipment and installations which will meet this requirement. The guidelines of this document address electric wiring, equipment, and systems installed in hazardous (classified) locations and contain specific provisions for the following: wiring methods, wiring connections, conductor insulation, flexible cords, sealing and drainage, transformers, capacitors, switches, circuit breakers, fuses, motor controllers, receptacles, attachment plugs, meters, relays, instruments, resistors, generators, motors, lighting fixtures, storage battery charging equipment, electric cranes, electric hoists and similar equipment, utilization equipment, signaling systems, alarm systems, remote control systems, local loud speaker and communication systems, ventilation piping, live parts, lightning surge protection, and grounding. Compliance with these guidelines will constitute one means, but not the only means, of compliance with this paragraph.

(c) Conduits. All conduits shall be threaded and shall be made wrench-tight. Where it is impractical to make a threaded joint tight, a bonding jumper shall be utilized.

[61 FR 5507, Feb. 13, 1996]

§ 1926.408 Special systems

(a) Systems over 600 volts, nominal. Paragraphs (a)(1) through (a)(4) of this section contain general requirements for all circuits and equipment operated at over 600 volts.

(a)(1) Wiring methods for fixed installations—

FIGURE 6.6 (*Continued*)

(a)(1)(i) Above ground. Above-ground conductors shall be installed in rigid metal conduit, in intermediate metal conduit, in cable trays, in cablebus, in other suitable raceways, or as open runs of metal-clad cable designed for the use and purpose. However, open runs of non-metallic-sheathed cable or of bare conductors or busbars may be installed in locations which are accessible only to qualified persons. Metallic shielding components, such as tapes, wires, or braids for conductors, shall be grounded. Open runs of insulated wires and cables having a bare lead sheath or a braided outer covering shall be supported in a manner designed to prevent physical damage to the braid or sheath.

§ **1926.408(a)(1)(ii)**

(a)(1)(ii) Installations emerging from the ground. Conductors emerging from the ground shall be enclosed in raceways. Raceways installed on poles shall be of rigid metal conduit, intermediate metal conduit, PVC schedule 80 or equivalent extending from the ground line up to a point 8 feet (2.44 m) above finished grade. Conductors entering a building shall be protected by an enclosure from the ground line to the point of entrance. Metallic enclosures shall be grounded.

(a)(2) Interrupting and isolating devices—
(a)(2)(i) Circuit breakers. Circuit breakers located indoors shall consist of metal-enclosed or fire-resistant, cell-mounted units. In locations accessible only to qualified personnel, open mounting of circuit breakers is permitted. A means of indicating the open and closed position of circuit breakers shall be provided.

(a)(2)(ii) Fused cutouts. Fused cutouts installed in buildings or transformer vaults shall be of a type identified for the purpose. They shall be readily accessible for fuse replacement.

(a)(2)(iii) Equipment isolating means. A means shall be provided to completely isolate equipment for inspection and repairs. Isolating means which are not designed to interrupt the load current of the circuit shall be either interlocked with a circuit interrupter or provided with a sign warning against opening them under load.

§ **1926.408(a)(3)**

(a)(3) Mobile and portable equipment—
(a)(3)(i) Power cable connections to mobile machines. A metallic enclosure shall be provided on the mobile machine for enclosing the terminals of the power cable. The enclosure shall include provisions for a solid connection for the ground wire(s) terminal to ground effec-

tively the machine frame. The method of cable termination used shall prevent any strain or pull on the cable from stressing the electrical connections. The enclosure shall have provision for locking so only authorized qualified persons may open it and shall be marked with a sign warning of the presence of energized parts.

(a)(3)(ii) Guarding live parts. All energized switching and control parts shall be enclosed in effectively grounded metal cabinets or enclosures. Circuit breakers and protective equipment shall have the operating means projecting through the metal cabinet or enclosure so these units can be reset without locked doors being opened. Enclosures and metal cabinets shall be locked so that only authorized qualified persons have access and shall be marked with a sign warning of the presence of energized parts. Collector ring assemblies on revolving-type machines (shovels, draglines, etc.) shall be guarded.

(a)(4) Tunnel installations—
(a)(4)(i) Application. The provisions of this paragraph apply to installation and use of high-voltage power distribution and utilization equipment which is associated with tunnels and which is portable and/or mobile, such as substations, trailers, cars, mobile shovels, draglines, hoists, drills, dredges, compressors, pumps, conveyors, and underground excavators.

(a)(4)(ii) Conductors. Conductors in tunnels shall be installed in one or more of the following:
(a)(4)(ii)(A) Metal conduit or other metal raceway,

§ **1926.408(a)(4)(ii)(B)**

(a)(4)(ii)(B) Type MC cable, or
(a)(4)(ii)(C) Other suitable multiconductor cable.

Conductors shall also be so located or guarded as to protect them from physical damage. Multiconductor portable cable may supply mobile equipment. An equipment grounding conductor shall be run with circuit conductors inside the metal raceway or inside the multiconductor cable jacket. The equipment grounding conductor may be insulated or bare.

(a)(4)(iii) Guarding live parts. Bare terminals of transformers, switches, motor controllers, and other equipment shall be enclosed to prevent accidental contact with energized parts. Enclosures for use in tunnels shall be drip-proof, weatherproof, or submersible as required by the environmental conditions.

(a)(4)(iv) Disconnecting means. A disconnecting means that simultaneously opens all

FIGURE 6.6   *(Continued)*

ungrounded conductors shall be installed at each transformer or motor location.

(a)(4)(v) Grounding and bonding. All nonenergized metal parts of electric equipment and metal raceways and cable sheaths shall be grounded and bonded to all metal pipes and rails at the portal and at intervals not exceeding 1000 feet (305 m) throughout the tunnel.

§ 1926.408(b)

(b) Class 1, Class 2, and Class 3 remote control, signaling, and power-limited circuits—

(b)(1) Classification. Class 1, Class 2, or Class 3 remote control, signaling, or power-limited circuits are characterized by their usage and electrical power limitation which differentiates them from light and power circuits. These circuits are classified in accordance with their respective voltage and power limitations as summarized in paragraphs (b)(1)(i) through (b)(1)(iii) of this section.

(b)(1)(i) Class 1 circuits—

(b)(1)(i)(A) A Class 1 power-limited circuit is supplied from a source having a rated output of not more than 30 volts and 1000 volt-amperes.

(b)(1)(i)(B) A Class 1 remote control circuit or a Class 1 signaling circuit has a voltage which does not exceed 600 volts; however, the power output of the source need not be limited.

(b)(1)(ii) Class 2 and Class 3 circuits—

(b)(1)(ii)(A) Power for Class 2 and Class 3 circuits is limited either inherently (in which no overcurrent protection is required) or by a combination of a power source and overcurrent protection.

(b)(1)(ii)(B) The maximum circuit voltage is 150 volts AC or DC for a Class 2 inherently limited power source, and 100 volts AC or DC for a Class 3 inherently limited power source.

(b)(1)(ii)(C) The maximum circuit voltage is 30 volts AC and 60 volts DC for a Class 2 power source limited by overcurrent protection, and 150 volts AC or DC for a Class 3 power source limited by overcurrent protection.

§ 1926.408(b)(1)(iii)

(b)(1)(iii) Application. The maximum circuit voltages in paragraphs (b)(1)(i) and (b)(1)(ii) of this section apply to sinusoidal AC or continuous DC power sources, and where wet contact occurrence is not likely.

(b)(2) Marking. A Class 2 or Class 3 power supply unit shall not be used unless it is durably marked where plainly visible to indicate the class of supply and its electrical rating.

(c) Communications systems—

(c)(1) Scope. These provisions for communication systems apply to such systems as central-station-connected and non-central-station-connected telephone circuits, radio receiving and transmitting equipment, and outside wiring for fire and burglar alarm, and similar central station systems. These installations need not comply with the provisions of 1926.403 through 1926.408(b), except 1926.404(c)(1)(ii) and 1926.407.

(c)(2) Protective devices—

(c)(2)(i) Circuits exposed to power conductors. Communication circuits so located as to be exposed to accidental contact with light or power conductors operating at over 300 volts shall have each circuit so exposed provided with an approved protector.

§ 1926.408(c)(2)(ii)

(c)(2)(ii) Antenna lead-ins. Each conductor of a lead-in from an outdoor antenna shall be provided with an antenna discharge unit or other means that will drain static charges from the antenna system.

(c)(3) Conductor location—

(c)(3)(i) Outside of buildings—

(c)(3)(i)(A) Receiving distribution lead-in or aerial-drop cables attached to buildings and lead-in conductors to radio transmitters shall be so installed as to avoid the possibility of accidental contact with electric light or power conductors.

(c)(3)(i)(B) The clearance between lead-in conductors and any lightning protection conductors shall not be less than 6 feet (1.83 m).

(c)(3)(ii) On poles. Where practicable, communication conductors on poles shall be located below the light or power conductors. Communications conductors shall not be attached to a crossarm that carries light or power conductors.

(c)(3)(iii) Inside of buildings. Indoor antennas, lead-ins, and other communication conductors attached as open conductors to the inside of buildings shall be located at least 2 inches (50.8 mm) from conductors of any light or power or Class 1 circuits unless a special and equally protective method of conductor separation is employed.

§ 1926.408(c)(4)

(c)(4) Equipment location. Outdoor metal structures supporting antennas, as well as self-supporting antennas such as vertical rods or dipole structures, shall be located as far

FIGURE 6.6 *(Continued)*

away from overhead conductors of electric light and power circuits of over 150 volts to ground as necessary to avoid the possibility of the antenna or structure falling into or making accidental contact with such circuits.

(c)(5) Grounding—

(c)(5)(i) Lead-in conductors. If exposed to contact with electric light or power conductors, the metal sheath of aerial cables entering buildings shall be grounded or shall be interrupted close to the entrance to the building by an insulating joint or equivalent device. Where protective devices are used, they shall be grounded.

(c)(5)(ii) Antenna structures. Masts and metal structures supporting antennas shall be perma-nently and effectively grounded without splice or connection in the grounding conductor.

(c)(5)(iii) Equipment enclosures. Transmitters shall be enclosed in a metal frame or grill or separated from the operating space by a barrier, all metallic parts of which are effectively connected to ground. All external metal handles and controls accessible to the operating personnel shall be effectively grounded. Unpowered equipment and enclosures shall be considered grounded where connected to an attached coaxial cable with an effectively grounded metallic shield.

[61 FR 5507, Feb. 13, 1996]

FIGURE 6.6   (*Continued*)

1926.416 General requirements.

(a) Protection of employees—

(a)(1) No employer shall permit an employee to work in such proximity to any part of an electric power circuit that the employee could contact the electric power circuit in the course of work, unless the employee is protected against electric shock by deenergizing the circuit and grounding it or by guarding it effectively by insulation or other means.

(a)(2) In work areas where the exact location of underground electric powerlines is unknown, employees using jack-hammers, bars, or other hand tools which may contact a line shall be provided with insulated protective gloves.

(a)(3) Before work is begun the employer shall ascertain by inquiry or direct observation, or by instruments, whether any part of an energized electric power circuit, exposed or concealed, is so located that the performance of the work may bring any person, tool, or machine into physical or electrical contact with the electric power circuit. The employer shall post and maintain proper warning signs where such a circuit exists. The employer shall advise employees of the location of such lines, the hazards involved, and the protective measures to be taken.

§ 1926.416(b)

(b) Passageways and open spaces—

(b)(1) Barriers or other means of guarding shall be provided to ensure that workspace for electrical equipment will not be used as a passageway during periods when energized parts of electrical equipment are exposed.

(b)(2) Working spaces, walkways, and similar locations shall be kept clear of cords so as not to create a hazard to employees.

(c) Load ratings. In existing installations, no changes in circuit protection shall be made to increase the load in excess of the load rating of the circuit wiring.

(d) Fuses. When fuses are installed or removed with one or both terminals energized, special tools insulated for the voltage shall be used.

(e) Cords and cables.

(e)(1) Worn or frayed electric cords or cables shall not be used.

(e)(2) Extension cords shall not be fastened with staples, hung from nails, or suspended by wire.

[58 FR 35179, June 30, 1993; 61 FR 9227, March 7, 1996; 61 FR 41738, August 12, 1996]

§ 1926.417 Lockout and tagging of circuits.

(a) Controls. Controls that are to be deactivated during the course of work on energized or deenergized equipment or circuits shall be tagged.

(b) Equipment and circuits. Equipment or circuits that are deenergized shall be rendered inoperative and shall have tags attached at all points where such equipment or circuits can be energized.

(c) Tags. Tags shall be placed to identify plainly the equipment or circuits being worked on.

[58 FR 35181, June 30, 1993; 61 FR 9227, March 7, 1996; 61 FR 41738, August 12, 1996]

FIGURE 6.7 Safety-Related Work Practices for Construction Workers (OSHA-CFR, Title 29, Part 1926, Paragraphs 416–417).

§ 1926.431 Maintenance of equipment.

The employer shall ensure that all wiring components and utilization equipment in hazardous locations are maintained in a dust-tight, dust-ignition-proof, or explosion-proof condition, as appropriate. There shall be no loose or missing screws, gaskets, threaded connections, seals, or other impairments to a tight condition.

§ 1926.432 Environmental deterioration of equipment.

(a) Deteriorating agents—

(a)(1) Unless identified for use in the operating environment, no conductors or equipment shall be located:

(a)(1)(i) In damp or wet locations;

(a)(1)(ii) Where exposed to gases, fumes, vapors, liquids, or other agents having a deteriorating effect on the conductors or equipment; or

(a)(1)(iii) Where exposed to excessive temperatures.

(a)(2) Control equipment, utilization equipment, and busways approved for use in dry locations only shall be protected against damage from the weather during building construction.

(b) Protection against corrosion. Metal raceways, cable armor, boxes, cable sheathing, cabinets, elbows, couplings, fittings, supports, and support hardware shall be of materials appropriate for the environment in which they are to be installed.

FIGURE 6.8   Safety-Related Maintenance and Environmental Considerations (OSHA-CFR, Title 29, Part 1926, Paragraphs 431–432).

§ 1926.441 Batteries and battery charging.

(a) General requirements—

(a)(1) Batteries of the unsealed type shall be located in enclosures with outside vents or in well ventilated rooms and shall be arranged so as to prevent the escape of fumes, gases, or electrolyte spray into other areas.

(a)(2) Ventilation shall be provided to ensure diffusion of the gases from the battery and to prevent the accumulation of an explosive mixture.

(a)(3) Racks and trays shall be substantial and shall be treated to make them resistant to the electrolyte.

(a)(4) Floors shall be of acid resistant construction unless protected from acid accumulations.

(a)(5) Face shields, aprons, and rubber gloves shall be provided for workers handling acids or batteries.

(a)(6) Facilities for quick drenching of the eyes and body shall be provided within 25 feet (7.62 m) of battery handling areas.

§ 1926.441(a)(7)

(a)(7) Facilities shall be provided for flushing and neutralizing spilled electrolyte and for fire protection.

(b) Charging—

(b)(1) Battery charging installations shall be located in areas designated for that purpose.

(b)(2) Charging apparatus shall be protected from damage by trucks.

(b)(3) When batteries are being charged, the vent caps shall be kept in place to avoid electrolyte spray. Vent caps shall be maintained in functioning condition.

FIGURE 6.9   Safety Requirements for Special Equipment (OSHA-CFR, Title 29, Part 1926, Paragraph 441).

§ 1926.449 Definitions applicable to this subpart.

The definitions given in this section apply to the terms used in Subpart K. The definitions given here for "approved" and "qualified person" apply, instead of the definitions given in 1926.32, to the use of these terms in Subpart K.

Acceptable. An installation or equipment is acceptable to the Assistant Secretary of Labor, and approved within the meaning of this Subpart K:

(a) If it is accepted, or certified, or listed, or labeled, or otherwise determined to be safe by a qualified testing laboratory capable of determining the suitability of materials and equipment for installation and use in accordance with this standard; or

(b) With respect to an installation or equipment of a kind which no qualified testing laboratory accepts, certifies, lists, labels, or determines to be safe, if it is inspected or tested by another Federal agency, or by a State, municipal, or other local authority responsible for enforcing occupational safety provisions of the National Electrical Code, and found in compliance with those provisions; or

(c) With respect to custom-made equipment or related installations which are designed, fabricated for, and intended for use by a particular customer, if it is determined to be safe for its intended use by its manufacturer on the basis of test data which the employer keeps and makes available for inspection to the Assistant Secretary and his authorized representatives.

Accepted. An installation is "accepted" if it has been inspected and found to be safe by a qualified testing laboratory.

Accessible. (As applied to wiring methods.) Capable of being removed or exposed without damaging the building structure or finish, or not permanently closed in by the structure or finish of the building. (See "concealed" and "exposed.")

Accessible. (As applied to equipment.) Admitting close approach; not guarded by locked doors, elevation, or other effective means. (See "Readily accessible.")

Ampacity. The current in amperes a conductor can carry continuously under the conditions of use without exceeding its temperature rating.

Appliances. Utilization equipment, generally other than industrial, normally built in standardized sizes or types, which is installed or connected as a unit to perform one or more functions.

Approved. Acceptable to the authority enforcing this Subpart. The authority enforcing this Subpart is the Assistant Secretary of Labor for Occupational Safety and Health. The definition of "acceptable" indicates what is acceptable to the Assistant Secretary of Labor, and therefore approved within the meaning of this Subpart.

Askarel. A generic term for a group of nonflammable synthetic chlorinated hydrocarbons used as electrical insulating media. Askarels of various compositional types are used. Under arcing conditions the gases produced, while consisting predominantly of noncombustible hydrogen chloride, can include varying amounts of combustible gases depending upon the askarel type.

Attachment plug (Plug cap)(Cap). A device which, by insertion in a receptacle, establishes connection between the conductors of the attached flexible cord and the conductors connected permanently to the receptacle.

Automatic. Self-acting, operating by its own mechanism when actuated by some impersonal influence, as for example, a change in current strength, pressure, temperature, or mechanical configuration.

Bare conductor. See *Conductor.*

Bonding. The permanent joining of metallic parts to form an electrically conductive path which will assure electrical continuity and the capacity to conduct safely any current likely to be imposed.

Bonding jumper. A reliable conductor to assure the required electrical conductivity between metal parts required to be electrically connected.

Branch circuit. The circuit conductors between the final overcurrent device protecting the circuit and the outlet(s).

Building. A structure which stands alone or which is cut off from adjoining structures by fire walls with all openings therein protected by approved fire doors.

Cabinet. An enclosure designed either for surface or flush mounting, and provided with a frame, mat, or trim in which a swinging door or doors are or may be hung.

Certified. Equipment is "certified" if it:

(a) Has been tested and found by a qualified testing laboratory to meet applicable test standards or to be safe for use in a specified manner, and

(b) Is of a kind whose production is periodically inspected by a qualified testing laboratory. Certified equipment must bear a label, tag, or other record of certification.

FIGURE 6.10 Definitions Applying to Part 1926, Subpart K (OSHA-CFR, Title 29, Part 1926, Paragraph 449).

Circuit breaker—(a) (600 volts nominal, or less.) A device designed to open and close a circuit by nonautomatic means and to open the circuit automatically on a predetermined overcurrent without injury to itself when properly applied within its rating.

(b) (Over 600 volts, nominal.) A switching device capable of making, carrying, and breaking currents under normal circuit conditions, and also making, carrying for a specified time, and breaking currents under specified abnormal circuit conditions, such as those of short circuit.

Class I locations. Class I locations are those in which flammable gases or vapors are or may be present in the air in quantities sufficient to produce explosive or ignitable mixtures. Class I locations include the following:

(a) Class I, Division 1. A Class I, Division 1 location is a location:

(1) In which ignitable concentrations of flammable gases or vapors may exist under normal operating conditions; or

(2) In which ignitable concentrations of such gases or vapors may exist frequently because of repair or maintenance operations or because of leakage; or

(3) In which breakdown or faulty operation of equipment or processes might release ignitable concentrations of flammable gases or vapors, and might also cause simultaneous failure of electric equipment.

NOTE: This classification usually includes locations where volatile flammable liquids or liquefied flammable gases are transferred from one container to another; interiors of spray booths and areas in the vicinity of spraying and painting operations where volatile flammable solvents are used; locations containing open tanks or vats of volatile flammable liquids; drying rooms or compartments for the evaporation of flammable solvents; inadequately ventilated pump rooms for flammable gas or for volatile flammable liquids; and all other locations where ignitable concentrations of flammable vapors or gases are likely to occur in the course of normal operations.

(b) Class I, Division 2. A Class I, Division 2 location is a location:

(1) In which volatile flammable liquids or flammable gases are handled, processed, or used, but in which the hazardous liquids, vapors, or gases will normally be confined within closed containers or closed systems from which they can escape only in case of accidental rupture or breakdown of such containers or

systems, or in case of abnormal operation of equipment; or

(2) In which ignitable concentrations of gases or vapors are normally prevented by positive mechanical ventilation, and which might become hazardous through failure or abnormal operations of the ventilating equipment; or

(3) That is adjacent to a Class I, Division 1 location, and to which ignitable concentrations of gases or vapors might occasionally be communicated unless such communication is prevented by adequate positive-pressure ventilation from a source of clean air, and effective safeguards against ventilation failure are provided.

NOTE: This classification usually includes locations where volatile flammable liquids or flammable gases or vapors are used, but which would become hazardous only in case of an accident or of some unusual operating condition. The quantity of flammable material that might escape in case of accident, the adequacy of ventilating equipment, the total area involved, and the record of the industry or business with respect to explosions or fires are all factors that merit consideration in determining the classification and extent of each location.

Piping without valves, checks, meters, and similar devices would not ordinarily introduce a hazardous condition even though used for flammable liquids or gases. Locations used for the storage of flammable liquids or of liquefied or compressed gases in sealed containers would not normally be considered hazardous unless also subject to other hazardous conditions.

Electrical conduits and their associated enclosures separated from process fluids by a single seal or barrier are classed as a Division 2 location if the outside of the conduit and enclosures is a nonhazardous location.

Class II locations. Class II locations are those that are hazardous because of the presence of combustible dust. Class II locations include the following:

(a) Class II, Division 1. A Class II, Division 1 location is a location:

(1) In which combustible dust is or may be in suspension in the air under normal operating conditions, in quantities sufficient to produce explosive or ignitable mixtures; or

(2) Where mechanical failure or abnormal operation of machinery or equipment might cause such explosive or ignitable mixtures to be produced, and might also provide a source of ignition through simultaneous failure of electric

FIGURE 6.10   *(Continued)*

equipment, operation of protection devices, or from other causes, or

(3) In which combustible dusts of an electrically conductive nature may be present.

NOTE: Combustible dusts which are electrically nonconductive include dusts produced in the handling and processing of grain and grain products, pulverized sugar and cocoa, dried egg and milk powders, pulverized spices, starch and pastes, potato and woodflour, oil meal from beans and seed, dried hay, and other organic materials which may produce combustible dusts when processed or handled. Dusts containing magnesium or aluminum are particularly hazardous and the use of extreme caution is necessary to avoid ignition and explosion.

(b) Class II, Division 2. A Class II, Division 2 location is a location in which:

(1) Combustible dust will not normally be in suspension in the air in quantities sufficient to produce explosive or ignitable mixtures, and dust accumulations are normally insufficient to interfere with the normal operation of electrical equipment or other apparatus; or

(2) Dust may be in suspension in the air as a result of infrequent malfunctioning of handling or processing equipment, and dust accumulations resulting therefrom may be ignitable by abnormal operation or failure of electrical equipment or other apparatus.

NOTE: This classification includes locations where dangerous concentrations of suspended dust would not be likely but where dust accumulations might form on or in the vicinity of electric equipment. These areas may contain equipment from which appreciable quantities of dust would escape under abnormal operating conditions or be adjacent to a Class II Division 1 location, as described above, into which an explosive or ignitable concentration of dust may be put into suspension under abnormal operating conditions.

Class III locations. Class III locations are those that are hazardous because of the presence of easily ignitable fibers or flyings but in which such fibers or flyings are not likely to be in suspension in the air in quantities sufficient to produce ignitable mixtures. Class III locations include the following:

(a) Class III, Division 1. A Class III, Division 1 location is a location in which easily ignitable fibers or materials producing combustible flyings are handled, manufactured, or used.

NOTE: Easily ignitable fibers and flyings include rayon, cotton (including cotton linters and cotton waste), sisal or henequen, istle, jute, hemp, tow,

cocoa fiber, oakum, baled waste kapok, Spanish moss, excelsior, sawdust, woodchips, and other material of similar nature.

(b) Class III, Division 2. A Class III, Division 2 location is a location in which easily ignitable fibers are stored or handled, except in process of manufacture.

Collector ring. A collector ring is an assembly of slip rings for transferring electrical energy from a stationary to a rotating member.

Concealed. Rendered inaccessible by the structure or finish of the building. Wires in concealed raceways are considered concealed, even though they may become accessible by withdrawing them. [See "Accessible. (As applied to wiring methods.)"]

Conductor—(a) Bare. A conductor having no covering or electrical insulation whatsoever.

(b) *Covered.* A conductor encased within material of composition or thickness that is not recognized as electrical insulation.

(c) *Insulated.* A conductor encased within material of composition and thickness that is recognized as electrical insulation.

Controller. A device or group of devices that serves to govern, in some predetermined manner, the electric power delivered to the apparatus to which it is connected.

Covered conductor. See *Conductor.*

Cutout. (Over 600 volts, nominal.) An assembly of a fuse support with either a fuseholder, fuse carrier, or disconnecting blade. The fuseholder or fuse carrier may include a conducting element (fuse link), or may act as the disconnecting blade by the inclusion of a nonfusible member.

Cutout box. An enclosure designed for surface mounting and having swinging doors or covers secured directly to and telescoping with the walls of the box proper. (See *Cabinet.*)

Damp location. See *Location.*

Dead front. Without live parts exposed to a person on the operating side of the equipment.

Device. A unit of an electrical system which is intended to carry but not utilize electric energy.

Disconnecting means. A device, or group of devices, or other means by which the conductors of a circuit can be disconnected from their source of supply.

Disconnecting (or Isolating) switch. (Over 600 volts, nominal.) A mechanical switching device used for isolating a circuit or equipment from a source of power.

Dry location. See *Location.*

Enclosed. Surrounded by a case, housing, fence or walls which will prevent persons from accidentally contacting energized parts.

FIGURE 6.10 *(Continued)*

Enclosure. The case or housing of apparatus, or the fence or walls surrounding an installation to prevent personnel from accidentally contacting energized parts, or to protect the equipment from physical damage.

Equipment. A general term including material, fittings, devices, appliances, fixtures, apparatus, and the like, used as a part of, or in connection with, an electrical installation.

Equipment grounding conductor. See *Grounding conductor, equipment.*

Explosion-proof apparatus. Apparatus enclosed in a case that is capable of withstanding an explosion of a specified gas or vapor which may occur within it and of preventing the ignition of a specified gas or vapor surrounding the enclosure by sparks, flashes, or explosion of the gas or vapor within, and which operates at such an external temperature that it will not ignite a surrounding flammable atmosphere.

Exposed. (As applied to live parts.) Capable of being inadvertently touched or approached nearer than a safe distance by a person. It is applied to parts not suitably guarded, isolated, or insulated. (See *Accessible* and *Concealed.*)

Exposed. (As applied to wiring methods.) On or attached to the surface or behind panels designed to allow access. [See *Accessible.* (As applied to wiring methods.)"]

Exposed. (For the purposes of 1926.408(d), Communications systems.) Where the circuit is in such a position that in case of failure of supports or insulation, contact with another circuit may result.

Externally operable. Capable of being operated without exposing the operator to contact with live parts.

Feeder. All circuit conductors between the service equipment, or the generator switchboard of an isolated plant, and the final branch-circuit overcurrent device.

Festoon lighting. A string of outdoor lights suspended between two points more than 15 feet (4.57 m) apart.

Fitting. An accessory such as a locknut, bushing, or other part of a wiring system that is intended primarily to perform a mechanical rather than an electrical function.

Fuse. (Over 600 volts, nominal.) An overcurrent protective device with a circuit opening fusible part that is heated and severed by the passage of overcurrent through it. A fuse comprises all the parts that form a unit capable of performing the prescribed functions. It may or may not be the complete device necessary to connect it into an electrical circuit.

Ground. A conducting connection, whether intentional or accidental, between an electrical circuit or equipment and the earth, or to some conducting body that serves in place of the earth.

Grounded. Connected to earth or to some conducting body that serves in place of the earth.

Grounded, effectively (Over 600 volts, nominal.) Permanently connected to earth through a ground connection of sufficiently low impedance and having sufficient ampacity that ground fault current which may occur cannot build up to voltages dangerous to personnel.

Grounded conductor. A system or circuit conductor that is intentionally grounded.

Grounding conductor. A conductor used to connect equipment or the grounded circuit of a wiring system to a grounding electrode or electrodes.

Grounding conductor, equipment. The conductor used to connect the noncurrent-carrying metal parts of equipment, raceways, and other enclosures to the system grounded conductor and/or the grounding electrode conductor at the service equipment or at the source of a separately derived system.

Grounding electrode conductor. The conductor used to connect the grounding electrode to the equipment grounding conductor and/or to the grounded conductor of the circuit at the service equipment or at the source of a separately derived system.

Ground-fault circuit interrupter. A device for the protection of personnel that functions to deenergize a circuit or portion thereof within an established period of time when a current to ground exceeds some predetermined value that is less than that required to operate the overcurrent protective device of the supply circuit.

Guarded. Covered, shielded, fenced, enclosed, or otherwise protected by means of suitable covers, casings, barriers, rails, screens, mats, or platforms to remove the likelihood of approach to a point of danger or contact by persons or objects.

Hoistway. Any shaftway, hatchway, well hole, or other vertical opening or space in which an elevator or dumbwaiter is designed to operate.

Identified (conductors or terminals). Identified, as used in reference to a conductor or its terminal, means that such conductor or terminal can be recognized as grounded.

Identified (for the use). Recognized as suitable for the specific purpose, function, use, environment, application, etc. where described as a requirement in this standard. Suitability of

FIGURE 6.10 *(Continued)*

equipment for a specific purpose, environment, or application is determined by a qualified testing laboratory where such identification includes labeling or listing.

Insulated conductor. See *Conductor.*

Interrupter switch. (Over 600 volts, nominal.) A switch capable of making, carrying, and interrupting specified currents.

Intrinsically safe equipment and associated wiring. Equipment and associated wiring in which any spark or thermal effect, produced either normally or in specified fault conditions, is incapable, under certain prescribed test conditions, of causing ignition of a mixture of flammable or combustible material in air in its most easily ignitable concentration.

Isolated. Not readily accessible to persons unless special means for access are used.

Isolated power system. A system comprising an isolating transformer or its equivalent, a line isolation monitor, and its ungrounded circuit conductors.

Labeled. Equipment or materials to which has been attached a label, symbol or other identifying mark of a qualified testing laboratory which indicates compliance with appropriate standards or performance in a specified manner.

Lighting outlet. An outlet intended for the direct connection of a lampholder, a lighting fixture, or a pendant cord terminating in a lampholder.

Listed. Equipment or materials included in a list published by a qualified testing laboratory whose listing states either that the equipment or material meets appropriate standards or has been tested and found suitable for use in a specified manner.

Location—(a) Damp location. Partially protected locations under canopies, marquees, roofed open porches, and like locations, and interior locations subject to moderate degrees of moisture, such as some basements.

(b) *Dry location.* A location not normally subject to dampness or wetness. A location classified as dry may be temporarily subject to dampness or wetness, as in the case of a building under construction.

(c) *Wet location.* Installations underground or in concrete slabs or masonry in direct contact with the earth, and locations subject to saturation with water or other liquids, such as locations exposed to weather and unprotected.

Mobile X-ray. X-ray equipment mounted on a permanent base with wheels and/or casters for moving while completely assembled.

Motor control center. An assembly of one or more enclosed sections having a common power bus and principally containing motor control units.

Outlet. A point on the wiring system at which current is taken to supply utilization equipment.

Overcurrent. Any current in excess of the rated current of equipment or the ampacity of a conductor. It may result from overload (see definition), short circuit, or ground fault. A current in excess of rating may be accommodated by certain equipment and conductors for a given set of conditions. Hence the rules for overcurrent protection are specific for particular situations.

Overload. Operation of equipment in excess of normal, full load rating, or of a conductor in excess of rated ampacity which, when it persists for a sufficient length of time, would cause damage or dangerous overheating. A fault, such as a short circuit or ground fault, is not an overload. (See "Overcurrent.")

Panelboard. A single panel or group of panel units designed for assembly in the form of a single panel; including buses, automatic overcurrent devices, and with or without switches for the control of light, heat, or power circuits; designed to be placed in a cabinet or cutout box placed in or against a wall or partition and accessible only from the front. (See "Switchboard.")

Portable X-ray. X-ray equipment designed to be hand-carried.

Power fuse. (Over 600 volts, nominal.) See "Fuse."

Power outlet. An enclosed assembly which may include receptacles, circuit breakers, fuseholders, fused switches, buses and watt-hour meter mounting means; intended to serve as a means for distributing power required to operate mobile or temporarily installed equipment.

Premises wiring system. That interior and exterior wiring, including power, lighting, control, and signal circuit wiring together with all of its associated hardware, fittings, and wiring devices, both permanently and temporarily installed, which extends from the load end of the service drop, or load end of the service lateral conductors to the outlet(s). Such wiring does not include wiring internal to appliances, fixtures, motors, controllers, motor control centers, and similar equipment.

Qualified person. One familiar with the construction and operation of the equipment and the hazards involved.

Qualified testing laboratory. A properly equipped and staffed testing laboratory which

FIGURE 6.10 (*Continued*)

has capabilities for and which provides the following services:

(a) Experimental testing for safety of specified items of equipment and materials referred to in this standard to determine compliance with appropriate test standards or performance in a specified manner;

(b) Inspecting the run of such items of equipment and materials at factories for product evaluation to assure compliance with the test standards;

(c) Service-value determinations through field inspections to monitor the proper use of labels on products and with authority for recall of the label in the event a hazardous product is installed;

(d) Employing a controlled procedure for identifying the listed and/or labeled equipment or materials tested; and

(e) Rendering creditable reports or findings that are objective and without bias of the tests and test methods employed.

Raceway. A channel designed expressly for holding wires, cables, or busbars, with additional functions as permitted in this subpart. Raceways may be of metal or insulating material, and the term includes rigid metal conduit, rigid nonmetallic conduit, intermediate metal conduit, liquidtight flexible metal conduit, flexible metallic tubing, flexible metal conduit, electrical metallic tubing, underfloor raceways, cellular concrete floor raceways, cellular metal floor raceways, surface raceways, wireways, and busways.

Readily accessible. Capable of being reached quickly for operation, renewal, or inspections, without requiring those to whom ready access is requisite to climb over or remove obstacles or to resort to portable ladders, chairs, etc. (See "Accessible.")

Receptacle. A receptacle is a contact device installed at the outlet for the connection of a single attachment plug. A single receptacle is a single contact device with no other contact device on the same yoke. A multiple receptacle is a single device containing two or more receptacles.

Receptacle outlet. An outlet where one or more receptacles are installed.

Remote-control circuit. Any electric circuit that controls any other circuit through a relay or an equivalent device.

Sealable equipment. Equipment enclosed in a case or cabinet that is provided with a means of sealing or locking so that live parts cannot be made accessible without opening the enclosure. The equipment may or may not be operable without opening the enclosure.

Separately derived system. A premises wiring system whose power is derived from generator, transformer, or converter windings and has no direct electrical connection, including a solidly connected grounded circuit conductor, to supply conductors originating in another system.

Service. The conductors and equipment for delivering energy from the electricity supply system to the wiring system of the premises served.

Service conductors. The supply conductors that extend from the street main or from transformers to the service equipment of the premises supplied.

Service drop. The overhead service conductors from the last pole or other aerial support to and including the splices, if any, connecting to the service-entrance conductors at the building or other structure.

Service-entrance conductors, overhead system. The service conductors between the terminals of the service equipment and a point usually outside the building, clear of building walls, where joined by tap or splice to the service drop.

Service-entrance conductors, underground system. The service conductors between the terminals of the service equipment and the point of connection to the service lateral. Where service equipment is located outside the building walls, there may be no service-entrance conductors, or they may be entirely outside the building.

Service equipment. The necessary equipment, usually consisting of a circuit breaker or switch and fuses, and their accessories, located near the point of entrance of supply conductors to a building or other structure, or an otherwise defined area, and intended to constitute the main control and means of cutoff of the supply.

Service raceway. The raceway that encloses the service-entrance conductors.

Signaling circuit. Any electric circuit that energizes signaling equipment.

Switchboard. A large single panel, frame, or assembly of panels which have switches, buses, instruments, overcurrent and other protective devices mounted on the face or back or both. Switchboards are generally accessible from the rear as well as from the front and are not intended to be installed in cabinets. (See *Panelboard.*)

Switches—(a) *General-use switch.* A switch intended for use in general distribution and branch circuits. It is rated in amperes, and it is capable of interrupting its rated current at its rated voltage.

(b) *General-use snap switch.* A form of general-use switch so constructed that it can be

FIGURE 6.10   (*Continued*)

installed in flush device boxes or on outlet box covers, or otherwise used in conjunction with wiring systems recognized by this subpart.

(c) *Isolating switch.* A switch intended for isolating an electric circuit from the source of power. It has no interrupting rating, and it is intended to be operated only after the circuit has been opened by some other means.

(d) *Motor-circuit switch.* A switch, rated in horsepower, capable of interrupting the maximum operating overload current of a motor of the same horsepower rating as the switch at the rated voltage.

Switching devices. (Over 600 volts, nominal.) Devices designed to close and/or open one or more electric circuits. Included in this category are circuit breakers, cutouts, disconnecting (or isolating) switches, disconnecting means, and interrupter switches.

Transportable X-ray. X-ray equipment installed in a vehicle or that may readily be disassembled for transport in a vehicle.

Utilization equipment. Utilization equipment means equipment which utilizes electric energy for mechanical, chemical, heating, lighting, or similar useful purpose.

Utilization system. A utilization system is a system which provides electric power and light for employee workplaces, and includes the premises wiring system and utilization equipment.

Ventilated. Provided with a means to permit circulation of air sufficient to remove an excess of heat, fumes, or vapors.

Volatile flammable liquid. A flammable liquid having a flash point below 38 degrees C (100 degrees F) or whose temperature is above its flash point, or a Class II combustible liquid having a vapor pressure not exceeding 40 psia (276 kPa) at 38 deg. C (100 deg. F) whose temperature is above its flash point.

Voltage. (Of a circuit.) The greatest root-mean-square (effective) difference of potential between any two conductors of the circuit concerned.

Voltage, nominal. A nominal value assigned to a circuit or system for the purpose of conveniently designating its voltage class (as 120/240, 480Y/277, 600, etc.). The actual voltage at which a circuit operates can vary from the nominal within a range that permits satisfactory operation of equipment.

Voltage to ground. For grounded circuits, the voltage between the given conductor and that point or conductor of the circuit that is grounded; for ungrounded circuits, the greatest voltage between the given conductor and any other conductor of the circuit.

Watertight. So constructed that moisture will not enter the enclosure.

Weatherproof. So constructed or protected that exposure to the weather will not interfere with successful operation. Rainproof, raintight, or watertight equipment can fulfill the requirements for weatherproof where varying weather conditions other than wetness, such as snow, ice, dust, or temperature extremes, are not a factor.

Wet location. See *Location.*

FIGURE 6.10   (*Continued*)

CHAPTER 7

ACCIDENT PREVENTION, ACCIDENT INVESTIGATION, RESCUE, AND FIRST AID

ACCIDENT PREVENTION

No matter how carefully a system is engineered, no matter how carefully employees perform their tasks, and no matter how well trained employees are in the recognition and avoidance of hazards, accidents still happen. This section provides a general approach that may be employed to reduce the number and severity of accidents. Four basic steps—employee responsibility, safe installations, safe work practices, and employee training—combine to create the type of safe work environment that should be the goal of every facility.

Individual Responsibility

The person most responsible for your own personal safety is you. No set of regulations, rules, or procedures can ever replace common sense in the workplace. This statement should not be construed to mean an employer has no responsibility to provide the safest practical work environment, nor does it mean that the injured person is "at fault" in a legal sense. Determining fault for accidents is, in part, a legal problem and is beyond the scope of this handbook.

Time after time, accident investigations reveal that the injured person was the last link in the chain. If the injured person had only been wearing appropriate safety equipment or following proper procedures, or if he or she had only checked one last time, the accident never would have occurred.

Table 7.1 lists five behavioral approaches that will help to improve the safety of all employees. To make certain employees have both the responsibility and the authority to carry out the five steps listed in Table 7.1, employers should adopt a policy similar to the one listed in Table 7.2. Simply putting such a statement in a safety handbook is insufficient. An employer must believe in this principle and must "put teeth into it." Such a credo provides the absolute maximum in individual employee responsibility and authority and will maximize the safety performance of the organization.

Installation Safety

Design. Proper design of electrical systems is composed of three parts—selection, installation, and calibration.

TABLE 7.1   Employee Safety Behavior

- Determine the nature and extent of hazards before starting a job.
- Each employee should be satisfied that conditions are safe before beginning work on any job or any part of a job.
- All employees should be thoroughly familiar with and should consistently use the work procedures and the safety equipment required for the performance of the job at hand.
- While working, each employee should consider the effects of each step and do nothing that might endanger themselves or others.
- Each employee should be thoroughly familiar with emergency procedures.

- *Selection.*   Electric equipment should be selected and applied conservatively. That is, maximum ratings must be well in excess of the quantities to which they will be exposed in the power system. To help with this, manufacturing organizations such as the National Electrical Manufacturer's Association (NEMA) have established ratings for equipment that ensure member companies only manufacture the highest-quality equipment. Equipment is tested per manufacturer's procedures by independent laboratories such as the Underwriter's Laboratory (UL). Equipment that is rated and labeled by such organizations should be used in electrical systems to help ensure safety. OSHA, NEC,* and NESC requirements should be considered as minimum criteria for safe selection. With increasing cost consciousness, many companies are opting for the installation of recycled rather than brand-new equipment. While the selection and use of such equipment can be a financial advantage, at least two criteria should be considered in the purchase:

  1. Even though used, the equipment should have been originally manufactured by a reputable, professional firm.
  2. The recycled equipment should be thoroughly reconditioned by a professional recycling company such as those represented by the Professional Electrical Apparatus Recyclers League.
  3. Engineering studies such as short-circuit analysis and coordination studies must be performed to ascertain that the recycled equipment has adequate ratings for the system in which it is being placed. Since the system short circuit values may change with time, this requirement is true even if the recycled equipment is a direct replacement for the previous equipment.

- *Installation.*   Equipment should be installed in a safe and *sensible* manner. Adequate work spaces for safety clearance should be allowed, safety barriers should be provided when necessary, and electrical installations should never be mixed with areas that are used for general public access.

- *Calibration.*   Equipment always should be properly calibrated. For example, protective devices should be calibrated so that they will operate for the minimum abnormal system

---

* National Electrical Code and NEC are registered trademarks of the National Fire Protection Association.

TABLE 7.2   Recommended Safety Credo

If it cannot be done safely, it need not be done!

---

condition. Equipment that is improperly calibrated can result in accidents as though the equipment had been improperly selected to begin with. Calibration is also a two-step process:

1. Proper engineering should be performed by professional engineers to ensure that the selected calibration settings are suitable for the application. The starting point for such a system is in the performance of an appropriate suite of power system studies, described later in the section.
2. Proper testing and physical setting of the devices should be carried out to ensure that the equipment is capable of performing when called upon. Such settings should be executed by professional technicians who are certified to perform this work. Organizations such as the International Electrical Testing Association (NETA) have been formed to ensure quality control on the education and performance of electrical technicians.

Electrical Protective Devices.   Protective devices such as circuit breakers, fuses, and switches must be capable of interrupting the currents to which they will be subjected. The National Electrical Code has numerous passages that require proper sizing of protective devices.

Maintenance.   Improperly maintained equipment is hazardous. For example, circuit breakers can explode violently if not properly maintained. Equipment should be periodically inspected and tested. If deficiencies are observed, the equipment must be repaired, adjusted, or replaced as required. Properly trained and certified technicians and mechanics should be used for such work. As mentioned earlier, national organizations such as NETA can be used to provide qualified personnel.

Operating Schemes.   Many personal injury accidents could be prevented if systems were all designed for safe operating procedures. The following are some of the key design concepts that should be used for any new or refurbished electrical system:

1. Arc-resistant switchgear should be employed. In addition to reducing the possibility of electrical arcs and resulting blasts, arc-resistant switchgear is designed to contain the blast and blast products. Such switchgear also has pressure relief systems that will redirect the release of arc and blast products and energy in directions that are safe for workers.
2. Remote operating controls should be used for all circuit breakers, switches, and other such control devices. This relatively simple and inexpensive design technique allows workers to operate equipment while stationed remotely from the possible results of failure. Whether supervisory control systems or simple remotely placed, hard-wired control switches are used, the distance will allow the workers to operate the equipment in relative safety.
3. Control panels should be designed with protective barriers to prevent shock hazard and contain arcs and arc products. For example, many industrial control panels are fed via 480 V circuit breakers. This breaker, in turn, feeds a step-down transformer. The control voltage output of the transformer feeds the various control circuit throughout the panel. If the 480 V portions of the circuit are designed with appropriate protective covers, workers in the panel need to protect themselves only from the electrical hazards presented by the 120 V circuits.

Power System Studies

Power system studies are engineering procedures that must be performed. There are three distinct times in the life of a power system when such studies should be performed:

1. During the initial design of the system to ensure that all selected equipment will perform properly during the day-to-day operation of the facility.

2. Anytime a significant change occurs such as replacement of an existing piece of equipment or addition of new equipment or circuits.

3. At intervals during the life of the system to identify any possible changes that may have occurred externally. For example, electric utility supplies are changing constantly. Such changes can have a profound effect on the short-circuit currents and normal operating voltage levels.

The studies described in this section should be considered safety-related procedures. If they are not performed equipment may be improperly applied and can fail, putting personnel at risk.

ANSI/IEEE standard Std-399 (*IEEE Recommended Practice for Industrial and Commercial Power Systems Analysis*) identifies 11 recommended power systems studies. Of these 11, the short-circuit analysis and protective device coordination study are among the most critical with respect to electrical safety. As described later in this section, such studies are required by some industry standards.

The following paragraphs briefly describe these studies and provide basic information about their importance and implementation. The reader is referred to the most recent edition of ANSI/IEEE standard Std-399 for details. Also note that these procedures are safety-related; they should be performed only by engineers and technicians with the education and experience to do them correctly.

Load Flow Analysis.   This type of study determines the voltage, current, reactive power, active power, and power factor in an operating power system. It is performed using computer software and can be set up to analyze contingencies of any type. Such studies are used to size and select equipment and will alert system operators to possible hazardous or poor operating conditions.

Stability Analysis.   Stability in a power system is defined as either transient or steady state. Steady state stability is the ability of a power system to maintain synchronism between machines within the system following relatively slow load changes. Transient stability is the ability to maintain synchronism after short-term events occur, such as switching and short circuits.

Stability studies are safety-related in that they will allow the power system operator to continue safe operation even when the system is exposed to abnormal or excessive events.

Motor-Starting Analysis.   When large motors are started, the high current surges and voltage dips can cause malfunction or failure of other system components. A motor-starting analysis uses a computer program to model the behavior of the system when the motor starts. This can be used to size the power system equipment to prevent outages, and the hazards associated with them will be reduced or eliminated.

Harmonic Analysis.   Harmonics and other types of power quality problems can cause premature equipment failure, malfunctions, and other types of hazardous conditions. Such failures can include heating and failure of rotating machinery and power system capacitors and their associated hazards. A harmonic analysis is performed to pinpoint the sources of harmonic distortion and to determine the solutions to such problems.

Switching Transients Analysis.   When certain loads are switched and/or when switches are malfunctioning, failures can occur, which can put employees at risk. A switching transients study determines the magnitude of such transients and allows the system engineers to develop solutions.

Reliability Analysis.   Reliability is usually expressed as the frequency of interruptions and expected number of interruptions in a year of system operation. When properly applied, the results of a reliability study can be used to make sure the system operates as continuously as possible. This means that workers will not be exposed to the hazards of working on the system to repair it.

Cable Ampacity Analysis. Cable ampacity studies determine the ampacity (current-carrying capacity) of power cables in the power system. Properly selecting power cables and sizing them will help to minimize unexpected failures.

Ground Mat Analysis. The subject of system grounding is covered in detail in Chap. 4. One of the most important pieces of equipment in the grounding system is the ground mat. A properly designed and installed ground mat will reduce step and touch voltages and provide a much safer environment for the worker. According to IEEE standard Std-399, at least five factors need to be considered in the ground mat analysis:

1. Fault current magnitude and duration
2. Geometry of the grounding system
3. Soil resistivity
4. Probability of contact
5. Humans factors such as
   (a) Body resistance
   (b) Standard assumptions on physical conditions of the individual

Short-Circuit Analysis and Protective Device Coordination. A short-circuit study determines the magnitude of the currents that flow for faults placed at various buses throughout the power system. This information is used to determine interrupting requirements for fuses and circuit breakers and to set trip points for the overcurrent devices.

A coordination study is performed to make certain the overcurrent devices in a system will trip selectively. Selective tripping means that only the nearest upstream device to the short circuit trips to clear the circuit.

The two studies, taken together, are used to properly select and calibrate the protective devices used in the power system. The information that they provide is used for the following purposes:

- Fuses and circuit breakers are selected so that they are capable of interrupting the maximum fault current that will flow through them.

- Instantaneous elements are selected so that they will respond (or in some cases not respond) to the short-circuit currents that will flow through them.

- Time curves and instantaneous settings are selected so that the nearest upstream device to the short circuit is the one that operates to clear the fault.

- Protective devices are selected so that the short circuit is cleared in a minimum amount of time with as little collateral damage as possible.

- Protective devices are selected so that fault currents that flow through cables and transformers will not cause thermal or mechanical damage to those pieces of equipment.

Each of these points is critical to the safe and economical operation of a power system. For example, if circuit breakers or fuses are incapable of interrupting fault currents, they may explode violently, injuring personnel in the area. If the wrong protective device operates, a "small" outage may expand to include an entire plant. If a transformer overload relay is too slow, the transformer may be damaged by excessive temperature rise. The only way that such malfunctions can be avoided is to perform a short-circuit analysis and a coordination study and then to select and set protective devices according to their results.

The NEC is the principal source of regulation in the area of electrical installation and design requirements for industrial and commercial facilities. The NEC has several sections that are pertinent. Table 7.3 reproduces a few of these sections.

In addition, ANSI/NFPA 70B, Electrical Equipment Maintenance, also has a section that applies. This is reproduced in Table 7.4. The only way to comply with these requirements is to ensure that a short-circuit analysis and a coordination study are performed for the power system.

Few would deny that such studies should be performed during the design phase of an electrical power system. But how about later, as the system ages? Several things happen to require the performance and/or reevaluation of these studies for an existing system:

- Electric utilities constantly add capacity. Your utility may have had a 200,000-kilovoltampere (kVA) fault capacity when the plant was new 20 years ago. Now, however, the utility's capacity may have doubled or even tripled. Such changes can cause fault duties to rise above the ratings of marginal interrupting equipment.

- Many plants are beginning to internally generate electricity. This generation adds to the fault capacity of the system.

TABLE 7.3 NEC Requirements for Short-Circuit Analyses and Coordination Studies

Location in NEC	Item
Definitions	Interrupting Rating. The highest current at rated voltage that a device is intended to interrupt under standard test conditions.
Article 110	110-9. Interrupting Rating. Equipment intended to break current at fault levels shall have an interrupting rating sufficient for the nominal circuit voltage and the current that is available at the line terminals of the equipment.
	Equipment intended to break current at other than fault levels shall have an interrupting rating at nominal circuit voltage sufficient for the current that must be interrupted.
	110-10. Circuit Impedance and Other Characteristics. The over current protective devices, the total impedance, the component short-circuit current ratings, and other characteristics of the circuit to be protected shall be selected and coordinated to permit the circuit-protective devices used to clear a fault to do so without extensive damage to the electrical components of the circuit. This fault shall be assumed to be either between two or more circuit conductors, or between any circuit conductor and the grounding conductor or enclosing metal raceway. Listed products applied in accordance with their listing shall be considered to meet the requirements of this section.
Article 240	240-12. Electrical System Coordination. Where an orderly shutdown is required to minimize hazard(s) to personnel and equipment, a system of coordination based on the following two conditions shall be permitted:
	(1) Coordinated short-circuit protection
	(2) Overload indication based on monitoring systems or devices.
	For the purposes of this section, *coordination* is defined as properly localizing a fault condition to restrict outages to the equipment affected, accomplished by the choice of selective fault-protective devices.

TABLE 7.4 ANSI/NFPA 70B Requirements for Short-Circuit Analyses and Coordination Studies

ANSI/NFPA 70B	Item
Paragraph 7-4.3	An up-to-date short-circuit and coordination study is essential for safety of personnel and equipment. It is necessary to analyze the momentary and interrupting rating requirements of the protective devices. That is, will the circuit breaker or fuse safely interrupt the fault or explode in attempting to perform this function?
Para 7-4.3.1	Another phase of the study is that of developing the application of the protective devices to realize minimum equipment damage and the least disturbance to the system in the event of a fault.

- Operating procedures may have changed. A bus tie circuit breaker that was normally open may now be normally closed. Such a change can greatly increase fault capacity.

- Technical standards can change. For example, in 1985 the protection requirements for liquid-filled transformers changed. Studies showed that many transformers were being mechanically damaged by high current through faults. The protection requirements became more stringent for such installations. Although the standards do not require existing systems to be changed, would it not make sense to at least review your system? The protection changes might be minimal.

- New installations or plant modernization may add capacity and other coordination streams to the system. For example, coordination studies require that the main breaker coordinate with the largest feeder device. If a larger feeder device is added later, the coordination study must be reviewed.

In general, short-circuit analyses and coordination studies should be reviewed at least every 5 years. These studies should be performed by a registered professional engineer. Many consulting firms have the ability and the experience to perform them; however, since short-circuit analyses and coordination studies are specialized types of engineering services, not all architect and engineering firms have the experience to do them. Closely review the qualifications of the firm that you retain.

ANSI/IEEE standard Std-399, Power Systems Analysis, is the standard that covers most of the engineering studies that are key to the proper design and performance of an electrical power system. The *Brown Book* is part of the IEEE color book series and should be referenced when you are deciding what studies to perform and how to perform them.

Proper selection, sizing, and calibration of the protective devices in a power system directly affect safety, efficiency, and economics. Common sense and regulatory requirements dictate that a short-circuit analysis and coordination study should be performed.

Arc-Flash Study.   The arc-flash study is performed to determine the incident energy to which workers will be exposed in the event that an electrical arc occurs. This study is described in some detail in Chap. 3.

FIRST AID

This handbook provides general coverage of the subject with expanded information on handling electrical injuries. Potential first aid givers should remember four very important points. Before an accident happens:

- Obtain hands-on first aid training for yourself and all employees. Such training may be obtained from the American Heart Association, the American Red Cross, or local sources such as fire departments or police departments.

If an accident does happen:

- Act quickly!!! You may be the only person that can prevent a death.

- Do not administer first aid that you are not qualified to administer. Injuries can be aggravated by improperly administered first aid.

- Get qualified medical help quickly. Paramedics and emergency medical technicians are trained to provide emergency first aid and should be summoned as soon as possible.

TABLE 7.5   General First Aid Procedure

- Act quickly.
- Survey the situation.
- Develop a plan.
- Assess the victim's condition.
- Summon help if needed.
- Move the victim only if danger is imminent.
- Administer required first aid:
  Shock
  Electrical burns

This handbook is not intended to be used as a first aid training manual. Table 7.5 summarizes the first aid steps that are discussed in the following sections.

General First Aid

Act Quickly.   *Remember—you may be the only person between the victim and death.* Whatever you do, do it quickly. This does not imply that you should act impetuously. Your actions should be planned and methodical, but you should not waste any time. Do not attempt to perform procedures for which you have no training or experience. Improperly applied procedures can be deadly.

Survey the Situation.   Remember that your purpose as a first aid giver is to help the problem, not contribute to it. If you are injured in the process of administering first aid, you cannot help the victim. If your preliminary assessment indicates that you need to wear safety clothing, put it on first, then administer aid. Table 7.6 lists key points that should be checked before you rush in.

Develop a Plan.   After the initial survey of the situation, develop the plan of attack. The specifics of any given situation will vary; however, the following guidelines should be used:

- If the victim is in immediate danger, he or she should be moved to a safe position. (See the next section on moving the victim and later sections on rescue techniques.)
- If the victim is nonresponsive, assess his or her condition and respond accordingly. (See the later section on assessing the victim's condition.)

TABLE 7.6   First Aid Checklist

- Is the circuit still energized?
- Is the victim contacting the circuit?
- Are noxious gases or materials present that may cause injury?
- Is fire present or possible?

TABLE 7.7 What to Do If the Victim Is Responsive

- Ask the victim what is wrong.
- Assess the victim's condition and treat the injuries as best as possible.
- Treat the worst injuries first.
- When the victim is out of immediate danger, or if you are unable to help because the injuries are beyond your abilities, summon help.
- Attend to the victim(s) and keep them safe until help arrives.
- When help arrives, give the first aid workers your assessment of the situation and stand by to help.

- If the victim is responsive, make him or her as comfortable as possible and summon aid. Do not abandon the victim until aid has arrived.

- Constantly monitor the condition of the victim. Electric shock can cause delayed failures and irregularities of heart rhythm.

Assess the Victim's Condition. The procedures to be used in administering first aid depend on the condition of the victim. If the victim is responsive, no action may be required. Table 7.7 lists the procedures to perform if the victim is awake and responsive.

If the victim is not responsive, you must perform a "hands-on" assessment of his or her condition. Table 7.8 lists the ABCs of first aid. This memory device can help the first aid giver to remember the proper procedure when examining a nonresponsive victim.

One of the biggest surprises to those who have not worked with accident victims is that the trauma of the accident can induce severe bleeding through the mouth and/or vomiting. Be prepared for these conditions before working with an injured person. When you have prepared yourself for this situation, begin the ABCs

TABLE 7.8 The ABCs of First Aid

Airway
Breathing
Circulation
Doctor

- **A**—Check the victim's **A**irway. Figure 7.1 illustrates the correct way to clear an injured person's airway. Remember to avoid moving the victim and to keep the victim's spine straight to avoid aggravating an injury. *Caution:* An accident victim may suffer from involuntary muscular reflexes and other such spasms. The strongest muscle in the human body is the jaw. Because of this, rescue workers should put their fingers into the victim's mouth only when absolutely necessary.

  Start by opening the victim's mouth as shown in Fig. 7.1. Search the mouth for foreign matter or other objects that may be blocking the air passage. Many times the victim's tongue may be blocking the air passage. To fix this problem, put your hand behind the victim's neck, gently pull the jaw forward, and, if required, carefully tilt the head back. If the air passage is clear and the victim is still not breathing, you should perform resuscitation.

- **B**—Check the victim's **B**reathing. First check to see if the victim is breathing. This can be done by observing his or her chest to see if it is moving. Then place your ear close to the victim's mouth and nose and listen carefully. If the victim is breathing but choking or gurgling sounds are heard clear the victim's airway.

- **C**—Check the victim's **C**irculation. Circulation should be checked by feeling for the victim's pulse at the carotid artery as shown in Fig. 7.2. To find the carotid artery, place your fingertips gently on the victim's larynx. Gently slide the fingers down into the groove between the windpipe and the muscle at the back of the neck. The carotid

TO OPEN AIRWAY

CLEAR MOUTH

TILT HEAD BACK

OBSTRUCTED   OPENED

FIGURE 7.1   Clearing the airway of an injured worker.

artery is located in this area. Gently feel for the pulse. Table 7.9 shows the steps to take for the various combinations of problems that may be found.

- **D**—Summon the **D**octor. After the victim's condition has been stabilized, summon help. If the resuscitation efforts are proving unsuccessful, the first aid giver may want to summon more qualified assistance even though the victim is not yet stabilized.

Summon Help If Needed.   One of the most difficult decisions is to determine when to summon help. If help is not summoned soon enough, the victim may die. On the other hand, if the first aid giver leaves to summon help, the victim may die. No concrete rules can be given here; however, the following guidelines may help:

- Relieve any immediate danger to the victim before summoning help.
- Perform the ABCs before summoning help.
- If the victim is not breathing or has no pulse, perform resuscitation before summoning help.
- If anyone else is in the area, yell or call for help while performing the preliminary accident assessment.

FIGURE 7.2   Checking the circulation.

Remember that the first aid giver is in charge of the victim until more qualified help arrives. Do not abandon the victim if immediate aid is required.

Move the Victim If Danger Is Imminent.   Unless they are in imminent danger, accident victims should be moved only when necessary and only by personnel who are qualified to move them. A victim of violent injury, such as a fall, may have spinal or other internal injuries. Moving such a victim could cause increased problems including paralysis or even

TABLE 7.9 How to Handle Unresponsive Victims

Breathing—pulse normal	Make victim comfortable. If help has not been summoned, do so and stand by until it arrives.
No breathing—pulse normal	Perform mouth-to-mouth resuscitation until breathing is restored or until help arrives and takes over.
Breathing normal—no pulse	Perform heart-lung resuscitation (CPR) until pulse is restored or until help arrives and takes over.
No breathing—no pulse	Perform heart-lung resuscitation (CPR) until pulse is restored or until help arrives and takes over.

death. Moving an injury victim is discussed in detail in the "Rescue Techniques" section in this chapter.

First Aid for Electric Shock.   Electric shock is one of the most difficult of all injuries to diagnose. In some cases, even if the injury is fatal, no external signs may be visible. Table 7.10 lists some of the clues and symptoms that may be present when a victim has received an electric shock.

Many prospective first aid givers are themselves injured when they contact an energized wire or a victim who is still in contact with an energized wire. Table 7.11 lists the precautions for working on or around accident victims who may be in contact with live wires. After cutting the power to the circuits or removing the victim from contact, if the victim is responsive and shows no signs of breathing or heart problems, the procedures listed in Table 7.12 should be followed. After cutting the power to the circuits or removing the victim from contact, if the victim is nonresponsive, the procedures listed in Table 7.13 should be followed.

First Aid for Electrical Burns.   Electrical burns may be internal and/or external. External burns are caused by the intense heat of the electric arc coupled with the current flow, while internal burns are caused by the current flow heating the tissue. Internal burns are virtually impossible to diagnose in the field. The symptoms of internal electrical burns are identical to the symptoms caused by severe electric shock. In addition to the symptoms described in Table 7.10, the victim may also experience significant pain caused by the damaged tissue. External burns are similar to thermally induced burns caused by fire or other heat sources.

For both internal and external burns, the first aid techniques are similar to those given in Tables 7.12 and 7.13.

One extremely important additional procedure for external burn victims is to cool the burns as quickly as possible. A shower or other source of clean, cool water can be used. Since the treatment of burns is a very specialized medical procedure, qualified medical help should be obtained as quickly as possible.

TABLE 7.10 Typical Symptoms of Electric Shock

- Victim may lose consciousness. This may occur at the moment of contact; however, it can also occur later.
- Victim has a weak or irregular pulse.
- Victim has trouble breathing or has stopped breathing.
- Small burns may appear at the entry and exit points of the electric current.

TABLE 7.11   Precautions for Performing First Aid on an Electric Shock Victim

---

- Do not touch any energized wires with any part of your body or with any conductive tools or equipment.
- Do not touch a victim who is still in contact with an energized wire with any part of your body or with conductive tools or equipment.
- Do not try to move any energized wires unless you are qualified to do so. Qualified in this instance means that you are trained in the performance of such a procedure and are able to avoid electrical hazards.

Resuscitation (Artificial Respiration)

Breathing trauma is one of the two very serious symptoms that result from severe electric shock or internal burns. When breathing is stopped or made irregular by electricity, it must be restored by resuscitation. Over the years many different types of resuscitation have been developed. Mouth-to-mouth resuscitation is the current preferred technique.

In order to be most effective, resuscitation must be started as soon as possible after breathing has ceased. Figure 7.3 shows a curve that approximates the possibility of success plotted against elapsed time before the start of resuscitation. Figure 7.4 illustrates a currently accepted procedure for the performance of artificial respiration. *Caution:* With the proliferation of communicable diseases such as acquired immunodeficiency syndrome (AIDS) and hepatitis, the decision of whether to perform life-saving mouth-to-mouth resuscitation has become much more difficult. Instruments such as breathing tubes should be used to protect the victim and the rescuer.

Heart-Lung Resuscitation

Heart-lung resuscitation is also called cardiopulmonary resuscitation (CPR). This technique should be applied when the victim has no pulse and no respiration. The procedure should be started as soon as possible to maximize the probability of successfully restoring full function to the victim.

Remember that while CPR is being performed, the first aid giver is actually pumping blood and breathing for the victim. Do not give up until qualified medical personnel say to.

CPR has two separate, but equally important parts; rescue breathing and chest compressions. The rescue breathing is performed to force air into an unbreathing victim's lungs. The chest compressions are used to compress the heart and force blood through the victim's circulatory system. Before CPR is performed the ABC's described earlier in this chapter should be checked. See Table 7.8 and associated text.

TABLE 7.12   First Aid Procedures for Conscious Electric Shock Victims Who Exhibit No Symptoms

---

- Keep the victim still and quiet. Remember that heart and respiratory problems can be delayed in electric shock victims.
- Monitor the victim's condition for at least 1 hour.
- If the victim continues to show no symptoms, take them to a doctor for a thorough examination.

TABLE 7.13   First Aid Procedures for Unconscious Electric Shock Victims with Symptoms

- Check the ABCs. If the victim is not breathing or has heart irregularities, perform resuscitation as described later in this handbook.
- If wounds are evident, cover them with sterile dressings.
- If external burns are evident, they should be cooled using clear, pure water.
- Try to cool burns with sterile compresses.
- Immediately seek medical aid.

Rescue Breathing (Mouth-to-Mouth Resuscitation).   If the preliminary assessment indicates that the victim is not breathing, mouth-to-mouth resuscitation should be performed as follows:

1. Open the airway, clear the mouth, and tilt the head back, as shown in Fig. 7.1.
2. Gently pinch the victim's nose to seal it closed, as shown in Fig. 7.4.
3. Take a deep breath and seal your lips around the victim's mouth creating an air-tight seal.
4. Gently blow into the victim's mouth for 1½ to 2 seconds. While blowing, watch to be sure the victim's chest is rising.
5. Stop blowing and allow the victim's lungs to exhale.
6. Repeat steps 4 and 5.
7. Stop the procedure and watch the victim for 5 to 10 seconds to determine if the victim is breathing on his/her own.
8. If the victim is not breathing normally, repeat steps 4 and 5 until the victim starts breathing normally or until qualified help arrives to relieve you.

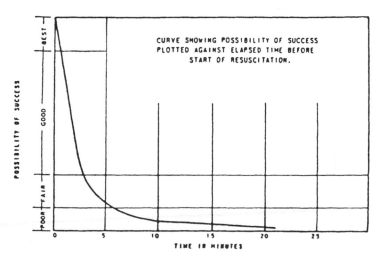

FIGURE 7.3   Elapsed time versus possibility of success for artificial respiration.

FIGURE 7.4 The rescuer gently pinches the nose closed with the thumb and index finger of the hand on the forehead breathes into the victim's mouth.

Chest Compressions. If the victim has no pulse, chest compressions should accompany the rescue breathing. The following steps should be used:

1. Properly position your hands on the victim's chest (See Figs. 7.5, 7.6, and 7.7)
   a. Kneel facing the victim's chest. Find the correct hand position by sliding your fingers up the rib cage to the breastbone as shown in Fig. 7.5)
   b. Place your middle finger in the notch and the index finger next to it on the lower end of the breastbone.
   c. Place your other hand beside the two fingers.
   d. Place the hand used to locate the notch on top of the other hand. Lace your fingers together. Be sure to keep your fingers off of the victim's chest. Position your shoulders over your hands with your elbows locked and your arms straight. (See Fig. 7.6 for alternates to this hand position.)
2. Compress the breastbone 1½ to 2 inches, as shown in Fig. 7.7.
3. Repeat step 2 fifteen times.
4. Stop and give the victim two rescue breaths as described under rescue breathing above. (Note that steps 2, 3, and 4 should take approximately 15 seconds.)
5. Recheck the victim's pulse for 5 seconds.
6. If no pulse is found, repeat steps 1 through 5.
7. Continue this process until:
   a. Another trained person takes over CPR from you.
   b. EMS personnel arrive and take over care of the victim.
   c. You are too exhausted and unable to continue.
   d. The scene becomes unsafe.

RESCUE TECHNIQUES

General Rescue Procedures

The first priority in any emergency is to remove living victims from the danger area if they cannot escape themselves. This procedure is called *rescue.* In some instances the rescuer will be risking his or her life in order to rescue a victim. The decision to risk one's life is a personal one and cannot be regulated by any sort of standard procedures. In any case, good judgment must be exercised by the rescuer. Remember that becoming a second victim does not help anyone.

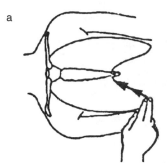

Kneel facing the victim's chest. Find the correct hand position by sliding your fingers up the rib cage to the breastbone.

Place your middle finger in the notch and the index finger next to it on the lower end of the breastbone.

Place your other hand beside the two fingers.

Place hand used to locate notch on top of the other hand. Lace the fingers together. Keep fingers off of the chest. Position shoulders over hands, with elbows locked and arms straight.

FIGURE 7.5 Steps a, b, c, and d illustrate the procedure to properly position your hands prior to perorming CPR.

FIGURE 7.6   Alternate positions for hands while applying CPR.

The American Red Cross method of rescue is shown in Table 7.14. Notice that the steps are virtually identical to those employed in preparing for first aid procedures. Rescue is, in fact, one of the preliminary steps in the performance of first aid.

After the victim has been removed from the hazardous area (in the case of an electrical accident, *do not touch the victim* until electric circuits have been de-energized and/or the victim has been removed from contact of the energized circuits) and after first aid has been administered, the accident area should be secured and made safe for other persons. *Do not disturb* the accident area beyond what is necessary to protect other persons. An accident investigation should be carried out by qualified personnel to determine the nature and cause of the accident and what corrective actions need to be taken to prevent the accident from occurring again.

The following outline details the generalized procedures defined by Table 7.14. This outline was taken from the AVO Multi-Amp Institute textbook entitled *High Voltage Rescue Techniques.*

1. Survey the scene. A quick survey (1 min or less—time is critical) to enable the rescuer to develop a mental plan. Consider the following:

FIGURE 7.7   The weight of the rescuer's upper body is used to compress the chest of the victim.

   a. Anticipating conditions.
     (1) Probable life-threatening hazards
     (2) Observed location of victim
     (3) Observed condition of victim
   b. Facts of the situation.
     (1) Time and conditions
     (2) Situations—What happened?
     (3) Contributing factors
   c. Weighing facts.
     (1) Rescue requirements
     (2) Accomplish rescue alone?
     (3) Equipment and manpower needed/ available
     (4) Additional assistance now or later
     (5) If electrical contact, get de-energized
   d. Determining procedures.
     (1) Immediate action
     (2) Course of action
   e. Initial action decision. May be made before a complete survey. May determine a future course of action.

TABLE 7.14 Primary Rescue Steps

---

1. Survey the scene.
   a. Develop an action plan.
2. Primary survey.
   a. Locate the victim.
   b. Assess the victim's condition.
   c. Try to arouse the victim.
   d. Move the victim to a safe place. *Caution*—do not jostle the victim any more than is necessary when moving. See later sections for a discussion of proper movement techniques.
   e. Apply first aid as soon as the victim is in a safe location.
3. Call for emergency medical assistance.

---

2. Primary survey plan activated. From the initial decision and action until the victim is being transported to medical facilities, the following would be considered:
   a. Victim location
      (1) Accessibility of victim.
      (2) Electric circuit contact.
      (3) Remove circuit contacts.
      (4) Try to arouse victim.
      (5) Help needed/called.
      (6) Assess condition of victim (ABCs plus broken bones, burns, twisted torso, etc.).
   b. Emergency aid required
      (1) What and where to give emergency aid?
      (2) How to move victim to safety
      (3) Emergency aid given
3. Call emergency medical service.
   a. How to move victim to medical aid.
   b. Continue emergency aid as required.
4. Secondary survey.
   a. Secure the area.
      (1) Do not damage or disturb physical evidence.
      (2) Isolate hazards.
      (3) Keep untrained personnel away.
   b. What happened.
      (1) Ask victim.
      (2) Ask witnesses.
      (3) Examine physical evidence.
      (4) Photographs, sketches.
      (5) Preserve physical evidence.
   c. Reports.
      (1) Written records
      (2) Notification
   d. Follow up.
      (1) Reports distributed
      (2) Action taken to prevent future similar accidents

Note: The action taken during the actual physical rescue must be started and accomplished very rapidly so that the victim is out of further danger and emergency aid may begin quickly. See Fig. 7.3. Accident investigation techniques are discussed later in this chapter.

Elevated Rescue

The physical techniques that are used for any type of rescue should be learned and practiced in a controlled environment under the direction of experienced, qualified training personnel.

Figure 7.8 graphically depicts a generally accepted method for performing an elevated, pole-top rescue. Note that this rescue technique requires a skilled line person with climbing ability. Do not attempt such a rescue unless you have had the requisite training. All rescue techniques should be accomplished using the basic rescue approach outline, which is illustrated in the previous section.

YOU MAY HAVE TO HELP A MAN ON A POLE REACH THE GROUND SAFELY WHEN HE—

- BECOMES ILL
- IS INJURED
- LOSES CONSCIOUSNESS

YOU MUST KNOW—

- WHEN HE NEEDS HELP
- WHEN AND WHY TIME IS CRITICAL
- THE APPROVED METHOD OF LOWERING

Evaluate THE SITUATION
CALL TO MAN ON POLE

IF HE DOES NOT ANSWER OR APPEARS STUNNED OR DAZED

- PREPARE TO GO TO HIS AID

TIME IS EXTREMELY IMPORTANT

BASIC STEPS IN
POLE TOP RESCUE

Evaluate THE SITUATION
Provide FOR YOUR PROTECTION
Climb TO RESCUE POSITION
Determine INJURED'S CONDITION

Then
IF NECESSARY—

- GIVE FIRST AID
- LOWER INJURED
- GIVE FOLLOW-UP CARE
- CALL FOR HELP

Provide
FOR YOUR PROTECTION YOUR SAFETY IS VITAL TO THE RESCUE

—PERSONAL TOOLS AND RUBBER GLOVES (RUBBER SLEEVES, IF REQUIRED)

Check
✓ **EXTRA RUBBER GOODS?**
✓ **LIVE LINE TOOLS?**
✓ **PHYSICAL CONDITION OF POLE?**
  ✓ DAMAGED CONDUCTORS, EQUIPMENT?
  ✓ FIRE ON POLE?
  ✓ BROKEN POLE?
✓ **HAND LINE ON POLE AND IN GOOD CONDITION?**

FIGURE 7.8   Illustrated method for elevated, pole-top rescue.

Climb TO RESCUE POSITION
- CLIMB CAREFULLY

POSITION YOURSELF—
- TO INSURE YOUR SAFETY
- TO CLEAR THE INJURED FROM HAZARD
- TO DETERMINE THE INJURED'S CONDITION
- TO RENDER AID AS REQUIRED
- TO START MOUTH-TO-MOUTH, IF REQUIRED
- TO LOWER INJURED, IF NECESSARY

THE BEST POSITION WILL USUALLY BE— SLIGHTLY ABOVE THE INJURED

IF THE INJURED IS
CONSCIOUS
- TIME MAY NO LONGER BE CRITICAL
- GIVE NECESSARY FIRST AID ON POLE
- REASSURE THE INJURED
- HELP HIM DESCEND POLE
- GIVE FIRST AID ON GROUND
- CALL FOR HELP, IF NECESSARY

Determine
THE INJURED'S CONDITION

HE MAY BE . . .
- **CONSCIOUS**

- **UNCONSCIOUS BUT BREATHING**

- **UNCONSCIOUS NOT BREATHING**

- **UNCONSCIOUS NOT BREATHING AND HEART STOPPED**

IF THE INJURED IS
UNCONSCIOUS BUT BREATHING
- **WATCH HIM CLOSELY IN CASE BREATHING STOPS**
- **LOWER HIM TO GROUND**
- **GIVE FIRST AID ON GROUND**
- **CALL FOR HELP**

FIGURE 7.8 *(Continued)*

IF THE INJURED IS

UNCONSCIOUS
AND NOT BREATHING

- **PROVIDE AN OPEN AIRWAY**

- **GIVE HIM 5 OR 6 QUICK BREATHS**

- **IF HE RESPONDS . . .**

CONTINUE
MOUTH-TO-MOUTH
UNTIL HE IS
BREATHING
WITHOUT
HELP

**IF PUPIL DOES NOT CONTRACT
AND SKIN COLOR IS BAD**

HEART HAS STOPPED

- **PREPARE TO LOWER HIM IMMEDIATELY**

- **GIVE 5 OR 6 MORE QUICK BREATHS JUST BEFORE LOWERING**

- **LOWER HIM TO THE GROUND**

- **START HEART-LUNG RESUSCITATION**

- **CALL FOR HELP**

FIGURE 7.8   (*Continued*)

IF HE DOES NOT RESPOND TO THE
FIRST 5 OR 6 QUICK BREATHS

✔ CHECK SKIN COLOR

✔ CHECK FOR PUPIL DILATION

NORMAL PUPIL   DILATED PUPIL

**IF PUPIL CONTRACTS
AND COLOR IS GOOD**
CONTINUE MOUTH-TO-MOUTH UNTIL HE
IS BREATHING WITHOUT HELP

Then

- HELP HIM DESCEND POLE

- WATCH HIM CLOSELY—HIS BREATHING MAY STOP AGAIN

- GIVE ANY ADDITIONAL FIRST AID NEEDED

- CALL FOR HELP

THE
METHOD
OF
LOWERING AN
INJURED MAN
IS

* SAFE
* SIMPLE
* AVAILABLE

RESCUER

Position **HAND LINE**

EQUIPMENT NEEDED

OVER ARM OR OTHER PART OF STRUCTURE

- 1/2 INCH HAND LINE

PROCEDURE . . .

- *Position* HAND LINE
- *Tie* INJURED
- *Remove* SLACK IN HAND LINE
- *Take* FIRM GRIP ON FALL LINE
- *Cut* INJURED'S SAFETY STRAP

  Lower INJURED

VICTIM   FALL LINE

- **POSITION LINE FOR CLEAR PATH TO GROUND**

Note **USUALLY BEST 2 OR 3 FEET FROM POLE**

FIGURE 7.8 *(Continued)*

SHORT END OF LINE IS WRAPPED AROUND FALL LINE TWICE. (TWO WRAPS AROUND FALL LINE.)

RESCUER TIES HAND LINE AROUND
VICTIM'S CHEST USING THREE HALF
HITCHES.

Tie INJURED—

**PASS LINE AROUND INJURED'S
CHEST**

TIE THREE HALF-HITCHES

- **KNOT IN FRONT**
- **NEAR ONE ARM PIT**
- **HIGH ON CHEST**
- **SNUG KNOT**

Remove SLACK IN HAND-LINE

- **ONE RESCUER**—REMOVES
  SLACK ON POLE

- **TWO RESCUERS**—MAN ON
  GROUND REMOVES SLACK

IMPORTANT
GIVE 5 OR 6 QUICK BREATHS
. . . IF NECESSARY . . .

THEN

Take FIRM GRIP ON FALL LINE

ONE RESCUER—HOLDS FALL LINE
WITH ONE HAND

TWO RESCUERS—MAN ON GROUND
HOLDS FALL LINE

FIGURE 7.8   (*Continued*)

Cut INJURED'S SAFETY STRAP

**CUT STRAP ON SIDE
OPPOSITE DESIRED SWING**

Caution

**DO NOT CUT YOUR OWN SAFETY
STRAP OR THE HAND LINE**

RIGHT

WRONG

Lower INJURED

ONE RESCUER
- **GUIDE LOAD LINE WITH
  ONE HAND**
- **CONTROL RATE OF DESCENT
  WITH THE OTHER HAND**

TWO RESCUERS
- **MAN ON THE POLE GUIDES
  THE LOAD LINE**
- **MAN ON THE GROUND
  CONTROLS RATE OF DESCENT**

**IF A CONSCIOUS MAN IS BEING ASSISTED
IN CLIMBING DOWN...
THE ONLY DIFFERENCE IS
THAT ENOUGH SLACK IS FED INTO THE
LINE TO PERMIT HIM CLIMBING FREEDOM**

**ONE MAN
RESCUE**

**TWO OR MORE
MAN RESCUE**

**RESCUES DIFFER ONLY IN
CONTROL OF THE FALL LINE**

Remember

**THE APPROVED METHOD OF
LOWERING AN INJURED
MAN IS . . .**

- *Position* HAND LINE
- *Tie* INJURED
- *Remove* SLACK IN HAND LINE
- *Take* FIRM GRIP ON FALL LINE
- *Cut* INJURED'S SAFETY STRAP
- *Lower* INJURED

FIGURE 7.8   *(Continued)*

IF POLE DOES NOT HAVE CROSSARM, RESCUER PLACES HAND-LINE OVER FIBERGLASS BRACKET INSULATOR SUPPORT, OR OTHER SUBSTANTIAL PIECE OF EQUIPMENT SUCH AS A SECONDARY RACK, NEUTRAL BRACKET, OR GUY WIRE ATTACHMENT, STRONG ENOUGH TO SUPPORT THE WEIGHT OF THE INJURED. SHORT END OF LINE IS WRAPPED AROUND FALL LINE TWICE AND TIED AROUND VICTIM'S CHEST USING THREE HALF HITCHES. INJURED'S SAFETY STRAP IS CUT AND RESCUER LOWERS INJURED TO THE GROUND.

LINE MUST BE REMOVED AND
VICTIM EASED ON
TO GROUND

LAY VICTIM ON HIS BACK AND OBSERVE IF VICTIM IS CONSCIOUS. IF VICTIM IS CONSCIOUS, TIME MAY NO LONGER BE CRITICAL. GIVE NECESSARY FIRST AID. CALL FOR HELP.

FIGURE 7.8   *(Continued)*

IF THE INJURED IS UNCONSCIOUS AND NOT BREATHING, PROVIDE AN OPEN AIR-WAY. GIVE HIM FIVE OR SIX QUICK BREATHS. IF HE RESPONDS, CONTINUE MOUTH-TO-MOUTH RESUSCITATION UNTIL HE IS BREATHING WITHOUT HELP. IF NO PULSE IS PRESENT, PUPIL DOES NOT CONTRACT, AND SKIN COLOR IS NOT NORMAL, HIS HEART HAS STOPPED. RESTORE CIRCULATION. USE HEART-LUNG RESUSCITATION.

BUCKET TRUCK
RESCUE
TIME
IS
CRITICAL

EQUIP PORTION OF INSULATED BOOM
OF TRUCK WITH ROPE BLOCKS
DESIGNED FOR HOT-LINE WORK.

FIGURE 7.8   (*Continued*)

STRAP IS PLACED AROUND INSULATED
BOOM APPROXIMATELY TEN FEET
FROM BUCKET TO SUPPORT ROPE
BLOCKS. BLOCKS ARE HELD TAUT ON
BOOM FROM STRAP TO TOP OF BOOM.

RESCUER ON THE GROUND EVALUATES
THE CONDITIONS WHEN AN EMERGENCY
ARISES. THE BUCKET IS LOWERED
USING THE LOWER CONTROLS.
OBSTACLES IN THE PATH OF THE
BUCKET MUST BE AVOIDED.

FIGURE 7.8   (*Continued*)

HOOK ON ROPE BLOCKS IS ENGAGED
IN A RING ON THE LINEMAN'S SAFETY
STRAP. SAFETY STRAP IS RELEASED
FROM BOOM OF TRUCK.

ROPE BLOCKS ARE DRAWN TAUT BY
RESCUER ON THE GROUND.
UNCONSCIOUS VICTIM IS RAISED OUT
OF BUCKET WITH ROPE BLOCKS.

FIGURE 7.8   (*Continued*)

**RESCUER EASES VICTIM ON TO THE
GROUND. CARE SHOULD BE TAKEN
TO PROTECT THE INJURED VICTIM
FROM FURTHER INJURY.**

**RELEASE ROPE
BLOCKS
FROM
VICTIM.**

FIGURE 7.8   (*Continued*)

BASKET OR BUCKET ON AERIAL
DEVICE MAY BE CONSTRUCTED
TO TILT AFTER BEING RELEASED,
ELIMINATING THE NEED FOR
SPECIAL RIGGING TO REMOVE AN
INJURED PERSON FROM THE BASKET
OR BUCKET.

LAY VICTIM ON HIS BACK AND
OBSERVE IF VICTIM IS CONSCIOUS.
IF VICTIM IS CONSCIOUS, TIME MAY
NO LONGER BE CRITICAL. GIVE
NECESSARY FIRST AID. CALL
FOR HELP.

IF THE INJURED IS UNCONSCIOUS
AND NOT BREATHING, PROVIDE AN
OPEN AIRWAY. GIVE HIM FIVE OR
SIX QUICK BREATHS. IF HE RESPONDS,
CONTINUE MOUTH-TO-MOUTH
RESUSCITATION UNTIL HE IS
BREATHING WITHOUT HELP.

FIGURE 7.8   (*Continued*)

FIGURE 7.9   Vault atmosphere monitoring.

Confined-Space Rescue

Confined-space rescue is made more complex by two basic problems. First, the confined space may become filled with gases that are not breathable or, worse, are toxic and therefore hazardous to human life. Second, the confined space does not allow for free motion of rescue personnel. This means that rescue workers may not be able to easily get a grip on the victim.

Atmosphere Checking in Confined-Space Rescues.   Before the rescue is attempted, the atmosphere must be tested for the presence of adequate supplies of oxygen, excessive quantities of toxic gases, and excessive quantities of combustible gases. Figure 7.9 shows a typical monitoring setup for an underground confined-space area. If the rescuer must descend into a confined space that does not have adequate oxygen concentrations or has excessive quantities of toxic or combustible gases, the rescuer must wear respiration gear.

Oxygen levels must be no less than 19.5 percent and no more than 23.5 percent. Table 7.15 lists the potential effects of insufficient oxygen. The permissible levels of toxic gas vary depending on the type of gas and the individual. Generally toxic gases should be kept to levels that are less than 50 parts per million (ppm) by volume. Tables 7.16 and 7.17 list the potential effects that carbon monoxide (CO) and hydrogen sulfide (H_2S) will have.

Securing the Victim to the Rescue Line.   The method used to secure a victim to the line depends upon the logistics of the given rescue. The nature of the injuries and the space available for the rescue both affect the method used. The simplest way is to tie a hand line around

TABLE 7.15   Potential Effects of an Oxygen-Deficient Atmosphere

Oxygen content (% by volume)	Effects and symptoms* (at atmosphere pressure)
19.5	Minimum permissible oxygen level for normal functioning.
15–19	Decreased ability to work strenuously. May impair coordination and may induce early symptoms in persons with coronary, pulmonary, or circulatory problems.
12–14	Respiration increases in exertion; pulse increases, impaired coordination, perception, judgment.
10–12	Respiration further increases in rate and depth, poor judgment, lips blue.
8–10	Mental failure, fainting, unconsciousness, ashen face, blueness of lips, nausea, and vomiting.
6–8	Eight min, 100% fatal; 6 min, 50% fatal; 4–5 min, recovery with treatment.
4–6	Coma in 40 s, convulsions, respiration ceases, death.

* The effects and symptoms at a given oxygen content are guidelines and will vary with the individual's state of health and physical activity.

TABLE 7.16 Potential Effects of Carbon Monoxide Exposure

PPM*	Effects and symptoms[†]	Time
35	Permissible exposure level limit	8 h
200	Slight headache	3 h
400	Headache, discomfort	2 h
600	Headache, discomfort	1 h
1000–2000	Confusion, headache, nausea	2 h
1000–2000	Tendency to stagger	$1\frac{1}{4}$ h
1000–2000	Slight palpitation of heart	30 min
2000–2500	Unconsciousness	30 min
4000	Fatal	Less than 1 h

* PPM—parts per million; volume measurement of gas concentration.
[†] The effects and symptoms at a given are guidelines and will vary with the individual's state of health and physical activity.

the victim using three half-hitches as shown in Fig. 7.10. Figure 7.11 shows a preferred method using a padded harness designed specifically for such rescue efforts. If space allows or if the victim's injuries require it, a full body rope harness may be rigged as shown in Fig. 7.12. Figure 7.13 shows a victim being lifted using the body rope harness. Sometimes a victim's injuries may be so severe that they must be securely immobilized before being removed from the accident area. Both of the devices shown in Fig. 7.14 may be used.

Retrieving the Victim.   After the victim has been secured to the rescue line, he or she must be raised from the confined space to an open area where treatment may be administered. A variety of different methods have been developed to raise victims. Figures 7.15 through 7.19 show various types of rigs that may be used as pulley points to retrieve victims.

The key to a successful victim retrieval is teamwork and adequate planning. Remember that human beings can weigh as much as 300 or 400 lb. Also remember that unconscious victims cannot assist in their own retrieval. Always be certain that the rigging is strong enough.

Ground-Level Rescue

The Problem.   Ground-level victims do not require rescue as often as elevated or confined-space victims. Occasionally, however, victims do need retrieval, especially from metal-clad switchgear assemblies. Walk-in metal-clad switchgear has circuit breaker cubicles that may represent a limited space problem. They also have doorways that must be traversed.

TABLE 7.17 Potential Effects of Hydrogen Sulfide Exposure

PPM	Effects and symptoms	Time
10	Permissible exposure level limit	8 h
50–100	Mild eye irritation, mild respiratory irritation	1 h
200–300	Marked eye irritation, marked respiratory irritation	1 h
500–700	Unconsciousness, death	$\frac{1}{4}$–1 h
1000 or more	Unconsciousness, death	Minutes

PASS LINE AROUND INJURED'S CHEST

TIE THREE HALF-HITCHES

- **KNOT IN FRONT**
- **NEAR ONE ARM PIT**
- **HIGH ON CHEST**
- **SNUG KNOT**

FIGURE 7.10 Securing the victim using a hand line.

FIGURE 7.11 Using a padded safety harness to secure the rescue line to the victim. (*Courtesy AVO, International.*)

FIGURE 7.12 Full body rope harness. (*Courtesy AVO, International.*)

FIGURE 7.13 Full body rope harness being used to lift the victim. (*Courtesy AVO, International.*)

STOKES NAVY STRETCHER

(a)

SKED SYSTEM MODEL SK-200

(b)

FIGURE 7.14   (*a*) Stokes navy stretcher. (*b*) SKED system model SK-200. (*Courtesy AVO, International.*)

FIGURE 7.15   Tripod hoist operation. (*Courtesy AVO, International.*)

FIGURE 7.16 Quad pod retractable lifeline. (*Courtesy AVO, International.*)

In addition to metal-clad gear problems, victims may be endangered by fire, chemical release, or any number of other similar hazards. Regardless of where the rescue is taking place, the general rescue procedures should be the starting point for all rescues.

Retrieving the Victim. Since moving a victim can induce additional injuries or aggravate existing injuries, always make certain that moving the victim is absolutely necessary. If the victim must be moved, immobilizing him or her using a stretcher or other technique is preferred. However, if time is critical, one of the following carries may be employed.

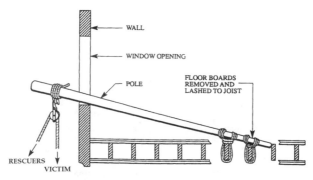

FIGURE 7.17 Gin pole hoist. (*Courtesy AVO, International.*)

FIGURE 7.18   Rope-tied A-frame and tripod hoists. (*Courtesy AVO, International.*)

The method used to move the victim depends on the nature and severity of the hazard to which he or she is exposed. For example, Figs. 7.20 and 7.21 illustrate two methods that may be used to remove a victim from a fire area. In these types of situations, smoke and heat may require that the rescuer stay low to the ground.

If only one person is available for the rescue, and the victim is conscious and mobile, the pack-strap carry (Fig. 7.22) may be employed. The firefighter's carry (Fig. 7.23) may be used if the victim is not conscious.

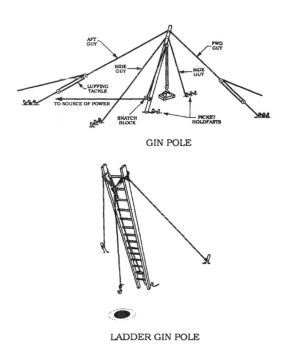

FIGURE 7.19   Gin pole hoists. (*Courtesy AVO, International.*)

FIGURE 7.20 Clothes drag. (*Courtesy AVO, International.*)

FIGURE 7.21 Blanket drag. (*Courtesy AVO, International.*)

FIGURE 7.22 Pack-strap carry (conscious victim). (*Courtesy AVO, International.*)

FIGURE 7.23 Firefighter's carry. (*Courtesy AVO, International.*)

If two people are available for the rescue, any of the carries shown in Figs. 7.24 through 7.26 may be used. The carry shown in Fig. 7.25 generally requires that the victim be conscious.

ACCIDENT INVESTIGATION

Purpose

When accidents occur, they must be investigated thoroughly. An investigation serves two major purposes as shown in Table 7.18. Note that accident investigations should be performed only by qualified professionals who are experienced in such matters. Litigation can often ensue and result in corporate, personal, or product liability; errors and omissions; insurance subrogation; or other claims.

FIGURE 7.24 Carry by extremities. (*Courtesy AVO, International.*)

Accident Prevention.   Accidents occur for a reason, and sometimes the reason is not obvious. A thorough investigation can reveal underlying causes. Employers must then act to eliminate those causes so such accidents will not occur again.

Litigation.   Modern society is becoming increasingly litigious. Evidence gathered during an accident investigation can be used in court to either defend or prosecute a lawsuit. Another aspect of our litigious society is the tendency of companies and individuals to "keep everything secret until the attorney approves it." Sadly, this approach may seem necessary to prevent frivolous lawsuits and/or to protect assets. It should be remembered by all parties

FIGURE 7.25   Two-person seat carry. (*Courtesy AVO, International.*)

FIGURE 7.26 Chair carry. (*Courtesy AVO, International.*)

involved in an accident investigation that the understanding of the accident and prevention of a recurrence should be the most important issues. Ultimately, the value of a single human life exceeds any corporate asset.

General Rules

Several general rules for carrying out accident investigations are listed in Table 7.19. Many investigations have been hampered by well-meaning, inexperienced individuals who remove or alter evidence in the interests of cleaning up the area. Remember that only absolutely necessary modifications should be made at the accident scene, and even these should be documented, preferably with photographs and/or videotapes.

One of the major mistakes made by employers is to wait for the lawsuit before an accident investigation begins. Employers owe it to themselves and their employees to investigate accidents completely and thoroughly, as soon after the accident as possible. This is especially true of witness interviews. Time quickly alters or destroys memories of traumatic events; therefore, people must be interviewed immediately.

Whenever possible, *retain an expert*. Forensic engineering firms specialize in the investigation and analysis of accidents. Such organizations have a wealth of factual information available to them. They can determine causes in cases where inexperienced investigators can only guess. For information on forensic consultants contact the American Academy of Forensic Sciences or the Canadian Society of Forensic Sciences.

Data Gathering

Site Survey.   The gathering of data should start with a survey of the accident scene. If the basic rules have been followed, the scene will be virtually unchanged. The investigator should follow the steps outlined in Table 7.20.

The first visual inspection can produce amazing insights into the nature and causes of the accident. Things to look for include

* Scorch and burn marks
* Damaged equipment

TABLE 7.18   Reasons for Performing Accident Investigations

* Future accidents may be prevented by elimination of accident causes.
* Evidence gathered during the investigation may be required in litigation.

TABLE 7.19 General Rules for Performance of an Accident Investigation

---

- Investigations should be performed only by qualified, experienced personnel.
- Until the investigation is complete, the accident scene should be modified only to the extent necessary to eliminate immediate hazards to personnel.
- No evidence should be destroyed or thrown away for at least 5 to 10 years after the accident.
- Photograph everything from many different angles.
- Document everything that is observed or taken in interviews.
- Perform the investigation as soon as possible. Remember that human memory is volatile.

- Evidence of safety equipment
- Position of victim and or objects

The photographs or videotape, preferably both, can permanently store vital evidence. Because of the two-dimensional nature of photography, shots should be taken from all angles including from above to get a complete perspective. A scale reference can be included in the scene. An audio or written record should be made to clearly define the nature of each photograph, where it was taken, and any specific information that should be noted.

The final detailed inspection and sketch can be handmade, clearly labeled, and dimensioned. Dimensions are particularly important since they may prove or disprove the possibility of any given event during the accident. The sketch should also reference photographs or videotape evidence that was taken.

Witness Interviews.   Witness interviews help to establish what, how, where, and when. However, the investigator must remember that of all the evidence he or she will gather, eyewitness accounts are the least reliable. Human beings tend to be influenced by their own personal backgrounds and prejudices. Within a few hours, or even minutes, of the accident, eye witnesses' memories start to modify. Witnesses should be interviewed as soon as possible and on audiotape if possible.

The cautions in the previous paragraph are especially true of the accident victim. The human mind will literally shut off to prevent the agony associated with an accident. In case after case, victims remember nothing regarding several hours or even days before and after the accident. Information supplied by a victim should always be viewed with skepticism. The more traumatic the injury, the less credible will be the victim's statement.

When interviewing witnesses, the following points should be followed:

- When possible, conduct the interview at the accident site. The familiar surroundings may help to arouse memories that would be lost.
- Prepare a series of questions to be asked before the interview, and start with these questions. If the witness seems more comfortable with a freewheeling statement, allow it; however, avoid rambling and nonrelevant statements since these will only cloud the results.

TABLE 7.20 Steps for an Accident Site Survey

- Visually inspect the area, making notes about general observations.
- Photograph or videotape the scene.
- Make a detailed inspection and develop a sketch of the entire scene.

- Try to keep the witness focused on the facts of the case. Blame, fault, and other such subjective conclusions should be left to the analysis stage of the investigation.
- When the interview is completed, the interview notes should be carefully stored and dated. If you make a summary of the interview, it should be stored with the notes; however, it should not replace the interview notes.
- Arrange for a follow-up interview to deal with questionable or conflicting accounts.

Physical Evidence.   One of the most frequent mistakes made by inexperienced investigators is the alteration, destruction, or loss of physical evidence. A scorched or melted wire, a bit of flaked insulation, a piece of torn cloth, or a damaged circuit breaker can all be extremely important evidence.

Physical evidence should be documented in the preliminary stages of the investigation by the use of photographs, sketches, and written notes. After it has been documented, it should be carefully marked and stored. Physical evidence should be kept for a minimum of 5 years and possibly as long as 10 years depending upon the circumstances of the accident.

Accident Analysis

After the evidence has been gathered and the interviews have been taken, the analysis phase begins. The actual steps taken in this portion depend to a large extent on the specifics of the accident itself. However, at least two major steps may be involved.

Experimentation.   Actual experimentation includes tests that are performed to evaluate or simulate certain conditions during an accident. For example, pork can be subjected to electric currents to evaluate the degree of burning that occurs for various currents, electric arcs can be staged with manikins to determine the degree of thermal burning that may occur, and air gaps can be checked to see how far an arc will jump in given circumstances. Experimentation is a detailed and sophisticated procedure and should be performed only by organizations with the necessary expertise.

Report.   A report is developed based on a comprehensive analysis of all the evidence and experimental data that has been gathered. The report should contain, at a minimum, the sections listed in Table 7.21.

- *Overview.*   This section should include an executive summary type of description of the events leading up to the accident, the accident itself, and the accident aftermath. Only very general information should be included, such as the number of injured persons, the basic cause of the injuries, and the generalized suggestions for preventing such a problem in the future.
- *Step-by-step description of the accident.*   Here the investigator takes the reader through each step of the accident from beginning to end. All the physical evidence is discussed,

TABLE 7.21   Major Elements of the Final Report of an Accident Investigation

- Overview
- Step-by-step description of the accident
- Conclusion
- Appendices

and eyewitness accounts are integrated into this portion of the report. All relevant information should be included, such as the activities of the victim prior to the accident that may have affected his or her behavior, the specific actions of the victim during the accident, the safety equipment that was in use at the time, the actions of witnesses, the duration of the electrical contact, the points on the body that were contacted, and other such details.

- *Conclusion.* The investigator should provide an evaluation of the accident including such information as the steps that could be taken to prevent similar accidents in the future, changes in safety equipment or rules that may be indicated, changes in employee training that may be required, and/or changes in work procedures that may be indicated. Since this is an engineering report, the investigator should avoid comments about who may be responsible for the accident. Such issues are better left to the legal system.

- *Appendices.* All the collected data can be included for reference. Remember that others may draw different conclusions than that of the investigator and should, therefore, be given the advantage of being able to examine all the evidence that has been collected. In addition, the evidence must be present to support the conclusions drawn. *Do not* include the originals of photographs and videotapes. Only copies should be included. The originals should be kept in a secure location such as a locked file cabinet or a safe-deposit box. Of course, physical evidence should not be included except at a formal presentation of the report. However, photographs of physical evidence can be included.

Please note that any and all content of such a report will be subject to insurance contract directives; insurance adjusters instructions; attorney-client privilege; powers granted to worker's compensation investigators; and powers granted to federal, state, and provincial agencies that have jurisdiction in the investigations.

CHAPTER 8

MEDICAL ASPECTS OF ELECTRICAL TRAUMA

INTRODUCTION

This handbook serves as a reference on how to be safe around electrical energy. In this chapter, the consequences of an electrical safety failure are presented from a medical point of view.

As noted in the Preface, each year many people are killed or injured by electrical energy. In the vast majority of these events, the individuals involved never realized how their time on the job might change their lives.

STATISTICAL SURVEY

Numbers give a statistical picture of known electrical injuries and fatalities. Published U.S. National Institute of Occupational Safety and Health (NIOSH), analyses of the U.S. Department of Labor (DOL), Bureau of Labor Statistics (BLS), and data on electrical incidents by Cawley and Homce show, from 1992 to 1998, in the United States, 32,309 employees were injured in electrical events resulting in days away from work, while 2,267 people died because of an occupational contact exposure to power frequency electrical energy.[1]

Cawley and Homce further analyzed 623 fatalities, and found that the incident voltage in 87 percent of the fatality reports was noted as 15,000 V or less, with 32 percent of the fatalities noted during work on 0-600 V (Table 8.1). Overhead power lines were an electrical source in about 40 percent of the deaths.

From 1980–1995, in the United States, electrocution accounted for 7.1 percent of occupational deaths among men, and 0.7 percent among women.[2] Among young workers, electrocution was the cause of 12 percent of all workplace deaths, and the third leading cause of work-related deaths among 16- and 17-year olds, after motor vehicle deaths and workplace homicide.[3]

While comprising about 7 percent of the U.S. workforce, construction workers sustained 44 percent of electrical fatalities.[1] Figure 8.1 illustrates possible electrical exposure hazards on a construction work site with a crane truck and overhead power line nearby[4]. In the illustration, electrical current flows from the power line through the truck to contacts between the truck tires and earth, ("step potential" hazard) and the truck and worker ("touch potential" hazard) while the worker is standing on the earth.

Focusing on electrical fatalities among U.S. construction workers, the researchers Ore & Casini studied 1980–1991 data, showing the age distribution of those construction

TABLE 8.1 Source Voltages Recorded in 623 Occupational Fatality Narratives

Cases (%)	No. deaths	Source voltages (V)
32	198	0–600
	50	480
	18	440
	24	277
	36	220/240
	42	110/120
8	50	601–5,000
35	215	5,001–10,000
13	81	10,001–15,000
13	79	Over 15,001

Source: Courtesy Cawley and Homce, 2003.

workers killed in electrical fatalities (Fig. 8.2), their job titles (Fig. 8.3), and the time of day when fatalities occurred (Fig. 8.4).[5] The data shows that a worker is vulnerable to an electrical failure at any age, regardless of his/her job title. The fatality time of day pattern reflects the increasing activity typically seen on a construction sight from start of day to end of day shift; an interpretation of this figure is that the more workers on a job site, the more likely a fatal incident is to occur.

FIGURE 8.1 Overhead power line contact by a construction bucket crane. Worker electrical exposure to an overhead power line in a construction scenario with a haul truck contacting an overhead power line, courtesy of James Cawley, NIOSH. Overhead power lines account for about 40% U.S. occupational electrical fatalities. (Use with permission J. Cawley, G.T. Homce, H.K. Sacks, and M.R. Yenachek, "Protecting Workers from Electrocution Caused by Contact of Cranes, Haul Trucks, and Drill Rigs with Overhead Powerlines: A New Approach, "Abstracts of the National Occupational Injury Research Symposium (NOIRS), Pittsburgh 2000.)

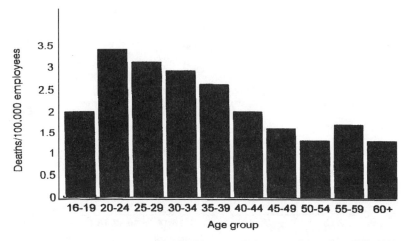

FIGURE 8.2 Distribution of electrical fatalities among U.S. construction workers, 1980–1991. Adapted from Figure 2 from Ore & Casini's 1996 report, this illustration shows the number of deaths per 100,000 workers by age groups. Courtesy of T. Ore, and V. Casini, "Electrical Fatalities Among U.S. Construction Workers," *JOEM*, vol. 38, no. 6, pp. 587–592, 1996.

Considering construction fatalities from another angle, when U.S. construction crane fatalities for 1984–1994 were studied by Suruda and colleagues, 196 of 502 fatalities, or 39 percent, involved electrocution.[6]

Often the question is asked, "Do electrical fatalities or injuries mostly occur in people new to their jobs?" This is a question that tries to sort out differences in the fatality trends

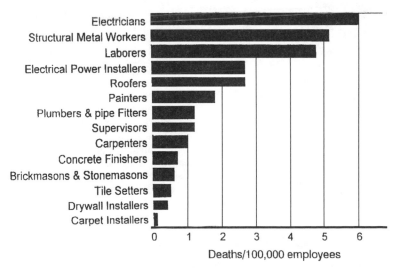

FIGURE 8.3 Job titles of electrical fatalities among U.S. construction workers, 1980–1991. Courtesy of T. Ore, and V. Casini, "Electrical Fatalities Among U.S. Construction Workers, *JOEM*, vol. 38, no. 6, pp. 587–592, 1996.

24 hour clock

FIGURE 8.4 Time of day of electrical fatalities among U.S. construction workers, 1980–1991. This figure shows the per centage of U.S. construction electrical fatalities compared across a 24 hour period. Adapted from Ore & Casini. Courtesy of T. Ore, and V. Casini, "Electrical Fatalities Among U.S. Construction Workers," *JOEM*, vol. 38, no. 6, pp. 587–592, 1996.

in relationship to the worker's job experience, rather than the worker's age while on the job. International Brotherhood of Electrical Worker (IBEW) data in part addresses this question. Power-line-transmission and distribution-worker fatalities for 1971–1987 were previously reported by seniority on the job (Fig. 8.5). Note that higher seniority status does not confer protection from an electrical fatality.

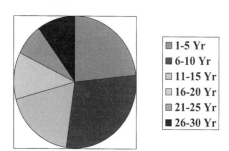

| ■ 1-5 Yr |
| ■ 6-10 Yr |
| □ 11-15 Yr |
| □ 16-20 Yr |
| ▨ 21-25 Yr |
| ■ 26-30 Yr |

FIGURE 8.5 Power line transmission and distribution worker fatality experience by seniority, 1975–1988.
This figure shows the distribution of power line fatalities in IBEW linemen form 1975-1988 by years of seniority. (*Courtesy IBEW Transmission and Distribution Worker Data.*)

Non-Occupational Electrical Trauma

Compared to workplace electrical trauma, non-occupational electrical injury and fatality statistics are not collected routinely, limiting injury and fatality estimates. Typically non-occupational electrical injuries are sustained in home or public settings, including households, medical facilities, police response sites, and educational and recreational locations (Table 8.2).[7,8,9,10,11]

Fatality and Injury Related Costs

Over 1.6 million U.S. employees work in the specialty trades, as shown by 2000 U.S. Census data (Table 8.3); the annual value of U.S. construction put in place by type of construction (Table 8.4) is shown for 2002.[12] According to a survey research from the Integrated Benefits

TABLE 8.2 Non-Occupational Electrical Trauma Scenarios

Setting	Source	Fatality/injury description
Homes	Receptacles	3,900 injuries annually, with 33% young children inserting metal objects[7]
Clinical/Hospital	Iatrogenic	Pectoral muscle damage after defibrillation[8]
Police Action	Stun guns	Associated with more than 50 fatalities since 2000[9]
Educational/Recreational	Swim pools	Electrocutions and Electric Shocks[10]
	Vending	Electrocutions, shocks[11]

TABLE 8.3 Construction Specialty Fabricators and Suppliers, 1997

NAICS code	Description special trade contractors	Establishments	Dollar value of business done ($millions)	Annual payroll ($millions)	Paid employees
2351	Plumbing, heating, and air-conditioning contractors	84,876	88,427.4	25,720.2	788,930
2353	Electrical contractors	61,414	64,915.1	21,680.0	641,984
23591	Structural steel erection contractors	4,238	8,152.7	2,387.1	72,301
23592	Glass and glazing contractors	4,713	4,045.5	1,051.6	35,823
23594	Wrecking and demolition contractors	1,541	2,304.0	592.2	18,820
23595	Building equipment and other machinery installation contractors	4,488	9,342.9	3,148.0	75,501
	Total	161,270	177,187.7	54,579.0	1,633,359

Source: U.S. Census Bureau, 2000a. Sums may not add to totals due to independent rounding. Taken from Gallaher and O'Conner's report for the U.S. Dept. of Commerce Technology Administration National Institute of Standards and Technology Table 1-8, August, 2004 [Gallaher and O'Conner, 2004].

TABLE 8.4. Value of U.S. Construction Put in Place, 2002

Type of construction	($Millions)
Office	37,578
Hotels	10,285
Religious	8,217
Other Commercial	55,879
Educational	68,995
Hospital and Institutional	22,366
Public Housing & Redevelopment	5,507
Industrial	18,486
Electric Light and Power	24,789
Public Utilities	54,370
Military	18,284
All Other Nonresidential	49,362
Total	374,118

Source: U.S. Census Bureau, 2004b. Taken from Gallaher and O'Conner's report for the U.S. Dept. of Commerce Technology Administration National Institute of Standards and Technology Table 1-1, August, 2004 [Gallaher and O'Conner, 2004].

Institute, including a traditional view of paid benefits, in 2002, workers compensation costs for work-related injuries and illnesses were estimated at 9 percent of health costs.[13] If this 9 percent estimate is used along side of the total health costs for the electrical contractors annual payroll (for example, 1997 figures of $21.7 billion as shown in Table 8.3[12]), the financial losses from workplace injuries and illnesses can be placed in context. Drawing on the most recently available national data from 1993, depending on the state, the annualized U.S. workers compensation cost may range from $573 to $1,979 per worker employed.[14]

ELECTRICAL EVENTS

An electrical event may scale from an almost unnoticeable release of energy, like localized heating in a conductor or a very brief light flash in a low voltage installation, to a massive explosion from an escalating arcing fault in a multiphase high voltage system.

Consider when the electrical energy release in an event is relatively small and slow, as for example, with ohmic heating in a fluorescent fixture's ballast. The electrical hazard can lead to a potentially fatal consequence if it serves as an ignition source in the presence of combustible dust, vapor, or materials. Fires and explosions from combustion of these hazardous materials can follow.

Another scenario can occur when an employee comes in an unprotected contact with an electrical shock hazard. Then electrical energy can flow from the hazard to the employee, essentially as if the employee has become a part of the source circuit. The employee's connection to the circuit can be "in series," as when current flows "into" or "through" the body; or the connection to the circuit can be "in parallel," as when current flows over the employee without creating a potential electrical gradient across specific tissues in the body. Complex connection in both series and parallel paths may occur as well.

A third scenario is when an employee or equipment the employee is using enters an air gap between conductors. This may happen initially intentionally as, for example, in the high voltage scenario of live-line work. Another possibility is that the employee or their equipment may enter the air gap between conductors initially accidentally, as with the operation of equipment too near to a power line. As a result of entry into an air gap between conductors, a conductive channel can open between the source and the employee or their tool, resulting in an electrical arc fault. With an electrical arc fault, a large amount of electrical energy can flow in an ultrashort time, creating an event that transforms electrical energy into multiple forms of energy flowing from the source into the electrical event space.

As suggested in the chapter on the hazards of electricity, an electrical event can be formally described using:

- the equipment voltage and involved current.

- the occurrence of an outage resulting in disruption of service, manufacturing, or industrial process; or

- the destructive effects of the energy released in the incident.

As earlier discussions about the National Electrical Code (NEC) have indicated, hazard rating of electrical installations is variable, depending on the location of the installation as well as on the surrounding operations and conditions. While there is consensus on how to rate electrical installations, there is no widely agreed upon scheme to rate an event's electrical failure severity.

As a general guide, the National Fire Protection Association's (NFPA) Failure Mode and Effects Analysis (FMEA) can be used as a qualitative rating from 0 to 4, where 0 is for an event where the hazards result in no effect; a rating of 1 suggests a slight effect; a rating of

2 is for a moderate effect; a rating of 3 is for an extreme effect; and a rating of 4 suggests a severe effect.

If an electrical event produces a fire, the severity of the fire's results may be rated. For example, Table 8.5 summarizes Noon's suggestions for fire scene ratings that can be assigned during investigation based on inferences using the materials present.[15]

Electrocution and Electrical Fatalities

The term *electrocution* refers to an electrical event with electrical current exposure that results in death. The implication is that the current flow has caused an electrical shock with subsequent death.

"Electrical accident fatality" is a general use phrase seen in news reports meaning either electrocution, or death resulting at the time of the electrical accident. This phrase may include fatalities associated shock or other forms of energy released at the time of the electrical accident, in particular those causing physical changes including burns, blast effects, and radiation damage.

"Electrical injury mortality" is a medical statistics phrase which suggests that persons who were injured in an electrical accident lived long enough to receive medical care for their injuries, but the medical care was not followed by survival.

It's important to appreciate that an electrical event can produce a fatality or injury even when there is no electrical current flow to the victim or electrical shock. This might be the situation, for example, when a victim is caught in an electrical ignition fire, explosion, or blast. In this type of scenario, the "root cause" of the accident is electrical, but the mechanism of death or injury is from thermal, acoustic, radiation, or blast exposure related to electrothermal chemical (ETC) combustion.

Another way employees can be killed or injured after an electrical event is that they are surprised by an energized source, either through a spark, like a static "zap" to exposed skin, or through a noise, like a sharp "gunshot" type sound close to the head. The surprise can lead to an unintended body movement which might be characterized as a "startle response." If the startle occurs at the top of a ladder or scaffold, the direct mechanism of death or injury can be through a fall. If the startle occurs in proximity to other energized equipment that is

TABLE 8.5 Severity of Fire Investigation Scenes

Rating	Description
10	materials are gone, wholly burned away
9	materials are mostly gone, some residual
8	materials are partially gone, recognizable residual
7	materials are burned all over but shape intact
6	materials are mostly burned
5	materials are partially burned
4	materials are slightly burned
3	materials are heat damaged from nearby fire
2	materials are heavily smoke damaged
1	materials are slightly smoke damaged
0	materials exhibit no significant fire damage

Source: Courtesy R. Noon. Suggested by R. Noon in *Engineering analysis of fires and explosions*. Boca Raton: CRC Press, 1995. p. 130.

moving, the direct mechanism of death or injury can be with a body part being caught in or by the moving equipment.

Fatal and non-fatal electrical incidents share three characteristics:

1. The *unintentional* exposure of employees to electrical energy;

2. Compliance *failure* in at least one aspect of electrical design, installation, policies, procedures, practices, or personal protection; and

3. Energy *transfer* to exposed employees in some combination of electrical, thermal, radiation, acoustic (pressure), mechanical, light, kinetic, or potential energy.

What is the difference between fatal and non-fatal electrical incidents? The answer depends in part on whether the question is asked hypothetically, like in a "what if" planning scenario; or whether the question is asked retrospectively after a traumatic accident has occurred.

Hypothetically, based on human physical and biological characteristics, we know that a fatal electrical event transfers a greater amount of energy to its victim than a non-fatal situation. This knowledge about the fatal risk of energy transfer underlies the use of equipment designs (for example, required doors, specified space clearances, venting systems on equipment to discharge combustion products, "umbilical corded" controls, infrared monitoring ports for doors closed heat monitoring) and barrier protection (such as PPE, including leather gloves, flash suits, safety glasses, face shields, long sticks, extended handles, and flame resistant clothing) discussed in this handbook.

By reducing the amount of possible energy transfer during an unintentional electrical exposure, strategies including equipment design and barrier protection can increase the likelihood of survival after an electrical incident.

Retrospectively, if two people are present in an electrical incident when one dies and the other survives, the difference in survival may come down to nuances in the victims' innate individual differences and their spatial and temporal relationship to the electrical hazard at the time of the energy release, transformation, and transfer. Medical and legal privacy protections tend to reduce accessibility to accident details, so systematic information is lacking about how various scenarios unfold.

Generally, there is a lethal exposure "dose" for different forms of energy that can result in death. When multiple forms of energy are involved in an electrical event, multiple lethal or sublethal doses of energy may flow from the event, transformed from the electrical hazard source, and transferred to nearby employees may result in highly variable damage to the body.

MEDICAL ASPECTS

While the structural or environmental consequences of an electrical event may be rated through schemes like FMEA, the medical consequences of an electrical incident can not be readily predicted from a rating or event description. This is a frustrating aspect to understanding how survivors look or feel after an electrical exposure.

In order to understand the fatalities or injuries seen as health effects following an electrical failure, there are three central questions:

1. How much energy was actually released during the electrical event?

2. What forms of energy were actually released during the electrical event?

3. What parts of the released energy were actually transferred to the affected worker?

A schematic of an electrical incident with photographs of an experimental simulation of a "real world" scenario are shown in Fig. 8.6.[16,17] As the illustration suggests, an electrical

FIGURE 8.6 Schematic of electrical event energy transformation and transfer with photographs of experimental simulation of a "Real World" scenario.

Industrial scenario with electrical energy input to a wall panel box while worker is standing nearby. At the moment of the electrical incident event, electrical energy transformation occurs. Uncontrolled energy release follows, resulting in the transfer of electrical and other energy forms to the space and nearby worker. Simulation photographs parallel schematic illustration. A staged test with a mannequin industrial worker standing nearby a panel box is sown under high voltage laboratory conditions with 480 V and approximately 22,000 A, available. An electrical failure is staged to show typical workplace scenario. [Schematic is adapted from J.E. Bowen, M.W. Wactor, G.W. Miller, and M. Capelli-Schellpfeffer, "Catch the Wave, "*IEEE Industry Applications Magazine*, vol. 10, no. 4, pp. 59–67, July–August 2004. Photographs by the author are from source documentation reported in R. Jones, D.P. Liggett, M. Capelli-Schellpfeffer, R.E. Downey, S. Jamil, T. Macalady, L.B. McClung, V.J. Saporita, L.F. Saunders, and A. Smith, "Staged Tests Increase Awareness Arc-Flash Hazards in Electrical Equipment," *IEEE Transactions of Industry Applications Society*, vol. 36, no. 2, pp. 659–667, March–April 2000.

incident unfolds over time. Again, when an electrical fault occurs, electrical energy is transformed and transferred, leading to multiple possible exposures.

Voltage ratings, available fault current estimates, and incident energy calculations permit a guess at the amount of energy which may be involved in an electrical incident. Practically, when more than one person is present at an electrical incident, more than one health effect from the incident is possible, ranging from no effect on those present to fatality.

Although it may seem unreasonable to employers and coworkers, the same exposure can and often does produce different levels of injury to individuals involved in the same event. Figure 8.7 illustrates the possible traumatic outcomes.

People can vary in their susceptibility to any single exposure (i.e., electricity, heat, radiation, noise, and pressure) because of their individual differences, like size, positioning, and use of personal protection. Moreover, an exposure itself can vary, depending first, on the event timing and, second, on the positions in space where the hazard is released relative to the affected employees.

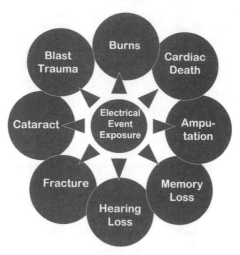

FIGURE 8.7   Possible trauma following an electrical event.
This bubble diagram illustrates the kinds of medical problems that may following the exposure to an electrical event. The body's response to the transferred energy from the event in part depends on the form of energy, the dose, and the duration of the exposure.

There is much scientific knowledge about how the body responds to direct electromechanical contact with 50 to 60 Hz electricity generated for electrical power, as discussed in Chap. 1. In fatality and injury scenarios where electrical shock is the dominant exposure, the tissues' vulnerability to electrical forces is the key.

Non-Electrical Effects in Electrical Events

As power density increases across more compact spaces, there is more fault energy available in electrical installations. With installations designed to be compact, there are smaller spatial cushions and consequently less physical or spatial barrier protection emerging in innovative designs. The result is that there may be less time and less physical distance through which an unintentional release of energy can be dissipated during an electrical event. With higher power density and smaller, more compact spaces, there is an increased risk of collateral damage.

High power density in a confined space can have bomblike potential. As a consequence, an electrical event can behave like closed blast explosion. For example, the potential ignition of 1 MW is roughly equivalent to the potential ignition of one stick of dynamite. One stick of dynamite is roughly 1/3 lb of TNT equivalent. So, a 100 MW scenario can be represented as loaded with 100 sticks of dynamite. Or, a 300 MW scenario can be represented as loaded with 100 lbs of TNT.

A closed space with an electrical event creates the possibility of closed blast health effects. At a pole top in a vacant field, the space is more like an open area suggesting an electrical scenario at pole top that is more likely to cause injuries similar to those seen with a free blast. Blast trauma may include destruction from shrapnel, confined space air contamination, acoustic or pressure effects, and acceleration-deceleration brain effects.

In brief, when an electrical event involves more than a direct electromechanical current exposure, the traumatic consequences can vary, as the bubble diagram suggests in Fig. 8.7.

Survivor Experience.   As suggested in Table 8.6, an electrical event may unfold in fractions of a second. In electrical scenarios, a full second is a "long time," suggesting a 60 cycle event. Compare this to the investigation of an electrical incident which may take weeks to months. The litigation accompanying such an event typically takes years. Note that the logistics of work do not permit a site to go "on hold" while these activities are underway.

Obviously, survivors experience an electrical event on a different time frame than their workplace. Practically, within a few work shifts, an electrical accident site may be operating routinely. In contrast, the survivor experience may unfold from the moment of the electrical event through minutes of triage to years of rehabilitation.

Worker Reflexes.   At the scene of an electrical event, survivors rely on their reflexes and senses. As discussed in Chap. 11, to see, think, and then respond to a scenario demanding a choice requires at least 1.5 seconds. Physical responses to an unfolding electrical event in a shorter amount of time occur with reflexes, such as eye blinks in response to flash, or quick hand withdrawal from a hot surface.

Sometimes the injured survivor can "walk away" from the electrical event. However, this immediate postincident behavior may be deceptive, given the possible medical consequences of electrical trauma.

Instead of walking away, if a non-fatal electrical contact is made, a person may collapse with the loss of regular heart function. As the heart's pumping action becomes inefficient, the blood supply to the lungs and brain are severely limited, reducing the delivery of oxygen and glucose (the body's fuel). When oxygen and glucose are not available because of circulation failure, a survivor can not breathe or remain upright or awake. This experience may be later recalled by a survivor as "near death" or "out of body." At this time, successful cardiopulmonary resuscitation (CPR) is essential to rescue the affected individual.

TABLE 8.6   Timing of Workplace and Survivor Experiences After Electrical Events

Experience	Timing
Hazard	
Event	Fractions of milliseconds
Investigation	Weeks to months
Litigation	Years
Survivor	
Worker reflexes	Seconds
Triage & medical evacuation	Event to Hour 1
Stabilization & initial evaluation	Hour 1 to Hour 8
Medical & surgical intervention	Hour 8 to Hour 24
Hospitalization	Day 2 to Day 9
Discharge to outpatient services	Day 10 to Day 30
Rehabilitation	Day 31 to Day 89
Return to work planning	Day 90 to Day 119
Re-entry to employment	Month 4 to Month 60
Plateau in Recovery	Beyond 5 Years

Once help arrives at the scene, an electrical incident survivor is routinely questioned. As first responders, emergency medical technicians, or paramedics become involved, the survivors are expected to tell these clinicians a "history" of their difficulties.

Survivors may need to explain how they were injured in order to get a medical evaluation. This may be difficult to do if the electrical accident has occurred quickly. For example, a 30 cycle event in a 60 Hz electrical power system occurs in 0.5 seconds. A scenario that takes a half second *to complete* is faster than most people can even recognize, much less remember in detail.

Triage and Medical Evacuation.   Medical triage, evacuation, and treatment requires minutes, hours, and longer, as discussed below. In the emergency or first response period, American Burn Association (ABA) Advanced Burn Life Support guidelines outline triage, initial resuscitation, and emergency transport protocols. On admission to the emergency medical services, physical examination, laboratory studies, and diagnostic radiology evaluation are prioritized depending on the patient's clinical situation.

When the survivor can not remember accident details there is little information available to clinicians for rigorous decision making about the exposure circumstances, possible mechanisms of injury, and need to obtain extensive laboratory studies or diagnostic imaging.

If in coma or heavily medicated, the survivor may not be able to coherently answer questions like:

• What happened in terms of the energy like volts, amps, cycles?
• Was there a blast or an explosion?
• Was the space small or confined?
• Did you fall or lose consciousness?
• Were you wearing protection?
• Did you need CPR?
• Was anyone else hurt or killed at the scene?

Further confusion between a survivor who becomes a patient and clinicians may arise when the circumstances around the injury event simply can not be immediately known. For example, when a work situation is believed to be "safely de-energized" but leads to an electrical accident, engineers may investigate over weeks to months to establish the "root causes" and available energy to the fault. During that investigation time, the electrical incident survivor may be repeatedly asked by clinicians, "What happened? Why didn't you turn off the power?" From the survivors' perspective, each episode of questioning by clinicians is an opportunity for self-doubt about their accident recollection.

Commonly, survivors report that they were doing a manual task or handling equipment when caught in an electrical event. Upper extremities, that is the body from the fingertips to the shoulders, are involved in most trauma because of how work is done.

From a clinician's viewpoint, knowledge of the electrical event's voltage conditions alone is of limited helpfulness because the extent of survivor's electrical injury is a function of the power frequency energy transfer due to a combination of the current density in the tissue and the contact duration.

Similarly, from a medical perspective, the heat, radiation, and pressure damage to the body are predicated on the efficiency of the electro-mechanical and electro-chemical coupling between the hazardous source, event space, and the body. This information is not usually available to the medical team.

Physically, when an extremity is exposed to electrical current, the skeletal muscle in the current path through the extremity can be significantly injured from the electrical effect, extreme heat, and damage to proteins. At the electrical accident site, first responders may see minimal skin wounds, or they might notice profound charring and vaporization of tissue. Even in the absence of impressive wounds, the survivor's risk of amputation can be high.

After a severe electrical contact injury, a survivor's injured muscles can swell massively, resulting in a situation referred to as "compartment syndrome." In a compartment syndrome, the blood supply to local muscle is literally squeezed by swelling. With this swelling, oxygen supply to the muscle can be "choked." Without oxygen from the blood supply, muscle tissue dies. If muscle tissue dies, proteins are lost as waste and filtered by the kidneys. These proteins can block the kidneys and result in kidney or renal failure.

Other significant features of severe electrical exposure include disturbances in the survivor's heart electrical pattern, called refractory cardiac arrhythmias. Further serious problems may include:

- Neck and back fractures from the electrical current exposure, falls, or blast.
- Disruption in the blood serum's balance of sodium, potassium, and chloride.
- Transient lifeless signs, including a situation referred to as keraunoparalysis.
- Shock lung, perforated ear drums, and ruptured gastrointestinal organs subsequent to blast trauma.

In high voltage electrical incidents, current passing through a limb across opposite sides of a joint can set up large skin or "trans-dermal" potentials, resulting in burns called "kissing" wounds. The most frequent sites of this trauma include:

- At the armpit or axilla, with injury between the limb and skin of the chest.
- At the leg by the space in back of the knee, where the skin of the upper and lower leg can meet when the survivor is caught in a crouched position.

Other skin burns may result in the following ways:

- With ohmic or Joule heating at the electrical contact points between the survivor and the electrical hazard.
- From the ignition of clothing burning in place against the survivor's skin.
- At unprotected skin exposed in the extremes of heat generated by electrical arcing.

Stabilization and Initial Evaluation

Severely electrically injured victims are difficult to resuscitate and the early phase of medical management can be very complex. Survivors are best served by admission to a burn trauma center where staffing is experienced in the management of patients with massive loss of tissue from electrical current damage and burns.

From the survivor's perspective, after a severely traumatic accident, a breathing tube, intravenous fluids, and massive pain control medicines obliterate awareness of all that happens in the early treatment period.

Because nerves are very sensitive to electrical forces, even minor electrical exposures may cause temporary nerve dysfunction in the affected arms or upper extremities. Pain may dominate, or nerve symptoms referred to with the medical terms anesthesia (no feeling), parathesia (some feelings), or dysesthesia (prickly, pin-prick, or knife-like feelings) may occur. In this situation the survivor may experience loss of feeling or "funny" feelings in the areas affected by electrical shock. These problems are referred to as "stunning" and hamper how effective hand grip is in terms of strength, endurance, and reliability. Stunning usually completely resolves within hours or days.

Although less common, electrical contacts may also lead to a nerve problem called temporary autonomic nervous system dysfunction, which may show up in the patient as specifically reflex sympathetic dystrophy (RSD), and hypertension or high blood pressure.

Less quickly resolved for a survivor are the classic complaints of burning pain and exquisite sensitivity which may develop within hours after an electrical injury. Swelling (called edema) and excessive sweats may then manifest during the next three to six months. This maybe followed in six to twelve months with loss of muscle mass, nails, and hair. Osteoporosis may develop in the limb. To evaluate this problem, clinicians conduct diagnostic nerve tests and treat with injections of medication near the nerves to produce a "nerve blockade."

Medical and Surgical Intervention

For a survivor with a severe electrical exposure, more intensive medical and surgical care are needed. The diagnostic evaluation of muscle and nerve injury unfolds while the patient's clinical condition is being stabilized. A goal is to distinguish central nervous system damage from peripheral nerve injury. Key steps are summarized in Table 8.7, adapted from Danielson et al.[18]

When an injury appears less severe, diagnostic testing in a hospital may still be necessary because the health changes resulting from electrical exposure may not be obvious for hours. An employee who "walks away" from an electrical event may not want to draw more attention to their accident by seeking or attending a medical evaluation. Yet because the body damage may not be "detectable" by the eye, an exposed employee should always be encouraged to have a medical check up.

Hospitalization. For the survivor who may have been involved in a serious occupational electrical event, during days or weeks of hospitalization the uncertainty in answering a clinicians' questions (eg., "What happened?" "Why did this happen?" "How were you hurt?") can hamper their medical treatment and amplify their distress at their injury.

Moreover, an experienced, knowledgeable, and competent employee can be made to feel ignorant, uninformative, manipulative, or uncooperative. After surviving a workplace electrical accident, the employee is essentially moved from the occupational environment, where he may be considered "a good worker" or an "expert at his job," to the hospital environment and the medical care delivery system, where the survivor may only be known by reference to a medical record number, a date of admission, and a treating physician's name. Concurrently, for family or coworkers who have not been hurt, the hospitalization experience can be alienating and frightening.

Discharge to Outpatient Services. As a consequence of an electrical event, the health effects can range from the absence of any external physical signs to severe multiple trauma. Even without a visible physical injury like burns or electrical contact wounds, survivors may report medical complaints. These complaints may seem unrelated to the electrical event sometimes because they show up hours, days, or weeks after the occurrence, or because they seem more or less severe than the event itself.

This situation can go from bad to worse when the facts of the injury scenario do emerge, and a survivor learns that he has experienced a near fatal situation. Hearing comments like "You are lucky to be alive!" can disintegrate the worker's belief in their professional competence. Employees who work around electricity don't survive on luck. Worse is the fact that having a near death accident doesn't "feel" lucky to most.

In particular, electricians especially hold to an occupational identity and set off beliefs that suggest workplace policies, procedures, planning, and personal protection can provide for safe occupational activity. When a worker's identity or beliefs are threatened, anxiety, adjustment reactions, or more serious psychiatric issues should be expected. If the facts of

TABLE 8.7 Determination of Muscle and Peripheral Nerve Injury[18]

Diagnostic	Description
Determination of the exposure circumstances	Determination of the electrical current exposure circumstances is fundamental to understanding the survivor's pattern of injury. Exposures may involve direct mechanical contact with an energized surface by an unprotected body part, or may occur through electrical arc current flow. Personal protection equipment such as industrial garments and gloves, face shields, safety glasses, and arc enclosures may moderate the intensity of the incidental exposure. For example, safety gloves may minimize skin wounds; or safety glasses may protect the eyes from the ultraviolet radiation released by an electrical arc. The duration of the electrical current flow, release of multiple forms of energy at time of the current flow, and geographic location of the event may increase the potential severity of the patient's trauma.
Documentation of the physical findings	Physical findings may include open wounds, traumatic amputation, fracture dislocations, impaired neuromuscular function, or edema. Physical examination supported by laboratory studies for the serum markers of muscle damage (e.g., serum enzymes, urine myoglobin) provide correlative information for radiology and electro-physiology diagnostic studies. When external wounds are not observed, a high degree of clinical suspicion of occult neurologic injury is advised.
Radiology studies	Radiographic and nuclear medicine examination can evaluate nervous system structural damage and confirm fractures, local tissue edema and focal areas of inflammation. For example, in the acute period, the localization of tissue edema in an upper extremity is an early sign of muscle injury.
Electro-physiology studies	Electro-physiology studies are guided by the patient's history and physical examination. Muscle weakness, easy fatigue, loss of endurance, paralysis, or pain are indications for electromyography, and peripheral nerve conduction studies (sensory and motor). Further evaluation may be needed with peripheral nerve refractory period spectroscopy and magnetic resonance spectroscopy. Neuro-physiologic evaluation may also be indicated by additional complaints. For example, history of loss of consciousness or seizures, central nervous system complaints such as headache, impaired memory, attentional problems, personality changes or cognitive changes suggests electro-encephalography (EEG) and electro-cardiography (EKG). A history of temporary or persistent loss of hearing, ringing in the ears, or change in memory to verbal communication are indications for auditory evoked potentials (AEP). A change in visual acuity or visual disturbance are indications for visual evoked potentials (VEP). Cardiac rhythm disturbances, syncope, chest pain, and easy fatigue are indications for electro-cardiography (EKG).

the injury scenario emerge as attributing blame for electrical incident on the survivor, then guilt should be added to the list of issues the worker may need to confront.

Besides difficulty in remembering event details because of the event's occurrence speed, a worker who does not have an easily noticed physical injury may still have problems which suggest a brain or nerve injury. These problems directly affect employability, and may be described as weakness, pain, headache, memory changes, disorientation, slowing of mental processes, agitation, or confusion. Personality changes such as irritability, moodiness, nightmares, difficulty sleeping, or feelings of depression or post-traumatic stress may also occur. These changes can interfere with personal and workplace relationships, straining a worker's support network.

REHABILITATION FOCUS AND RETURN TO WORK PLANNING

Rehabilitation and return to work (RTW) may take weeks to months. With massive trauma, years may be necessary to return a survivor to active employment. For example, a seriously injured high-voltage electrical shock survivor can be reemployed, however the rehabilitation period may extend months beyond the time of the patient's acute hospitalization and surgical management.

Successful rehabilitation often does not lead to the return of the patients to their preinjury job. The disability for electrical incident survivors is disproportionate to the incidence of this preventable condition. This means, for relatively few injuries, there is a relatively high frequency of permanent disability. With the typical youth of those injured in electrical accidents and the loss of potential productivity in economic terms, the health effects of electrical incidents carry significant costs for victims, their families, and their employers.

Re-Entry to Employment Settings

When workers have difficulty in the use of their limbs after an electrical incident, their security in completing tasks (like exerting a forceful grip, climbing a ladder, using to hands to lift a load, assisting a coworker in a hazardous activity) may be unacceptably compromised. This difficulty can drastically reduce employment options. Workers may wish to return to their job, but face resistance in this step from their family because of their concerns for reinjury. When the electrical incident is felt to be the victim's "fault," coworkers may harbor undisclosed resentments, acting as a roadblock to reentry to employment.

Plateau in Recovery

The complexities of electrical injury rehabilitation are often under-appreciated by the medical community. As a general guide, three elements are essential to successful recovery following an electrical incident: team, time, and talk. The involvement of an experienced occupational rehabilitation team is needed to avoid the catastrophe of repeated failure to return to work, depression, and loss of self-respect that may evolve when injured workers fear for their livelihood. Time is necessary for the survivor to go through the healing stages following repeated surgeries, possible amputation, and long hospitalizations. Talk or excellent communication is critical to maintaining the employee's relationships with family, friends, coworkers, and caregivers.

Research suggests electrical trauma survival is associated with significant functional impairment. In a landmark retrospective study of employees of a national electrical energy company, Gourbiere and colleagues reviewed the electrical trauma survival experience during 1970–1989 for a workforce between 100,000 and 120,000 during this period. Electrical burns affected 2,080 workers. Of these, 515 patients or 25 percent were noted to have postinjury problems, including:

- 63 percent burn related, with amputations in 5 percent
- 18 percent neuropsychiatric

- 12 percent sensory
- 5 percent orthopedic
- 1 percent cardiovascular

Sense organ problems were also noted, including:

- Vision related changes due to conjunctivitis, keratitis, and cataracts.
- Auditory sequella, with conductive or sensorineural hearing loss, tinnitus, and vertigo.
- Anosmia, or the loss of sense of smell.

In 59 of the 515 patients, disability was considered serious, with impairment rating 31 to 100 percent.[19]

Return to work success of electrical trauma patients is variable. While some electrical trauma patients are able to return to their preinjury activities without functional impairment, others are not. Postinjury job loss can be seen with the survivors who are "owner operators" of their business, as well as with employees from larger organizations. While their postinjury medical care may not interfere with their company's core business, after accident a survivor's job may not exist due to competition from other vendors, contract completion, restaffing, or workforce reductions.

In one study, all patients had peripheral electrical contacts, with no evidence on history or examination of direct mechanical electrical contact with the head. Forty-four patients were injured on the job and nine were injured during non-vocational activities. At the time of follow-up study interviews, 30 (56.6 percent) patients were working again, 18 (44.0 percent) patients were unemployed or retired, 1 patient was deceased, and 4 patients could not be contacted.[20]

If the survivors wish to return to work, and their preinjury job is not available, the clinician can be influential in assisting placement into new work. To limit the potentially destructive effect of financial uncertainty on the survivor and their personal or family situation, when a clinician is advising on placement in new work, a vocational evaluation is recommended as soon as possible to establish the local employment area's market for the survivor's skills set, occupational abilities, and training.

For example, driving is a common requirement in service jobs where travel is necessary to deliver tasks to client sites, like construction locations, factories, or customer businesses. Under U.S. Department of Transportation (DOT) rules, a person can not be medically certified for a commercial driver's license if there is a recent history of loss of consciousness, vision less than 40/40 in each eye and both eyes, certain cardiac conditions, and certain medications.[21] Where amputations are present, demonstration must be made of the performance in driving a vehicle that may fall under DOT rules. When the patient can not drive a DOT regulated vehicle, designed work support is necessary to allow access to job sites and equipment.

Another aspect of industrial work is the need for bimanual secure grip. To climb a ladder, lift a load, or handle a heavy tool, secure, forceful bimanual grip which can be reliably and repeatedly used is an essential function of the job. When finger or hand amputation is a consequence of electrical trauma, supported work to offset the limitation of secure bimanual grip is needed.

At a certain point after an electrical trauma, survivors reach a plateau in their recovery. It can take years longer than the time to plateau after recovery from a serious illness like pneumonia or a heart attack. With consideration for numerous surgeries, retraining for employment, and mental health care, plateaus are assessed on an individual basis. Nevertheless, a common characteristic is that the survivors rarely go "back" to the life they had before their incident. A survivor typically goes "forward" by living differently. The differences can be small, as with an attitude change, or dramatic, as with a job change. The bottom line for the survivor is that, after electrical trauma life may never be the same.

REFERENCES

1. J. Cawley, and G. Homce, "Overview of Electrical Injuries, 1992–1998," and "Recommendations for Future Research," *Journal of Safety Research*, vol. 34, pp. 241–248, October 2003.

2. NIOSH, *Worker Health Chartbook*, DHHS (NIOSH), Publication No. 2000-127, Figure 2-6, p. 33, September 2000.

3. D.N. Castillo, *NIOSH Alert: Preventing Death and Injuries of Adolescent Workers*, U.S. Department of Health and Human Services, Public Health Service, Centers for Disease Control and Prevention, National Institute for Occupational Safety and Health, DHHS (NIOSH) Publication No. 95-125, Cincinati, OH, 1995.

4. J. Cawley, G.T. Homce, H.K. Sacks, and M.R. Yencheck, "Protecting Workers from Electrocution Caused by Contact of Cranes, Haul Trucks, and Drill Rigs with Overhead Powerlines: A New Approach," Abstracts of the National Occupational Injury Research Symposium (NOIRS), Pittsburgh, 2000.

5. T. Ore, and V. Casini, "Electrical Fatalities Among U.S. Construction Workers," (Please see Figures 2, 3, and 5) *JOEM*, vol. 38, no. 6, pp. 587–592, 1996.

6. A. Suruda, D. Liu, M. Egger, and D. Lilliquist, "Fatal Injuries in The United States Construction Industry Involving Cranes 1984–1994," *JOEM*, vol. 41, no. 12, pp. 1052–1058, 1999.

7. Consumer Product Safety Commission, "Electrical Receptacle Outlet Injuries," CPSC Document, no. 524, 2004. URL: www.cpsc.gov/cpscpub/pubs/524.html

8. U. Vogel, T. Wanner, and B. Bultmann, "Extensive Pectoral Muscle Necrosis After Defibrillation: Nonthermal Skeletal Muscle Damage Caused by Electroporation," *Intensive Care Medicine*, vol. 24, pp. 743–745, 1998.

9. W. Andrews, *TASER Danger*? Reported by CBS News, 2004. Retrieved in print format from www.cbsnews.com/stories/main648859.shtml

10. Consumer Product Safety Commission, "Don't Swim with Shocks: Electrical Safety In and Around Pools, Hot Tubs, and Spas," May 13, 2003. URL: www.cpsc.gov/cpscpub.pubs/519.pdf

11. Consumer Product Safety Commission, Standard Submission to Prevent Electrocutions and Non-Fatal Electrical Shocks in Vending Machines Based on CPSC 7 Investigations of Fatal and Non-Fatal Serious Injuries, 2004. URL: www.cpsc.gov/volstd/gfci/422.xx.pdf

12. M.P. Gallaher, and A.C. O'Connoer, "Cost Analysis of Inadequate Interoperability in the U.S. Capital Facilities Industry," NIST GCR 04-867, U.S. Department of Commerce Technology Administration National Institute of Standards and Technology, August 2004.

13. T. Parry, M.S. Schweitzer, and J.D. Molmen, IBI Integrated Benefits Institute, "The Business Case for Managing Health and Productivity from IBI's Full-Cost Benchmarking Program," p. 2, June 2004. URL: www.acoem.org

14. G. Waehrer, J.P. Leigh, D. Cassady, and T.R. Miller, "Costs of Occupational Injury and Illness Across States, *JOEM*, vol. 46, no. 10, pp. 1064–1095, 2004.

15. R. Noon, *Engineering Analysis of Fires and Explosions*, CRC Press, Boca Raton, Florida, p. 130, 1995.

16. J.E. Bowen, M.W. Wactor, G.W. Miller, and M. Capelli-Schellpfeffer, "Catch the Wave," *IEEE Industry Applications Magazine*, vol. 10, no. 4, pp. 59–67, July–August 2004.

17. R. Jones, D.P. Liggett, M. Capelli-Schellpfeffer, R.E. Downey, S. Jamil, T. Macalady, L.B. McClung, V.J. Saporita, L.F. Saunders, and A. Smith, "Staged Tests Increase Awareness of Arc-Flash Hazards in Electrical Equipment," *IEEE Transactions of Industry Applications Society*, vol. 36, no. 2, pp. 659–667, March–April 2000.

18. J.R. Danielson, M. Capelli-Schellpfeffer, and R.C. Lee. " Upper Extremity Electrical Injurys" *Hand Clinics*, vol. 16, no. 2, pp. 225–234, May 2000.

19. E. Gourbiere, J.P. Corbut, and Y. Bazin, "Functional Consequence of Electrical Injury. In *Electrical Injury: A Multi-disciplinary Approach to Therapy, Prevention, and Rehabilitation*," in: R.C. Lee, M. Capelli-Schellpfefferl, and K.M. Kelley, *NYAS*, vol. 720, pp. 259–271, 1994.

20. N. Pliskin, M. Capelli-Schellpfeffer, A. Malina, R. Law, K.M. Kelley, and R.C. Lee, "Neuro-psychological Sequelae of Electrical Shock," *J Trauma*, vol. 44, no. 4, pp. 709–715, 1998.

21. R.M. Gerbo, "What Criteria are Disqualifying for a Driver Undergoing a Department of Transportation Medical Evaluation, in: J.J. Schwerha, *JOEM*, vol. 46, no. 7, pp. 755–756, 2004.

CHAPTER 9
LOW-VOLTAGE SAFETY SYNOPSIS

INTRODUCTION

Each year 120-V circuits cause more deaths and injuries than circuits of all other voltage levels combined. Such low-voltage circuits are extremely hazardous for two reasons. First, low-voltage circuits are the most common. Because they are the final distribution voltage, 240-V and 120-V circuits are used throughout residential, commercial, industrial, and utility systems.

The second reason for the extreme danger of low-voltage circuits is user apathy. Comments such as "It can't hurt you, it's only 120 volts," are heard all too often. Reference to Table 9.1 shows that 120-V circuits can produce currents through the human body that can easily reach fibrillation levels. Consider a perspiring worker using a metal electric drill with one foot immersed in water. Table 9.1 clearly shows that under such conditions, a worker can be subjected to a lethal shock. Furthermore, if sustained for a sufficient period, 120-V contact can create severe burns.

A 480-V circuit is more than four times as lethal as a 120-V circuit. A 480-V circuit has sufficient energy to sustain arcing faults and to create severe blast conditions. This chapter summarizes some of the safety-related concerns that apply to low-voltage circuits—that is, circuits of 1000 V AC and less and 250 V DC and less. Generally, workers should treat low-voltage circuits with the same degree of respect afforded medium- and high-voltage circuits. Refer to Chaps. 2 and 3 for detailed information. Note that some of the safety-related information covered in this chapter also applies to medium- and high-voltage systems. Where necessary, the information is repeated in Chap. 10.

Electrical safety requirements when working on or near electronic circuits can be a problem. For example, the dictates of electrical safety would seem to require that circuit parts and workers be insulated from one another. In some electronic circuits, however, the prevention of static electricity damage requires that the worker be grounded. Also, many workers develop a sense of false security believing that 12,000 V in an electronic circuit is somehow less hazardous than 12,000 V in a power system. This chapter will also present and explain the electrical safety procedures to be employed when working on or near electronic circuits.

LOW-VOLTAGE EQUIPMENT

Hand tools and extension cords are the most commonly used pieces of low-voltage equipment. Each of these items is responsible for hundreds of injuries and deaths each year.

TABLE 9.1 Possible Current Flow in 120-V Circuit

Body part	Resistance, Ω
Wet hand around drill handle	500
Foot immersed in water	100
Internal resistance of body	200
Total resistance	800

Total current flow possible in 120-V circuit = $I = (120/800) = 150$ mA

The following sections summarize the types of usage procedures that should be employed with such equipment.

Extension Cords

Flexible cord sets (extension cords) are used to extend the reach of the power cord for low-voltage-operated tools and equipment. They typically have a male plug at one end and a female receptacle at the other. The tool's power plug is inserted into the female receptacle, and the extension cord's male plug is inserted into the power source or into yet another extension cord's female receptacle. Extension cords can be used to supply power to tools that are located many feet or yards away from the source of power.

Extension cords can be extremely hazardous if not used properly. The following precautions should always be observed when using extension cords:

- Closely inspect extension cords before each use. Table 9.2 lists the types of items that should be looked for during the inspection. (Note that Table 9.2 applies to both extension cords and portable tools.)
- Never use an extension cord or power cord to lift or support a tool.
- Make certain the ground connection is complete from one end of the cord to the other.
- Never alter the plug or receptacle on an extension cord. This applies especially to altering the ground connection.
- If the extension cord is of the locking or twist-locking type, the plugs should be securely locked before using the cord.
- Do not use an extension cord in wet or hazardous environments unless it is rated for such service by the manufacturer. If extension cords are used in wet or hazardous environments, insulating safety equipment such as rubber gloves with appropriate leather protectors should be used.

TABLE 9.2 Visual Inspection Points for Extension Cords and Cord-Connected Tools

- Missing, corroded, or damaged prongs on connecting plugs
- Frayed, worn, or missing insulation
- Improperly exposed conductors
- Loose screws or other poorly made electrical connections
- Missing or incorrectly sized fuses
- Damaged or cracked cases
- Burns or scorch marks

- Only personnel who are authorized and trained in the use of extension cords should be allowed to use them.

- Extension cords should be subjected to the electrical tests outlined in Table 9.3. If the cord fails any of the listed tests, it should be replaced or repaired.

- When not in use, extension cords should be carefully rolled and stored in such a way that they cannot be damaged.

- Extension cords should never be used for permanent power illustrations. If power be needed for more than the short duration, permanent electrical installation should be used. The maximum allowed is generally 90 days for seasonal types of installations, such as Christmas lights and other types of decorations.

- All temporary power connections in industrial and commercial facilities should be supplied from a circuit protected by a GFCI. (See Chap. 3 for more information on GFCIs.)

An extension cord that exhibits any of the visual inspection problems listed in Table 9.2 or that does not pass the tests listed in Table 9.3 should be removed from service until it can be repaired or replaced.

Plug- and Cord-Connected Equipment

Drills, circular saws, sanders, and other such electrically operated equipment fall into this category. These types of devices are powered via a male plug located at the end of a flexible cord. The flexible cord, in turn, connects to the operating circuits of the tool.

Hand tools generally fall into one of two categories—the standard metal tools with ground, and the double-insulated type of tool. The metallic tool power cord is a standard three-conductor type with a hot wire, neutral wire, and the safety ground. The hot wire and the neutral wire form the actual power circuit of the tool. The safety ground is connected to the metallic frame of the tool. In the event that the hot wire is short-circuited to the case of the tool, the safety ground wire forms a continuous, low impedance path back to the service box. Figure 9.1 is a pictorial diagram of such an assembly.

Even the relatively low impedance path provided by the safety ground wire can create lethal voltages from the hand to the feet under some circumstances. The double-insulated tool does not employ a metallic case. Rather, its case is made of a high-strength, nonconductive plastic or composite material. The power supply for such tools is a two-conductor power cord with no safety ground. Since the tool case is nonconductive, the user is protected by both the case and the normal insulation of the electric circuit.

TABLE 9.3 Recommended Periodic Tests and Test Results for Extension Cords

Test	Description	Pass/fail criteria
Ground continuity test	25 A minimum is passed through the cord's ground circuit.	Voltage drop across the cord should not exceed 2.5 V.
Insulation breakdown	High voltage is applied to the cord's insulation system and the leakage current or insulation resistance is measured (3000 V maximum direct current applied).	Leakage current no more than 6 μA @ 3000 V (500 MΩ).

FIGURE 9.1 Typical metallic case hand tool.

Power tools should be subject to the same precautions as those outlined for extension cords in the previous section. Specifically the following should be observed:

- Closely inspect tools before each use. Table 9.2 lists the types of items that should be looked for during the inspection. (Note that Table 9.2 applies to both extension cords and portable tools.)

- Never use the tool power cord to lift the tool. If the tool must be lifted, tie a hand line or rope to the tool.

- Make certain the tool's ground connection is complete. For metal-cased tools, the safety ground connection is the difference between life and death when an internal short circuit occurs.

- Never alter the plug or receptacle on a tool. This applies especially to altering the ground connection.

- If a tool employs a twist-lock or locking type of plug, it should be securely fastened before the tool is used.

- Do not use a tool in wet or hazardous environments unless it is rated for such service by the manufacturer.

- If tools are used in wet or hazardous environments, insulating safety equipment such as rubber gloves with appropriate leather protectors should be worn.

- Only personnel who are authorized and trained in the use of power tools should be allowed to use them.

- Cord-connected tools should be subjected to the electrical tests outlined in Table 9.4. If the tool fails any of the listed tests, it should be replaced or repaired.

- When not in use, tools should be carefully stored in such a way that they cannot be damaged.

- All temporary power connections in industrial and commercial facilities should be supplied from a circuit protected by a GFCI. (See Chap. 3 for more information on GFCIs.)

TABLE 9.4 Recommended Periodic Tests and Recommended Results for Cord-Powered Tools

Test	Description	Pass/fail criteria
Ground continuity test	25 A minimum is passed through the cord's ground circuit.	Voltage drop across the cord should not exceed 2.5 V.
Insulation breakdown	High voltage is applied to the cord's insulation system and the leakage current or insulation resistance is measured (3000 V maximum direct current applied)*.	Leakage current not more than 6 mA @ 3000 V (500 MΩ).
Leakage test	Measures the current (0 to 10 mA) that would flow through the operator if he or she were to provide a path to ground at normal operating voltage.	
Operational check	Operates the tool to verify proper operation and to indicate operating current.	Operating current should be within nominal nameplate values.

* Refer to tool manufacturer's directions for allowed maximum test voltages.

Any tool that does not meet the pass/fail criteria listed in Table 9.4 should be removed from service until it can be repaired or replaced.

Current Transformers

The safety hazards of current transformers are identical for low-, medium-, and high-voltage circuits. Refer to Chap. 10 for a detailed coverage of the nature of current transformer hazards and methods for protecting workers from those hazards.

GROUNDING LOW-VOLTAGE SYSTEMS

The subject of electrical grounding is a complex one. The following sections focus on the grounding concepts and requirements of low-voltage systems as they relate to safety. For more complete coverage of electrical grounding, refer to the various ANSI/IEEE standards including IEEE Guide for Safety in AC Substation Grounding—ANSI/IEEE standard 80 and IEEE Recommended Practice for Grounding of Industrial and Commercial Power Systems—ANSI/IEEE standard 142.

What Is a Ground?

A *ground* is an electrically conducting connection between equipment or an electric circuit and the earth, or to some conducting body that serves in place of the earth. If a ground is properly made, the earth or conducting body and the circuit or system will all maintain the same relative voltage.

Figure 9.2a shows an electrical system that is not grounded. In such a situation, a voltage will exist between the ground and some of the metallic components of the power

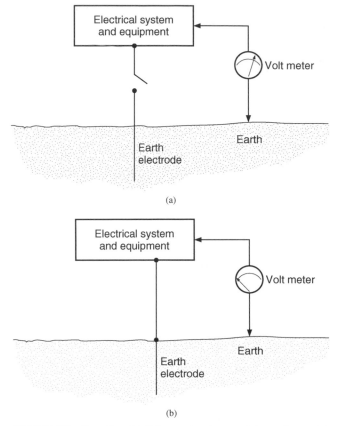

FIGURE 9.2   Grounding. (*a*) Before system is grounded, a voltage exists between the system and the earth; (*b*) when ground connection is made, no voltage exists.

system. In Fig. 9.2*b,* the earth connection has been made. When the system is grounded, the voltage is reduced to zero between the previously energized sections of the system.

Bonding Versus Grounding

Bonding is the permanent joining of metallic parts to form a continuous, conductive path. Since the earth is generally not a good conductor, bonding is used to provide a low-impedance metallic path between all metallic parts. This, in essence, bypasses the earth and overcomes its relatively high impedance. A good grounding system is a combination of solid connections between metallic parts and the earth as well as between all metallic parts.

Voltage Hazards

Figure 9.3 illustrates the four standard voltage hazards that are associated with and relieved by proper system grounding. If a system is ungrounded and the nonenergized metallic parts

FIGURE 9.3 Four voltage hazards related to system grounding. (*Courtesy Institute of Electrical and Electronic Engineers.*)

become energized, the metallic parts will have a measurable voltage between themselves and ground. If the system is grounded by driving ground electrodes into the earth, current will flow from the rods into the earth. At each ground electrode, the voltage will rise relative to the remote reference point. The voltage will drop off away from the electrodes and peak at each electrode. This situation creates the four types of voltage hazards as follows:

- *Step voltage.* As a worker steps across the ground, the front foot will be at a different potential than the rear foot. This effect is caused by the voltage gradient created by the ground electrodes (Fig. 9.3). Step voltage can easily reach lethal levels. It can be mitigated by increasing the number of grounding electrodes, increasing bonding between the metal parts, and employing a grounding grid.

- *Touch voltage.* Because of the voltage gradient, the voltage of the earth only a short distance from the grounded metallic equipment will be different than the equipment. Thus, if a worker touches the grounded equipment, his or her feet will be at a different potential than his or her hands. This voltage can be lethal. Touch voltage is mitigated by increasing the number of grounding electrodes, increasing bonding between metal parts, and employing a grounding grid.

- *Mesh voltage.* Mesh voltage is the worst case of touch voltage. Cause, effect, and mitigation methods are identical to those described previously under Touch voltage.

- *Transferred voltage.* Because metal parts have a much lower impedance than the earth, the voltage drop between remote locations is lower on the metallic connections than the earth. This means that the earth may be at a significantly different voltage than the metallic connections. Transferred voltages are particularly noticeable on neutral wires that are grounded at the service point and nowhere else. Transferred potential can be mitigated by providing the entire area with a ground grid; however, this solution is infeasible in any but the smallest systems. Transferred voltage must be mitigated by avoiding contact with conductors from remote locations and/or using rubber insulating gloves.

System Grounds

What Is a System Ground?   A *system ground* is the connection of one of the conductors to the earth. Such a connection is accomplished by connecting an electric wire to the selected system conductor and the grounding electrode.

Why Are Systems Grounded?   Power systems have conductors grounded for a variety of safety and operational reasons including

- Grounded systems provide sufficient short-circuit current for efficient operation of protective equipment.
- Grounded systems are less prone to transient overvoltages, which can cause insulation failures.
- Grounded systems are generally more easily protected from lightning.
- Solidly grounded systems are less prone to resonant conditions, which can cause equipment and insulation failures.

What Systems Must Be Grounded?   The NEC requires that both AC and DC systems be grounded. Table 9.5 summarizes low-voltage DC grounding requirements, and Table 9.6 summarizes low-voltage AC grounding requirements.

How Are Systems Grounded?   Electrical systems are grounded by connecting one of the electric conductors to earth. The conductor chosen and the location of the ground are determined as part of the engineering design for the system. Figures 9.4 through 9.7 illustrate four different low-voltage circuits and how they are grounded. Note that in some cases, the point of ground is determined by regulatory requirements such as the NEC.

TABLE 9.5   DC Circuits That Require Grounding

Type of circuit	Circuits to be grounded	Exceptions/comments
Two-wire DC	All	Systems less than 50 V or greater than 300 V between conductors need not be grounded. Limited-area industrial systems with ground detectors need not be grounded. Certain rectifier-derived DC systems do not need to be grounded. DC fire-protective signaling circuits with no more than 0.030 A may not need to be grounded.
Three-wire DC	All	The neutral wire is grounded.

Note: See the current edition of the National Electrical Code for details.

TABLE 9.6 AC Circuits That Require Grounding

Type of circuit	Circuits to be grounded	Exceptions/comments
AC systems 50 to 1000 V	If maximum voltage to ground is less than 50 V Three-phase, 4-wire wye, if neutral is a circuit conductor Three-phase, 4-wire delta with midpoint grounded on one leg In some cases when grounded service conductor is uninsulated	The National Electrical Code has many exceptions to these grounding requirements.

Note: See the current edition of the National Electrical Code for details.

The wire(s) that must be grounded depend on the voltage level and the system application. For DC systems, one of the conductors or the neutral wire is to be grounded depending on the type of system (Table 9.5). For AC systems, the NEC specifies five different locations for the selection of the grounding point for AC systems. Table 9.7 identifies each of the five locations.

Systems with voltages between 480 and 1000 V phase to phase may be grounded through an impedance to limit the amount of fault current. In the circuit shown in Fig. 9.8, the resistor will oppose current flow between the earth and the phase wires. For example, assume that the phase A wire falls to the earth. The circuit that is formed will be composed of the power system's phase-to-neutral voltage (277 V) impressed across the series combination of the power system's impedances plus the grounding resistor. If the grounding resistor is properly sized, it will limit the fault current to any maximum value that is chosen during the design.

By limiting the amount of fault current, resistance-grounded systems provide a somewhat higher level of safety. However, the engineer making the design decision to ground through a resistor must take into consideration more variables than just safety. Protective systems, ground-fault current, voltage transients, and many other such concerns must be considered before the engineer decides to use a resistance-grounded system.

FIGURE 9.4 Grounding a 120-V single-phase circuit.

FIGURE 9.5 Grounding a 240/120-V single-phase, three-wire circuit.

FIGURE 9.6   Grounding a 208/120-V three-phase, four-wire system.

FIGURE 9.7   Grounding a 240/120-V, three-phase, four-wire system.

TABLE 9.7   Typical Grounding Requirements for Low-Voltage Systems

Type of premises wiring circuit	Location of ground
One-phase, 2-wire	Either conductor (Fig. 9.4)
One-phase, 3-wire	The neutral conductor (Fig. 9.5)
Multiphase with one wire common to all phases	The common conductor (not illustrated)
Multiphase systems required on grounded phase	One of the phase conductors (not illustrated)
Three-phase, 4-wire circuits	The neutral conductor (Fig. 9.6)
240/120, 3-phase, 4-wire	The center point of the grounded leg (Fig. 9.7)

FIGURE 9.8   Resistance grounded 480/277-V, three-phase, four-wire system.

Equipment Grounds

What Is an Equipment Ground?   An *equipment ground* is an electrically conductive connection between the metallic parts of equipment and the earth. For example, transformer cases and cores are connected to the earth—this connection is called an equipment ground. Note that the metallic parts of equipment are grounded and bonded together. This bonding serves to reduce the voltage potential between all metallic parts and the earth.

Why Is Equipment Grounded?   The equipment ground is one of the most important safety aspects of grounding. Workers are constantly in contact with transformer shells, raceways, conduits, switchgear frames, and all the other conductive, non–current-carrying parts. Proper equipment grounding and bonding ensures that the voltage to which workers will be subjected is kept to a minimum. Proper bonding and grounding mitigates touch and step voltages.

How Is Equipment Grounded?   The NEC requires that the path to ground from circuits, equipment, and metal enclosures must meet the conditions listed in Table 9.8. The NEC does not allow the earth to be the sole equipment grounding conductor. Simply setting metal equipment on the earth is insufficient. Equipment must be connected to the earth via metal electrodes and conductors.

   Equipment used in grounded systems is grounded by bonding the equipment grounding conductor to the grounded service conductor and the grounding electrode conductor. That is, the equipment ground is connected to the system ground. Equipment used in ungrounded systems is grounded by bonding the equipment grounding conductor to the grounding electrode conductor.

What Equipment Must Be Grounded?   Tables 9.9 through 9.12 list the types of equipment that must be grounded according to the NEC. Always refer to the current edition of the NEC for up-to-date information.

TABLE 9.8   Equipment Grounding Requirements

- Must be permanent and continuous
- Must have the capacity to conduct any fault current likely to be imposed on it
- Must have sufficiently low impedance to limit the voltage to ground and to facilitate the operation of the circuit protective devices

---

TABLE 9.9   Grounding Requirements for Equipment Fastened in Place or Connected by Permanent Wiring

Must ground	Exceptions
• Equipment within 8 ft (2.44 m) horizontally or 5 ft (1.52 m) vertically of ground or grounded metal objects subject to human contact • Equipment located in wet or damp locations • Equipment in electrical contact with other metallic objects • Equipment located in classified hazardous locations • Equipment that is supplied by metal-clad, metal-sheathed, metal-raceway, or other wiring method which provides an equipment ground • Equipment that operates with any terminal in excess of 150 V to ground	• Enclosures for switches or circuit breakers used for other than service equipment and accessible to qualified persons only. • Metal frames of electrically heated appliances; exempted by special permission, in which case the frames shall be permanently and effectively insulated from ground. • Distribution apparatus, such as transformer and capacitor cases, mounted on wooded poles, at a height exceeding 8 ft (2.44 m) above ground or grade level. • Listed equipment protected by a system of double insulation, or its equivalent, shall not be required to be grounded. Such equipment must be distinctively marked.

TABLE 9.10   Grounding Requirements for Equipment Fastened in Place or Connected by Permanent Wiring

Must ground	Exceptions
Switchboard frames and structures	Frames of 2-wire DC switchboards where effectively insulated
Pipe organs (generator and motor frames)	Where the generator is effectively insulated from ground and from the motor driving it
Motor frames	
Enclosures for motor controllers	Enclosures attached to ungrounded portable equipment Lined covers of snap switches
Elevators and cranes	
Garages, theaters, and motion picture studios	Pendant lamp-holders supplied by circuits less than 150 V to ground
Electric signs	
Motion picture projection equipment	
Remote-control, signaling, and fire-protective signaling circuits	
Lighting fixtures	
Motor-operated water pumps	

TABLE 9.11 Nonelectric Equipment Grounding Requirements

Must ground
Cranes
Elevator cars
Electric elevators
Metal partitions
Mobile homes and recreational vehicles

TABLE 9.12 Grounding Requirements for Equipment Connected by Cord and Plug

Must ground	Exceptions
In classified hazardous locations	
In systems that are operated in excess of 150 V to ground	Motors, where guarded.
	Metal frames of electrically heated appliances, exempted by special permission, in which case the frames shall be permanently and effectively insulated from ground.
	Listed equipment protected by a system of double insulation, or its equivalent, shall not be required to be grounded. Where such a system is employed, the equipment shall be distinctively marked.
In residential occupancies • Refrigerators, freezers, and air conditioners • Clothes washing/drying machines, sump pumps, aquariums • Handheld motor-operated tools including snow blowers, hedge clippers, etc. • Portable hand lamps	Listed equipment protected by a system of double insulation, or its equivalent, shall not be required to be grounded. Where such a system is employed, the equipment shall be distinctively marked.
In other than residential occupancies • Refrigerators, freezers, air conditioners • Clothes washing/drying machines, electronic computers/data-processing equipment, sump pumps, aquariums • Handheld, motor-operated tools including hedge clippers, lawn mowers, snow blowers, etc. • Cord- and plug-connected appliances used in damp or wet locations • Tools likely to be used in wet or conductive locations • Portable hand lamps	Tools and hand lamps in wet locations when supplied through an isolating transformer with an ungrounded secondary of not over 50 V. Listed equipment protected by a system of double insulation, or its equivalent, shall not be required to be grounded. Where such a system is employed, the equipment shall be distinctively marked.

Ground Fault Circuit Interrupters

GFCIs are described in detail in Chap. 2. Although few mandatory standards require universal applications of GFCI devices, prudence and common sense suggest that they should be applied in all industrial/commercial environments. Their sensitivity and operating speed (5 mA, 25 m/s) make them the only type of protective device that is capable of being used to protect human lives.

SAFETY EQUIPMENT

Overview

Because of the delusion that "low voltage can't hurt you," many workers do not use safety equipment when working on or near energized, low-voltage conductors. As described previously in this chapter, the amount of available energy in the system determines its lethal effects, not the voltage. Low-voltage systems are extremely hazardous and should be treated with the respect they deserve. The following sections describe the types of safety equipment that should be worn when working on or near energized, low-voltage conductors. The recommendations given are minimum recommendations. If additional or more stringent protection is desired or required, it should be worn. Refer to the tables in Chap. 3 for specific recommendations.

Hard Hats

Protective headgear for persons working on or near energized, low-voltage circuits should provide both mechanical and electrical protection. Since the ANSI class C helmet provides no electrical protection, class C helmets should not be worn. Workers should be supplied with and should wear either ANSI class G or class E helmets. If workers are never required to work around high-voltage circuits, the ANSI class G helmet may be used. If, however, workers are required to work around both high- and low-voltage circuits, they should be supplied with and should wear ANSI class E helmets. Table 9.13 summarizes the characteristics of the three ANSI classes. Note that class G and E were formerly class A and B, respectively.

TABLE 9.13 Summary of the Characteristics for ANSI Class C, E, and G Hard Hats

Class	Description	Comments
G	Reduce the impact of falling objects and reduce danger of contact with exposed, low-voltage conductors. Representative sample shells are proof-tested at 2200 V phase to ground.	Recommended to be worn by personnel working around only low-voltage circuits
E	Reduce the impact of falling objects and reduce danger of contact with exposed high-voltage conductors. Representative sample shells are proof-tested at 20,000 V phase to ground.	Recommended to be worn by personnel working around high- and low-voltage circuits
C	Intended to reduce the force of impact of falling objects. This class offers no electrical protection.	Should not be worn by personnel working on or around energized conductors of any voltage

Eye Protection

Even low-voltage systems are capable of producing extremely powerful and hazardous electric arcs and blasts. This is especially true of 480-V and 575-V systems. Because of this, eye protection for electrical workers should provide protection against heat and optical radiation. ANSI standard Z87.1-1989 provides a selection chart as well as a chart that illustrates the various protection options available. The selection chart is reproduced in this handbook as Table 9.14, and eye protection options are shown in Fig. 9.9.

Electrical workers should consider using eye protection that combines both heat and optical radiation protection. Table 9.14 and Fig. 9.9 show several different types of equipment that will provide such protection including types B, C, D, E, and F. If arcing and molten splashing is possible, as with open-door switching or racking of circuit breakers, the type N eye protection should be employed.

NOTES:
(1) Care shall be taken to recognize the possibility of multiple and simultaneous exposure to a variety of hazards. Adequate protection against the highest level of each of the hazards must be provided.
(2) Operations involving heat may also involve optical radiation. Protection from both hazards shall be provided.
(3) Faceshields shall only be worn over primary eye protection.
(4) Filter lenses shall meet the requirements for shade designations in Table 1.
(5) Persons whose vision requires the use of prescription (Rx) lenses shall wear either protective devices fitted with prescription (Rx) lenses or protective devices designed to be worn over regular prescription (Rx) eyewear.

(6) Wearers of contact lenses shall also be required to wear appropriate covering eye and face protection devices in a hazardous environment. It should be recognized that dusty and/or chemical environments may represent an additional hazard to contact lens wearers.
(7) Caution should be exercised in the use of metal frame protective devices in electrical hazard areas.
(8) Refer to Section 6.5, Special Purpose Lenses.
(9) Welding helmets or handshields shall be used only over primary eye protection.
(10) Non-sideshield spectacles are available for frontal protection only.

FIGURE 9.9 Eye protection devices. (Refer to Table 9.14 for selection criteria.) (*Courtesy American National Standards Institute.*)

TABLE 9.14 Eye Protection Selection Chart

	Assessment SEE NOTE (1)	Protector type[a]	Protectors	Limitations	Not recommended
IMPACT	Chipping, grinding, machining, masonry work, riveting, and sanding	B, C, D, E, F, G, H, I, J, K, L, N	Spectacles, goggles, face shields SEE NOTES (1) (3) (5) (6) (10). For severe exposure, add N.	Protective devices do not provide unlimited protection. SEE NOTE (7).	Protectors that do not provide protection from side exposure. Filter or tinted lenses that restrict light transmittance, unless it is determined that a glare hazard exists. Refer to Optical Radiation.
HEAT	Hot sparks	B, C, D, E, F, G, H, I, J, K, L, *N	Face shields, goggles, spectacles *For severe exposure, add N. SEE NOTES (2) (3).	Spectacles, cup and cover type goggles do not provide unlimited facial protection.	Protectors that do not provide protection from side exposure.
	Splash from molten metals	*N	*Face shields worn over goggles H, K SEE NOTES (2) (3).	SEE NOTE (2).	
	High temperature exposure	N	Screen face shields, reflective face shields SEE NOTES (2) (3).		
CHEMICAL	Splash	G, H, K	Goggles, eyecup and cover types	SEE NOTE (3).	Spectacles, welding helmets, hand shields.
		*N	*For severe exposure, add N.	Ventilation should be adequate but well protected from splash entry.	
	Irritating mists	G	Special-purpose goggles	SEE NOTE (3).	
DUST	Nuisance dust	G, H, K	Goggles, eyecup and cover types	Atmospheric conditions and the restricted ventilation of the protector can cause lenses to fog. Frequent cleaning may be required.	

OPTICAL RADIATION

		Typical filter lens shade	Protectors	
Welding:			SEE NOTE (9).	
Electric Arc	O, P, Q	10-14	Welding helmets or welding shields	Protection from optical radiation is directly related to filter lens density. SEE NOTE (4). Select the darkest shade that allows adequate task performance.
Welding:		SEE NOTE (9).		
Gas	J, K, L, M, N, O, P, Q	4-8	Welding goggles or welding face shield	
Cutting		3-6		
Torch brazing		3-4		SEE NOTE (3).
Torch soldering	B, C, D, E, F, N	1.5-3	Spectacles or welding face shield	
Glare	A, B	Spectacle SEE NOTES (9) (10).		Shaded or special-purpose lenses, as suitable. SEE NOTE (8).
				Protectors that do not provide protection from optical radiation. SEE NOTE (4).

[a] Refer to Fig. 9.9 for protector types.

NOTE: For NOTES referred to in this table, see text accompanying Fig. 9.9.

Source: Courtesy of American National Standards Institute.

TABLE 9.15 Recommended Arc Protection and Clothing for Persons Working on or Near Energized Low-Voltage Conductors

Description of work	Type of clothing
Routine work on or close to energized conductors (circuits above 50 V to ground)	Flame retardant work clothing
Open box switching and/or fuse removal (circuits 208 V phase to phase and higher)	Flame retardant work clothing and/or flash suits
Installation or removal of low-voltage circuit breakers and/or motor starters with energized bus (circuits 208 V phase to phase or higher)	Flame retardant work clothing and/or flash suits

* Head, eye, and insulating protection should also be worn. See Chap. 2 for additional information on the use of various types of protective clothing.

† See Chap. 3 for calculating the required weight of protective clothing.

Arc Protection

Low-voltage systems can create and sustain significant electric arcs accompanied by electric blast. Employees performing work in 575-V, 480-V, or 208-V phase-to-phase systems should wear the type of clothing listed in Table 9.15. Refer to Chap. 3 for methods that can be used to calculate flash clothing weights.

Rubber Insulating Equipment

The introduction of the so-called low-voltage rubber glove has made the use of insulating protection much more convenient than in the past. Personnel working on or near energized low-voltage conductors should wear low-voltage rubber gloves with appropriate leather protectors. Such gloves will have red ANSI labels and are rated for use in circuits of up to and including 1000 V rms.

Voltage-Testing Devices

Proximity or contact testers intended for use in low-voltage circuits should be used to test circuits and to verify that they are de-energized and safe to work on. Voltage-testing devices are

FIGURE 9.10   Volt-stick proximity voltage sensor for use on circuits up to 600 V. (*Courtesy Santronics, Inc.*)

FIGURE 9.11 Safety voltage/continuity tester for circuits of 600 V AC or DC and less. (*Courtesy Ideal Industries, Inc.*)

described in Chap. 2. Examples of low-voltage–measuring instruments are shown in Figs. 9.10 through 9.12. Workers should use receptacle and GFCI testers such as that shown in Fig. 9.13. These testers are especially important for the verification of the ground path in an operation duplex receptacle.

FIGURE 9.12 Digital readout contact type safety voltmeter for circuits of 1000 V AC and DC or less. (*Courtesy Tegam, Inc.*)

FIGURE 9.13 Receptacle and GFCI tester. (*Courtesy Direct Safety Company, Phoenix, Arizona.*)

SAFETY PROCEDURES

General

The general procedures described in Chap. 3 should be used on all circuits in excess of 50 V to ground. The following sections describe key procedures that apply to low-voltage circuits.

Approach Distances

Although approach distances are laid out in a number of OSHA references, the NFPA 70E provides a much more useful and practical concept for approach distances. The following paragraphs identify methods that may be used to determine approach distances. Refer to Chap. 3—Fig. 3.29 and Table 3.15—for reference to the terms used in the following section.

Crossing the Limited Approach Boundary.   Unqualified workers are not allowed to cross the limited approach boundary under any circumstances. Qualified workers may cross the limited approach boundary if they are qualified to perform the work.

Requirements for Crossing the Restricted Approach Boundary.   In order to cross the restricted approach boundary, the following criteria must be met:

• The worker must be qualified to do the work.

• There must be a plan in place that is documented and approved by the employer.

• The worker must be certain that no part of the body crosses the prohibited approach boundary.

• The worker must work to minimize the risk that may be caused by inadvertent movement by keeping as much of the body out of the restricted space as possible. Allow only protected body parts to enter the restricted space as necessary to complete the work.

• Personal protective equipment must be used appropriate for the hazards of the exposed energized conductor.

Requirements for Crossing the Prohibited Approach Boundary.   NFPA 70E considers crossing the prohibited approach boundary to be the same as working on or contacting an energized conductor. To cross into the prohibited space, the following requirements must be met:

• The worker must have specified training required to work on energized conductors or circuit parts.

• There must be a plan in place that is documented and approved by the employer.

• A complete risk analysis must be performed.

• Authorized management must review and approve the plan and the risk analysis.

• Personal protective equipment must be used appropriate for the hazards of the exposed energized conductor.

Voltage Measurement

The voltage-measurement techniques defined in Chap. 3 should be employed on circuits of all voltages, including 1000 V and below. Note that although the OSHA standards do not require the three-step measurement process for circuits of below 600 V, best safety practice does, call for instrument checks both before and after the actual circuit measurement.

TABLE 9.16 Typical Situations That May Allow Reclosing of a Protective Device
That Has Operated

- The faulted section of the system is found and repaired.
- The nature of the protective device makes it clear that no hazard is present. For example, if the device that operated is an overload type of device, it may be safe to reclose.
- The reclosing operation can be made in such a way that the workers are not exposed to additional hazard. For example, if the reclosing operation can be made by remote control, and if all personnel are kept away from all parts of the circuit, it may be reclosed.

Locking and Tagging

Lockout-tagout and energy control procedures apply to circuits of all voltage levels. Refer to Chap. 3 for detailed discussions of lockout-tagout procedures.

Closing Protective Devices After Operation

Workers should never reclose any protective device after it has operated, until it has been determined that it is safe to do so. Several criteria may be used to determine whether it is safe to reclose the protective device (Table 9.16). Other conditions may or may not indicate a safe reclosing situation.

ELECTRICAL SAFETY AROUND ELECTRONIC CIRCUITS

Modern technology requires many persons to work on or near electronic circuitry. Such circuits can provide special or unusual hazards. The following sections provide information about the nature of the hazard and some specific procedures that may be used by workers to enhance their personal safety.

The Nature of the Hazard

Frequencies. The relationship of frequency to electrical hazards is discussed in Chap. 1. Generally, the following points apply:

- DC currents and AC currents up to approximately 100 Hz seem to affect the body in a very similar manner. For all practical purposes, when working around a DC circuit, the worker should use the same types of procedures as when working around power system frequencies.
- Above 100 Hz, the threshold of perception increases. Between 10 and 100 kHz, the threshold increases from 10 to 100 mA.

Capacitive Discharges. According to the NFPA 70E, the following are true with respect to capacitor discharges:

- A current caused by the discharge of a 1-μF, 10,000-V capacitor may cause ventricular fibrillation.
- A current caused by the discharge of a 20-μF, 10,000-V capacitor will probably cause ventricular fibrillation.

Specific Hazards of Electronic Equipment.   Although seemingly harmless, electronic circuits present a number of hazards including

1. Electrical shock from 120-, 240-, or 480-V AC power supplies
2. High power supply voltages
3. Possible shock and burn hazards caused by radio frequency (RF) fields on or around antennas and antenna transmission lines
4. RF energy–induced voltages
5. Ionizing (x-radiation) hazards from magnetrons, klystrons, thyratrons, CRTs, and other such devices
6. Nonionizing RF radiation hazards from
   (a) Radar equipment
   (b) Radio communication equipment
   (c) Satellite earth-transmitters
   (d) Industrial scientific and medical equipment
   (e) RF induction heaters and dielectric heaters
   (f) Industrial microwave heaters and diathermy radiators

Special Safety Precautions

The following methods are offered in addition to the other safety equipment and procedures that are discussed throughout this handbook.

AC and DC Power Supplies.   The nature of these hazards is similar to the hazards that are discussed throughout Chaps. 3 and 9. One piece of equipment that is finding increasing use in protecting workers from these types of hazards is the PVC sheeting that can be placed over the exposed circuit parts. This PVC material provides an insulating blanket for voltages up to 1000 V and will allow the worker to perform the necessary tasks in the equipment.

Protection from Shock and Burn Caused by RF Energy on Antennas and/or Transmission Lines.   Avoidance of contact is the best possible protection for this type of hazard. Transmitting equipment should always be disabled before workers are allowed to approach antennas or transmission lines.

Electrical Shock Caused by RF-Induced Voltage.   Electric shocks from contacting metallic objects that have induced RF voltages on them can be dangerous in at least two ways:

1. The surprise effect of the shock can cause the victim to fall from a ladder or other elevated location.
2. RF discharge can cause ventricular fibrillation under the right circumstances.

Three methods can be used to protect personnel from induced RF voltages:

1. De-energize the RF circuits to eliminate the energy.
2. Use insulating barriers to isolate the metal objects from the worker.
3. Ground and bond all non–current-carrying metal parts such as chassis, cabinets, covers, and so on. Proper RF ground wires must be very short compared to the wavelength of the RF. If a solid ground cannot be reached because of distance, a counterpoise type of ground can be employed. The design of such a ground is beyond the scope of this handbook. The reader should refer to one of the many engineering texts available.

Radiation (Ionizing and Nonionizing Hazards).   The best methods for protecting workers from this type of hazard are:

1. De-energize the circuit so that the worker is not exposed to the radiation.
2. Protect the worker from the radiation by using appropriate shielding.

STATIONARY BATTERY SAFETY

Introduction

Stationary batteries (Fig. 9.14) are used for various types of stand-by and emergency power requirements throughout electrical power systems. Batteries are usually connected to the power system as shown in Fig. 9.15. Because of their construction and energy capacity, batteries offer a special type of safety hazard that includes chemical, electrical, and explosive hazards.

FIGURE 9.14 Partial view of a typical stationary battery installation.

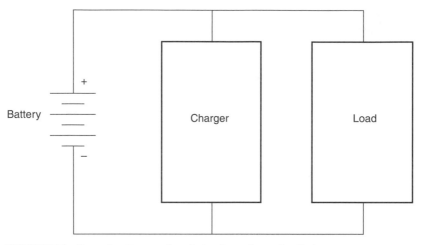

FIGURE 9.15   Connection diagram of a typical stationary battery installation.

Basic Battery Construction

Stationary batteries operate on the basic principle of galvanic action; that is, two dissimilar materials will produce a voltage when they are put close together. While the basic chemistry of stationary batteries may vary and change, the two most common types in use today are the lead-acid and the nickel-cadmium types.

Lead-Acid Batteries.   The lead-acid battery uses lead (Pb) and lead peroxide (PbO_2) for its negative and positive plates, respectively. The electrolyte in which the plates are immersed is a solution of sulfuric acid (H_2SO_4) and water (H_2O). The basic chemical reaction is shown in Eq. 9.1.

$$PbO_2 + Pb + 2H_2SO_4 \underset{\text{Discharging}}{\overset{\text{Charging}}{\rightleftharpoons}} 2PbSO_4 + 2H_2O$$

Modern lead-acid batteries are constructed in one of two general formats:

1. Vented cell batteries (also known as flooded cells) are a mature technology in which the plates are completely immersed in the electrolyte. The containers are generally open to the atmosphere with flame arresters used to minimize the chance of explosion or fire.
2. Valve-regulated cell (VRLA) is also called the starved electrolyte cell. Such a cell is essentially sealed except for the presence of a relief valve. The electrolyte is restrained internally either by gelling the material or by insertion of a fiber mat. The VRLA is, in its essence, lead-acid technology.

NiCad (Nickel-Cadmium) Batteries.   The NiCad batteries used for stationary applications use nickel hydrate [$Ni(OH)_3$] for the positive plate and cadmium (Cd) for the negative plate. The electrolyte commonly used in the NiCad battery is potassium hydroxide (KOH). The basic chemical action of the NiCad battery is shown in Eq. 9.2. Notice that the electrolyte does not take part in the chemical reaction.

$$2Ni(OH)_3 + Cd \underset{\text{Discharging}}{\overset{\text{Charging}}{\rightleftharpoons}} 2Ni(OH)_2 + Cd(OH)_2$$

TABLE 9.17 Recommended Safety Procedures/Equipment for Stationary Batteries

Hazard	Protection procedure/equipment
Electrical	• Low-voltage rubber gloves (Class 00 or Class 0). • Insulated tools. • Arc protection (face shield and flame-retardant clothing, minimum).

Chemical	• Chemical protective apron.
	• Chemically protective face shield and goggles.
	• Chemically resistant gloves.
	• Safety shoes.
	• Ample supply of pure water.
	• Eye and body wash station.
	• Neutralizing solution. (Use with caution and only with the approval of the battery manufacturer.)
	  • NiCad—7 oz boric acid/gal H_2O.
	  • Lead-Acid—1 lb baking soda/gal H_2O.

Explosion	• Be sure that battery room is adequately ventilated. Typically, hydrogen concentrations should be kept to less than 1%. • Use nonsparking insulated tools. • A Class C fire extinguisher should be immediately available.

Safety Hazards of Stationary Batteries

Electrical Hazards. Stationary batteries have sufficient stored energy to represent both shock and arcing hazards. Additionally, the high current capacity of stationary batteries can cause extremely dangerous heat. Severe burns have been caused by the high battery currents through personal jewelry (such as wedding rings) and tools.

Chemical Hazards. The electrolytes from both of the major types of batteries are destructive to human tissues. Although not normally in strong concentrations, the sulfuric acid and potassium hydroxide solutions can destroy eye tissue and cause serious burns on more hardy locations.

Explosion Hazards. Explosions of stationary batteries result from two different sources:

1. Excessive heat from ambient conditions, excessive charging, or excessive discharging can cause a cell to explode if it cannot properly vent. This hazard exists to some degree for all batteries; however, it tends to be more of an issue with VRLAs and NiCads.

2. The chemical reaction during charging of a lead-acid battery is not 100 percent efficient. In fact, if the battery is charged too quickly, not all of the hydrogen will find a sulfate radical with which it can combine. This can cause the release of hydrogen to the air. Concentrations of hydrogen in air of more than 4 percent or 5 percent volume can explode violently.

Battery Safety Procedures

Electrical Safety. Table 9.17 lists the minimum safety procedures and equipment that should be available to personnel working on or near stationary batteries. Refer to manufacturer's instructions for more specific recommendations.

CHAPTER 10

MEDIUM- AND HIGH-VOLTAGE SAFETY SYNOPSIS

INTRODUCTION

High-voltage systems* have three uniquely hazardous characteristics. First, high-voltage systems are very high energy systems. This means that they easily establish and maintain hazardous arc and blast conditions. Second, most workers do not have much experience working on or around such systems. Personnel ignorance can lead to very serious accidents and injuries. And finally, high voltages can puncture through the keratin skin layer, virtually eliminating the only significant skin resistance. Lack of knowledge can be corrected by personnel training. Chapter 13 covers the topic of personnel training in great detail.

The energy content and skin-puncturing characteristics of high-voltage systems are inherent to their nature. Safety precautions must be strictly observed to reduce the possibility of accidents, and safety equipment must be worn to reduce the severity of accidents.

This chapter highlights some of the key or critical safety concepts related to high-voltage systems. Many of the types of safety precautions used in high-voltage systems are identical to those used in low-voltage systems. Such precautions have been repeated in this chapter.

HIGH-VOLTAGE EQUIPMENT

Current Transformers

Note: The material in this section applies to current transformers used in low-, medium-, and high-voltage circuits.

Current transformers (CTs) (Fig. 10.1) are used to reduce primary current levels to lower values that are usable by instruments such as meters and protective relays. In doing this the primary winding of the current transformer is connected in series with the system load current and the secondary winding is connected to the instruments (Fig. 10.2).

---

* ANSI/IEEE standard 141 defines low-voltage systems as those which are less than 1000 V, medium-voltage systems as those equal to or greater than 1000 V and less than 100,000 V, and high-voltage systems as those equal to or greater than 100,000 V. To simplify notation, this chapter will refer to all systems of 1000 V or higher as *high-voltage* systems.

FIGURE 10.1   Typical current transformers. (*Courtesy AVO Training Institute.*)

If the secondary of an energized transformer is open-circuited, an extremely high voltage appears at the secondary (Fig. 10.3). Depending on the type of transformer and the conditions at the time of the open circuit, this high voltage can create shock, arc and blast hazards. Current transformers can explode violently when their secondary circuits are opened.

Caution: Never open-circuit the secondary of an energized current transformer. A CT secondary circuit should always be terminated with rated load or a short circuit.

When working on or near CT circuits, personnel should always be aware of the hazard and should wear rubber gloves, arc protective clothing, and face shields if working on the wiring for the circuit. The primary circuit should be de-energized before the secondary wiring is opened.

FIGURE 10.2   Schematic diagram of a current transformer connection.

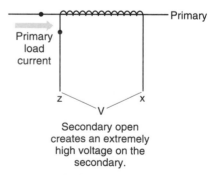

FIGURE 10.3   Open circuit on secondary of current transformer.

GROUNDING SYSTEMS OF OVER 1000 V

The subject of electrical grounding is a complex one. The following sections focus on the grounding concepts and requirements of high-voltage systems as they relate to safety. For more complete coverage of electrical grounding refer to the various ANSI/IEEE standards including IEEE Guide for Safety in AC Substation Grounding—ANSI/IEEE standard 80 and IEEE Recommended Practice for Grounding of Industrial and Commercial Power Systems—ANSI/IEEE standard 142. Chapter 4 covers grounding in detail.

What Is a Ground?

A ground is an electrically conducting connection between equipment or an electric circuit and the earth, or to some conducting body that serves in place of the earth. If properly made, the earth or conducting body and the circuit or system will all maintain the same relative voltage.

Figure 10.4*a* shows an electrical system that is not grounded. In such a situation, a voltage will exist between the ground and some of the metallic components of the power

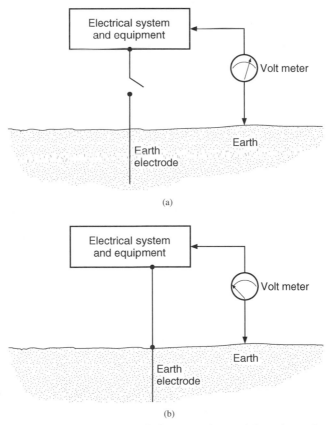

(a)

(b)

FIGURE 10.4 Grounding. (*a*) Before system is grounded, a voltage exists between the system and the earth; (*b*) when ground connection is made, no voltage exists.

FIGURE 10.5   Four voltage hazards related to system grounding. (*Courtesy Institute of Electrical and Electronic Engineers.*)

system. In Fig. 10.4*b*, the earth connection has been made. When the system is grounded, the voltage is reduced to zero between the previously energized sections of the system.

Bonding Versus Grounding

Bonding is the permanent joining of metallic parts to form a continuous, conductive path. Since the earth is generally not a good conductor, bonding is used to provide a low-impedance metallic path between all metallic parts. This, in essence, bypasses the earth and overcomes its relatively high impedance. A good grounding system is a combination of solid connections between metallic parts and the earth as well as between all metallic parts.

Voltage Hazards

Figure 10.5 illustrates the four standard voltage hazards that are associated with and relieved by proper system grounding. If a system is ungrounded and the nonenergized metallic parts become energized, the metallic parts will have a measurable voltage between themselves and ground. If the system is grounded by driving ground electrodes into the earth, current will flow from the rods into the earth. At each ground electrode the voltage will rise relative to the remote reference point. The voltage will drop off away from the electrodes and peak at each electrode. This situation creates the four types of voltage hazards as follows:

- *Step voltage.*   As a worker steps across the ground, the front foot will be at a different potential than the rear foot. This effect is caused by the voltage gradient created by the ground electrodes (Fig. 10.5). Step voltage can easily reach lethal levels. Step voltage can be mitigated by increasing the number of grounding electrodes, increasing bonding between the metal parts, and employing a grounding grid.

- *Touch voltage.*   Because of the voltage gradient, the voltage of the earth, only a short distance from the grounded metallic equipment, will be different than the equipment. Thus if a worker touches the grounded equipment, his or her feet will be at a different potential than his or her hands. This voltage can be lethal. Touch voltage is mitigated by increasing the number of grounding electrodes, increasing bonding between metal parts, and employing a grounding grid.

- *Mesh voltage.*   Mesh voltage is the worst case of touch voltage. Cause, effect, and mitigation methods are identical to those described previously under Touch voltage.

- *Transferred voltage.*   Because metal parts have a much lower impedance than the earth, the voltage drop between remote locations is lower on the metallic connections than the earth. This means that the earth may be at a significantly different voltage than the metallic connections. Transferred voltages are particularly noticeable on neutral wires, which are grounded at the service point and nowhere else. Transferred potential can be mitigated by providing the entire area with a ground grid; however, this solution is infeasible in any but the smallest systems. Transferred voltage must be mitigated by avoiding contact with conductors from remote locations and/or using rubber insulating gloves.

System Grounds

What Is a System Ground?   A system ground is the connection of one of the conductors to the earth. Such a connection is accomplished by connecting an electric wire to the selected system conductor and the grounding electrode.

Why Are Systems Grounded?   Power systems have conductors grounded for a variety of safety and operational reasons including

- Grounded systems provide sufficient short-circuit current for efficient operation of protective equipment.

- Grounded systems are less prone to transient overvoltages that can cause insulation failures.

- Grounded systems are generally more easily protected from lightning.

- Solidly grounded systems are less prone to resonant conditions that can cause equipment and insulation failures.

What Systems Must Be Grounded?   The NEC allows the grounding of permanently installed electrical utilization systems and requires the grounding of mobile electrical supply systems. If the mobile supply system is supplied by a high-voltage delta system, the NEC requires that a ground be derived using a grounding transformer. The National Electrical Safety Code requires that electric utility supply systems be grounded.

How Are Systems Grounded?   Electrical systems are grounded by connecting one of the electrical conductors to earth. The conductor chosen and the location of the ground are determined as part of the engineering design for the system. With systems that have delta transformer or generator windings, the neutral, and therefore the ground point, can be derived by using a grounding transformer. Figure 10.6 shows a grounded wye-delta transformer used to derive a system neutral and ground. Figure 10.7 shows a zigzag transformer used for the same purpose.

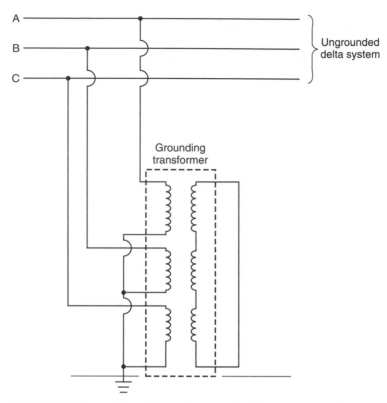

FIGURE 10.6   Grounded wye-delta transformer used to derive a system ground.

The zigzag transformer (Fig. 10.7) is the most commonly used method for deriving grounds on high-voltage systems. The windings consist of six equal parts, each designed for one-third of the line-to-line voltage; two of these parts are placed on each leg and connected as shown in Fig. 10.7. In the case of a ground fault on any line, the ground current

FIGURE 10.7   Grounding autotransformer with interconnected Y (also called zigzag windings).

TABLE 10.1 Equipment Grounding Requirements

- Must be permanent and continuous
- Must have the capacity to conduct any fault current likely to be imposed on it
- Must have sufficiently low impedance to limit the voltage to ground and to facilitate the operation of the circuit protective devices

flows equally in the three legs of the autotransformer, and the interconnection offers the minimum impedance to the flow of the single-phase fault current.

Equipment Grounds

What Is an Equipment Ground?   An *equipment ground* is an electrically conductive connection between the metallic parts of equipment and the earth. For example, transformer cases and cores are connected to the earth—this connection is called an equipment ground. Note that the metallic parts of equipment are grounded and bonded together. This bonding serves to reduce the voltage potential between all metallic parts and the earth.

Why Is Equipment Grounded?   The equipment ground is one of the most important aspects of grounding as far as safety is concerned. Workers are constantly in contact with transformer shells, raceways, conduits, switchgear frames, and all the other conductive, non–current-carrying parts. Proper equipment grounding and bonding ensures that the voltage to which workers will be subjected is kept to a minimum. Proper bonding and grounding mitigates touch and step voltages.

How Is Equipment Grounded?   The NEC requires that the path to ground from circuits, equipment, and metal enclosures must meet the conditions listed in Table 10.1. Equipment used in grounded systems is grounded by bonding the equipment grounding conductor to the grounded service conductor and the grounding electrode conductor. That is, the equipment ground is connected to the system ground. Equipment used in ungrounded systems is grounded by bonding the equipment grounding conductor to the grounding electrode conductor.

What Equipment Must Be Grounded?   All non–current-carrying metal parts of fixed, portable, and mobile high-voltage equipment and associated fences, housings, enclosures, and supporting structures shall be grounded.

The NEC does allow two exceptions to this rule: High-voltage equipment does not have to be grounded if

1. It is isolated from ground and located so as to prevent any person who can make contact with ground from contacting the metal parts when the equipment is energized

2. It is certain pole-mounted distribution equipment that is exempted

Refer to the current edition of the NEC for detailed information.

SAFETY EQUIPMENT

Overview

Since the amount of available energy in the system determines its lethal effects, and since high-voltage systems are generally very high energy systems, high-voltage systems are extremely hazardous and should be treated with the utmost respect.

The following sections describe the types of safety equipment that should be worn by persons working on or near energized, high-voltage conductors. The recommendations given are minimum recommendations. If additional or more stringent protection is desired or required, it should be worn. Refer to the tables in Chap. 3 for specific recommendations.

Hard Hats

Protective headgear for persons working on or near energized, low-voltage circuits should provide both mechanical and electrical protection. Personnel working on or near energized high-voltage circuits should be supplied with and should wear ANSI class E helmets. ANSI class G and class C helmets are not acceptable. The ANSI class G helmet is not rated or tested for use in high-voltage circuits and the ANSI class G helmet provides no electrical insulation at all. Table 10.2 summarizes the characteristics of the three ANSI classes. Note that class G and E were formerly class A and B, respectively.

Eye Protection

High-voltage systems are capable of producing extremely powerful and hazardous electric arcs and blasts; therefore, eye protection for electrical workers should provide protection against heat and optical radiation. ANSI standard Z87.1 provides a selection chart as well as a chart that illustrates the various protection options available. The selection chart is reproduced in this chapter as Table 10.3 and eye protection options are shown in Fig. 10.8.

Electrical workers should consider using eye protection that combines both heat and optical radiation protection. Table 10.3 and Fig. 10.8 show several different types of protection that will provide such protection including types B, C, D, E, and F. If arcing and molten splashing are possible, as with open-door switching or racking of circuit breakers, the type N eye protection should be employed.

TABLE 10.2 Summary of the Characteristics for ANSI Class C, G, and E Hard Hats

Class	Description	Comments
C	Intended to reduce the force of impact of falling objects. This class offers no electrical protection.	Should not be worn by personnel working on or around energized conductors of any voltage
E	Reduce the impact of falling objects and reduce danger of contact with exposed high-voltage conductors. Representative sample shells are proof-tested at 20,000 V phase to ground.	Recommended to be worn by personnel working around high- and low-voltage circuits
G	Reduce the impact of falling objects and reduce danger of contact with exposed, low-voltage conductors. Representative sample shells are proof-tested at 2200 V phase to ground.	Recommended to be worn by personnel working around only low-voltage circuits

FIGURE 10.8   Eye protection devices. Refer to Table 9.3 for selection criteria.

Arc Protection

High-voltage systems create and sustain significant electric arcs accompanied by electric blast. Workers performing routine work on or around energized, high-voltage conductors should wear flame-retardant work clothing or its equivalent with a rating of $4\frac{1}{2}$ oz/yd or more. Workers operating open-air switches or performing open-door switching procedures should wear flame-retardant flash suits or their equivalent with a rating of 10 oz/yd or more. Refer to Chap. 3 for methods that can be used to calculate required flash clothing weights.

Rubber Insulating Equipment

Rubber insulating equipment and appropriate leather protectors should be used by personnel who are working on or around energized, high-voltage conductors. Rubber gloves, rubber sleeves, rubber blankets, line hose, and other such protective devices should be employed any time workers must approach such conductors. Rubber insulating equipment used around high-voltage systems must have a rating sufficient for the voltage of the system. ANSI class 1, 2, 3, or 4 rubber insulating goods should be employed as required.

Voltage-Testing Devices

Proximity or contact testers intended for use in high-voltage circuits should be used to test circuits and to verify that they are de-energized and safe to work on. Voltage-testing

TABLE 10.3 Eye Protection Selection Chart

	Assessment SEE NOTE (1)	Protector type†	Protectors	Limitations	Not recommended
IMPACT	Chipping, grinding, machining, masonry work, riveting, and sanding	B, C, D, E, F, G, H, I, J, K, L, N	Flying fragments, objects, large chips, particles, sand, dirt, etc. → Spectacles, goggles, face shields SEE NOTES (1) (3) (5) (6) (10). For severe exposure, add N.	Protective devices do not provide unlimited protection. SEE NOTE (7).	Protectors that do not provide protection from side exposure. Filter or tinted lenses that restrict light transmittance, unless it is determined that a glare hazard exists. Refer to Optical Radiation.
HEAT	Furnace operations, pouring, casting, hot dipping, gas cutting, and welding — Hot sparks	B, C, D, E, F, G, H, I, J, K, L, *N	Face shields, goggles, spectacles *For severe exposure, add N. SEE NOTES (2) (3).	Spectacles, cup and cover type goggles do not provide unlimited facial protection.	Protectors that do not provide protection from side exposure.
	Splash from molten metals	*N	*Face shields worn over goggles H, K SEE NOTES (2) (3).	SEE NOTE (2).	
	High temperature exposure	N	Screen face shields, reflective face shields SEE NOTES (2) (3).	SEE NOTE (3).	
CHEMICAL	Acid and chemicals handling, degreasing, plating — Splash	G, H, K; *N	Goggles, eyecup and cover types *For severe exposure, add N.	Ventilation should be adequate but well protected from splash entry.	Spectacles, welding helmets, hand shields.
	Irritating mists	G	Special-purpose goggles	SEE NOTE (3).	
DUST	Woodworking, buffing, general dusty conditions — Nuisance dust	G, H, K	Goggles, eyecup and cover types	Atmospheric conditions and the restricted ventilation of the protector can cause lenses to fog. Frequent cleaning may be required.	

OPTICAL RADIATION

	Protector types[†]	Typical filter lens shade	Protectors	Comments
Welding:			SEE NOTE (9).	
Electric Arc	O, P, Q	10-14	Welding helmets or welding shields	Protection from optical radiation is directly related to filter lens density. SEE NOTE (4). Select the darkest shade that allows adequate task performance. Protectors that do not provide protection from optical radiation. SEE NOTE (4).
Welding:		SEE NOTE (9).		
Gas	J, K, L, M, N, O, P, Q	4-8	Welding goggles or welding face shield	
Cutting		3-6		
Torch brazing		3-4		
Torch soldering	B, C, D, E, F, N	1.5-3	Spectacles or welding face shield	SEE NOTE (3).
Glare	A, B	Spectacle SEE NOTES (9) (10).		Shaded or special-purpose lenses, as suitable. SEE NOTE (8).

[†] Refer to Fig. 10.8 for protector types.
NOTE: For NOTES referred to in this table, see text accompanying Fig. 9.9.
Source: Courtesy of American National Standards Institute.

10.11

FIGURE 10.9 "TIC" tracer proximity voltage sensor for use on circuits up to 35,000 V. (*Courtesy TIF Electronics.*)

devices are described in Chap. 2. An example of a high-voltage–measuring instrument is shown in Fig. 10.9. ASTM standard F 1796 is the standard defining proximity testers.

SAFETY PROCEDURES

General

The general procedures described in Chap. 3 should be used on all high-voltage circuits. The following sections describe key procedures that apply to high-voltage circuits.

Approach Distances

Although approach distances are laid out in a number of OSHA references, the NFPA 70E provides a much more useful and practical concept for approach distances. The following paragraphs identify methods that may be used to determine approach distances. Refer to Chap. 3—Figure 3.29 and Table 3.15—for reference to the terms used in the following section.

Crossing the Limited Approach Boundary. Unqual-ified workers are not allowed to cross the limited approach boundary under any circumstances. Qualified workers may cross the limited approach boundary if they are qualified to perform the work.

Requirements for Crossing the Restricted Approach Boundary. In order to cross the restricted approach boundary, the following criteria must be met:

- The worker must be qualified to do the work.
- There must be a plan in place that is documented and approved by the employer.
- The worker must be certain that no part of the body crosses the restricted approach boundary.
- The worker must work to minimize the risk that may be caused by inadvertent movement by keeping as much of the body out of the restricted space as possible. Allow only protected body parts to enter the restricted space as necessary to complete the work.
- Personal protective equipment must be used appropriate for the hazards of the exposed energized conductor.

Requirements for Crossing the Prohibited Approach Boundary. NFPA 70E considers crossing the prohibited approach boundary to be the same as working on or contacting an energized conductor. To cross into the prohibited space, the following requirements must be met:

- The worker must have specified training required to work on energized conductors or circuit parts.
- There must be a plan in place that is documented and approved by the employer.
- A complete risk analysis must be performed.
- Authorized management must review and approve the plan and the risk analysis.
- Personal protective equipment must be used appropriate for the hazards of the exposed energized conductor.

Voltage Measurement

The voltage-measurement techniques defined in Chap. 3 should be employed on circuits of all voltages.

Locking and Tagging

Lockout-tagout and energy control procedures apply to circuits of all voltage levels. Refer to Chap. 3 for detailed discussions of lockout-tagout procedures.

Closing Protective Devices After Operation

Workers should never reclose any protective device after it has operated, until it has been determined that it is safe to do so. Several criteria may be used to determine whether it is safe to reclose the protective device (Table 10.4).

Other conditions may or may not indicate a safe reclosing situation. ANSI/IEEE standard 141 defines low-voltage systems as those that are less than 1000 V, medium-voltage systems as those equal to or greater than 1000 V and less than 100,000 V, and high-voltage systems as those equal to or greater than 100,000 V. To simplify this notation, this chapter will refer to all systems of 1000 V or higher as *high-voltage* systems.

TABLE 10.4 Typical Situations That May Allow Reclosing of a Protective Device That Has Operated

- The faulted section of the system is found and repaired.
- The nature of the protective device makes it clear that no hazard is present. For example, if the device that operated is an overload type of device, it may be safe to reclose.
- The reclosing operation can be made in such a way that the workers are not exposed to additional hazard. For example, if the reclosing operation can be made by remote control, and if all personnel are kept away from all parts of the circuit, it may be reclosed.

CHAPTER 11
HUMAN FACTORS IN ELECTRICAL SAFETY

INTRODUCTION

In this chapter, the topic of human factors is reviewed in the context of electrical safety. Background information is followed by examples for power systems. Brief discussion is also presented on the topics of visualization and cognitive ergonomics.

Background

The topic of human factors evolves from the obvious statement that, "People are not machines." In the early 19th century, during the early periods of industrialization, a persistent question was how to optimize the placement of people in "machine systems." Pre-dating electrical power, when jobs involved assembly line operations with repetitive tasks requiring uniformity and efficiency of physical movements between operations, materials, and employees, studies of "time-and-motion" and human "anthropomorphics," or "average body part sizes" were used by management to figure out problems, like "how far can an average worker reach to push or pull this piece" on the assembly line.

In the early 20th century, as electrification enabled industrialization to move beyond manufacturing across all economic sectors, the complexity of machine systems increased. More than one process could be pulled into a work environment, leading to multiple operations situated in geographic proximity, running concurrently, and producing output at high speeds.

With electrification, increasing industrial complexity, and faster production rates, the demands on people working with the machine systems grew. As suggested by Rasmussen[1] at this stage, human factors more commonly studied measures of human physical abilities along with measures of physical reliability. In this context, "knob-and-dial" ergonomics was a research focus, for example, identifying the correct size of knob for an employee to easily grasp doing a turning task, and measuring how often the knob would be incorrectly versus correctly turned.

Then, in the 1930s, Rasmussen notes that as automation was introduced into the chemical process industries, the link between machines and people often became indirect. The interface of automatic control engineering made the wide reach of individual employees possible over large scale and significantly hazardous production.

As production failures occurred, interest in human factors extended from human "physical" reliability to include the reliability people might show in understanding process information and executing successful decisions. This specialty later became known as *cognitive ergonomics*.

Then "human error" research developed as an aspect of human factors studies, with the goal being to break down or untangle causes of human physical and/or cognitive malfunction from machine or process malfunction in post-accident studies. In anticipation of installation design and development, human error probability, or human reliability assessment (HRA) also came into human factors consideration as part of operational engineering analysis.

In the mid-20th century, military experience in World War II aviation and the post-World War II growth in civilian aviation added more information about how humans and engineered systems could interact with potentially tragic results of plane crashes. By the early 1970s, generally accepted ideas about how to evaluate the fit between work and workers were maturing into an accepted professional approach within the industrial engineering community.

Concurrently, within the electrical power industry, the post-World War II development of the civilian nuclear power industry resulted in an active transfer of human factor's ideas into the management of civilian nuclear power operations. This development became more substantial following two major tragedies: the Unit 2 scram at the Three Mile Island Power Station (TMI-2) in Pennsylvania on March 28, 1979; and the explosion of Unit 4 of the Chernobyl Nuclear Power Station in the Ukrainian former Soviet Union on April 26, 1986.[2]

With TMI-2 and Chernobyl, nations learned that human error in complex operational environments producing electrical power could potentially result in global disaster. Investigations identified human factors in the causal chain of events leading to these disasters. This learning created additional focus on basic concepts of human performance.

Continuing to the present, the topics of human factors have moved beyond "human machine interactions"[3] to "human-systems interfaces" or HIS[4]. The term *human factors* may have two meanings. Human factors may refer to various traits or "elements of the human"[2] as individuals which should be considered for safe and effective results from engineered systems. Or, the term may mean the applied science technology relating fundamental human sciences (like anatomy, physiology, neuro-psychology) to industrial systems.[2]

To illustrate these definitions, consider that electrical work demands intense integration of visual and spatial information to successfully complete tasks. Anything that inhibits an employee's vision creates risk for an electrical event. Considering the two definitions of human factors applied to vision, as a human factor relating to the individual, vision may be understood as limited to the measurement of visual acuity (or accuracy of how well an employee sees a wall chart) to ensure that they have adequate vision ability to read a computer screen or see a stoplight; or the factor may be as complicated as the engineering advancement of vision assistance devices, such as these examples:

- Infrared cameras installed in vehicle dashboards for night driving
- Infrared intrusion detection systems
- Infrared monitors used in heat tracing as a safety monitoring strategy to identify excessive ohmic heating of installed conductors

For an individual, vision is a critically important human factor. About 80 percent of information about machines and systems comes to employees via their vision.[5] Note visual acuities go down as the complexity of the visual target increases. In other words, when there is more to look at, it is harder to see. Multiple physical factors influence visual acuity, including:

- Illumination
- Contrast
- Time of exposure to the visual target
- Color of the visual target and the target's background

The smallest detectable threshold for vision is 10^{-6} milliliter (mL). Vision ability is decreased with vibration, hypoxia or low oxygen, and motion of the visual target.

POWER SYSTEMS AND HUMAN FACTORS

As the previous comments suggest, the power industry has significantly contributed to understanding about human factors. Here further examples of human factors in the electrical industry are presented. In Table 11.1, selected workplace environments are identified.

The generation, transmission, distribution, design, installation, operation, and maintenance of electrical power happens in diverse situations. Given the diversity in workplaces, the human factors of relevance to electrical safety depend in part on where in the power system the employee may be working. An employee assigned to a control room at a nuclear power station can reasonably be expected to confront demands in his/her work that differ compared to an employee wiring a house under construction or doing maintenance in a cogeneration facility. So, it follows then that the human factors in a control room can be expected to differ from those at the construction site.

TABLE 11.1 Power Systems Environments and Human Factors Examples

Workplace	Human factors considerations

Generation	Dust, fumes, noise in fossil fuel environments
	Common features to control rooms,[11] including
	    Compact workstations using visual displays
	    Large overview displays
	    Increased cognitive workloads as staffing changes
	    Information multiplicity
	Virtual workspaces[4] with
	    Serial access to information & controls
	    More time spent on secondary tasks

Transmission	Live line work,[12] with
	    Helicopter approaches at high elevations
	    Moving parts (Helicopter rotars)
	    Required calculations: minimum approach distance
	Placing Person with Tools in Air Gap[13]

Distribution	Voltage protection personal equipment, including
	    Rubber goods
	    Extended tool handles
	Recognition of minimal power line approach distances
	At or above shoulder work requiring stressful postures
	Work in vaults or confined space requiring respirators
	Repetive motion and lifting heavy loads[14]

Construction	Crouching, kneeling, or reaching in various spaces Vibration in powered equipment Environmental temperature extremes Heavy equipment operation, with noise, moving parts Potential contact with power lines above and below

The influence of the environment on human factors is not isolated. Relating Hawkins'[6] "SHEL" illustration of this concept, Kawano[2] describes the acronym as:

- S – for software, including for example, computer software, paper documents or instructions, permits, or procedures
- H – for hardware, or the engineered or constructed aspects of the environment
- E – for the environment or ambient conditions, such as geography, meteorology, humidity, or altitude
- L – for "liveware," or the people involved directly or indirectly in the work situation

Kawano adds an "m" to Hawkins' SHEL acronym, to include *managing systems* as a dynamic and represented part of the model. As suggested by Hawkins and amended by Kawano, the m-SHEL model provides a step towards an integrated systems approach to human factors. The model offers another strategy for thinking about electrical safety so that the programmatic focus goes beyond the individual to include the technologies, engineering, interpersonal, and environmental contributions to the individual's experience in the system of production where they work.

Visualization

Rasmussen[1] has suggested that automation relieved employees of repetitive manual tasks; consequently, work with more decision-making or trouble-shooting content became common in the industrial setting. Today, information technology, including the use of computer-based approaches, has placed volumes of data at the center of employee tasks.

Tory and Möller[7] describe the methods available with computer visualization and graphics which can support perception and cognition, or the human abilities to sense (via eyes, ears, nose, taste, touch) and think. Table 11.2 gives electrical safety examples where these methods can serve as cognitive aids.

Cognitive Ergonomics

The early 21st century is notable for the promise of new horizons in research advancing cognitive ergonomics to fully benefit the industrial workplace. As brain science maps the

TABLE 11.2   Electrical Safety Examples of Cognitive Aids with Computer Visualization Technologies

Cognitive Challenge	Visualization method
Multiple permits per job step with multiple steps in a job	Reducing search via data grouping or structure
Monitoring for breaker trips, power surges, or faults	Perceptual monitoring using pre-attentive visual characteristics
Switching procedures in power outage or routine maintenance	Enhancing data recognition, abstraction, and aggregation

Note: Visualization methods are based on Table 1 of Tory and Möller's report,[7] which is adapted in part from work by Card, *et al. in Information Visualization: Using Vision to Think.* San Francisco: Morgan Kaufmann Publishers, 1999.

terrain of thought and mind-body interactions using scanning technologies like magnetic resonance imaging (MRI), positron emission tomography (PET), and photon emission computerized tomography (SPECT), data is accumulating to explain how people think in different scenarios.

A key insight from this research is that context is hugely influential over thought. Using the metaphor of the brain as a computer, one way to appreciate the role of context is to suggest that no person comes to his/her work with a "blank hard drive or screen." Rather people do their jobs with the advantage and disadvantage of what they see, hear, smell, and touch in their surrounding; what they believe; and how they physically feel. This information loads the "drive and screen" occupying space alongside of where "the programs run" or thought happens.

Cognitive ergonomics specifically addresses how to modify what occupies an employee's "hard drive and screen" when work is being done, to improve the way thought unfolds, and thereby, improve decisions.

Context can be appreciated not simply as a physical background to the employee. Context can also be in the form of background beliefs or work myths. The dictionary defines a myth as "an old traditional story or legend, especially one concerning fabulous or supernatural beings, giving expression to early beliefs, aspirations, and perceptions of a people and serving to explain natural phenomena or the origins of a people."[8]

Following are examples of myths which can affect electrical safety. Exploring these myths offers promise to improve electrical safety by reducing the chance that an employee will act based on an inaccurate background belief.

"I'm experienced, so I won't get injured."   This belief assumes that experiences protect against injury and death. However, as the statistics shown in Chap. 8 suggest, the employee who is most at risk of an electrical event is between the ages of 25 to 45 years and with accident-free years on the job. To the extent that an employee's experience dulls his/her awareness of the distinctive features present in the job he/she is immediately doing, his/her experience may create a false context for thinking about situational facts like:

- Task electrical configuration
- Personal protection and barrier needs
- Required resources (like people, equipment, and time) to complete the job

"Electrical accidents happen when an employee isn't paying attention."   This mythic belief is based on a logic that goes as follows: As long as an employee does pay attention, no accidents will occur; therefore, if an accident happens, an employee must have been inattentive.

As an explanation for electrical events, inattention is assumed by the comments like:

- "He wandered off in his thinking."
- "He was daydreaming on the job."
- "He was worried about something else."

However, through many debriefing of electrical accidents, engineers have come to appreciate modern electrical work unfolds in highly complicated situations. If an electrical safety failure occurs, typically more than one thing goes wrong. Multiple system faults or personnel errors contribute to the event. This understanding suggests that attention alone is not *sufficient* to prevent an electrical accident. Given the embedded complexity in the power system, multiple actions are required to preserve safety. How to sharpen or focus the actions of human attention is a research target of cognitive ergonomics.

Related to attention is the ability to hear. If an employee cannot hear instructions he is being given verbally, that employee can be expected to have difficulty in his/her focus on the instructions or attention to the verbal guidance given. Hearing as a limited human factor depends on sound. Sound pressures needed for hearing are a function of the material or media (for example, air) through which the sound or acoustic waves are propagating.[5] The threshold for hearing in the frequency range of the spoken word, from about 1,000 to 5,000 Hz, is about 20 μPascals (2.9×10^{-9} psi).

According to Chapanis, there are multiple aspects to a listener's ability to detect sound, including:

- The listener's age
- The listener's history of past or ongoing noise exposure
- Whether the listener is using one ear or two to hear, with the use on one ear (mono-aural listening) requiring 3 decibels (dB) more sound pressure
- The sound's acoustic frequency, expressed in hertz (Hz)
- The presence of competing sounds or masking
- The complexity of the bandwidth tones
- The sound's duration, with durations less than 200 mili-seconds (ms) requiring increased intensity (that is, as acoustic signal duration is halved, the intensity of the signal must double to be audible).

Returning to the M-SHEL model of human factors, the interaction of "liveware" or people depends on successful communication. To the extent that employees can not adequately communicate because they can not hear each other talk, safety is jeopardized. So the physical ability or individual human factor of hearing can affect the ability to communicate, which can affect attention, which is influenced in part by background belief or context.

"As long as I don't touch an electrical source, I won't get shocked." This belief correctly identifies the need to avoid exposure to shock hazard. However, mechanical contact is *not* necessary for an employee to be shocked. Returning to the M-SHEL model, the mythic belief here under-rates the role of environment in its influence over the individual human factors of body size and positioning in relationship to electrical conductivity, resistivity, and impedance.

Electricity is conducted along copper wires in power generation, transmission, and distribution. Depending on the current, when an employee's body comes sufficiently near to an electrical source, the charge that is carried by electrons in copper wire may be converted through an electrochemical reaction to charge conducted by the ions in the human body. Employee size and positioning, as well as meteorological conditions and geography play a role in whether electricity can cross an air gap by arcing, and flow through or around the employee. The resulting shock can be destructive, even fatal, if adequate personal protection is not being used.

During an electrical hazard exposure, human factors studies suggest how little opportunity the employee may be left with to react. Reaction times for responses to stimuli (like buzzing or light shocks), "word" information, or other prompts have been studied as a human factor. These times are studied in laboratory and "real world" scenarios. Generally, reaction times vary from person to person, and sometimes between successive trials by the same person.

Environmental stress, such as heat exhaustion, altitude sickness, or hyperbaric conditions (such as work in mining, undersea, or in certain medical facilities) can change mental efficiency and lengthen reactions. Responses are relatively slower when language is involved. For example,

- A printed word is registered in the reader's brain in about an eighth of a second (0.125 s)
- A spoken word is accessed by a listener's brain in about a fifth of a second (0.200 s), before the speaker has finished pronouncing the word
- The brain takes about a quarter second (0.250 s) to find a word to name an object, and another quarter of a second (0.250 s) to program the mouth and tongue to pronounce the name (total: 0.500 s).[9]

Psychologists have studied attention and voluntary action in responses, finding:

- An average physical movement (motor) response time of >0.600 s in healthy people tested; and
- An average non-motor response time of >1.050 s in adults asked to *say* verbs for printouts of words shown as *visually presented* nouns.

In Chap. 1, the thresholds for neuromuscular responses in response to electrical current were reviewed. In comparison to electrical responses, motor and non-motor responses to language stimuli generally take much longer.

The implication of this human performance limitation can be shown with this example: If a coworker needs 250 ms to process a spoken word like STOP or HELP, while an electrical incident like an arcing fault is unfolding, say in less than 6 cycles, there is going to be a mismatch between

- The amount of time the coworker needs to detect, process, and respond to the fault; and
- The amount of time where in maximum risk is present from the fault.

This time mismatch is even more pronounced when there is a need to sequence perception, thought, and response in a rapid amount of time, as in to:

- Sense physical stimuli
- Perceive and process information; and
- Act.

Age influences reaction time, with times slower for those older than 60 and younger than 15 years. Certain situational conditions generally slow reaction time and increase its variability for a person, including

- Sleep deprivation
- Fatigue
- Time of day
- Environmental extremes
- Alcohol or other drug use
- Medical problems
- Nutritional status

Through a number of research studies it has been found that on the job reaction time may be many time longer than what is found in a lab set up, especially for specific kinds of tasks requiring physical exertion and mental concentration,[10] such as is found in electrical work.

SUMMARY

Future protective human factors innovations in electrical safety will depend in part on maximizing opportunities to integrate the demands of power systems, environments, and advances in human factors studies, visualization research and cognitive ergonomics insights. Until then, employees must be prepared with knowledge of the limits of their physical and mental performance in the complex and potentially hazardous tasks required to deliver electrical power.

REFERENCES

1. J. Rasmussen, "Human Factors in High Risk Systems," IEEE Fourth Conference on Human Factors and Power Plants, Conference Record, Monterey, CA, pp. 43–48, 1988.
2. R. Kawano, Steps Toward the Realization of "Human-Centered Systems"-An Overview of the Human Factors Activities at TEPCO, In Global Perspectives of Human Factors in Power Generation, IEEE Sixth Conference on Human Factors and Power Plants, Conference Proceedings, Orlando, pp. 13/27–13/32, 1997.
3. W.H. Hawkins, "Where Does Human Factors Fit in R&D Organizations?" *IEEE Aerospace and Electronic Systems Magazine,* vol. 5, no. 9, pp. 31–33, September, 1990.
4. J. O'Hara, W.F. Stubler, J. Kramer, "Human Factors Considerations in Control Room Modernization: Trends and Personnel Performance Issues, In Global Perspectives of Human Factors in Power Generation," IEEE Sixth Conference on Human Factors and Power Plants, Conference Proceedings, Orlando, pp. 4/7–4/10, 1997.
5. A. Chapanis, *Human factors in systems engineering,* John Wiley & Sons, Inc., New york, pp. 186; 218–220; and 228, 1996.
6. F.H. Hawkins, *Human Factors in Flight,* Ashgate Publishing Limited Gower House, Hants, England, 1987.
7. M. Tory, and T. Mőller, "Human Factors in Visualization Research," *IEEE Transactions on Visualization and Computer Graphics,* vol. 10, no. 1, pp. 72–84, January/February, 2004.
8. *New Lexicon Webster's Dictionary of the English Language,* Lexicon Publications, Inc., New York, 1989, p. 660.
9. S. Pinker, *Words and Rules,* HarperCollins, New York, p. 335, 1999.
10. A.N. Beare, R.E. Dorris, and E.J. Kozinsky, "Response Times of Nuclear Plant Operations: Comparison of Field and Simulator Data." In *Proceedings of the Human Factors Society 26th Annual Meeting,* Human Factors Society, Santa Monica, CA, pp. 669–673, 1982.
11. J.W. Lee, I.S. Oh, H.C. Lee, Y.H. Lee, and B.S. Sim, "Human Factors Researches in KAERI for Nuclear Power Plants, In Global Perspectives of Human Factors in Power Generation," IEEE Sixth Conference on Human factors and Power Plants, Conference Proceedings, Orlando, pp. 13/11–13/16, 1997.
12. IEEE Task Force 15.07.05.05, "Recommendedd Practices for Helicopter Bonding Procedures for Live Line Work, *IEEE Transactions on Power Delivery,* vol. 15, no. 1, 333–349, January, 2000.
13. ESMOL Subcommittee 15.07, "Safety Considerations When Placing a Person With Tools in an Air Gap to Change Porcelain and Glass Insulators on Transmission Systems of 345 kV and Above, Using Ladder and Aerial Lift Methods," *IEEE Transactions on Power Delivery,* vol. 17, no. 3, pp. 805–808, July, 2002.
14. OSHA, "Ergonomics Solutions for Electrical Contractors," URL: www.osha.gov/SLTC/etools/electricalcontractors/index.html

CHAPTER 12
SAFETY MANAGEMENT AND ORGANIZATIONAL STRUCTURE

INTRODUCTION

An electrical safety program will only be effective if management makes a strong commitment to. This chapter will develop some of the key management concepts and procedures that must be present for a safety program to work. Of course, electrical safety is only part of an overall safety program; consequently, much of the material in this chapter is applicable to the entire safety effort.

The procedures introduced in this chapter should be applicable to all types of electrical installations. However, the effectiveness of any specific program must be determined by ongoing evaluations.

Safety organizations should be responsible to the very highest management levels and generally should not report to operations. Safety-related decisions should not be made by personnel with direct, bottom-line responsibility.

Additionally, the decisions made by legal counsel should be closely evaluated in terms of their effect on personnel safety. Many well-meaning attorneys and/or senior level managers are required to make decisions that will maximize shareholder returns and/or limit corporate liabilities. Such decisions are, unfortunately, not always consistent with long-term worker safety.

Problems also may be introduced by labor organizations in their attempt to secure the best overall package for their members. Care should be exercised to avoid using safety as a "bargaining chip." This caution applies to labor and management.

ELECTRICAL SAFETY PROGRAM STRUCTURE

Figure 12.1 is a suggested design for the overall structure of a company electrical safety program. Of course, such a structure must be integrated into the overall safety program; however, the unique needs of the electrical safety program should be included in the design in a manner similar to the one shown.

Each of the various elements of the design are described in the following sections.

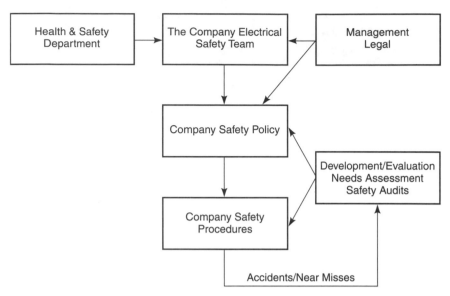

FIGURE 12.1   The electrical safety program structure.

ELECTRICAL SAFETY PROGRAM DEVELOPMENT

An electrical safety program is composed of a variety of procedures, techniques, rules, and methods. Each of these individual items must be developed independently; the sum of them then adds up to the overall safety program. The development of an entire program becomes the accumulation of all the individual procedures.

This does not mean that the development of a safety program is a hodgepodge effort. On the contrary, the whole procedure should be viewed as an engineering design problem. The program should be put together by corporate electrical safety team composed of management, safety, and technical personnel.

Some companies take regulatory and consensus standards such as the OSHA Electrical Safety Related Work Practices rule and the NFPA 70E and apply them almost verbatim for their safety policies, procedures, and/or rules. This is a very poor practice. Even the NFPA 70E standard should be looked at only as a set of minimum requirements. In addition to regulatory requirements, company policies and procedures must include specific local requirements and must be developed in a way that is consistent with the facility culture.

The Company Electrical Safety Team

Introduction.   One of the best overall safety concepts of the last part of the twentieth century was that of the worker *team.* Nowhere is the team of more use than in the safety arena. The following paragraphs describe the concept of the electrical safety team and provide enough information to serve as a template for the setup of such a body.

Note that the company health and safety department (HSD) has the ultimate responsibility for the safety program. The HSD should work closely with the team and review all procedures and policies developed by the electrical safety team. As described in the following, the HSD should have a representative on the team.

Structure.   Generally, the safety team should include the following representatives/members:

• Electrical workers	An electrical worker should serve as the chairperson of the team. In addition to the chairperson, sufficient members should be included from the workforce to ensure that the workers have a significant say in their overall safety program. Working supervisors should be eligible for these positions; however, supervision should never dominate the team.
• Health and safety professionals	At least one health and safety professional should be included on the team to advise and assist in the areas where his or her expertise apply.
• Management	A member of management should be present as an advisor. This person can direct the team with respect to company policies to avoid the team conflicting with company directives. The management representative should be as high a level as possible. The management representative should always be willing to consider modification of conflicting company policies in the interest of improved safety.
• Legal representation	Company counsel should be represented on the team; however, they should be there strictly in an advisory capacity.

Responsibilities.   The company electrical safety team (CEST) should have the following responsibilities and authorities:

• The CEST should have overall responsibility for the development, implementation, evaluation, and modification of the company electrical safety procedures. Either directly or through delegation, the CEST should develop the entire program, working closely with the health and safety professionals.

• The degree of authority vested in the CEST must be a matter of individual company policy. Generally, the CEST should have the maximum authority allowable under existing company structure.

• If the company has multiple locations, the CEST should appoint or sponsor elections for employee electrical safety teams (EEST) at each site. The local teams will participate in local accident reviews, evaluate procedures, and determine the best way to apply them locally, counsel employees, and implement and plan safety meetings.

• The CEST should have representation at the management level for the purpose of participating in the development of the company safety policy.

Employee Electrical Safety Teams.   Employee electrical safety teams should be put in place to perform the actual fieldwork and legwork required by the CEST. The EEST will participate in accident investigation, program development, and any other activities deemed necessary. Some companies appoint permanent employee electrical safety teams as part of their ongoing corporate structure.

Company Safety Policy

All safety programs should be underwritten by a company and/or departmental safety policy. Although specific policy statements must vary from industry to industry, all policies should contain the following key statements:

1. The company is committed to safe work practices.
2. At a minimum, all company safety policies and procedures shall comply with applicable federal, state, and local standards as well as recognized consensus standards.
3. Safety is the premier consideration in performing work.
4. Employees will be required to follow all company safety procedures.
5. If a job cannot be safely done, it need not be done.
6. Each individual employee is uniquely responsible for his or her own personal safety.
7. The cooperation of all personnel will be required to sustain the safety program.

Assessing the Need

The development or revision of an electrical safety program should begin with an evaluation of any existing programs. This initial survey should closely examine and catalog the number and types of electrical accidents. Investigators should also be creative in their analysis. That is, they should identify potential hazards as well as demonstrated ones.

One of the most effective ways to catalog safety hazards is to perform a safety audit. Safety audits are discussed later in this chapter.

Problems and Solutions

Based on the results of the needs assessment, accidents and potential problem areas should be cataloged into *cause* categories. The specific categories selected should be chosen to fit the industry. Good starting points are the seven categories used by the Occupational Safety and Health Administration in the development of the Safety-Related Work Practices Rule. These categories, illustrated in Table 12.1, can be used for most electrical accidents. The

TABLE 12.1 Categories to Classify Accidents or Potential Accidents

1. Use of equipment or material too close to exposed energized lines
    a. Vehicles (e.g., cranes and dumptrucks)
    b. Other mechanical equipment (e.g., augers and derricks)
    c. Tools and materials (e.g., ladders and tree limbs)
2. Failure to use electrical protective equipment
3. Assuming an unsafe position
4. Failure to de-energize (and lockout-tagout) equipment
5. Use of visibly defective electric equipment
6. Blind reaching, drilling, digging, etc.
7. No unsafe work practice or not enough information to classify

specific installation should develop additional categories or subcategories as required so that very few accidents fall into category 7.

Identification of the safety problems will inevitably lead to solution concepts. These solutions should take the form of specific plans and programs that can be implemented to the new or existing electrical safety program.

Program Implementation

After the analysis phase, the solutions can be integrated into the facility's safety program. The method used here will depend upon the facility; however, a reasonable starting point would be the development of an energy control program similar to that described in Chap. 3.

Examples

The following examples (Tables 12.2 through 12.5) will illustrate the steps just outlined. These examples are drawn from actual industry experiences and illustrate the specific kinds of problems that workers may face every day.

TABLE 12.2   Example of Fuse Removal Safety Procedure Development

Need assessment	In the last 5 years, 15 employees have been shocked while removing and/or replacing low-voltage cartridge fuses. Five of the injuries were extremely serious. No deaths have occurred.
Problem(s)	Review of the problems indicate that employees must, from time to time, work on such circuits when they are energized. Further investigation shows that employees are using channel lock type pliers to remove the fuses. Employees are not using rubber insulating gloves during such "hot" work.
Solutions	To resolve this problem, employees will be supplied with insulated tools that are specifically designed for the removal and replacement of fuses. Employees are already supplied with rubber insulating gloves.
Implementation	A procedure is added to the employee safety procedures manual, which specifies the following:
	When removing or replacing fuses, the bus which feeds them shall first be de-energized unless doing so introduces additional hazard or unless de-energizing requires a major plant outage.* Removing energized fuses is classified as hot work. Only qualified, trained personnel shall be allowed to remove and replace energized fuses.
	Low-voltage cartridge type fuses shall be removed using only approved, insulated fuse pullers.
	Employees shall wear at least the minimum required safety equipment when removing or replacing energized low-voltage, cartridge type fuses. Minimum safety equipment shall include hard hats, safety glasses or goggles, and electrical insulating rubber gloves approved for the voltage level encountered.
	The new procedure should be introduced at an employee safety training meeting. The complete procedure should be demonstrated and employees should be checked to make certain that they understand it.

*See Chap. 3 for a specific procedure to assess additional hazards and/or outages.

TABLE 12.3   Example of Flash Suit Safety Procedure Development

Need assessment	Companies of similar size and business, with comparably sized and arranged electrical systems have had problems with severe flash injuries. These same companies have implemented flash suit use during certain phases of their operations, and have exhibited a marked decrease in the number of injuries.
Problem(s)	Employees must, from time to time, perform open-door circuit breaker switching, remove motor starters from cubicles, and operate open-air disconnect switches. This exposes them to more than a normal electrical arc hazard.
Solutions	To resolve this problem, the company will supply and require the use of flame-retardant flash suits.
Implementation	A procedure is added to the employee safety procedures manual that specifies that employees shall be required to wear flame-retardant flash suits when performing open-door switching, motor starter removal or installation, or open-air switch operation. This requirement applies to all medium-voltage circuits and all low-voltage circuits with ampacities in excess of 100 A. This requirement is in addition to other safety equipment including hard hats, eye protection, and rubber gloves with protective leather covers.
	The new procedure should be introduced at an employee safety training meeting. The complete procedure should be demonstrated and employees should be checked to make certain that they understand it.

Electric Shock While Replacing Low-Voltage Fuses.   Assume that your electrical safety audit has identified several instances where employees were shocked during the removal and/or installation of 120-V fuses. Table 12.2 shows the problem, solution, and implementation required to provide a safer work environment.

Protection for Arc Injury During Switching Operations.   Not all safety procedures are developed based on local facility experience. In this example, we assume that the local

TABLE 12.4   Example of Recognition of Exposed Live Parts Procedure Development

Need assessment	An employee was working in a 480-V cabinet. His supervisor had previously told him that the cabinet was de-energized. The employee had looked at the cabinet and had measured the *obvious* energized locations. They were dead.
Problem(s)	There were exposed fasteners in the front of the cabinet that were not de-energized. They were not obvious and the employee missed them. The employee believed his supervisor and did not inspect the cabinet closely enough. The employee did not perform a thorough pre-work inspection to check for energized parts. He made several assumptions.
Solutions	Employees must be retrained to closely inspect for exposed energized parts.
Implementations	At a safety meeting, the entire incident was discussed and the problem brought to the forefront. All employees were reminded to inspect and check for energized parts no matter how certain they are that the system is safe.
	A training program was implemented in which the various specific pieces of equipment are identified and employees learn where all of the possible exposed parts are located. A formal procedure was added to require that employees carefully inspect equipment before they work on it.

TABLE 12.5 Example of Voltage Measurement Procedure Development

Need assessment	In addition to the problem discussed in the previous example (Table 12.4), other employees have been observed performing voltage measurements incorrectly.
Problem(s)	OSHA and other regulations require that voltage measurements must be made prior to working on a piece of de-energized equipment. All qualified electrical workers must be thoroughly trained in the proper and safe methods of voltage measurement.
Solutions	Employees are to be retrained in the proper methods of voltage measurement.
Implementations	The correct methods of voltage measurement were discussed and reinforced in employee safety meetings. Specific, hands-on training programs were implemented and all employees were tested to make certain that they were thoroughly familiar with the procedures.
	Additionally, the methods of selecting and inspecting measuring instruments was reinforced.

safety personnel have reviewed other industrial safety programs with respect to flash protection. They have discovered that other companies have experienced severe injuries during routine switching and racking operations. Furthermore, they discover that other companies have implemented successful programs that involve the use of flash suits. The number of flash-related injuries has dropped at these other companies. Table 12.3 was developed based on this analysis.

Recognizing Exposed Energized Parts. Unfortunately, many procedures are developed and modified based on an accident or injury. The example shown in Table 12.4 is one such example. In this example, the employee was qualified and trained; he simply made a mistake. In spite of this, a significant amount of effort must go into the continuous retraining of personnel.

Voltage Measurement. A significant part of the electrical worker's job is the measurement of electrical circuits to make certain that they are, in fact, de-energized. In the previous example, the worker did not recognize the exposed energized parts. Even if they are recognized, they must be measured to be certain of their condition. Table 12.5 is an example of voltage measurement procedure development.

Company Safety Procedures

The actual development and structure of company electrical safety procedures are specific to each company. The methods and format described in Chaps. 2 and 3 of this handbook should be referred to as a basis.

Also, the OSHA procedures have suggestions and formats for such plans.

Results Assessment

Few, if any, procedures will endure forever without modification. Personnel replacements, equipment changes, enhanced operating experience, and new safety innovations will require that any safety program be periodically reviewed and updated. The following criteria should be used to evaluate the need to revise an existing safety procedure:

1. Accidents and near misses
2. Employee suggestions
3. Employee electrical safety committee recommendations (see next section)
4. Changes in regulatory or conseusus standards

EMPLOYEE ELECTRICAL SAFETY TEAMS

Reason

Electrical safety programs that are developed and/or administered without employee input will be ineffective. Employees know their jobs better than anyone. They understand those procedures with which they are comfortable and those which seem wrong. An electrical safety program that is developed without complete employee involvement has very little chance of success.

Method

Employee safety teams provide the best method for ensuring employee participation in a safety program. Such teams should be composed of a minimum of three employees chosen by their peers. The team should have regularly scheduled meetings on a monthly basis—more often if required by accidents or emergency conditions. The meetings should be held on company time.

Employee safety teams should be intimately involved with the entire safety program from initial design to ongoing results evaluations. They should be officially sanctioned by management and should have direct access to executive management. A limited, but useful, travel budget should be made available to the team to allow them to attend seminars and visit other facilities to study other safety programs. Although the specific responsibilities of such a team may vary from one company to another, the following may be used as guidelines.

Safety Program Development. Because of their familiarity with their jobs, employees are in a unique position to evaluate safety needs. What appears to be a perfectly safe procedure to a layperson may be obviously unsafe to a skilled, qualified worker. Furthermore, employees will often talk to their peers more freely than to supervisors or management.

An employee safety team will be able to identify problems and solutions that outsiders might miss. Because of this, the safety team should be directly involved with the safety program development and/or modification described earlier in this chapter.

Safety Meetings. Employees should control all or at least part of their safety meetings. Legislative or regulatory requirements should be scheduled by management and safety personnel; however, other safety-related presentations should be generated by employees and administrated through the employee safety team.

Accident Investigation. The team should be involved in the investigation of accidents. The best way to ensure this involvement is to have the employee safety team appoint a member to serve on the accident investigation team. Accident investigation is a very demanding science and should not be performed by those unfamiliar with the information presented in Chap. 3.

Employee Training. The safety team should be allowed to review and recommend employee safety-related training. Employees who have been to outside safety training courses should be interviewed by the safety team. The team should then issue an annual recommendation for additional or modified training. This report should be considered a primary source of information when training budgets and schedules are produced.

SAFETY MEETINGS

The venerable safety meeting is one of the most universal of all safety vehicles. This one brief period combines training, program review, new concept presentation, and hazard notification all in one. The specific design and implementation of a safety meeting is necessarily dependent on the organization arrangement of the company. The following items highlight some of the key imperatives in the structure of a safety meeting.

Who Attends

Safety meetings should be attended by all personnel who are affected by the safety topics that will be discussed. Those who should attend electrical safety meetings include the electricians, electrical technicians, electronics technicians, electrical supervisors, electrical department management, safety personnel with electrical responsibility, and anyone else who may be exposed to any of the electrical hazards.

Safety meetings should be chaired by employees, preferably members of the employee safety team. Remember that the company's attitude toward proper safety can make an enormous difference in employee attitudes.

What Material Should Be Covered

The safety meeting should have a standard, but flexible, schedule. Table 12.6 shows the type of schedule that should be included. Flexibility is the key word. One of the most frequent complaints about safety meetings is that they always contain the same old messages, are presented the same old way, and use the same old films or videos. The program should be flexible and new. Employees should be consulted as to topics that they feel may be relevant.

TABLE 12.6 Typical Safety Meeting Schedule

- Welcome and Review of Topics to Be Covered
- Accident Report
    Summary of Accidents
    Analysis and Lessons Learned
    Procedure/Policy Modifications if Any
- Training Schedules and/or Safety Awards
- Meeting Topic Presentation
- Employee Safety Suggestions
- Questions/Answers
- Announcement of Next Meeting and Adjournment

Avoid using the safety meeting time for non–safety-related information. General company announcements should be made at general employee meetings, not at safety meetings.

Also remember that employees who are injured at home cannot report to work. Safety meetings should also address home safety issues. Topics such as safe use of appliances, house wiring safety, home lightning protection, and first aid can be useful to employees both on and off the job.

When Meetings Should Be Held

To accommodate work schedules, many companies hold their safety meetings at the beginning of the work day. Others may wait until the end of the day. Try moving the time of the meetings around. When the results of the meeting are evaluated, as described later, the most appropriate time will be that in which the greatest amount of information is imparted.

For real variety, some companies have found that continually moving the meeting times is effective. Some companies even hold meetings at night and invite spouses to attend. This approach allows the whole family to become involved in safety and allows the employer to present home safety topics more effectively.

Safety meetings should only be canceled when absolutely no other option is available. Major operational emergencies such as fires and major outages are the only acceptable reasons for canceling a safety meeting.

Where Meetings Should Be Held

Variety should also be exploited when choosing meeting locations. Most meetings will be held at the work site for economic reasons; however, some meeting topics might call for different locations. For example, a demonstration of the proper use of hot sticks might best be held at a substation area, while tool safety training might be most effective if presented in the shop. *Remember that when demonstrations are given, even more extreme safety precautions should be taken.*

Evening safety meetings can be held at private banquet rooms or restaurants with meeting facilities. The meeting can be integrated with a social occasion that will allow management personnel, safety personnel, employees, and families to discuss problems and share solutions on a more informal basis. When such meetings are held, they should remain serious, with safety as the premier topic.

How Long Meetings Should Be

Meetings held at the workplace should be kept to a maximum of 1 hour unless some special topic or presentation requires more time. Remember that a safety meeting is a training presentation. Adult training sessions are best kept to short segments. Thus, if a meeting requires more than the 1 hour time period, be certain that breaks are given every 45 minutes or so. Evening meetings may be somewhat longer if a dinner or other social event is integrated with the meeting.

Evaluation of Safety Meetings

Like all parts of a safety program, meetings should be constantly evaluated. Employee questionnaires and quizzes should be used to evaluate the quality of the meetings and the amount of information retained. These evaluations can be done on an anonymous basis so that employees feel free to share negative as well as positive comments. When tests or questionnaires identify problem areas, the meeting structure should be modified to correct the problem.

Follow-through. The safety program will not work if the company is not truly committed. When safety suggestions are made, the company should follow through and report back to the employee in writing. The employee safety committee should monitor suggestions and intervene if the company is not following through properly.

OUTAGE REPORTS

Safety hazards and electrical outages are frequently related to each other. Knowledge of the details that surround an electrical outage can be invaluable in pinpointing safety hazards. Outage reports can be used to

1. Target potential or existing safety hazards
2. Justify maintenance or additions to a power system to increase reliability
3. Gather data for designing better power systems or additions to other plants or facilities

Outage reports should include

• Time and length of outage
• Cause of outage if known
• Results of outage (costs, injuries, etc.)
• Suggestions to present such an outage in the future.

SAFETY AUDITS

Description

Depending on the intended extent, audits vary from short, simple inspections of specific problems or areas to large, companywide reviews that explore every facet of a safety program. Whatever their size or complexity, a safety audit is performed to review and assess the ongoing safety elements of a business. Audits are designed to identify the weaknesses and strengths of a safety program.

Process. A comprehensive audit starts with a review of upper-management safety awareness and attitudes and continues through the entire organization. During the procedure, every facet of the safety program is reviewed, cataloged, critiqued, and—if necessary—modified.

Attitudes. The attitudes of company personnel can make or break the effectiveness of a safety audit. Many employees tend to take safety audits very personally. A properly run safety audit should not be performed as an inquisition. Honest errors should be noted and corrected without hostile indictments or defensive recriminations. All personnel should be aware that a safety audit is only one tool used to make the workplace safer for everyone.

Audits Versus Inspections. Some references differentiate between an audit and an inspection. In such references, audits are portrayed as major, comprehensive efforts that include all facets of the safety program. Inspections, on the other hand, are smaller, more concise efforts that may assess only one small area or procedure.

While there may be some psychological advantage in separating the two terms, remember that both are intended to identify, and subsequently eliminate, existing or potential safety hazards.

Purposes

The single most important purpose of a safety audit is to identify and eliminate safety hazards. There are, however, additional side benefits and purposes that are realized including

1. Employee morale is improved as safety problems are eliminated.
2. Audits provide a dynamic record of safety performance.
3. A positive cycle of safety improvement is created. That is, departments that get good scores tend to work harder to maintain their high scores. Departments that do not do as well tend to work harder to improve for the next audit.
4. Managers at all levels are made aware of many safety problems and procedures that they might not otherwise discover. For example, an audit that includes analysis of OSHA compliance may make managers aware of OSHA rules that were previously unknown.
5. When performed as a "family" procedure, a safety audit can actually enhance employer-employee relations. Employees respect management teams who truly put employee safety first.

Procedure

The safety audit should review and evaluate each element of the safety procedure. Site-inspection trips, personnel interviews, task observation, and documentation review are among the methods that may be used. Each portion of the company that is reviewed will have its own specific concerns. Note that the items presented in the following sections are intended for example only. Specific facilities may require more or different audit criteria.

Facilities.   Checkpoints for the electrical physical plant include the following:

1. Is the electric equipment kept clean?
2. Are safety exits clearly marked and unblocked by wire reels, ladders, and other such equipment?
3. Are relay flags and other such indicators kept reset?
4. Is explosion-proof equipment properly maintained and sealed?
5. Are all electrical grounds in place and secure?

Employees.   Properly trained, alert employees are at the heart of any safety program. An individual employee is always the person most responsible for his or her own personal safety. The following points are typical of the types of checks that should be made to audit the safety awareness of individual employees.

1. Are employees adequately trained?
2. Are employees familiar with their safety handbook?
3. Do employees know where all electrical safety equipment is located?

4. Can employees successfully perform safety-related procedures such as safety grounding and voltage measurement?
5. Are personnel familiar with the electrical safety one-line diagram?
6. Are employees familiar with the safety techniques that are unique to the various voltage levels to which they will be exposed?

Management. Good safety practice starts at the top. If management includes itself in the audit process, employees will accept that they too must be reviewed. Management involvement is more than just window dressing, however. Other management audit points include the following:

1. Does the facility have a formal, written safety policy?
2. Are safety rules enforced uniformly?
3. Is there an employee safety committee to which management defers when technical and day-to-day decisions must be made?
4. Does management promote both on-the-job and off-the-job safety?
5. Does management provide, at company expense, CPR and other such training?

Safety Equipment. Various rules and regulations require the use of approved electrical safety equipment. The following points are typical of safety equipment checks during an audit.

1. Is safety equipment readily available for all personnel?
2. Have all rubber goods been tested within the required period?
3. Are meters and instruments mechanically sound and electrically operable?
4. Are test equipment fuses properly sized?
5. Are safety interlocks operational and not bypassed?
6. Are safety signs, tags, warning tapes, and other such warning devices readily available to all employees?

Safety Procedures. Energy control and other such safety procedures must be properly developed and implemented. A few of the key points for procedures include

1. Are written switching orders required for planned outages?
2. Is a written copy of the lockout-tagout procedure available to all personnel?
3. Are all employees trained in electrical safety grounding?
4. Are safety procedures followed for each and every operation or are they bypassed in the interest of production?

Documentation. Safety electrical one-line diagrams and other such documents are too frequently ignored and allowed to fall behind or remain inaccurate. The following key points should be checked for safety-related documentation:

1. Are the electrical portions of the company safety handbook up-to-date and consistent with currently accepted safety practice?
2. Are copies of all safety standards and practices readily available to all personnel?
3. Is the safety electrical one-line diagram up to date, accurate, and legible?
4. Are accident reporting forms, test sheets, procedures, and other such operational aids readily available?

The Audit Team

Audit teams should be composed of safety personnel, management personnel, and employees. Because of their familiarity with safety auditing techniques, safety personnel should, generally, be the *team leaders*. The audit and assessment tasks should be shared equally by all team members.

The number of team members will, of course, depend upon the size of the facility or facilities being audited. A minimum of three members is recommended even for very small facilities. This will allow representations for safety, management, and employees.

Some companies use the employee safety team to spearhead safety audits. The safety team assigns audit personnel, reviews audit results, and continues to follow up on changes.

Audit Tools

The principal tool of the safety audit is the audit form (Figs. 12.2 to 12.7). A form of this type should be developed for every area involved in the audit. In addition to such forms, audit team members should be supplied with copies of safety procedures, standards, and other such documents. Team members can thus evaluate whether the procedures and standards are being followed or not. Note that the management and employee questionnaires can be used for statistical analysis.

Follow-Up

Emergency or life-threatening problems should be corrected immediately. Problems that are left uncorrected continue to put employees and equipment at risk. Moreover, major audits that find problems which are not corrected send very negative signals to employees.

Follow-up to correct a problem should occur within a very short time after the results of the audit. Major safety problems should be corrected immediately. Action on less serious problems should begin within 1 month of the audit.

When audit results call for additional employee training, major system modification, or other such long-term expenditures, management should develop the plan within 1 month. No one expects major expenditures to be performed on a short-term basis; however, safety considerations must be met in a timely and effective manner.

The audit report should be written as soon as the audit is complete. The timetable for the correction of observed problems should be based on three criteria as follows:

1. How serious is the problem? If an accident caused by this problem would result in very severe injuries or high levels of damage, the problem should be corrected immediately. This criterion is the most important of the three, and it should outweigh any other considerations.
2. What is the likelihood that an accident will occur as a result of the problem noted in the audit? If there is little possibility of an accident, correction of the problem may be relegated to a lower status.
3. How much will it cost to fix the problem?

Internal Versus External Audits

Although companies usually employ their own personnel in the performance of a safety audit, sometimes an external firm may be used. Each approach has its own advantages and

Outdoor Substations

Company _____

Location _____

Substation _____

Date _____ Auditors _____

General Conditions

Item	Rating	Comments
Properly fenced		
Restricted access		
Fence grounding		
Rock and gravel (no vegetation)		
Corrosion-free		
Proper clearances		
"Danger High Voltage" signs		
Control room properly ventilated		

Additional observations and comments:

(a)

FIGURE 12.2 Facilities audit form. (*a*) Outdoor substations; (*b*) indoor substation/electrical room; (*c*) miscellaneous outdoor equipment; (*d*) power transformers; (*e*) metal-clad switchgear; (*f*) Motor control centers; (*g*) battery stations; (*h*) panel boards. (*Courtesy Cadick Corporation.*)

Indoor Substation/Electrical Room

Company _____

Location _____

Substation/Area _____

Date _____ Auditors _____

General Conditions

Item	Rating	Comments
Restricted access		
"Danger High Voltage" signs		
Improper material storage		
Room properly ventilated		
Proper clearances		
Voltage level markings		
General housekeeping		
Corrosion-free		
Fire extinguishers		
Clear egress		
Additional observations and comments:		

(b)

FIGURE 12.2 *(Continued)*

Miscellaneous Outdoor Equipment

Company _____

Location _____

Substation _____

Date _____ Auditors _____

Item	Rating	Comments
Lightning arresters		
Air disconnect switches		
Capacitor banks		
Oil circuit breakers		
Reclosers		
Fused disconnects		
CTs and PTs		
Control transformers		
Additional observations and comments:		

(c)

FIGURE 12.2 (*Continued*)

Power Transformers

Company _____

Location _____

Substation/Area _____ Designation(s) _____

Date _____ Auditors _____

Item	Rating	Comments
Insulators		
Gauges operable		
Proper clearances		
No signs of leaks		
Equipment grounding		
Grounding plates		
Neutral ground resistors		
Proper ventilation		
Type of maintenance program		
PCB stickers		

Additional observations and comments:

(d)

FIGURE 12.2 (*Continued*)

Metal-Clad Switchgear

Company _____

Location _____

Substation/Area _____ Designation(s) _____

Date _____ Auditors _____

Item	Rating	Comments
Door and cover properly secured		
Proper ventilation		
Proper working space & clearances		
Panel identification markings		
Circuit identification markings		
Voltage level markings		
Protective device indicators		
Type of maintenance program		
Clear egress		
Equipment grounding		
Panel meters operative		
Additional observations and comments:		

(e)

FIGURE 12.2 (*Continued*)

Motor Control Centers

Company _____

Location _____

Substation/Area _____   Designation(s) _____

Date _____   Auditors _____

Item	Rating	Comments
Door and cover properly secured		
Proper ventilation		
Proper working space & clearances		
Panel identification markings		
Circuit identification markings		
Voltage level markings		
Indicators		
Type of maintenance program		
Clear egress		
Equipment grounding		
Panel meters operative		
Additional observations and comments:		

(f)

FIGURE 12.2   *(Continued)*

Battery Stations

Company _____

Location _____

Substation/Area _____ Designation(s) _____

Date _____ Auditors _____

Item	Rating	Comments
Restricted access		
Proper ventilation		
Eyewash station		
Neutralizing solution		
Proper working space & clearances		
Electrolyte level		
Signs of out-gassing		
Voltage level markings		
Corrosion-free		
Type of maintenance program		
Clear egress		
Equipment grounding		
Availability of battery PPE		
Battery charger		
Additional observations and comments:		

(g)

FIGURE 12.2 (*Continued*)

Panel Boards

Company _____

Location _____

Substation/Area _____ Designation(s) _____

Date _____ Auditors _____

Item	Rating	Comments
Door and cover properly secured		
Proper ventilation		
Proper working space & clearances		
Panel identification markings		
Circuit identification markings		
Voltage level markings		
Type of maintenance program		
Equipment grounding		
Additional observations and comments:		

(h)

FIGURE 12.2 (*Continued*)

Company/Location: Date: Auditor: Page - 1

Item	Accessibility/Availability	Properly Rated	Tested or inspected — Interval (Months)	Tested or inspected — Yes or No	Functional	Physical Condition (1 = OK, 2 = Marginal, 3 = Bad) — Insulation	Connectors/Cables	Cleanliness	Case	Proper Fusing	Proper Storage (Yes or No) — Dry	UV Protected	Proper Containers	Physical Protection	Comments
Rubber Insulating Good															
Rubber Gloves			6												
Leather Protectors															
Rubber Blankets			12												
Rubber Sleeves			12												
Line Hose															
Rubber Covers															
Rubber Mats															

(a)

FIGURE 12.3 Safety equipment audit form. (a) Page 1; (b) Page 2; (c) Page 3. (*Courtesy Cadick Corporation.*)

12.23

Item	Accessibility/Availability	Properly Rated	Tested or inspected			Physical Condition (1 = OK, 2 = Marginal, 3 = Bad)					Proper Storage (Yes or No)				Comments
			Interval (Months)	Yes or No	Functional	Insulation	Connectors/Cables	Cleanliness	Case	Proper Fusing	Dry	UV Protected	Proper Containers	Physical Protection	
Eye Protection															
Safety Glasses															
Face Shields															
Thermal Protection															
Thermal Work Clothes															
Flash Suits															
Fire Blankets															
Warning Devices															
Barricades/Barrier Tape															
Signs															

(b)

FIGURE 12.3 (*Continued*)

12.24

Company/Location: Date: Auditor:

Item	Accessibility/Availability	Properly Rated	Tested or inspected — Interval (Months)	Tested or inspected — Yes or No	Functional	Physical Condition (1 = OK, 2 = Marginal, 3 = Bad) — Insulation	Connectors/Cables	Cleanliness	Case	Proper Fusing	Proper Storage (Yes or No) — Dry	UV Protected	Proper Containers	Physical Protection	Comments
Measuring Instruments															
Voltage Testers															
Ground Measurement Equip.															
Current Measuring Devices															
Thermographic Devices															
Other Safety Equipment															
Safety Grounds															
Hard Hats															
Hot Sticks			24												
Insulated Tools															
Ladders															
Foot Protection															

(c)

FIGURE 12.3 *(Continued)*

Company/Location: Date: Auditor:

Item	Existence (Y or N)		Review		Utilization 1 = Always 2 = Sometimes 3 = Never	Accessibility	Comments
	Written	Oral	Interval	Last Reviewed			
First Aid and CPR							
Accident Investigation							
Electrical Firefighting							
Protective Clothing							
Emergency Comm.							
Energy Control Procedures							
Switching Procedures							
Lockout/Tagout							
Voltage Measurement							
Safety Grounding							

FIGURE 12.4 Safety procedures audit form. (*Courtesy Cadick Corporation.*)

Company/location:		Date:		Auditors:	
Documentation	Document exists 1 = Yes 2 = Some 3 = No	Current revision 1 = Yes 2 = Some 3 = No	Accessibility 1 = Yes 2 = Some 3 = No	Legibility 1 = Good 2 = Marginal 3 = Unacceptable	Comments
Electrical safety manual					
Safety procedures					
Safety equipment records					
Safety equipment manuals					
Accident investigation					
Engineering — One-lines					
Three-lines					
Schematics					
Wiring diagrams					
SCA and coordination					
Maintenance records — PM records					
Protective device tests					
Oil & gas analysis					
Grounding tests					
Battery tests					

FIGURE 12.5 Documentation audit form. (*Courtesy Cadick Corporation.*)

12.27

Electrical Safety Program

Employee questionnaire—page 1	Company/location:

This interview form is anonymous. Please feel free to answer honestly.

The answers that you provide on this form are critical to the successful implementation and maintenance of a Zero Incidents Safety Program. Please provide your reaction to each statement by checking the appropriate box. You may be asked for an interview to further expand on your answers.

	Strongly Agree	Agree	Disagree	Strongly Disagree
1. I am not able to perform my job safely because I have not been provided with the right safety equipment.	☐	☐	☐	☐
2. It is not appropriate to point out my coworkers' at-risk behavior.	☐	☐	☐	☐
3. Many of the safety procedures we are supposed to follow are unnecessary.	☐	☐	☐	☐
4. I follow the correct safety procedures and use the appropriate equipment.	☐	☐	☐	☐
5. Management at my company supports a company, sponsored electrical safety training program.	☐	☐	☐	☐
6. Management at my company supports the electrical safety program.	☐	☐	☐	☐
7. I would not hesitate to report one of my coworkers who was exhibiting at-risk behavior.	☐	☐	☐	☐
8. I am provided with all the necessary safety procedures to allow me to perform my job safely.	☐	☐	☐	☐
9. My supervisor encourages productivity at the expense of safety.	☐	☐	☐	☐
10. My company says they support the electrical safety program, but in reality they discourage anything that detracts from production.	☐	☐	☐	☐
11. My supervisor supports the electrical safety program.	☐	☐	☐	☐
12. My coworkers do not follow the correct safety procedures.	☐	☐	☐	☐
13. I enjoy my job.	☐	☐	☐	☐
14. I know what to do in an electrical emergency.	☐	☐	☐	☐
15. My team has adequate skills to perform their job safely and efficiently.	☐	☐	☐	☐
16. My company does not provide adequate skills training.	☐	☐	☐	☐

(a)

FIGURE 12.6 Employee audit questionnaire. (*a*) Page 1; (*b*) Page 2; (*c*) Page 3; (*d*) Page 4. (*Courtesy Cadick Corporation.*)

Electrical Safety Program

Employee questionnaire—page 2	Company/location:
This interview form is anonymous. Please feel free to answer honestly.	

The answers that you provide on this form are critical to the successful implementation and maintenance of a Zero Incidents Safety Program. Please provide your reaction to each statement by checking the appropriate box. You may be asked for an interview to further expand on your answers.

	Strongly Agree	Agree	Disagree	Strongly Disagree
1. I am skeptical about the ultimate value of our electrical safety program.	□	□	□	□
2. The concept of "zero incidents" is unrealistic.	□	□	□	□
3. On occasion electrical safety must be compromised for production.	□	□	□	□
4. The concept of "zero incidents" is fine in theory but not achievable at a reasonable cost.	□	□	□	□
5. I am aware of incidents where safety has been compromised for production.	□	□	□	□
6. I believe the company is committed to correcting, in a timely manner, deficiencies that are identified by this audit.	□	□	□	□
7. All my coworkers have adequate electrical safety and skills training.	□	□	□	□
8. I endorse a "zero incidents" electrical safety program.	□	□	□	□
9. Electrical safety should never be compromised for production.	□	□	□	□
10. We have adequate electrical personal protective equipment.	□	□	□	□

(b)

FIGURE 12.6   (*Continued*)

Electrical Safety Program

Employee questionnaire—page 3	Company/location:

This interview form is anonymous. Please feel free to answer honestly.

The following questions are designed to determine your knowledge of the various safety-related equipment and procedures. Please check the box that best describes your understanding of the subject matter.

	In-depth Understanding	Competent	Need Additional Training	Totally Unfamiliar
11. Company safety handbook (electrical section)	□	□	□	□
12. Switching procedures for plant distribution system	□	□	□	□
13. Voltage measurement procedures	□	□	□	□
14. Use of drawings (one-lines, schematics, etc.)	□	□	□	□
15. Lockout/tagout procedures	□	□	□	□
16. Use of temporary safety grounds	□	□	□	□
17. First aid and CPR	□	□	□	□
18. Personal protective equipment usage (tools and clothing)	□	□	□	□
19. Electrical firefighting	□	□	□	□
20. Hazard analysis/recognition	□	□	□	□
21. Tool usage	□	□	□	□
22. Familiarity with job safety analysis	□	□	□	□

- Please fill out the form on the next page -

(c)

FIGURE 12.6   (*Continued*)

Electrical Safety Program

Employee questionnaire—page 4 | Company/location:

This interview form is anonymous. Please feel free to answer honestly.

Type of training	Training received	Training current	How trained (check all appropriate)		Comments
			Classroom	OJT	
Print reading					
Troubleshooting					
National Electrical Code					
CPR & first aid					
Electrical firefighting					
Protective clothing					
Hazard risk analysis					
Confined space					
Fall restraint					
Switching procedures					
Lockout/tagout					
Voltage measurement					
Safety grounding					
Equipment specific training					

Safety procedures and use of equipment

Directions: Please enter a (Y)es or (N)o into each box.
Training is considered current if it is done within 10 years except CPR, which is 1 year.

(d)

FIGURE 12.6 (*Continued*)

Electrical Safety Program

Management questionnaire—page 1	Company/location:
This interview form is anonymous. Please feel free to answer honestly.	

The answers that you provide on this form are critical to the successful implementation and maintenance of a Zero Incidents Safety Program. Please provide your reaction to each statement by checking the appropriate box. You may be asked for an interview to further expand on your answers.

Columns: Strongly Agree / Agree / Disagree / Strongly Disagree

1. Our electrical employees are not able to perform their job safely because they have not been provided with the right safety equipment. ☐ ☐ ☐ ☐
2. It is not appropriate for an employee to point out coworkers' at-risk behavior. ☐ ☐ ☐ ☐
3. Many of the safety procedures we are supposed to follow are unnecessary. ☐ ☐ ☐ ☐
4. Our employees follow the correct safety procedures and use the appropriate equipment. ☐ ☐ ☐ ☐
5. I support a company-sponsored electrical safety training program. ☐ ☐ ☐ ☐
6. I support the electrical safety program. ☐ ☐ ☐ ☐
7. I would not hesitate to counsel an employee who was exhibiting at-risk behavior. ☐ ☐ ☐ ☐
8. The company provides all the necessary safety procedures to allow our workers to perform their jobs safely. ☐ ☐ ☐ ☐
9. Our supervisors have encouraged productivity at the expense of safety. ☐ ☐ ☐ ☐
10. Our company says they support the electrical safety program, but in reality they discourage anything that detracts from production. ☐ ☐ ☐ ☐
11. All supervisors support the electrical safety program. ☐ ☐ ☐ ☐
12. Our employees know what to do in an electrical emergency. ☐ ☐ ☐ ☐
13. Our electrical team has adequate skills to perform their job safely and efficiently. ☐ ☐ ☐ ☐
14. Our company does not provide adequate skills training. ☐ ☐ ☐ ☐
15. I am skeptical about the ultimate value of our electrical safety program. ☐ ☐ ☐ ☐

(a)

FIGURE 12.7   Management audit questionnaire. (*a*) Page 1; (*b*) Page 2. (*Courtesy Cadick Corporation.*)

Electrical Safety Program

Management questionnaire—page 2	Company/location:
This interview form is anonymous. Please feel free to answer honestly.	

The answers that you provide on this form are critical to the successful implementation and maintenance of a Zero Incidents Safety Program. Please provide your reaction to each statement by checking the appropriate box. You may be asked for an interview to further expand on your answers.

Strongly Agree Agree Disagree Strongly Disagree

16. The concept of "zero incidents" is unrealistic. ☐ ☐ ☐ ☐

17. On occasion electrical safety must be compromised for production. ☐ ☐ ☐ ☐

18. The concept of "zero incidents" is fine in theory but not achievable at a reasonable cost. ☐ ☐ ☐ ☐

19. I am aware of incidents where safety has been compromised for production. ☐ ☐ ☐ ☐

20. I am committed to correcting, in a timely manner, deficiencies that are identified by this audit. ☐ ☐ ☐ ☐

21. All of my employees have adequate electrical safety and skills training. ☐ ☐ ☐ ☐

22. I endorse a "zero incidents" electrical safety program. ☐ ☐ ☐ ☐

23. Electrical safety should never be compromised for production. ☐ ☐ ☐ ☐

24. All of my employees have adequate electrical personal protective equipment. ☐ ☐ ☐ ☐

(b)

FIGURE 12.7 (*Continued*)

disadvantages. Three principal reasons may call for the use of an external consultant in the performance of a safety audit:

1. Consultants perform such audits on a routine basis and, therefore, tend to be more efficient than inexperienced employees.
2. Consultants tend to be dispassionate and therefore more objective in their analysis.
3. In-house personnel may not be available because of work schedules.

Of course, the use of consultants has some disadvantages as well, such as the following:

1. Consultants lack familiarity with their clients' in-house systems and conditions.
2. Employees may not speak as freely with an outsider as they do with their own coworkers.

CHAPTER 13

SAFETY TRAINING METHODS AND SYSTEMS

INTRODUCTION

"Here's your rubber gloves. Go follow Joe." For much of the second half of the twentieth century, this was the ultimate safety training method for many facilities. Fortunately, safety training professionals have managed to get two critical messages across:

1. Bad training is worse than no training at all.
2. Good training is both an art and a science.

This chapter defines "good" training and shows how it can augment a safety program. The information can be used by employers and consultants alike. Consultants can use the information to help set up training programs for their clients. Employers can use the information to evaluate training consultants and/or take preliminary steps to set up their own training programs. *Note:* This information is applicable to all types of technical training. Refer to the American Society of Training and Development (www.astd.org) for training details.

Safety Training Definitions

For the purposes in this chapter, *training* is defined as a formal process used to generate a desired positive change in the behavior of an individual or group of individuals. Notice the use of the word "formal." Training must be "on purpose." Never allow training to just happen. Organize it, develop objectives, and administer.

Training is not the same thing as education, although training may well include education. Education is used to add to the cognitive skills and abilities of an individual or individuals. In other words, education teaches you how to think, while training modifies your behavior.

Behavioral modification is the core of training; however, there are many different types of behavioral modification. Three examples follow.

1. Many training companies offer training courses wherein they attempt to teach general electrical safety procedures. While these courses are generally excellent, they do not directly address the specific procedures of each student's company. Rather, the courses emphasize the importance of good safety and show clients what should be done. The behavioral change that occurs is an increase in the safety effort. Such an increase will lead to many benefits; however, the student will need additional safety training in his or her own specific job requirements.

2. Many in-house training programs teach the safety aspects of the maintenance or operation of a range of specific types of equipment. For example, a four-day program which teaches the maintenance of low-voltage power circuit breakers will spend great amounts of time discussing and showing the safety aspects of the particular equipment. This type of behavior modification is relatively specific in that it teaches detailed procedures for a class of equipment—low-voltage power circuit breakers. The desired behavior is the use of proper safety techniques to ensure that students will safely maintain their breakers.

3. A manufacturer's training program on a specific piece of equipment is a third example. When the manufacturer of an uninterruptible power supply (UPS) offers a course on the operation and maintenance of their UPS system, very specific safety procedures are taught. In this instance, the manufacturer knows that good maintenance and operation will result in fewer problems. However, the manufacturer also knows that trained personnel must perform the maintenance safely. Here, the desired modification is reduced to a relatively few, very specific procedures.

In each of the examples just given, the training programs should be prefaced with educational objectives. An *educational objective* is a specific, measurable behavioral change required of the student. For example, in a lockout-tagout course, an objective might be as follows: "The student is expected to determine that a given system has been de-energized by correctly performing the three-step voltage-measurement process." A complete list of objectives should precede every module in a training program. The use of objectives allows the trainer and the trainee to be measured accurately and objectively. All the training discussed in this chapter will be objective-based training.

Training Myths

The electrical safety industry has at least two major training myths. These myths hurt or in some cases eliminate the implementation of good training programs.

Myth #1: My People Can Learn by Just "Doing Their Job." This is the ultimate in "go follow Joe" thinking. The fact is that today's electrical systems are too complex to permit self-training unless it is very well organized and documented. The best result from such an approach will be slow, very expensive learning curves. The more likely result will be employees who learn improperly and make potentially grave errors.

Misconceptions are among the biggest problems for professional trainers. Adult learners have substantially more trouble unlearning a concept than they do learning one. While experience is definitely the best teacher, experience should be prefaced with a properly organized and presented training experience.

Of course, well-designed self-training programs are an excellent source of training. The key word here, however, is *designed*. Simply turning an employee loose with an instruction book is not effective.

Myth #2: My Supervisors Can Train My People. Training is a part of every supervisor's job, but it should not be allowed to become the only part for three basic reasons. First, in today's highly regulated workplace, a supervisor has his or her hands full with the day-to-day requirements of supervising. They simply do not have the time to put in the detailed effort required for good training. Second, good technical or supervision skills do not necessarily include good training skills. The ability to train well is an art. The third consideration is the "person with a briefcase" syndrome. Employees tend to listen closely and learn from a trainer from outside the company. The feeling is that this person brings a fresh viewpoint to the job. Frankly, this belief has some merit. The outside viewpoint can often instill skills and ideas that are not available from inside.

Conclusion

1. Insurance companies are increasing safety requirements for their insured companies.
2. The technology of power systems continues to spiral upward.
3. Systems grow at an ever increasing rate.
4. Regulatory requirements continue to increase in number and complexity.

All these facts demand that employees be trained. Good training is available from a variety of sources.

ELEMENTS OF A GOOD TRAINING PROGRAM

If you do not know what it is made of, it is hard to build. Because of this fact, many companies, even with strong training commitment, have poor training programs. From a safety training standpoint, three types of training can be defined—*classroom, on-the-job,* and *self-training.* A complete adult training package must include all three of these methods assembled in a formal, planned program. Each method has certain critical elements, some of them shared.

Classroom Training

Classrooms. When most of us think of classrooms, we think of the rooms where we went to grade school . . . the small, cramped desks with books underneath the seat, chalk dust, and the spinsterly schoolmarm. Adult classrooms need to be designed for adults.

- The classroom should be spacious but not huge. The students should not be cramped together and should have sufficient table or desk space to spread out their texts and other materials.
- The classroom should be well lit and quiet. Dim lighting is more conducive to sleeping than learning, and noise from a nearby production facility can completely destroy the learning process.
- The classroom should be as far from the students' work area as possible. This serves two purposes.
  1. A remote location tends to focus the student on learning.
  2. Interruptions by well-meaning supervisors are minimized.
- All required audiovisual equipment should be readily available. Screens should be positioned for easy viewing and controls should be readily accessible to the instructor. Avoid using a white board for a screen due to the glare.

Laboratories. The adult learner must learn by doing. For safety training, learning by doing means laboratory sessions and/or on-the-job training. Many of the design features mentioned for classrooms also apply to laboratories. However, there are a few special considerations.

1. Avoid using the same room for both laboratory and classroom. A room large enough for both is usually too large for effective classroom presentation. Too small a room, on the other hand, crowds the students during lecture and laboratory. Equipment needed to train students in procedures should be placed in positions where it does not interfere with the lecture presentation.

2. Initial training on safety procedures should be done on de-energized equipment.

3. Students should work in groups to help support each other.

Materials.   Student materials can easily make or break a training program. Although a complete material design treatise is beyond the scope of this book, a few critical considerations follow:

1. All student text chapters should be prefaced by concise, measurable training objectives. See the Safety Training Definitions model earlier in this chapter.

2. Illustrations should be legible and clear. Avoid photocopies when possible. Also, put illustrations on separate pages which may be folded out and reviewed while the student reads the text. Nothing detracts more than flipping back and forth from text to illustration.

3. Students should have copies of all visual aids used by the instructor.

4. Laboratories should have lab booklets which clearly define the lab objectives and have all required information at the students' fingertips.

5. Avoid using manufacturer's instruction literature for texts. While these types of documents make excellent reference books, they are not designed for training purposes.

Instructors.   No one has really defined what makes the difference between a good instructor and a mediocre one. Some of the most technically qualified individuals make poor instructors while others are excellent. Some excellent public speakers are very poor instructors, while others are quite good. A few points are common to all good instructors.

1. Good instructors *earn* the respect of their students by being honest with them. No student expects an instructor to know everything there is to know about a subject. An instructor who admits that he or she does not know has taken a giant step toward earning respect.

2. Instructors must have *field* experience in the area that they are teaching. Learning a subject from a textbook or an instruction manual does not provide sufficient depth. (Unless, of course, the material is theoretical.)

3. Instructors must be given adequate time to prepare for a presentation. Telling an instructor on Friday that a course needs to be taught on the following Monday will lead to poor training. Even a course an instructor has taught many times before should be thoroughly reviewed before each session.

4. Avoid using a student's supervisor for the instructor. A supervisor will often expect the student to learn *because* he or she is the supervisor. In addition, the student may feel uncomfortable asking questions.

Frequency of Training.   How often to train is an extremely complex problem. Management must balance the often-conflicting requirements of production schedules, budgetary constraints, insurance regulations, union contracts, and common sense with the needs of the individual employee to know what he or she is doing.

Overshadowing all this, however, is one basic fact—the employee who is not learning is moving backward. Technology and regulations are changing constantly; therefore, employees must be trained constantly. In general, the following points can be made:

1. Employees in technical positions should be constantly engaged in a training program. Two or even three short courses per year are not unreasonable or unwarranted. The courses should be intensive, and related directly to the job requirements of the employee.

2. Employee training should be scheduled strategically to occur before job requirements. In other words, an employee charged with cable testing during an outage should go to a cable testing school a few weeks prior to the outage. Few of us can remember a complex technical procedure for two or three years.

3. Training should be provided at least as often as required by regulatory requirements. Remember that standards and regulations only call out minimum requirements.

4. Regulatory requirements must be met. For example, OSHA rules require annual review of lockout/tagout. Training is required if problems are noted.

On-the-Job Training (OJT)

On-the-job training will be discussed in great detail in a later section. As far as the elements of laboratories, instructors, materials, and frequency of training are concerned, virtually all the previous discussion applies with the following additions:

1. The laboratory for OJT is the whole plant or facility. The student learns while actually on the job.

2. In OJT, the supervisor is often the most logical choice for an instructor. In this role, however, the instructor does not usually get involved in a direct one-on-one training session. Rather the instructor gives the student the objectives for each OJT session and then evaluates progress.

3. Good training is done on purpose. A complete training program will have OJT materials which include objectives, reading materials, and examples of what is to be done. In addition the student should have a record card (see Fig. 13.1) (called a *qual* card in the military) which identifies each area of responsibility. When the student has performed satisfactorily, the instructor checks off on the card to indicate successful completion.

FIGURE 13.1 A student qualification sheet.

4. Since employees are on the job every day, they are (theoretically) learning every day. This is not, however, true OJT. Remember that all good training is done on purpose. Typically a student should be given one specific OJT assignment at least once per week. An instructor (supervisor) should observe the successful completion of the assignment, and the employee should be given credit for the job.

Self-training

Most well-intentioned self-training fails simply because it is not organized. An adult must know what is expected before he or she can perform. In this sense then, self-training has much in common with OJT, and virtually all the comments which apply to OJT also apply to self-training.

The most important element of self-training is materials. The material for self-training must be written very carefully. Small, self-supporting modules which are prefaced with clear objectives should serve as the core for self-training materials. Each module should be ended with self-progress quizzes. After the student completes each module or related group of modules, an examination should be administered by the instructor.

Conclusion

When confronted with requirements for training, such as those previously outlined, many companies are staggered by what appears to be a very expensive investment. The investment made in training is returned many times over in improved safety, morale, and productivity. Furthermore, a training program that does not have the proper elements will probably not be good training, and poor training is worse than no training at all.

ON-THE-JOB TRAINING

The major difference between 25 years of experience and 1 year of experience repeated 25 times is OJT. Good, well-planned, on-purpose OJT is the heart and soul of an adult learning program. Adults must learn by doing, and the workplace is the best laboratory. The three major segments of OJT include *setup, implementation,* and *evaluation.*

You must involve the employees in the development of this program. Their input and feedback will be critical every step of the way. The employee safety committee can also be used as a training committee and should oversee every aspect of the training program development and presentation.

Note that safety topics are especially suitable for presentation in an OJT format. The best way to memorize critical safety procedures is while actually performing them under field conditions. Note also, however, that special supervision must be made when personnel are learning safety procedures with energized equipment.

Setup

As noted previously, all training must be properly planned. Good training implies that it is being done on purpose. The setup portion is the most critical of all and must be properly executed because it serves as the foundation.

Step 1.   Define the employees' job functions. This step (often called a job analysis) is critical to all training but especially to OJT. Since OJT takes place while actually on the job, you must know what your employees do before you can train them. This is a simple concept, but it is surprising how many organizations miss this step. The performance of the job analysis can be as simple as asking your employees to make a list of what they do. The analysis of time sheets is another good approach. A final source of information is to ask your supervisors to list job functions for employees.

Notice that interviewing supervisors is the last of the three approaches. Unfortunately, company management and supervision are not always as closely in tune with day-to-day employee activities as they should be. This is not always true, but too often it is.

At the conclusion of this step, you should have a complete listing of the major jobs performed by your employees; for example, one major job is to remove motor starters from motor control centers with energized buses.

To simplify this step, concentrate on major job functions. Detail can be added as the OJT program develops over the years.

Step 2. Isolate the job functions that are practical and measurable. For example, you may have one task such as troubleshooting the south plant UPS system. While troubleshooting is a trainable skill, it is very difficult to measure troubleshooting ability in an OJT environment. The example given in the previous step (e.g., removal of starters from motor control centers) is measurable and practical. Initially only the practical and measurable skills should be included in the OJT program. As your program's sophistication increases, you may wish to add more cognitive skills.

As each job function is determined, it should be identified with a simple numbering system. For example, the starter removal might be job function number 1; cable testing, job function number 2; and so on. This makes it possible to keep records of which job functions employees have received training in.

Step 3. Break each job function into tasks, subtasks, and detailed procedures. Using relay testing as an example, we might develop several tasks including the de-energizing of the cubicle, proper use of safety equipment, mechanically removing the starter, and so forth. The safety subtasks would include the proper voltage measurement, inspection of rubber gloves, application of safety grounds, and so on. The detailed procedures can be developed either in house or from manufacturer's literature.

Step 4. Develop training objectives for the tasks and subtasks. The objectives should have the measurement criteria contained within them. The following examples use the motor starter removal as its subject.

Task

1. The employee shall independently remove a low-voltage motor starter from its cubicle. The job shall be performed using all applicable company safety standards.

    Subtasks
    a. The employee shall perform a safety inspection for a set of low-voltage rubber gloves.
    b. The employee shall wear a flash suit, previously inspected rubber gloves, hard hat, and face shield.
    c. The employee shall stop the motor and de-energize the starter module. He or she shall demonstrate proper procedure and stand in a safe work location while performing these steps.

Here we have shown only one of the objectives; there would be several more, one for each safety task and one for the mechanical procedures.

Step 5. Develop the written materials necessary to support the program. The following should be developed as a minimum:

1. Complete set of written objectives.
2. Written, detailed description of each task. To accomplish this portion, manufacturer's literature may augment in-house–developed materials.
3. Diagrams, illustrations, or examples.
4. Employee record cards. The cards should have space for the employee's name, classification, and other demographic data. The majority of the card should contain space for completed job functions (by identification code), time of completion, and supervisor's initials indicating successful completion.

Implementation

The implementation phase of an OJT program should be a simple extension of the setup phase.

Step 1. Select several of the job functions to be used as trial runs of the OJT program. At the same time, select an experienced group of employees to receive the initial training. Develop the record cards for each employee.

Step 2. Thoroughly indoctrinate each of the test employees in the program and give them advance knowledge of what is expected. Of course, if you have been properly including the employees in the development phase, this step will be unnecessary.

Step 3. Introduce the training one job function at a time. Scheduling should be made by working the program into the normal work schedule. That is, when plant scheduling requires relay testing, use the opportunity to provide the OJT.

Step 4. Have the employee's supervisor go over the task with the employee and make certain the employee thoroughly understands what is required of him or her. Remember, this is training and the supervisor is the instructor in this case. The employee should perform the work, with the instructor giving guidance. The supervisor should watch the employee in the task and make certain all tasks are performed to within measurement criteria.

Step 5. After each training session is completed, thoroughly evaluate the results. See the next section for a more detailed description of evaluation.

Step 6. Gradually involve each existing and new employee in the OJT program. Anytime an employee is called upon to perform a new job function or task, the OJT program should be implemented and the employee's record card should be updated accordingly.

Evaluation

Step 1. The results each employee obtains should be measured against the criteria established during the setup phase. Do not make any assumptions based on just one or two employees. Poor performance by several employees could mean the program is bad, or it could mean that more classroom training is required.

Step 2. Interview the employees in a confidential manner. Give them a chance to critique the entire training program. All aspects—materials, implementation, objectives—should be reviewed.

Step 3. Interview the supervisor to get his or her feedback.

Step 4. Based on the first three steps, modify the program as required.

Step 5. Every two years review the program from top to bottom.

Conclusion

One major misstep taken by many organizations when they enter the world of OJT is to forget the key word—*training*. On-the-job training is not an on-the-job test. Rather it is on-the-job *training*. Remember that and do not forget to provide the necessary field instruction as the program progresses.

In general, OJT will not stand alone. It must be accompanied by good classroom instruction and/or self-training properly implemented and monitored. As stated in the previous section, OJT can be the most cost-effective of all the parts of a training program.

TRAINING CONSULTANTS AND VENDORS

Only very large companies can justify having their own, completely self-supporting, in-house training organizations. Even when in-house organizations exist, they frequently use outside vendors for support and special programs. In other words, you should use training consultants and vendors for the same reasons you use any consultant or vendor—when time, personnel, expertise, or economics make it impossible to perform the task yourself.

Training consultants fall roughly into three categories—those who provide canned programs, those who develop and present tailored programs, and those who provide training analysis. Of course, some companies fit into more than one of the categories.

Canned Programs and Materials

Programs. Companies which provide the short one-week, manufacturer's type of seminars have proliferated since the early 1960s. Training programs are available in electrical safety, circuit breaker maintenance, relay testing, uninterruptible power supplies, programmable logic controllers, and a host of other such topics. Usually these courses are from three days to two weeks in duration. They are normally presented either in a training center or at a hotel with adequate meeting facilities. Some of them are equipment-specific, while others are generic in nature. Pricing varies, but these seminars generally offer a very economical approach when you have only one or two employees to train.

Learning by mail programs, called correspondence schools, also fall into this general category. Some companies have been providing programmed mail-order instruction for decades. The correspondence schools are especially useful for vocational-type training at a more generic level.

The vocational technical schools (votechs), both private and public, fill yet another need in this arena. Many companies are requiring two-year associate's degrees or the equivalent for employment in certain technical areas. The votech schools provide such programs. Most of the votechs have evening courses, thus allowing students to attend in their free time.

Each of these three types of programs has its own advantages. The one-week seminar is very intensive and provides training in very specialized areas in a hurry. The correspondence schools are very effective for students who have schedules that make attending classes difficult. The votech types of programs are excellent for those students who wish to start at the beginning and learn the entire discipline.

Materials. As a subcategory, some organizations will sell training materials such as programmed courses, vidcotapes, and/or interactive video programs. The principal difference between these materials and the correspondence school is that the materials may be used to present training to many people in one organization. These packages may also cover very specific equipment maintenance and/or operations.

Tailored Programs

When a company is truly committed to highly cost-effective training, they should opt for the tailored program. Often surprisingly affordable, tailoring may be as simple as taking a canned one-week course and teaching only about the equipment which is in use at the particular plant or company. Most of the large vendors are glad to do this sort of tailoring. Other vendors will develop entire tailored presentations to a client's specification.

This type of service is usually most useful when your organization has a large number of people to train in a given area or areas. For example, immediately prior to a major maintenance shutdown, you may wish to have five or ten of your maintenance personnel trained in cable testing and maintenance. Such training is usually much more affordable than sending individual employees to single presentations.

Some training consultants will develop completely tailored programs based on existing industry standard texts, for example, safety programs based on NFPA 70E or substation maintenance based on an electric equipment text. Such programs are extremely effective and are generally no more costly than those described above.

Training Analysis

Sometimes, a company may have no idea what its training needs are. This is less common than in years past, but it does still happen. Training consultants can be retained to provide an entire training systems analysis. The major areas in such a development include *needs analysis, job and task analysis,* and *curriculum development.*

Needs Analysis. Also called a front-end analysis, the needs analysis is the starting point for any training program. The overall method is covered in many industry standards such as those by ASTD. The general principle is quite simple. Desired performance is compared to actual performance for the workers under consideration. If the desired performance is below nominal, training is indicated.

Job and Task Analysis. This portion is usually a part of the needs or front-end analysis. The procedure involves defining each employee's job functions in measurable objective terms; breaking each job function into tasks and subtasks; and finally writing these job functions, tasks, and subtasks as measurable learning objectives. These learning objectives are then used as the basis for the training programs.

Curriculum Development. The actual curriculum is written and/or selected to fulfill the learning objectives developed during the job and task analysis. However, how this is accomplished may be quite flexible. For example, if a given group of employees has the responsibility to perform routine maintenance on the uninterruptible power supply, the learning objectives may be met by sending the group to one of the canned courses previously discussed.

Evaluating Training Vendors and Consultants

Evaluation of any training must be based on the learning objectives. In order to know how well we are doing we must know what we intended to do. Training must be objective-based and done on purpose. Proper evaluation demands that such an approach be used. With this in mind, the evaluation of vendor training can be made by using four simple criteria.

Does the Vendor Provide Properly Documented, Objective-Based Training? If the answer to this is no, do not use that vendor. Of course, this does not apply directly to some of the situations described in this book. However, even training analysis consultants can be checked for competency in and knowledge of objective-based learning.

Does the Vendor Listen Closely to Your Desires and Counsel You If He or She Believes That You Are Wrong? One of the most difficult things for any consultant to do is to argue with his or her client. Remember, the client pays the bills. Therefore, if your consultant listens, and then counsels (or argues) with you when he or she believes you are wrong, that consultant has your best interests at heart.

Does the Vendor Provide Only Qualified and Competent Personnel? Consultants are only as good as the people they employ. Take the time to quiz the individuals that your consultant provides. Ask for resumes and, where applicable, interview the individuals. Even qualified people will not be effective if they have a personality conflict with you.

Did the Program Meet the Objectives That Were Set Out at the Beginning? Do not depend on course critiques and/or personal evaluations of attendees. Give quizzes where applicable. Sit down with employees and have them review the material that they learned. Develop your own short quiz based on the course materials and administer it to your employees. The quiz can be administered orally, in written form, or as a field-type performance quiz. (For example, one task might be to "test this relay like they taught you at school.")

As a corollary to this last question, always insist that a vendor administer performance quizzes as part of a training program. Quizzes do three things:

1. They measure the student's performance.
2. They measure the trainer's performance.
3. They measure the course's success.

In some cases union contracts may make quizzing difficult or impossible; however, since the quizzes also measure the quality of the training that occurs, many unions are eager for their members to be tested. In this way, they know their members are receiving proper and adequate training treatment.

Conclusion

The use of training consultants is really no different in concept than the use of any other consultant or contractor. You should first determine what you need, monitor what you are getting, and, finally, evaluate what you got.

TRAINING PROGRAM SETUP—A STEP-BY-STEP METHOD

Introduction

Among the many difficulties faced by employers wishing to establish a viable electrical safety program is the need to validate that workers have the proper background, skills, and training as required by OSHA and NFPA 70E. Unfortunately the existing standards do not address detailed methods that allow employers to validate that their personnel are truly "trained on and familiar with the operation and the hazards of the equipment upon which they are working." The program described in this chapter uses long recognized industrial education procedures such as job-task analysis, skills assessment, and training needs analysis to develop and validate a program which will provide the highest level of compliance with existing regulatory and consensus standards.

Background

All of the modern regulatory and consensus standards allow only "qualified persons" to perform work that requires approaching energized conductors. The OSHA defines a qualified person as *"one familiar with the construction and operation of the equipment and the hazards involved."* (29CFR1910.399)

The OSHA also offers some additional clarification in several locations as shown in Table 13.1.

Unfortunately none of the regulations give clear methods for certifying that personnel are truly qualified in the OSHA sense. This can lead to situations wherein the employer and employee are caught in a "no-win" scenario.

For example, several years ago a worker was injured in an arc flash incident. This individual was a journeyman electrician with almost 25 years of direct, hands-on experience and training. He was working in an electrical switch that was identical to hundreds of others he had serviced over the years. The arc occurred when the worker allowed a grounded lead to contact exposed, energized conductors. These conductors were exposed via mounting hardware that was in a very "difficult-to-reach" location in the equipment.

In subsequent investigation it was determined that the worker did not know that the mounting hardware was directly connected to the energized bus through a plastic mounting piece. The design of the equipment predated the dead-front requirement of modern equipment. The injured worker had never realized that the exposed hardware was energized even when the switch was turned off.

The experience and the training history of the worker did not sway the OSHA investigator who correctly determined that the worker was, in fact, unqualified. He had not recognized and had not avoided the hazard.

This puts employers in the difficult position of declaring that their employees are qualified for a certain work, only to find out later that, in fact, the OSHA disagrees. Short of the OSHA certifying employees (a highly unlikely scenario), there seems to be no way to validate that employees are qualified.

Figure 13.2 illustrates the problem.

The regulatory standards define the term qualified, and give definitions and explanations as to what the term means, tasks the employer with the responsibility of determining who is and who is not qualified.

An accident involving a "qualified" employee occurs, which OSHA investigates. As a part of the evaluation, the OSHA investigator determines that one or more of the employees involved in the accident were not qualified. OSHA then issues fines and requires that the employer modify procedures and training.

TABLE 13.1   OSHA Sections on "Qualified Persons"

Location	Text
1910.399	Note 1: Whether an employee is considered to be a "qualified person" will depend upon various circumstances in the workplace. It is possible and, in fact, likely for an individual to be considered "qualified" with regard to certain equipment in the workplace, but "unqualified" as to other equipment. (See 1910.332(b)(3) for training requirements that specifically apply to qualified persons.)
1910.331	The provisions of 1910.331 through 1910.335 cover electrical safety work practices for both qualified persons (those who have training in avoiding the electrical hazards of working on or near exposed energized parts) . . .
1910.332	**1910.332(b)(3)** Additional requirements for qualified persons. "Qualified persons" (i.e. those permitted to work on or near exposed energized parts) shall, at a minimum, be trained in and familiar with the following: **1910.332(b)(3)(i)** The skills and techniques necessary to distinguish exposed live parts from other parts of electric equipment. **1910.332(b)(3)(ii)** The skills and techniques necessary to determine the nominal voltage of exposed live parts, and **1910.332(b)(3)(lii)** The clearance distances specified in 1910.333(c) and the corresponding voltages to which the qualified person will be exposed. Note 1: For the purposes of 1910.331 through 1910.335, a person must have the training required by paragraph (b)(3) of this section in order to be considered a qualified person. Note 2: Qualified persons whose work on energized equipment involves either direct contact or contact by means of tools or materials must also have the training needed to meet 1910.333(C)(2).

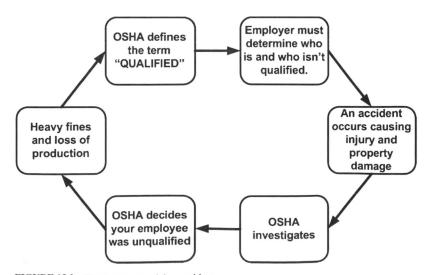

FIGURE 13.2   The regulatory training problem.

At this point, the circle starts again. The employer must again use the OSHA definitions to determine who is and who is not qualified, and so on.

A plan

The following paragraphs present a program that can be used to validate the training of electrical workers as well as organizing the overall electrical safety program. This program has been successfully deployed in many locations throughout the United States and continues to be introduced in a number of other plants.

This validation program uses a holistic approach by integrating well established engineering, technical, and educational methods. At its core, the overall strategy is based on the Instructional Systems Design (ISD) model which is shown in Fig. 13.3.

Note that this model uses a "scientific" format that should be familiar to those in the engineering and technical disciplines. Note also that sufficient time and effort must be made in each stage to ensure that subsequent steps will be completed effectively. For example, shortening the Analysis stage will create a disproportional increase in the time required for the design and development stages.

While the overall approach is designed to be a comprehensive methodology, it is flexible enough to be adjusted to fit the specific needs and situations found in a variety of different industrial, commercial, and utility cultures. The six major sections of the program are described in the following paragraphs.

Analyze

Needs Analysis.   Not all organizations are in need of a comprehensive training effort. Some, for example, may need only special skills training in a few isolated areas. A needs analysis is performed to determine what, if any, training is needed in the organization.

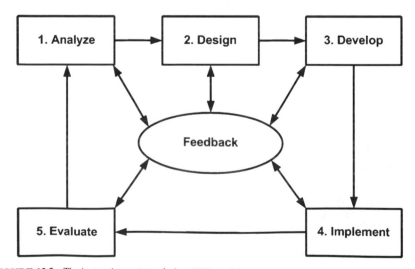

FIGURE 13.3 The instruction systems design (ISD) model.

You should consult three sources to assess your needs—*employees, supervisors and managers,* and *company performance.*

1. *Employees.* Formally poll your employees for training requirements. Usually they know their shortcomings better than anyone. The poll should be concise and sincere. Be prepared for a revelation because experience has shown that, more often than not, employers and employees have very different views about the state of their training.

2. *Supervisors and managers.* Extend the training poll to your management and supervision. Give them a chance to tell you where training deficiencies are causing problems. Their input should be solicited for their employees' and their own training needs. Do not forget to include yourself in this analysis. Employees at all levels may need training.

3. *Company performance.* An honest evaluation of company performance can be one of the best sources of training needs assessment. The most obvious example of this element is in the area of safety. Do some departments have a consistent record of safety problems? Such a department is an obvious target for safety training. Remember to involve your employees and supervisors in the review and analysis of the material. The people who are going to receive the training should be involved at all stages.

An electrical safety audit, as described in Chap. 12, can be used to satisfy all the requirements of the needs analysis. This comprehensive audit requires a minimum of two days for a small facility and a maximum of several days for a large facility. A combination of interviews, questionnaires, and site inspections combine to create a thorough assessment of the existing conditions. An exit presentation is held with key plant personnel to review critical findings. In most cases, severe hazards are discovered during the audit which can be corrected on the spot. A comprehensive report follows the audit within two weeks. This report details the findings and recommendations of the audit and is the first concrete step in the process.

Job and Task Analysis. After the needs analysis is complete, you should have a well-defined set of training needs. The next step is the performance of a job and task analysis (JTA) for each of the areas which require training. This critical step defines the actual composition of a given job. The JTA involves the following three steps.

1. *Employee interviews.* Each employee is interviewed, and a listing is made of the jobs and tasks that he or she performs. The listing should be very detailed to allow for the development of performance objectives.

2. *Supervisor and manager interviews.* The department supervisors are interviewed to obtain their input to the jobs and tasks performed by their employees. This listing is compared to the one developed by the employee interviews, and modifications are made as required. Many training problems disappear when this step is performed since supervisors and employees develop a common understanding of the task at hand.

3. *Job performance observation.* Employees are observed in the performance of their jobs. The job and task listings developed during the interviews are reviewed and refined to be consistent with actual performance. Remember that the actual performance of the given job is under question, so do not modify a JTA based on performance observation unless you are certain that the job is being performed correctly.

The arc-energy analysis. An arc-energy analysis which includes a short-circuit study, coordination study, and flash-hazard analysis is a critical part of the analysis stage and should be performed if it is not already available. Some companies may prefer to perform a "mini-study" before any of the other steps to determine the severity of the flash hazard at their location. The results of this study will determine the minimum PPE requirements for workers, provide the required flash-hazard warning labels for each location, and introduce overall electrical safety requirements to plant personnel.

Because of the perceived cost of such a study, many companies opt for the use of the so-called simplified method as described in NFPA70E, and discussed in Chap. 3. The authors of this book discourage this practice for several reasons including:

1. The simplified method requires knowledge of much of the same baseline data including short circuit duties and clearing times.

2. The simplified method frequently results in an over-dressed work force. In fact, if done conservatively, it should.

3. We have documented situations in which the simplified method actually results in under-dressing personnel. That is, even when correctly applied, the simplified method may give results that will not provide adequate protection. While this occurrence is rare, even one time could result in severe injuries or death.

4. While most will agree that the current arc-energy calculation methods are a work in progress, they still provide the best available information of the degree of exposure. We owe our workers our best possible effort in determining the risk and attendant protection methods.

Design

Performance Objective Design. Based upon the results of the JTA, a set of measurable performance objectives is developed. You may wish to develop training objectives in two tiers—*terminal objectives* and *enabling objectives.* Terminal objectives are broad-based objectives that represent the actual required behavior. Enabling objectives are "sub" objectives that define skills required to meet a given terminal objective. The following give examples.

- Terminal objectives. The employee will be able to test a Westinghouse induction disk overcurrent relay. The tests will be performed using the manufacturer's instruction leaflet, a transformer-loaded overcurrent relay test set and associated instructions, and the company-supplied test sheet. Test results shall be as defined in the relay instruction leaflet.
- Enabling objectives
  1. The employee will be able to describe the purpose and use of each of the controls on the transformer-loaded test set.
  2. The employee will be able to describe the pickup and timing adjustments on the Westinghouse induction disk overcurrent relay. (Add other objectives as defined from the JTA.)

Note the specificity of the objectives. The regulations and standards are very clear in the requirement that employees must be trained on the equipment with which they work.

Structure Design. The information generated during the needs analysis, JTA, and performance objectives development is reviewed, and an overall structure for the training effort is defined. Of course, not all companies are going to need a large, expensive training center with a staff of trainers.

- *In-house training.* If you define an extensive training need with many ongoing course requirements, you may need to consider the development and use of an in-house training system. The materials for such training can include both specifically developed and/or vendor-supplied materials. On-the-job training is usually a very important part of in-house training.

- *Vendor-supplied training.* Specific skills such as cable testing, relay testing, breaker testing, and instrument repair may best be supplied by training companies that specialize in these areas. In fact, in some cases an entire program may be implemented through commercially purchased training.
- *Self-study.* Many skills can be learned effectively through one of the many self-study programs that are commercially available. Even if a program of this type is not purchased, employees should be encouraged (monetarily if necessary) to put in some study time on their own.

Infrastructure Design

1. Training department organization
   a. The training department should report to the executive level of management. This is especially important for technical training departments.
   b. Training department supervisors should be technically qualified personnel. Furthermore, training supervisors should maintain a teaching schedule.
   c. Salaries for training department personnel should be equal to or exceed the salaries of equivalent level personnel in other departments.
2. Training facilities. The selection of a training facility will depend upon the quantity and nature of the training that you wish to conduct. If you plan to use a lot of outside vendors, you may be able to use their facilities. Conference rooms, local meeting centers, and hotel facilities may also be viable alternatives. Specialized laboratory needs may require the construction of facilities. As an alternative to this, you may consider using your company work areas for laboratories; however, this is not recommended as a general practice.

Develop

Overall Training Structure. In some cases the required training will be minimal and perhaps limited only to the specific qualification exercises necessary to assure that all workers have the requisite technical and safety skills to perform their work safely. However, it should always be recalled that OSHA requires skilled personnel to be familiar with the hazards as well as the operation of the equipment. This means that every worker must have a solid grounding in the hazards of electricity along with the various specific job requirements. This means that the final training program will be a mix of several different training scenarios including:

1. Prerequisite skills training from apprenticeship programs, public education, and/or validated on-the-job programs. The proof of such training may be a prerequisite for obtaining a job.
2. Formal classroom training to be attended by all personnel as they progress through their job structure. Such training may be either:
   a. Internally developed and presented classes or
   b. External training from vendors, vocational schools, or community colleges.
   c. Formal on-the-job training requiring demonstration of skills in the presence of a supervisor with an appropriate qualification form.

Whatever the mix of training, there is only one way to ensure that workers are truly qualified. They must demonstrate their competence in a controlled environment, under supervision. To be fully qualified, a worker must be trained on a switch of the exact type that he/she will encounter in the field. Trainees must demonstrate to a supervisor where all of the hazardous locations are and what steps/precautions should be taken to prevent an accident. Figure 13.1 is a section of an employee on-the-job training qualification form for technical skills. Such forms should be used along with examinations and field demonstrations to insure that workers are capable of performing all of the tasks to which they may be assigned.

Course Development.   Courses must be developed or purchased to fulfill the objectives defined during the setup step.

1. *Purchased materials.*   Both self-study and classroom materials are available for purchase from a variety of training vendors. Make certain that the purchased materials meet the objectives defined during the setup step.

2. *Developed materials.*   Although somewhat more expensive, developing company-specific course materials has some distinct advantages. In the first place, your employees will tend to identify with materials that are obviously site-specific; moreover, presentation time is optimized since all material is company-specific. Note that these materials may be developed either by in-house personnel or by consultants.

Implementation

The format of training implementation depends to a large extent on the structure of the facility and the training needs of the workers as determined in previous sections. However, the correct implementation is absolutely essential in the validation process. At each step the workers must demonstrate their knowledge of the operation and the hazards of the equipment that they work on; moreover, they must demonstrate their ability to plan ahead and avoid those hazards by using appropriate work practices, PPE, and other precautions as required.

Of course examinations and hands-on exercises will be developed based on the job task analysis and other assessments discussed previously. Any missed questions should be reviewed with the student until he/she understands why they missed them, and has demonstrated their ability to answer similar questions correctly. In other words, where safety is at risk, 70 percent is not a passing grade. In fact, anything less than 100 percent success is insufficient when safety is involved.

The actual presentation of the training program will depend upon the extent and type of training being implemented. Some courses may be presented on an ongoing basis. For example, skills updates such as cable testing, motor and generator maintenance, and the like may be repeated immediately before maintenance intervals that will require those skills. Safety training should be repeated frequently, especially in areas such as lifesaving and CPR.

Training in areas such as basic electricity, mathematics, and other such topics is usually performed only once per employee. Large companies may need to schedule such training on a regular basis to allow for turnover. Smaller companies, however, may opt to use self-training or correspondence courses for such topics.

Evaluation

Although evaluation is identified as a separate step, it should be an ongoing effort employed at every level of the training program. Five basic elements of the training program may be used to analyze its effectiveness—*test review, course critiques, employee interviews, course content,* and *needs analysis.*

Test Review. Examinations evaluate the student, the instructor, and the course. They should be administered for every training course. Tests should be reviewed for a variety of problems such as

- *Commonly missed questions.* If one question or group of questions is consistently missed by all students, you have a problem with the instructor and/or the course.
- *Consistently high or low grades.* If the average grades on a test are consistently high or low, the course is probably too easy or difficult, respectively. Another possible cause of such a problem is that the course material does not fulfill the course objectives.

Course Critiques. Carefully review the course critiques for trouble spots. Consistently high or low evaluations may be an indication of a problem. Take the occasional "sour grapes" critique with a grain of salt.

Employee Interviews. Periodically interview employees to keep your finger on their pulse. Make certain that the employees are learning and that the training program is accomplishing what you intended.

Course Content. Electrical technology is changing rapidly. Although basic electrical principles remain the same, testing technology changes rapidly. Periodically you must have your course materials reviewed for technical accuracy and relevance. Nothing frustrates students more than learning yesterday's technology.

Needs Analysis. Periodically update the performance of the needs analysis. Make certain that the desired behavioral objectives are being met. Ask the following questions:

1. Has company performance improved as desired?
2. Is employee and supervisor morale at the levels you wished to achieve?
3. Is your company safety record acceptable?

A Training Program Tracking Program. Keeping track of the feedback received during the various portions of the program is a monumental task, especially for very large companies. A training program tracking program should be implemented to document and verify all of the various training steps.

For example, during training implementation, records are stored in an electronic database. Forms, queries, and reports are structured to allow the user to search for information relating to employees, hazards, or departments. Individual status reports are quickly and easily accessible, and show each hazard for which the individual has completed training. Such record keeping is critical to optimize safety and to serve as proof of training should an OSHA investigation occur.

Whatever criteria were used to establish the original need for the training program must be evaluated. If the desired changes have occurred, then the program is meeting its goals.

Modification

Update and modify your training program based on your evaluation. Although this may seem obvious, a surprising number of companies evaluate their programs and then make no changes. An evaluation is useless if no changes are implemented.

INDEX

AAFS (*see* American Academy of Forensic Sciences)

ABA (Advanced Burn Life Support guidelines), 8.12

Abrasions (in rubber goods), 2.33, 2.35

AC (*see* Alternating current)

Access plates, for rotating equipment, 5.23

Accident investigation, 7.37–7.41
  accident analysis in, 7.40–7.41
  data gathering for, 7.38–7.40
  employee safety team in, 12.8–12.9
  general rules for, 7.38
  purpose of, 7.37, 7.38

Accident prevention, 7.1–7.7
  and first aid training, 7.7–7.8 (*See also* First Aid)
  individual responsibility for, 7.1, 7.2
  installation safety for, 7.1–7.3
  power system studies for, 7.3–7.7

Advanced Burn Life Support guidelines (ABA), 8.12

A-frame hoist, 7.35

Afterglow, 2.3

Age cracks (in rubber goods), 2.33, 2.36

AIEE (American Institute of Electrical Engineers), 6.3

Airline industry, RCM in, 5.5

Alternating current (AC):
  circuits requiring grounding, 9.9
  grounding of conductors for, 4.11
  power supplies for electronic circuitry, 9.22
  service-supplied, grounding of, 4.9–4.11
  and severity of shock, 1.4

American Academy of Forensic Sciences (AAFS), 6.13, 7.38

American Burn Association, 6.13

American Institute of Electrical Engineers (AIEE), 6.3

American National Standards Institute (ANSI), 2.14, 6.1–6.2, 6.13

American Red Cross rescue method, 7.16, 7.17

American Society for Testing and Materials (ASTM), 2.4, 2.16, 6.4–6.5, 6.13

American Society of Safety Engineers (ASSE), 6.5–6.6, 6.13

American Society of Training and Development, 6.13

ANSI (*see* American National Standards Institute)

ANSI C-2 (*see* National Electrical Safety Code)

ANSI standards:
  for electrical grounding, 9.5, 10.3
  for eye protection, 9.15
  for hard hats, 9.14, 10.8
  for power systems studies, 7.4
  for short-circuit analyses and coordination studies, 7.7

ANSI/NFPA 70 (*see* National Electrical Code)

ANSI/NFPA 70B (*see* Electrical Equipment Maintenance)

ANSI/NFPA 70E (*see* Standard for Electrical Safety in the Workplace)

Approach distances, 3.46–3.51
  barriers and warning signs for, 3.56, 3.60–3.61
  flash hazard distance, 3.46, 3.49–3.51
  with high- and medium-voltage systems and equipment, 10.12–10.13
  and illumination, 3.61
  with low-voltage systems and equipment, 9.20
  protective clothing for, 3.56–3.60
  shock hazard distance, 3.46–3.48
  terms related to, 3.46, 3.47

Arc (electric), 1.7–1.16
  blasts formed by, 1.16
  burns from, 1.16
  defined, 1.1, 1.8

Arc (electric) (*Cont.*):
  description of, 1.8–1.9
  effects of, 1.24
  energy input to, 1.12–1.13
  energy release from, 1.9, 1.12
  flash and thermal protection from, 2.2–2.13
  incident energy of, 1.14–1.16
  procedural protection strategy for, 1.24
  protective equipment for, 1.24, 2.2
  surface area of, 1.13–1.14
  and voltage magnitude, 1.4
  voltage of, 1.13
Arc energy input, 1.1
Arc incident energy, 1.1
Arc protection:
  calculations for, 3.51–3.62
  with high- and medium-voltage systems,
    10.9
  with low-voltage systems, 9.18
Arc Thermal Performance Value (ATPV):
  calculation of, 2.1
  defined, 2.1
  developing rating for, 2.9
  evaluation of, 2.3
Arc-energy analysis, 13.15–13.16
Arc-flash, 1.1
Arc-flash study, 7.7
Arc-in-a-box, 3.52
Arcing current, 1.12–1.13, 3.53, 3.54
Arc-resistant switchgear, 1.1–1.2
Area enclosures, maintenance of, 5.15
Artificial respiration, 7.12–7.14
ASSE (*see* American Society of Safety
    Engineers)
ASSE standards, 6.6
ASTM (*see* American Society of Testing and
    Materials)
ASTM standards, 2.5, 6.4–6.5, 6.19–6.21
ATPV (*see* Arc Thermal Performance Value)
Attendants, hazard, 2.50
Audiovisual proximity tester, 2.56
Audit teams, 12.14
Audits (*see* Safety audits)
Authorized controller (lockout/tagout), 3.29
Authorized employee (lockout/tagout), 3.29

Barricades:
  for temporary hazards, 2.50
Barrier tape, 2.49–2.51
Barriers:
  for approach distances, 3.56, 3.60–3.61
  for rotating equipment, 5.23

BASOFIL®, 2.7
Batteries, stationary, 9.23–9.25
Behavioral approaches to safety, 7.1, 7.2
Behavioral modification, 13.1–13.2
Blanket drag, 7.36
Blankets, rubber, 2.21–2.23, 2.29, 2.31
Blast (electric), 1.16–1.18
  defined, 1.2
  procedural protection strategy for, 1.24
  protective equipment for, 1.24
Blood vessels, 1.20
Body (*see* Human body)
Bonding, 4.18–4.19
  connection of equipment for, 4.8–4.9
  defined, 4.2
  of electrically conductive materials, 4.6–4.7
  of equipment, 4.6–4.7
  general requirements for, 4.2–4.9
  grounding vs., 9.6, 10.4
Bonding jumper:
  defined, 4.2
  equipment, 4.2
  main, 4.2
  and system grounding, 4.11–4.12
Breakers, locking devices for, 2.53, 2.54
Breathing, 1.7 (*See also* Pulmonary system)
Brown Book (IEEE), 7.7
Burns:
  from arcs, 1.16, 1.20, 1.21
  categories of, 1.16
  first aid for, 7.11–7.13
  heart damage from, 1.23
  from shock, 1.3
  sources of, 1.20

Cable ampacity analysis, 7.5
Cables, jumper, 2.66, 2.67, 2.70, 2.72, 2.73
Cadick, John, 5.10
Canadian Society of Forensic Sciences (CSFS),
    6.13, 7.38
Capacitive discharges, 9.21
Carbon monoxide exposure, 7.30, 7.31
Cardio-pulmonary resuscitation (CPR),
    7.12–7.16
CBM (*see* Condition Based Maintenance)
Cell wall damage, from shock, 1.3
CEST (*see* Company electrical safety team)
Chair carry, 7.38
Chemical bloom (in rubber goods), 2.33, 2.36
Chemically treated materials, 2.6
Chernobyl Nuclear Power Station, 11.2
Chest compressions, 7.14–7.16

Circuit breakers:
  failure of, 5.17
  low voltage, 5.18–5.20
  medium voltage, 5.20–5.21
  molded-case, 5.16–5.18
Clamps, jumper, 2.66–2.68, 4.9
Classroom training, 13.3 13.5
Cleaning, in Eight Step Maintenance Program, 5.12
Cleared circuits (*see* De-energized circuits)
Closed-door operation:
  of low-voltage switchgear, 3.10, 3.12, 3.14–3.15
  of medium-voltage switchgear, 3.10–3.11
  of motor starters, 3.21–3.23
Clothes drag, 7.36
Clothing:
  for approach distances, 3.56–3.60
  comparison of material characteristics, 2.7
  conductive, 3.61
  for crossing flash hazard boundary, 3.59
  flame-resistant materials for, 2.6–2.9
  flash suits, 2.11–2.13
  ignition temperatures and fiber types, 1.8
  non-flame-resistant materials for, 2.5
  thermal protective, 2.2–2.5, 2.9, 2.11
  work, 2.9–2.11
Cognitive ergonomics, 11.1, 11.4–11.7
Color splash (in rubber goods), 2.36, 2.37
Company electrical safety team (CEST), 12.2–12.4
Condition Based Maintenance (CBM), 5.9, 5.14
Condition-directed maintenance, 5.7
Conductive clothing, 3.61
Conductive materials, bonding of, 4.6–4.7
Conductors:
  in AC systems, 4.11
  grounding, 4.2, 4.11
  grounding electrode, 4.3
  maintenance of, 5.15
Confined work spaces:
  avoiding hazards in, 3.62
  rescues from, 7.30–7.36
Construction industry:
  electrical fatalities in, 8.1–8.5
  installation safety standards, 6.23, 6.165–6.184
  OSHA standard definitions for, 6.23, 6.188–6.194
  OSHA standards, 6.22–6.23
  safety standards for special equipment, 6.23, 6.187
  safety-related maintenance and environmental considerations standards, 6.23, 6.186
  safety-related work practices standards, 6.23, 6.185
Construction work clothing, 2.9–2.11
Consultants, training by, 13.9–13.11
Contact testers, 2.56–2.59
Contractor muscle, 1.2
Cord-connected equipment:
  grounding of, 4.19–4.21
  low-voltage safety with, 9.3–9.5
  OSHA standards for, 4.20
  (*See also* Extension cords)
Correspondence schools, 13.9
Cotton clothing materials, 2.5
Course critiques (training), 13.19
Covers, rubber, 2.22, 2.24–2.25
CPR (cardio-pulmonary resuscitation) (*see* Heart-lung resuscitation)
CSFA (*see* Canadian Society of Forensic Sciences)
CTs (*see* Current transformers)
Current transformers (CTs):
  in high- and medium-voltage safety, 10.1–10.2
  in low-voltage systems, 9.5
  in power protection scheme, 5.22
Current(s):
  alternating (*see* Alternating current)
  arcing, 1.12–1.13
  direct (*see* Direct current)
  objectionable, 4.8
  severity of shock and duration of, 1.3
  severity of shock and magnitude of, 1.5–1.7
  temporary, 4.8
Curriculum development, 13.10–13.11
Cuts (in rubber goods), 2.36, 2.37

DC (*see* Direct current)
Deaths:
  causes of, 1.25
  in construction industry, 8.1–8.5
  cost-related, 8.4–8.6
  from electrical events, 8.7–8.8
  statistical survey of, 8.1–8.6
De-energized (cleared) circuits:
  safety procedures and methods, 3.3–3.6
  steps before working with, 3.6
  verification of isolation, 3.28
De-energizing equipment, energy control rules for, 3.24–3.25
Depressions (in rubber goods), 2.37
Design of electrical systems, 7.1–7.3

Design safety standards, OSHA, 6.22, 6.24–6.47
Detergent cracks (in rubber goods), 2.37
Digital multimeter, 2.57, 2.58
Digital readout safety voltmeter, 2.57, 2.58
Direct current (DC):
  circuits requiring grounding, 9.8
  power supplies for electronic circuitry, 9.22
  and severity of shock, 1.4
DOT rules, 8.17
Double-insulated tools, 9.3
Duct systems, grounding of, 4.6, 4.7
DuPont Company, 2.6

E_{BT} (Energy break-through), 2.1
Economic case for electrical maintenance,
  5.3–5.4
Educational objectives, 13.2
EEST (*see* Employee electrical safety teams)
Effectively grounded, 4.2
The Eight Step Maintenance Program
  (John Cadick), 5.10–5.14
Electric arc (*see* Arc [electric])
Electric blast (*see* Blast [electric])
Electric shock (*see* Shock [electric])
Electrical accident fatality, 8.7
Electrical Equipment Maintenance—
  ANSI/NFPA 70B, 6.15–6.16
  for maintenance, 5.1
  for short-circuit analyses and coordination
    studies, 7.7
Electrical events, 8.6–8.8
  arc (*see* Arc [electric])
  blast (*see* Blast [electric])
  deaths from, 8.7–8.8
  electrocution, 8.7–8.8
  non-electrical effects in, 8.10–8.13
  physical responses to, 8.11, 8.12
  possible trauma following, 8.9, 8.10
  schematic of, 8.8–8.9
  shock (*see* Shock [electric])
  survivor experience, 8.11–8.13
Electrical hold, 1.7
  loss of control during, 1.21
  and muscular system, 1.22
Electrical injury mortality, 8.7
Electrical maintenance, 5.1–5.24
  area enclosures, 5.15
  conductors, 5.15
  economic case for, 5.3–5.4
  eight-step maintenance program, 5.10–5.14
  electrical insulation, 5.16
  enclosures, 5.15

frequency of, 5.14
fuses, 5.16
general requirements for, 5.14–5.15
low-voltage power circuit breakers, 5.18–5.20
medium voltage circuit breakers, 5.20–5.21
molded-case circuit breakers, 5.16–5.18
OSHA standards for, 6.23, 6.186
personal safety/protective equipment, 5.24
portable electric tools/equipment, 5.23–5.24
protective devices, 5.16
protective relays, 5.21–5.23
Reliability-Centered Maintenance, 5.4–5.10
rotating equipment, 5.23
safety-related case for, 5.1–5.3
Electrical paralysis, 1.21, 1.22
Electrical safety program, 12.1–12.34
  company electrical safety team, 12.2–12.4
  development of, 12.2–12.8
  employee safety teams, 12.3–12.4, 12.8–12.9
  outage reports, 12.11
  safety audits, 12.11–12.34
  safety meetings, 12.9–12.11
  structure of, 12.1–12.2
  in training program, 13.12–13.19
Electrical testing:
  for rubber-insulating equipment, 2.43–2.44
  of tools and extension cords, 3.63
Electrically conductive materials, bonding of,
  4.6–4.7
Electrician's safety kit, 2.78, 2.79
Electrocution, 8.7–8.8
  deaths from, 8.1
  defined, 1.2
Electronic circuitry:
  grounding of workers with, 9.1
  safety around, 9.21–9.23
Electroporation, 1.3, 1.4
Elevated rescue, 7.18–7.29
Embedded foreign matter (in rubber goods),
  2.37
Employee electrical safety teams (EEST),
  12.3–12.4, 12.8–12.9
Employee interviews (training evaluation), 13.19
Enabling training objectives, 13.16
Enclosed switches and disconnects, 3.17–3.18
Enclosures, maintenance of, 5.15
Energized, defined, 4.2
Energized circuits:
  costs requiring, 3.5
  decision tree for, 3.4, 3.5
  safety procedures and methods for, 3.3–3.6
  work requiring, 3.6

Energy, incident, 1.14
Energy break-through (E_{BT}), 2.1
Energy control programs, 3.23–3.26
Energy control rules:
  for de-energizing equipment, 3.24–3.25
  for procedures involving more than one
    person, 3.25–3.26
  for reenergizing equipment, 3.25
Energy input (arcs), 1.12–1.13
Energy release (arcs), 1.9, 1.12
Environmental issues, OSHA standards for,
  6.23, 6.186
Epidermis, 1.19, 1.20
Equipment:
  bonding of, 4.6–4.7
  construction, OSHA standards for, 6.23,
    6.187
  defined, 4.2
  improper maintenance of, 5.2–5.3
  portable, 5.23–5.24
  safety (*see* Safety equipment)
  (*See also specific types*)
Equipment grounding, 4.6, 4.19–4.26
  conductors for, 4.21–4.26
  cord- and plug-connected equipment,
    4.19–4.21
  and ferroresonance, 4.25–4.26
  ground-fault protection, 4.3
  grounding and bonding, 4.8
  for high- and medium-voltage systems, 10.7
  for low-voltage systems, 9.11–9.13
Equipotential grounding, 3.38, 3.43–3.44
Ergonomics, 11.1, 11.4–11.7
Evacuation, medical, 8.12
Evaluation:
  in Eight Step Maintenance Program,
    5.13–5.14
  of on-the-job training, 13.8–13.9
  of training programs, 13.11, 13.18–13.19
Experimentation (accident investigation), 7.40
Extension cords:
  low-voltage safety with, 9.2–9.3
  testing of, 3.63
Extensor muscle, 1.2, 1.21
Eye protection, 2.13, 2.15, 2.16
  for high- and medium-voltage systems,
    10.8–10.11
  for low-voltage systems, 9.15–9.17

Face shields, 2.12, 2.15
Facilities life cycle, impact of maintenance on,
  5.9–5.10

Failure:
  of circuit breakers, 5.17
  defined, 5.8
Failure Modes and Effects Analysis (FMEA),
  5.6, 8.6–8.7
Failure-finding maintenance, 5.7
Fatalities (*see* Deaths)
Fault path, performance of, 4.8
Ferroresonance, 4.25–4.26
Ferrules, jumper cable, 2.66, 2.69–2.73
Fibrillation, 1.2, 1.3, 1.7
Field testing:
  of rubber goods, 2.29–2.35
  of voltage testers, 2.62
Fire scene ratings, 8.7
Firefighter's carry, 7.36
First aid, 7.7–7.16
  ABCs of, 7.9–7.10
  for electric shock, 7.11–7.13
  for electrical burns, 7.11–7.13
  general procedures for, 7.8–7.11
  heart-lung resuscitation, 7.12–7.16
  resuscitation (artificial respiration), 7.12–7.14
First-degree burns, 1.16
Fittings, protection of, 4.9
Flame resistance (FR):
  defined, 2.1
  and energy break-through, 2.3
  evaluation of, 2.3
  for protective clothing, 2.4–2.5
Flame retardant, defined, 2.1
Flame-resistant materials, 2.6–2.9
Flammable, defined, 2.1
Flash (*see* Arc [electric])
Flash boundary, 2.1, 2.3
Flash hazard:
  calculating protection for, 3.51–3.62
  distance for, 3.46, 3.49–3.51
  IEEE Std 1584–2002 methods of calculating,
    3.53–3.55
  Lee method of calculating, 3.52
  NFPA 70E methods of calculating, 3.52, 3.53
  PPE for crossing boundary for, 3.55
  software solutions for calculating, 3.53–3.55
Flash line (in rubber goods), 2.40
Flash protection:
  safety equipment for, 2.2–2.13
  thermal performance evaluation, 2.3–2.4
  thermal protective clothing, 2.2, 2.3
Flash suits, 2.11–2.13
Flexible cord sets (*see* Extension cords)
Flexor muscles, 1.21

FMEA (*see* Failure Modes and Effects Analysis)
Form marks (in rubber goods), 2.37
FR (*see* Flame resistance)
FR Cotton, 2.7
FR Polyester, 2.7
Frequency:
  with electronic circuitry, 9.21
  and severity of shock, 1.3–1.4
FR9B®, 2.7
Full body rope harness, 7.32
Functional failure, 5.8
Fuses, maintenance of, 5.16

General industry standards (OSHA), 6.19, 6.22, 6.156–6.164
GFCIs (*see* Ground fault circuit interrupters)
Gin pole hoists, 7.34, 7.35
Gloves, rubber, 2.16–2.20, 2.29–2.35
Goggles, 2.15, 2.16
Ground, defined, 4.2, 9.5
Ground fault circuit interrupters (GFCIs), 2.74–2.76, 4.3, 4.20, 9.14
Ground mat analysis, 7.5
Ground rods/pipes, 4.12–4.15
Grounded:
  defined, 4.2
  effectively, 4.2
Ground-fault protection of equipment, 4.3
Grounding, 4.1–4.27
  bonding vs., 9.6, 10.4
  connection of equipment for, 4.8–4.9
  and electric shock hazard, 4.1–4.2
  of equipment, 4.19–4.26, 9.11–9.13, 10.7
  equipment used for (*see* Grounding equipment)
  general requirements for, 4.2–4.9
  of high- and medium-voltage systems, 10.3–10.7
  of low-voltage systems, 9.5–9.14
  need for, 2.63–2.64
  to prevent objectionable current, 4.8
  of systems, 4.9–4.19, 9.8–9.11, 10.5–10.6
Grounding conductor, 4.2
  equipment, 4.2
  for equipment grounding, 4.21–4.26
Grounding electrode conductor, 4.3, 4.14–4.16
Grounding electrode system, 4.3, 4.12–4.14
Grounding equipment, 2.63–2.74
  cables, 2.66, 2.67, 2.70, 2.72, 2.73
  clamps, 2.66–2.68
  ferrules, 2.66, 2.69–2.73
  installation and location of, 2.74

jumpers, 2.65–2.73
  and need for grounding, 2.63–2.64
  switches, 2.64, 2.65
Ground-level rescue, 7.31, 7.34–7.38
"Group" padlocks, 2.52
Guards, for rotating equipment, 5.23

Hand protection, 2.13
Hand tools:
  grounding of, 4.19–4.21
  insulated, 2.48–2.49
  low-voltage safety with, 9.3–9.5
Hard hats, 2.14, 2.16
  for high- and medium-voltage systems, 10.8
  for low-voltage systems, 9.14
Hard spots (in rubber goods), 2.37–2.39
Harmonic analysis, 7.4
Hazardous energy source control, 6.22, 6.139–6.155 (*See also* Lockout/tagout)
Hazardous environments, use of tools in, 3.63
Hazards of electricity, 1.1–1.25
  arc, 1.7–1.16
  blast, 1.16–1.18
  body parts affected by, 1.17, 1.19–1.23
  causes of injury and death, 1.23–1.25
  with electronic circuitry, 9.21–9.22
  field marking of, 3.65
  and improper maintenance of equipment, 5.2–5.3
  protective strategies, 1.25
  shock, 1.2–1.7
  with stationary batteries, 9.25
  voltage hazards, 9.6–9.8
Head protection, 2.13–2.16
Health and safety department (HSD), 12.2–12.3
Heap, Howard, 5.5, 5.8
Heart, 1.22–1.23
  burn damage to, 1.23
  fibrillation of, 1.2, 1.3, 1.7
  shock damage to, 1.23
Heart-lung resuscitation (CPR), 7.12–7.16
High- and medium-voltage safety, 10.1–10.13
  current transformers, 10.1–10.2
  equipment for, 10.7–10.12
  flash protection boundary, 3.50–3.51
  grounding, 10.3–10.7
  procedures for, 10.12–10.13
High Voltage Rescue Technique, 7.16, 7.17
Hoechst Celanese Corporation, 2.6
Horny layer, 1.19
  defined, 1.2
  and electrical resistance, 1.20

Hospitalization of victims, 8.14
Hot stick kit, 2.45–2.47
Hot sticks, 2.44–2.48, 3.38
Hot work (*see* Energized circuits)
HSD (*see* Health and safety department)
Human body:
   arc effects on, 1.8–1.12, 1.24
   body parts affected by electricity, 1.17,
      1.19–1.23
   heart, 1.22–1.23
   muscular system, 1.21–.122
   nervous system, 1.20, 1.21
   nominal resistance values for parts of, 1.6
   pulmonary system, 1.23
   response to current magnitudes, 1.7
   shock effects on, 1.2–1.3, 1.6, 1.23
   skin, 1.19–1.21
   tolerance to heat, second-degree burns, 3.50
Human factors in safety, 11.1–11.8
   background of, 11.1–11.3
   meanings of term, 11.2
   and power systems, 11.3–11.7
Hydrogen sulfide exposure, 7.30, 7.31

IAS Electrical Safety Workshop, 6.13
IEC (*see* International Electrotechnical
   Commission)
IEEE (*see* Institite of Electrical and Electronic
   Engineers)
IEEE standards:
   for calculating flash hazard, 3.53–3.55
   for conductor size, 4.17
   for electrical grounding, 9.5, 10.3
   for power systems studies, 7.4
Illumination of work areas, 3.61
Incident energy:
   of arcs, 1.14–1.16
   calculation of, 3.54–3.55
Indentations (in rubber goods), 2.37
Infrastructure design (for training), 13.17
In-house training, 13.16
Injury:
   causes of, 1.24–1.25
   cost-related, 8.4–8.6
   electrical injury mortality, 8.7
   to muscles and peripheral nerves, 8.15
   non-occupational, 8.4
   statistical survey of, 8.1–8.6
Inspection, 2.2
   in Eight Step Maintenance Program, 5.11–5.12
   of rubber-insulating equipment, 2.29–2.44
   safety audit vs., 12.11–12.12

timing of, 2.2
   of tools and test equipment, 3.63
   with tools/instruments, 5.11–5.12
   unaided, 5.11
   of voltage-measuring instruments, 2.59
Installation:
   accident prevention during, 7.1–7.3
Instantaneous trip units, 5.23
Institite of Electrical and Electronic Engineers
   (IEEE), 6.3, 6.13
Instructional Systems Design (ISD) model, 13.14
Instructors, training, 13.4
Instruments, voltage-measuring (*see* Voltage-
   measuring instruments)
Insulated tools, 2.48–2.49, 9.3
Insulation:
   electrical, 5.16
   insulated tools, 2.48–2.49, 9.3
   for jumper cables, 2.67
   maintenance of, 5.16
   rubber insulated equipment, 2.29–2.44 (*See
     also* Rubber insulating equipment)
InterNational Electrical Testing Association
   (NETA), 5.1, 5.12, 5.13, 6.13
International Electrotechnical Commission
   (IEC), 6.2, 6.13
International Organization for Standardization
   (ISO), 6.2, 6.13
Ionizing hazards, 9.22–9.23
ISD model, 13.14
ISO (*see* International Organization for
   Standardization)

Job and task analysis, 13.10, 13.15
Jumper, bonding, 4.2
Jumpers, safety grounding (safety grounds),
   2.65–2.73
   cables, 2.66, 2.67, 2.70, 2.72, 2.73
   clamps, 2.66–2.68
   ferrules, 2.66, 2.69–2.72
   selection of, 2.70, 2.72–2.73

Keratin, 1.20
KERMEL®, 2.6, 2.7
KEVLAR®, 2.7

Laboratory training, 13.3–13.4
Lead-acid batteries, 9.24
Leather protectors, 2.19
Lee, Raphael C., 1.3, 1.9, 1.14, 3.52
Lee Method, 1.14–1.15
"Let-go" threshold, 1.4

Line hose, 2.25–2.26, 2.29
Litigation, accident, 7.37, 7.38
Load flow analysis, 7.4
Lockout/tagout, 3.26–3.32
  applications for, 3.28
  control bypass for, 3.29–3.31
  control transfer for, 3.31
  employee responsibilities for, 3.27–3.28
  with high- and medium-voltage systems and
    equipment, 10.13
  isolation verification, 3.28
  locks without tags, 3.27
  with low-voltage systems and equipment, 9.21
  by nonemployees and contractors, 3.31
  normal removal of, 3.28–3.29
  OSHA standards for, 6.22
  procedural reviews for, 3.32
  removal of, 3.28–3.31
  rules for using locks and tags, 3.27
  safety ground application with, 3.29, 3.31
  sequence for, 3.28
  SHA standards for, 6.139–6.155
  tags without locks, 3.26–3.27
  training for, 3.31
  when/where to use, 2.53–2.55, 3.26
Locks and locking devices, 2.51–2.55 (*See also*
  Lockout/tagout)
Low voltmeter safety standards, 2.60, 2.61
Low-voltage measuring instruments, 2.62–2.63
Low-voltage power circuit breakers,
  maintenance of, 5.18–5.20
Low-voltage safety, 9.1–9.25
  arc hazard, 1.9
  current transformers, 9.5
  equipment for, 9.14–9.19
  extension cords, 9.2–9.3
  flash protection boundary, 3.49, 3.50
  grounding, 9.5–9.14
  on or near electronic circuitry, 9.21–9.23
  plug- and cord-connected equipment, 9.3–9.5
  procedures for, 9.20–9.21
  with stationary batteries, 9.23–9.25
Low-voltage switchgear:
  closed-door operation of, 3.10, 3.12, 3.14–3.15
  open-door operation of, 3.10, 3.15
  safety procedures/methods with, 3.11–3.15
Lubrication, in Eight Step Maintenance
  Program, 5.12–5.13

Main bonding jumper, 4.11–4.12
Maintenance, 7.3 (*See also* Electrical
  maintenance)

"Maintenance Testing Specifications for
  Electrical Power Distribution Equipment
  and Systems" (NETA), 5.1
Manufacturer's seminars, 13.9
Materials, training, 13.4, 13.10, 13.18
Mats, rubber, 2.20, 2.29
Medical aspects of electrical trauma, 8.8–8.17
  medical and surgical intervention, 8.14–8.16
  non-electrical effects in electrical events,
    8.10–8.13
  rehabilitation and return to work, 8.16
  stabilization and initial evaluation, 8.13–8.14
  statistical survey of known injuries/fatalities,
    8.1–8.6
Medical interventions, 8.14–8.16
Medium voltage circuit breakers, maintenance
  of, 5.20–5.21
Medium-voltage safety, 3.50–3.51, 10.1n. (*See
  also* High- and medium-voltage safety)
Medium-voltage switchgear:
  closed-door operation of, 3.10–3.11
  open-door operation of, 3.10, 3.11
  safety procedures/methods with, 3.7–3.11
Meetings, safety, 12.8–12.11
Mesh voltage hazards, 9.7, 10.5
Mold marks (in rubber goods), 2.37, 2.40
Molded-case circuit breakers:
  locking devices for, 2.53, 2.54
  maintenance of, 5.16–5.18
  safety procedures/methods with, 3.15–3.17
Motor starters:
  closed-door operation of, 3.21–3.23
  open-door operation of, 3.23
  safety procedures/methods with, 3.20–3.23
Motor-starting analysis, 7.4
Mouth-to-mouth resuscitation, 7.13–7.14
Multiple-lock devices, 2.52–2.53, 2.55
Muscles:
  contractor, 1.2
  electrical hold, 1.7
  extensor, 1.2, 1.21
  flexor, 1.21
  "Let-go" threshold, 1.4
Muscular system, 1.21
  determination of injuries to, 8.15
  shock damage to, 1.21–1.22

National Electrical Code (NEC)—ANSI/NFPA
  70, 2.75, 4.1, 6.14–6.15
  bonding, 4.18
  cord- and plug-connected equipment,
    4.19–4.20

equipment grounding conductors, 4.21–4.22, 4.25, 9.11, 9.12
equipment to be grounded, 4.19, 9.11–9.13
ground-fault protection for temporary wiring, 4.20
grounding electrode conductor installation, 4.15–4.16
grounding of electrical systems, 4.3–4.6
grounding service-supplied AC systems, 4.10–4.11
performance of fault path, 4.8
safety requirements, 5.3
National Electrical Manufacturer's Association, 6.13
National Electrical Safety Code (NESC)—ANSI C-2, 4.16, 6.12–6.14
National Electrical Safety Foundations (NESF), 6.13
National Fire Protection Association (NFPA), 6.3–6.4, 6.13, 8.6
Natural fibers, as non-flame-resistant, 2.5
NEC (*see* National Electrical Code)
Needs analysis, 13.10, 13.14, 13.15, 13.19
Neon light proximity tester, 2.55, 2.56
Nervous system, 1.3, 1.20, 1.21, 8.15
NESC (National Electrical Safety Code), 4.16
NESF (National Electrical Safety Foundations), 6.13
NETA (*see* InterNational Electrical Testing Association)
NFPA (*see* National Fire Protection Association)
NFPA 70 (*see* National Electrical Code)
NFPA 70B (*see* Electrical Equipment Maintenance)
NFPA 70E (*see* Standard for Electrical Safety in the Workplace)
NFPA standards system, 6.4
Nickel-Cadmium (NiCad) batteries, 9.24
Nicks (in rubber goods), 2.37, 2.40
NOMEX®, 2.7, 2.8
NOMEX IIIA, 2.6
Non-current-carrying metal parts, grounding of, 4.6, 4.19
Non-flame-resistant materials, 2.5
Nonionizing hazards, 9.22–9.23
Nowland, Stan, 5.5, 5.8
Nylon, 2.7

Objectionable current, 4.8
Occupational Safety and Health Act of 1970, 6.6, 6.11

Occupational Safety and Health Administration (OSHA), 4.1, 6.6–6.12, 6.13
electrical safety standards, 6.9
employee responsibilities and rights, 6.8–6.9
employer responsibilities and rights, 6.7–6.8
purpose of, 6.6–6.7
"qualified persons" defined by, 13.12, 13.13
safety regulations developed by (*See also* OSHA standards)
state OSHA organizations, 6.11, 6.12
technical consultation, 6.9, 6.11
training and education, 6.11
voluntary protection programs, 6.9, 6.11, 6.12
OJT (*see* On-the job training)
One-line diagrams, 2.77, 2.78
One-minute safety audit, 3.65–3.66
On-the-job training (OJT), 13.5–13.9
Open-air disconnects, 3.18–3.21
Open-door operation:
of low-voltage switchgear, 3.10, 3.15
of medium-voltage switchgear, 3.10, 3.11
of motor starters, 3.23
Operating procedures, for accident prevention, 7.3
OSHA (*see* Occupational Safety and Health Administration)
OSHA standards, 6.9, 6.10, 6.19, 6.22–6.194
for construction industry, 6.22–6.23
control of hazardous energy source, 6.22, 6.139–6.155
for cord- and plug-connected equipment, 4.20
definitions used in (construction), 6.23, 6.188–6.194
definitions used in (general), 6.22, 6.156–6.164
design safety standards, 6.22, 6.24–6.47
for general industry, 6.19, 6.22
for grounding of electrical equipment, 4.6
for grounding of electrical systems, 4.3–4.6
for grounding service-supplied AC systems, 4.9–4.11
installation safety requirements (construction), 6.23, 6.165–6.184
lockout/tagout, 6.22, 6.139–6.155
for maintenance, 5.1
for molded-case circuit breakers, 5.17–5.18
for performance of fault path, 4.8
power generation/transmission/distribution, 6.66–6.138
power generation/transmission/distribution, 6.22
safety requirements for special equipment (construction), 6.23, 6.187

OSHA standards (*Cont.*):
  safety-related maintenance and environmental considerations (construction), 6.23, 6.186
  safety-related work practices (construction), 6.23, 6.185
  safety-related work practices (general), 6.22, 6.48–6.65
  for voltage measurement, 9.20
OSHA Training Institute, 6.11
Outage reports, 12.11
Outpatient services for victims, 8.14–8.16
Ozone cracks (in rubber goods), 2.40, 2.41

Pack-strap carry, 7.36
Padlocks, 2.52, 2.53
Pain, 1.20
Panel voltmeters, 3.36
Panelboards, safety procedures/methods with, 3.15–3.17
Parting line (in rubber goods), 2.40
PBI (*see* Polybenzimidazole)
PBI Gold, 2.7
PCIC (Petroleum and Chemical Industry Committee), 6.13
PdM (*see* Predictive Maintenance)
PEARL (Professional Electrical Apparatus Recyclers League), 6.13
PEARL Reconditioning Standards, 5.17
Performance objective design, 13.16
Peripheral nerve injury, 8.15
Personal protective equipment (PPE):
  for crossing flash hazard boundary, 3.55, 3.56, 3.59
  maintenance of, 5.24
Petroleum and Chemical Industry Committee (PCIC), 6.13
PFR Rayon, 2.7
Phasing tester, 2.57, 2.59
Physical condition, severity of shock and, 1.3
Physical evidence (accidents), 7.40
Piping systems, grounding of, 4.6, 4.7
Pitting (in rubber goods), 2.40, 2.41
Planning, in Eight Step Maintenance Program, 5.11
Plasma, 1.8
  burns from, 1.20
  defined, 1.2
Plug-connected equipment:
  grounding of, 4.19–4.21
  low-voltage safety with, 9.3–9.5
  OSHA standards for, 4.20
PM (*see* Preventive Maintenance)

Pole-top rescue, 7.18–7.29
Policy, safety, 12.4
Polybenzimidazole (PBI), 2.6, 2.9
Portable electric tools/equipment, maintenance of, 5.23–5.24
Power generation/transmission/distribution, OSHA standards for, 6.22, 6.66–6.138
Power system studies, 7.3–7.7
Power systems:
  and human factors, 11.3–11.7
  switching of, 3.6–3.23
  voltage ratings of, 10.1n.
PPE (*see* Personal protective equipment)
Predictive Maintenance (PdM), 5.5, 5.7, 5.9
Pre-job briefings, 3.3
Preventive Maintenance (PM), 5.5, 5.7, 5.9
Privette, Alan, 3.55
Proactive Maintenance, 5.5
Professional Electrical Apparatus Recyclers League (PEARL), 6.13
Professional Electrical Apparatus Recyclers League (PEARL) Reconditioning Standards, 5.17
Prohibited approach boundary, 3.47, 3.48, 9.20
Protective clothing:
  for approach distances, 3.56–3.60
  characteristics of, 3.60
  for crossing flash hazard boundary, 3.59
Protective devices, 7.3
  coordination study for, 7.5–7.7
  with high- and medium-voltage systems and equipment, 10.13
  with low-voltage systems and equipment, 9.21
  maintenance of, 5.16
  (*See also* Personal protective equipment (PPE))
Protective relays, maintenance of, 5.21–5.23
Protective strategies, 1.24, 1.25 (*See also* Safety equipment; Safety procedures and methods)
Protex, 2.7
Protuberance (in rubber goods), 2.42
Proximity testers, 2.55–2.59
Pulmonary system, 1.23
Puncture (in rubber goods), 2.42

Quad pod retractable lifeline, 7.34
"Qualified persons," 13.12, 13.13

Racking a breaker, 3.7
Radiation burns, 1.20
Radiation hazards, 9.22–9.23

RCM (*see* Reliability-Centered Maintenance)
RCM White Paper (Al Rose), 5.4
Reactive maintenance, 5.5
Real-time Monitoring (RTM), 5.5
Record keeping, in Eight Step Maintenance
    Program, 5.13
Recovery plateaus, 8.16–8.17
Reenergizing equipment, energy control rules
    for, 3.25
Regulatory bodies, 6.1–6.12
    ANSI, 6.1–6.2
    ASSE, 6.5–6.6
    ASTM, 6.4–6.5
    IEEE, 6.3
    NFPA, 6.3–6.4
    OSHA, 6.6–6.12
Rehabilitation of victims, 8.16–8.17
Relays, protective, 5.21–5.23
Reliability analysis, 7.4–7.5
Reliability-Centered Maintenance (RCM),
    5.4–5.10
    and facilities life cycle, 5.9–5.10
    and failure, 5.8
    history of, 5.5
    in industry and utilities, 5.5–5.6
    maintenance actions in, 5.8–5.9
    primary principles of, 5.6–5.8
Remote operation, safety procedures/methods
    for, 3.7
Repair marks (in rubber goods), 2.42, 2.43
Reports, accident, 7.40, 7.41
Rescue breathing, 7.13–7.14
Rescue techniques, 7.14, 7.16–7.38
    confined-space rescue, 7.30–7.36
    elevated rescue, 7.18–7.29
    general procedures for, 7.14, 7.16–7.17
    ground-level rescue, 7.31, 7.34–7.38
Restricted approach boundary, 3.47, 3.48, 9.20
Resuscitation:
    artificial respiration, 7.12–7.14
    heart-lung, 7.12–7.16
Return to work (RTW), 8.16–8.17
Rose, Al, 5.4
Rotating equipment, maintenance of, 5.23
RTF (*see* Run-to-Failure)
RTM (Real-time Monitoring), 5.5
RTW (*see* Return to work)
Rubber insulating equipment, 2.15–2.44
    blankets, 2.21–2.23
    classifications of, 2.17
    covers, 2.22, 2.24–2.25
    electrical testing of, 2.43–2.44

gloves, 2.16–2.20
    for high- and medium-voltage systems, 10.9
    inspection and testing of, 2.29–2.44
    irregularities in, 2.33, 2.36–2.43
    labeling requirements for, 2.16, 2.19
    line hose, 2.25–2.26
    for low-voltage systems, 9.18
    mats, 2.20
    maximum-use voltages for, 2.17
    sleeves, 2.26–2.28
    test voltages for, 2.17
Runs (in rubber goods), 2.42
Run-to-Failure (RTF), 5.5, 5.8

Safety audits, 3.65–3.66, 12.11–12.34
    description of, 12.11–12.12
    follow-up for, 12.13
    inspections vs., 12.11–12.12
    internal vs. external, 12.14, 12.34
    procedure for, 12.12–12.14
    purposes of, 12.13
    team for, 12.14
    tools for, 12.14–12.33
Safety credo, 7.2
Safety electrical one-line diagram (SEOLD),
    2.77, 2.78
Safety equipment, 1.24, 2.1–2.79
    barriers, 2.49–2.51
    clothing materials, 2.4–2.5
    electrician's safety kit, 2.78, 2.79
    eye protection, 2.13, 2.15, 2.16
    flame-resistant materials, 2.6–2.9
    for flash and thermal protection, 2.2–2.13
    flash suits, 2.11–2.13
    general inspection/testing requirements for,
        2.2
    ground fault circuit interrupters, 2.74–2.76
    grounding equipment, 2.63–2.74
    hand protection, 2.13
    head protection, 2.13–2.16
    for high- and medium-voltage systems,
        10.7–10.12
    hot sticks, 2.44–2.48
    insulated tools, 2.48–2.49
    locks and locking devices, 2.51–2.55
    for low-voltage systems, 9.14–9.19
    maintenance of, 5.24
    non-flame-resistant materials, 2.5
    one-line diagrams, 2.77, 2.78
    rubber-insulating equipment, 2.15–2.44
    signs, 2.49–2.51
    tags, 2.51–2.55

Safety equipment (*Cont.*):
  thermal performance evaluation for, 2.3–2.4
  thermal protective clothing, 2.2, 2.3
  voltage-measuring instruments, 2.54–2.63
  work clothing, 2.9–2.11
  (*See also specific equipment*)
Safety glasses, 2.15
Safety grounds:
  application of, 3.38–3.43
  control of, 3.44–3.46
  with lockout/tagout, 3.29, 3.31
  placement of, 3.37–3.46
  removal of, 3.44
  (*See also* Jumpers, safety grounding)
Safety kit for electricians, 2.78, 2.79
Safety meetings, 12.8–12.11
Safety policy, 12.4
Safety procedures and methods, 1.24, 3.1–3.66
  approach distances, 3.46–3.51
  arc protection calculations, 3.51–3.62
  around electronic circuitry, 9.21–9.23
  development and structure of, 12.7
  for energized or de-energized circuits, 3.3–3.6
  energy control programs, 3.23–3.26
  field marking of potential hazards, 3.65
  flash hazard calculations, 3.46–3.62
  for high- and medium-voltage systems (*see* High- and medium-voltage safety)
  lockout/tagout, 3.26–3.32
  for low-voltage safety (*see* Low-voltage safety)
  one-minute safety audit, 3.65–3.66
  operating enclosed switches and disconnects, 3.17–3.18
  operating low-voltage switchgear, 3.11–3.15
  operating medium-voltage switchgear, 3.7–3.11
  operating molded-case breakers and panelboards, 3.15–3.17
  operating motor starters, 3.20–3.23
  operating open-air disconnects, 3.18–3.21
  pre-job briefings, 3.3
  remote operation, 3.7
  requiring flash suits, 2.13
  results assessment for, 12.7–12.8
  safety grounds, 3.37–3.46
  six-step safety method, 3.1–3.3
  with stationary batteries, 9.25
  for switching of power systems, 3.6–3.23
  tools and test equipment, 3.62–3.64
  voltage-measurement techniques, 3.32–3.37
Safety standards (*see* Standards)
Safety tags, 2.51–2.55

Safety teams:
  company electrical safety team, 12.2–12.4
  employee safety teams, 12.3–12.4, 12.8–12.9
Safety voltage measurement (*see* Voltage-measuring instruments)
Safety-related case for electrical maintenance, 5.1–5.3
Safety-related work practices (OSHA standards), 6.22, 6.48–6.65, 6.185
Scratches (in rubber goods), 2.33, 2.35, 2.37
Seat carry, 7.37
Second-degree burns, 1.16
Self-training, 13.6, 13.17
SEOLD (*see* Safety electrical one-line diagram)
Service enclosures, grounding of, 4.6, 4.7
SHEPP briefings, 3.3
Shock circuit, 1.2
Shock (electric), 1.2–1.7
  defined, 1.2
  description of, 1.2–1.3
  effects of, 1.23
  and equipment grounding, 4.1–4.2
  factors influencing severity of, 1.3–1.7
  first aid for, 7.11–7.13
  heart damage from, 1.23
  muscular system damage from, 1.21–1.22
  nervous system damage from, 1.20, 1.21
  procedural protection strategy for, 1.24
  protective equipment for, 1.24
  pulmonary system damage from, 1.23
Shock hazard distance, 3.46–3.48
Short-circuit analysis, 7.5–7.7
Shotgun hot sticks, 2.44–2.46
Signs:
  for approach distances, 3.56, 3.60–3.61
  safety, 2.49–2.51
Single-phase systems, voltage measurement in, 3.34, 3.35
Single-point grounding, 3.38
Site surveys (accidents), 7.38, 7.39
Six-step safety method, 3.1–3.3
SKED system, 7.33
Skin, 1.19–1.21
Skin breaks (in rubber goods), 2.42
Sleeves, rubber, 2.26–2.28
Snags (in rubber goods), 2.37, 2.40
Soft spots (in rubber goods), 2.43
Stability analysis, 7.4
Standard for Electrical Safety in the Workplace— ANSI/NFPA 70E, 5.2–5.4, 6.16–6.19
  approach distances, 9.20
  on capacitive discharges, 9.21

flash hazard calculation, 3.52, 3.53
as minimum requirements, 12.1
Standard universal fitting (hot sticks), 2.44
Standards:
 ASSE, 6.6
 ASTM, 6.19–6.21
 OSHA (*see* OSHA standards)
 (*See also specific equipment and procedures; specific standards*)
Stationary batteries, 9.23–9.25
Step voltage hazards, 9.7, 10.4–10.5
Step-down voltage transformers, 3.35, 3.36
Stokes navy stretcher, 7.33
Stoll curve, 2.1, 2.3
Stratum corneum, 1.19 (*See also* Horny layer)
Structure design (for training), 13.16–13.17
Super Valzer, 2.7
Surgical interventions, 8.14–8.16
Sweat glands, 1.20
Switches, safety grounding, 2.64, 2.65
Switchgear:
 arc-resistant, 1.1–1.2
 low-voltage, 3.11–3.15
 medium-voltage, 3.7–3.11
Switching of power systems, 3.6–3.23
 enclosed switches and disconnects, 3.17–3.18
 low-voltage switchgear, 3.11–3.15
 medium-voltage switchgear, 3.7–3.11
 molded-case breakers and panelboards, 3.15–3.17
 motor starters, 3.20–3.23
 open-air disconnects, 3.18–3.21
 remote operation, 3.7
 safety procedures/methods for, 3.6–3.23
Switching transients analysis, 7.4
Synthetic clothing materials, 2.5
Synthetic-cotton materials, 2.5
System grounding, 4.3–4.6, 4.9–4.19
 bonding, 4.18–4.19
 conductors, AC systems, 4.11
 defined, 9.8
 grounding conductor connection to electrodes, 4.16–4.18
 grounding electrode conductor, 4.14–4.16
 grounding electrode system, 4.12–4.13
 for high- and medium-voltage systems, 10.5–10.6
 for low-voltage systems, 9.8–9.11
 main bonding jumper, 4.11–4.12
 methods of, 4.9, 4.10
 purposes of, 4.9
 service-supplied AC systems, 4.9–4.11

Tags, 2.51–2.55 (*See also* Lockout/tagout)
"Tailgate meetings," 3.3
Tailored training programs, 13.10
Tears (in rubber goods), 2.43
Technora, 2.7
Telescoping hot sticks, 2.44, 2.46
Terminal boxes, 5.23
Terminal training objectives, 13.16
Test reviews, 13.19
Testing, 2.2
 determining intervals for, 5.14
 in Eight Step Maintenance Program, 5.13
 of extension cords, 3.63
 for flame resistance, 2.3
 of ground fault circuit interrupters, 2.76
 of hot sticks, 2.47, 2.48
 of molded-case circuit breakers, 5.16–5.17
 of rubber-insulating equipment, 2.29–2.44
 safety procedures/methods for, 3.62–3.64
 of tools, 3.63
Thermal burns, 1.20
Thermal performance evaluation, 2.3–2.4
Thermal protection:
 safety equipment for, 2.2–2.13
 usage standards for, 2.4
Thermal protective clothing, 2.2–2.5, 2.9, 2.11
Thermo-Man®, 2.8, 2.9
Third-degree burns, 1.16
Thone-Poulenc, 2.6
Three Mile Island Power Station, 11.2
Three-phase systems, voltage measurement in, 3.35, 3.36
Three-step voltage measurement process, 2.60–2.62, 3.34
Tightening, in Eight Step Maintenance Program, 5.12
Time-directed maintenance, 5.7
Tools:
 grounding of, 4.19–4.21
 insulated, 2.48–2.49, 9.3
 low-voltage safety with, 9.3–9.5
 portable, 5.23–5.24
 safety procedures/methods for, 3.62–3.64
Touch voltage hazards, 9.7, 10.5
Training, 13.1–13.19
 classroom, 13.3–13.5
 consultants and vendors offering, 13.9–13.11
 defined, 13.1
 of employee safety team, 12.9
 in first aid, 7.7–7.8
 for lockout/tagout, 3.31
 myths about, 13.2–13.3

Training (*Cont.*):
  on-the-job, 13.5–13.9
  program for, 13.12–13.19
  self-training, 13.6
Transferred voltage hazards, 9.7–9.8, 10.5
Transformers (*see* Current transformers)
TREVIRA®, 2.7
Triage, medical, 8.12
Tripod hoist, 7.33, 7.35
TWARON®, 2.7
Two-phase systems, voltage measurement in,
  3.34, 3.35
Two-point grounding, 3.38

United Airlines, 5.5
U.S. Department of Transportation (DOT) rules,
  8.17

Vendors, training by, 13.9–13.11, 13.17
VINEX®, 2.7
Visualization, 11.4
Vocational technical schools, 13.9–13.10
Voltage:
  of arcs, 1.13
  severity of shock and magnitude of, 1.4
Voltage hazards:
  with high- and medium-voltage systems,
    10.4–10.5
  with low-voltage systems, 9.6–9.8
Voltage measurement, 3.32–3.37
  with high- and medium-voltage systems and
    equipment, 10.13
  how to measure, 3.36–3.37

instrument condition for, 3.33–3.34
instrument selection for, 3.32–3.33
with low-voltage systems and equipment,
  9.20
three-step process for, 2.60–2.62, 3.34
what to measure, 3.34–3.36
Voltage-measuring instruments, 2.54–2.63,
  3.32–3.34
  condition of, 3.33–3.34
  contact testers, 2.56–2.59
  inspecting condition of, 2.59
  low voltmeter safety standards, 2.60, 2.61
  low-voltage, 2.62–2.63
  proximity testers, 2.55–2.59
  selecting, 2.57–2.59
  selection of, 3.32–3.33
  and three-step process, 2.60–2.62, 3.34
Voltage-testing devices:
  for high- and medium-voltage systems, 10.9,
    10.12
  for low-voltage systems, 9.18–9.19
Voluntary protection programs (VPPs), 6.9,
  6.11, 6.12
VPPs (*see* Voluntary protection programs)

Wall switches, locking devices for, 2.53, 2.54
Warning signs, 2.49–2.51, 3.56, 3.60–3.61
Wet environments, use of tools in, 3.63
Witness interviews (accidents), 7.39, 7.40
Wool clothing materials, 2.5
Work clothing, 2.9–2.11
Work practices, safety-related (OSHA
  standards), 6.22, 6.48–6.65, 6.185

ABOUT THE AUTHORS

John Cadick, P.E., is a registered professional engineer and the founder and president of the Cadick Corporation. Mr. Cadick has specialized for over three decades in electrical engineering, training, and management. His consulting firm, based in Garland, Texas, specializes in electrical engineering and training and works extensively in the areas of power system design and engineering studies, condition-based maintenance programs, and electrical safety. Prior to creating the Cadick Corporation and its predecessor, Cadick Professional Services, he held a number of technical and managerial positions with electric utilities, electrical testing companies, and consulting firms. In addition to his consultation work in the electrical power industry, Mr. Cadick is the author of *Cables and Wiring* and of numerous professional articles and technical papers.

Mary Capelli-Schellpfeffer, M.D., M.P.A., delivers outpatient medical care services to employees in occupational health service centers. Board-certified as a physician in general preventive medicine and public health, she is also a consultant to both the NJATC (National Joint Apprenticeship and Training Committee of National Electrical Contractors Association and International Brotherhood of Electrical Workers) and IEEE (Institute of Electrical and Electronics Engineers) Standards Committees. She lives in Chicago, Illinois.

Dennis K. Neitzel, a Certified Plant Engineer (C.P.E.), has over 38 years of experience in the electrical field. During eight years with the U.S. Air Force he served consecutively as an electrician, electrical shift supervisor, and quality control inspector and evaluator. Subsequent civilian employment included positions with Westinghouse Idaho Nuclear Company at the Idaho National Engineering and Environmental Laboratory where he advanced to Senior Project Engineer. Mr. Neitzel went to AVO Training Institute in 1989 and has progressed from Senior Training Specialist to Institute Director. He has served as a Principal Committee Member for the NFPA 70E standard since 1992 and is also serving as the IEEE Standard 902 Working Group Chairman. Mr. Neitzel has published numerous technical papers and magazine articles on electrical safety, holds a Master's Degree in Electrical Engineering Applied Sciences, and is a Certified Electrical Inspector.